**W9-BXU-102**

Common Functions and Graphs

Identity
$y = x$

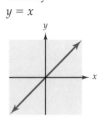

Square
$y = x^2$

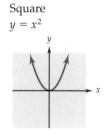

Cube
$y = x^3$

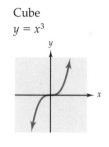

Absolute value
$y = |x|$

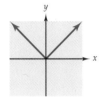

Reciprocal
$y = \dfrac{1}{x}$

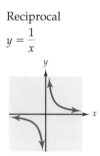

Square root
$y = \sqrt{x}$

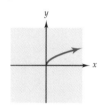

Cube root
$y = \sqrt[3]{x}$

Exponential
$y = b^x, b > 1$

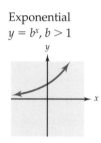

Logarithmic
$y = \log_b x, b > 1$

Common Models and Graphs

Linear
$y = ax + b$

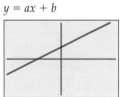

Quadratic
$y = ax^2 + bx + c$

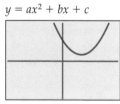

Cubic
$y = ax^3 + bx^2 + cx + d$

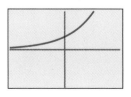

Exponential
$y = a \cdot b^x$

Logarithmic
$y = a + b \ln x$

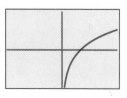

Logistic
$y = \dfrac{a}{1 + be^{-kx}}$

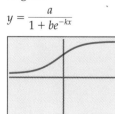

College Algebra

Visualizing and Determining Solutions

College Algebra

Visualizing and Determining Solutions

Elaine Hubbard
Kennesaw State University

Ronald D. Robinson

HOUGHTON MIFFLIN COMPANY   Boston   New York

Publisher: Charles Hartford
Editor-in-Chief: Jack Shira
Senior Associate Editor: Maureen Ross
Senior Project Editor: Nancy Blodget
Editorial Assistants: Joy Park, Kathy Yoon
Senior Production/Design Coordinator: Jennifer Waddell
Senior Manufacturing Coordinator: Sally Culler
Marketing Manager: Michael Busnach

Cover design: Walter Kopec

Cover image: Bruce Forster/Tony Stone Images

Printed in the U.S.A.

Library of Congress Catalog Card Number: 99-71965

ISBN: 0-395-81856-7

123456789-VH-03 02 01 00 99

CONTENTS

PREFACE

College Algebra: Visualizing and Determining Solutions is written for students who are meeting their core curriculum requirements in mathematics. The content includes the topics found in any mainstream college algebra course. Moreover, this text provides excellent preparation for such additional courses as statistics and business calculus.

The Approach

A major goal of *College Algebra: Visualizing and Determining Solutions* is to engage the students by promoting their active participation in the study of mathematics. The pedagogical features of the book are designed to achieve that goal.

Technology   The use of a graphing calculator is integrated throughout the text. Rather than compartmentalizing technology in special boxes or marginal activities, we employ the graphing calculator as an integral part of the presentation and as an essential aspect of the learning process.

Although solutions of equations and properties of functions, for example, can be determined with a graphing calculator, we prefer not to treat graphing and algebraic methods as equivalent. Rather, our focus is on the mathematics and the algebraic methods, with the technology serving as a guide to the development of concepts and as a visual aid for reaching conclusions. Therefore, when both approaches are presented, the graphing method leads to the algebraic method rather than one approach being an alternative to the other. In short, first we see, then we do.

Exploration   We believe that "discovery learning" is primarily a classroom activity. Students are best able to explore and discover when they are guided by the teaching skills of the instructor. Throughout this text, we have included classroom activities under the heading "Exploration." With leading questions and, usually, the visual assistance of the graphing calculator, the student is guided from a concrete set of circumstances to a general conclusion. All the questions are open-ended, thereby providing maximum flexibility in the use of these activities.

Developing the Concept   A parallel feature called "Developing the Concept" usually follows each "Exploration" or, less often, is a stand-alone activity. This feature is one that students are better able to use on their own. Structured like an example, "Developing the Concept" guides the student with leading questions, and answers them as well. The combination of exploring and developing the concept is a major effort to promote students' interaction and involvement in their own learning.

Communication Skills   Increased emphasis has been placed on writing and speaking in mathematics. When a student says, "I know the answer, I just don't know how to say it," the student probably does not know the answer.

xi

To help with this, we have included two features that are specifically designed to improve the student's ability to communicate about mathematics. One, "Speaking the Language," precedes the exercise sets (except in sections that are dedicated exclusively to applications) and helps the student to practice the use of vocabulary in context.

The other main communication feature is the generous number of writing exercises, which are identified with a ✏ icon. Although the answers to such writing exercises can often vary, a sample answer is given in the Answers section at the back of the book. Rather than assigning writing exercises, some instructors may prefer to use them as the basis of class discussion. Thus, writing exercises can promote both written and oral communication development. These exercises truly set this text apart from many others.

Exercises Most exercise sets are divided by the headings "Concepts and Skills," "Concept Extension," and "Applications." Sometimes the applications exercises have subheads to indicate the type of application that is included. All of these headings have been designed to assist the instructor in the preparation of assignments.

Exercises included in the "Concepts and Skills" are representative of the basic material of the section. Nearly all are illustrated by text examples.

Most exercise sets include a number of "Concept Extension" exercises designed to go somewhat beyond the basic concepts and skills discussed in the text section. These exercises are designed to synthesize concepts and to use previously learned skills in a new context.

Most sections contain numerous real-life and real-data applications, with source acknowledgments. Several sections are devoted exclusively to such exercises. These include a wide variety of subject areas and are designed to promote a greater understanding of the relevancy of mathematics.

Other Features

Definitions and Rules All definitions, rules, properties, and procedures are highlighted in colored boxes for easy reference.

Examples Numerous titled examples provide immediate illustrations of the concepts and techniques discussed in the sections. Comments helpful to the student are often included in the detailed solutions.

Graphs Visualization is provided by a large number of traditional coordinate plane graphs and calculator graphs that appear in both the exposition and the exercises. Calculator graphs are designed to resemble the displays of most calculator models.

Quick Reference With the exception of applications sections, a "Quick Reference" appears at the end of each section. These are detailed summaries of the rules, definitions, properties, and procedures discussed in the section. All are grouped by subsection and can be used by the student for reference and review.

Calculator Guide When a graphing calculator function is first introduced in the text, a key word ▢ icon appears along with a brief description of the function. Students can reference these key words in Appendix B, which is a calculator keystroke guide for the TI-83 Plus calculator.

Calculator Programs Appendix C contains sample calculator programs. Although these programs are written for the TI-83 Plus, the algorithms are described in detail, and they can be used to program any calculator model.

A Final Word

The methods and approaches used in *College Algebra: Visualizing and Determining Solutions* are based on our many years of classroom teaching experience and are time-tested for their effectiveness in promoting student success. We are pleased to share our experiences with you.

Supplements

Student Solutions Manual This manual includes solutions to the odd-numbered exercises in the text.

Graphing Calculator Guide The Graphing Calculator Keystroke Guide includes information for the TI-82, TI-83, and TI-83 Plus. A Key Word icon in the text alerts students to specific keystroke information in this supplement. The sequence of the Key Word explanations in the Guide corresponds to the sequence in the main text.

Complete Solutions Guide The Complete Solutions Guide includes detailed solutions to all Preamble Exercises, end-of-section exercises, and Chapter Review Exercises, including those in Chapter R. The Guide also includes answers to the Speaking the Language exercises, which appear before the section exercises in each chapter.

Test Item File This printed manual contains multiple-choice, open-ended, and writing questions. There is a set of questions for each section of the book.

Computerized Test Bank Available for both Windows and Macintosh.

Tutorial Software This tutorial software is algorithmically driven and interactive. Lessons and problems are presented in a lively manner. It includes student-friendly features such as: animated solution steps, extensive hints, and a glossary of key terms and definitions.

Videotapes Prepared by Dana Mosely, the videotapes provide a thorough review of concepts and worked-out examples to reinforce lessons within the text.

Acknowledgments

We would like to thank the following colleagues who reviewed the manuscript and made many helpful suggestions:

Judy Ahrens, *Pellissippi State Technical Community College*
Robert Bohac, *Northwest College*
Curtis Card, *Black Hills State University*
Joseph F. Cieply, *Elmhurst College*

John W. Coburn, *St. Louis Community College—Florissant Valley*
Victor M. Cornell, *Mesa Community College*
Elaine N. Daniels, *Salve Regina University*
Rob Farinelli, *Community College of Allegheny County*
Nicholas E. Geller, *Collin County Community College*
Betty Givan, *Eastern Kentucky University*
Allen Hesse, *Rochester Community College*
Jesse D. Parete, *Edison Community College*
Claudinna Rowley, *Johnson County Community College*
Fred Schineller, *Arizona State University*
John Seims, *Mesa Community College*
Joanne Shansky, *Milwaukee Area Technical College*
Stanley Smith, *Black Hills State University*
Lyndon Weberg, *University of Wisconsin—River Falls*

ELAINE HUBBARD
RONALD D. ROBINSON

OPENING FEATURES

Chapter Opener

Each chapter begins with a short introduction to a real-data application. Connecting mathematics with students' view of the world leads students to a better understanding of the practical nature of the discipline. The chapter opener also includes a helpful overview of the topics that will be covered in the chapter and a list of the section titles.

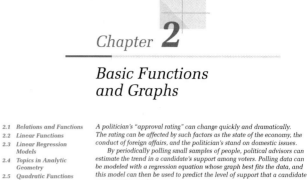

Chapter **2**

Basic Functions and Graphs

A politician's "approval rating" can change quickly and dramatically. The rating can be affected by such factors as the state of the economy, the conduct of foreign affairs, and the politician's stand on domestic issues.

By periodically polling small samples of people, political advisors can estimate the trend in a candidate's support among voters. Polling data can be modeled with a regression equation whose graph best fits the data, and this model can then be used to predict the level of support that a candidate can expect to receive.

In this chapter, we introduce the essential concept of a function, and we investigate the properties of the graphs of a variety of functions. In particular, you will learn how real data can be modeled with functions and how to judge the most appropriate model to use. Although many different models are available, our emphasis in this chapter is on linear and quadratic regression models.

This chapter provides you with a solid foundation for the important skills of analyzing, modeling, and interpreting real data.

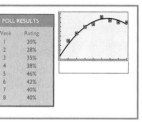

POLL RESULTS	
Week	Rating
1	20%
2	28%
3	35%
4	38%
5	46%
6	42%
7	40%
8	40%

111

2.1 *Relations and Functions*

■ *Relations* ■ *Functions* ■ *Function Notation and Evaluation*
■ *Domain and Range* ■ *Piecewise Functions* ■ *Applications*

In many real-world situations, the value of one quantity [depends on] the value of some other quantity. For example, the federal t[ax that you pay] depends on your income, the distance that you can driv[e in a car] depends on your speed, and your success in a college cou[rse depends] on the number of hours that you study.

Being able to describe how one variable affects another va[riable is an] important step in analyzing a problem. In this section, we consider how to use a *function* to describe the association of two sets of numbers. The topic of functions is central to the study of mathematics.

Relations

The bar graph in Figure 2.1 shows how advertising affects the profits of a small business.

FIGURE 2.1

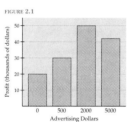

The graph shows that advertising does increase profits, but that beyond a certain level, the cost of too much advertising actually decreases profits. The advertising costs and corresponding profits shown in the bar graph can be written as a set of ordered pairs.

$$A = \{(0, 20,000), (500, 30,000), (2000, 50,000), (5000, 42,000)\}$$

Such a set is called a **relation.**

> *Definition of a Relation*
>
> A **relation** is a set of ordered pairs. The **domain** of a relation is the set of all first coordinates of the ordered pairs. The **range** is the set of all second coordinates.

Section Opener

Each section begins with a list of subsection titles, thus providing a brief outline of the material that follows.

VISUALIZING AND DETERMINING SOLUTIONS

These headings appear in examples
in which a graphing solution
approach is considered.

Visualizing the Solution

This first portion shows students
how to depict the problem using a
graphing calculator and how to
estimate a solution from this
visualization (often highlighting
any potential pitfalls from using
graphing technology).

Determining the Solution

The algebraic solution follows
the graphing solution and shows
students how to obtain an exact
solution with algebraic methods.
Together, these headings remind
students that graphical approaches
are used to best advantage when
coupled with algebraic approaches—
a "see first, then do" logical
progression.

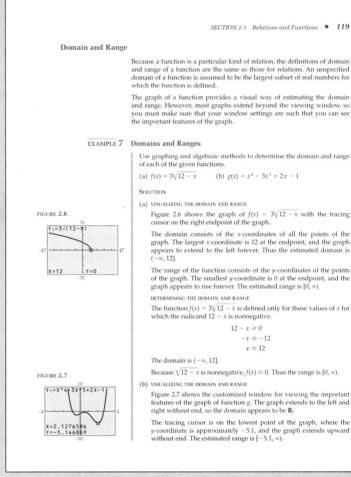

Domain and Range

Because a function is a particular kind of relation, the definitions of domain
and range of a function are the same as those for relations. An unspecified
domain of a function is assumed to be the largest subset of real numbers for
which the function is defined.

The graph of a function provides a visual way of estimating the domain
and range. However, most graphs extend beyond the viewing window, so
you must make sure that your window settings are such that you can see
the important features of the graph.

EXAMPLE 7   **Domains and Ranges**

Use graphing and algebraic methods to determine the domain and range
of each of the given functions.

(a) $f(x) = 3\sqrt{12 - x}$ (b) $g(x) = x^4 - 3x^3 + 2x - 1$

SOLUTION

(a) VISUALIZING THE DOMAIN AND RANGE

Figure 2.6 shows the graph of $f(x) = 3\sqrt{12 - x}$ with the tracing
cursor on the right endpoint of the graph.

The domain consists of the x-coordinates of all the points of the
graph. The largest x-coordinate is 12 at the endpoint, and the graph
appears to extend to the left forever. Thus the estimated domain is
$(-\infty, 12]$.

The range of the function consists of the y-coordinates of the points
of the graph. The smallest y-coordinate is 0 at the endpoint, and the
graph appears to rise forever. The estimated range is $[0, \infty)$.

DETERMINING THE DOMAIN AND RANGE

The function $f(x) = 3\sqrt{12 - x}$ is defined only for those values of x for
which the radicand $12 - x$ is nonnegative.

$$12 - x \geq 0$$
$$-x \geq -12$$
$$x \leq 12$$

The domain is $(-\infty, 12]$.

Because $\sqrt{12 - x}$ is nonnegative, $f(x) \geq 0$. Thus the range is $[0, \infty)$.

(b) VISUALIZING THE DOMAIN AND RANGE

Figure 2.7 shows the customized window for viewing the important
features of the graph of function g. The graph extends to the left and
right without end, so the domain appears to be **R**.

The tracing cursor is on the lowest point of the graph, where the
y-coordinate is approximately -5.1, and the graph extends upward
without end. The estimated range is $[-5.1, \infty)$.

FIGURE 2.6

FIGURE 2.7

REAL-DATA AND REAL-LIFE APPLICATIONS

Applications

We have included numerous real-life and real-data applications, with source acknowledgments. These include a wide variety of subject areas and are designed to promote a greater understanding of the relevancy of mathematics.

Some sections are devoted exclusively to examples and exercises involving real-life applications. Other sections contain dedicated blocks of such problems.

Real-Life Applications

Because the vertex of a vertical parabola is the highest or lowest point of the curve, we sometimes make use of a quadratic regression equation to estimate or predict a maximum or minimum value. "The regression equation has the form $y = ax^2 + bx + c$. Knowing a and b allows us to determine the coordinates of the vertex."

EXAMPLE 2   **Enrollment at Two-Year Colleges**

Figure 2.53 shows the fall enrollments (in millions) at two-year colleges for selected years in the period 1986–1996. (*Source*: National Center for Educational Statistics.)

FIGURE **2.53**

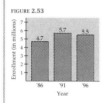

(a) Produce a quadratic regression equation to model the given data.
(b) According to your regression equation, in what year did enrollment peak?

SOLUTION

(a) We enter ordered pairs of the form (year, enrollment) in the statistics list. For convenience, we let the first coordinate be the number of years since 1986:

$$(0, 4.7), (5, 5.7), (10, 5.5)$$

Then we select the quadratic regression option and obtain the equation $y = -0.024x^2 + 0.32x + 4.7$.

(b) Note that $a = -0.024$ and $b = 0.32$. Because $a < 0$, the parabola opens downward and the function has a maximum (peak) value at the vertex. The first coordinate of the vertex is

$$-\frac{b}{2a} = -\frac{0.32}{2(-0.024)} = 6.\overline{6}$$

Thus we estimate the peak year to be 7 years after 1986, or 1993.

15. Cellular Phone Rates   In the early 1990s, only two cellular telephone carriers were allowed in each U.S. city. As shown in the bar graph, deregulation has increased competition and lowered rates. (*Source*: Herschel Shosteck Associates.)

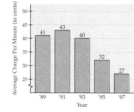

(a) Use a scatterplot of the data to help you decide what type of model is reasonable.
(b) Let x represent the number of years since 1990 and determine a quadratic regression equation to model the data. (Round coefficients to the nearest integer.)

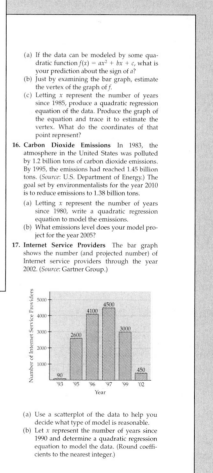

(a) If the data can be modeled by some quadratic function $f(x) = ax^2 + bx + c$, what is your prediction about the sign of a?
(b) Just by examining the bar graph, estimate the vertex of the graph of f.
(c) Letting x represent the number of years since 1985, produce a quadratic regression equation of the data. Produce the graph of the equation and trace it to estimate the vertex. What do the coordinates of that point represent?

16. Carbon Dioxide Emissions   In 1983, the atmosphere in the United States was polluted by 1.2 billion tons of carbon dioxide emissions. By 1995, the emissions had reached 1.45 billion tons. (*Source*: U.S. Department of Energy.) The goal set by environmentalists for the year 2010 is to reduce emissions to 1.38 billion tons.

(a) Letting x represent the number of years since 1980, write a quadratic regression equation to model the emissions.
(b) What emissions level does your model project for the year 2005?

17. Internet Service Providers   The bar graph shows the number (and projected number) of Internet service providers through the year 2002. (*Source*: Gartner Group.)

DISCOVERY LEARNING

Exploration

This pedagogical device provides opportunities for students to become active participants by exploring and discovering mathematical concepts. Classroom activities are accompanied by leading questions and, usually, the visual assistance of the graphing calculator. The student is guided from a concrete set of circumstances to a generalization. All the questions are open-ended, thereby providing maximum flexibility in the use of these activities as in-class discussion starters.

Developing the Concept

This is a parallel feature that immediately follows Exploration and that provides a similar scenario, but this feature also provides answers to leading questions. Because the features are separate and use slightly different scenarios, the Exploration is intact as a discovery learning exercise. The instructor can thus progress through the Exploration and Developing the Concept, or the Exploration can be bypassed entirely, with Developing the Concept serving as the sole instructional vehicle. Sometimes, Developing the Concept features appear by themselves to lead the reader to an understanding of the formula or theoretical concept.

WRITING AND SPEAKING IN MATHEMATICS

Increased emphasis has been placed on communicating in mathematics.

Speaking the Language

This feature precedes the exercise sets (except in sections that are dedicated exclusively to applications) and helps students think and communicate in the language of mathematics by reinforcing vocabulary and contextual meanings. Speaking the Language answers can be found in the Complete Solutions Guide.

Writing Exercises

The generous number of writing exercises are identified with a pencil icon. Although the answers to such writing exercises can often vary widely, a sample answer is given in the Answers section at the back of the book. Rather than assigning writing exercises, some instructors may prefer to use them as the basis of class discussion. Thus, writing exercises can promote both written and oral communication development.

CALCULATOR FEATURES

Key Words

Appearing at the initial point of use, Key Words briefly describe a pertinent graphing calculator function and alert students to related detailed coverage, including keystrokes, found in Appendix B and the accompanying Graphing Calculator Guide supplement. Selected keys from the TI-83 Plus are found inside the front cover.

The sequence of Key Words in the main text matches the sequence of Key Word explanations in the Graphing Calculator Keystroke Guide.

Graphs

Visualization is provided by a large number of traditional coordinate plane graphs and calculator graphs that appear in both the exposition and the exercises. Calculator graphs are designed to resemble the displays of most calculator models.

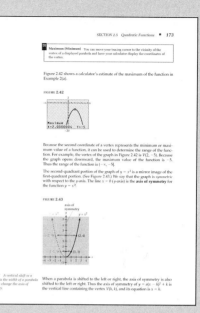

PEDAGOGICAL SUPPORT

Definitions, Properties, and Rules

All definitions, rules, properties, and procedures are highlighted in colored boxes for easy reference.

Examples/Solutions

All sections contain numerous, titled examples, many with multiple parts graded by difficulty. These Examples illustrate concepts, procedures, and techniques, and they reinforce the reasoning and critical thinking needed for problem solving. Detailed solutions include comments helpful to the student that justify the steps taken, explain their purpose, and identify properties and rules.

Notes *(see sample page above)*

Special remarks and cautionary notes offering additional insight appear throughout the text.

END-OF-SECTION FEATURES

A typical section ends with Quick Reference, Speaking the Language (see Writing and Speaking in Mathematics), and Section Exercises.

Quick Reference

Quick Reference appears at the end of all sections except those dealing exclusively with applications. These detailed summaries of the important rules, properties, and procedures are grouped by subsection for a handy reference and review tool.

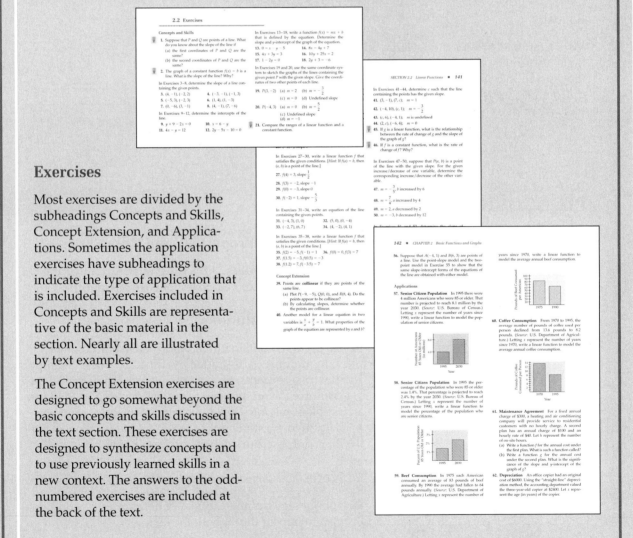

Exercises

Most exercises are divided by the subheadings Concepts and Skills, Concept Extension, and Applications. Sometimes the application exercises have subheadings to indicate the type of application that is included. Exercises included in Concepts and Skills are representative of the basic material in the section. Nearly all are illustrated by text examples.

The Concept Extension exercises are designed to go somewhat beyond the basic concepts and skills discussed in the text section. These exercises are designed to synthesize concepts and to use previously learned skills in a new context. The answers to the odd-numbered exercises are included at the back of the text.

END-OF-CHAPTER FEATURES

Review Exercises

These appear at the end of each chapter, organized by section. The answers to the odd-numbered exercises are included at the back of the text.

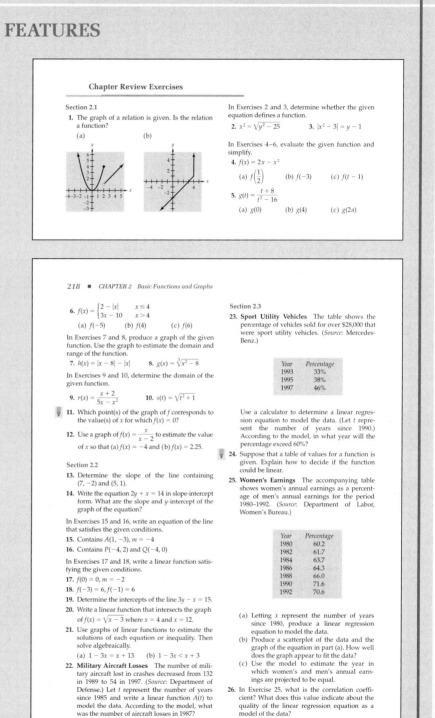

Chapter Review Exercises

Section 2.1

1. The graph of a relation is given. Is the relation a function?

(a)

(b)

In Exercises 2 and 3, determine whether the given equation defines a function.

2. $x^2 = \sqrt{y^2 - 25}$

3. $|x^2 - 3| = y - 1$

In Exercises 4–6, evaluate the given function and simplify.

4. $f(x) = 2x - x^2$

(a) $f\left(\frac{1}{2}\right)$ (b) $f(-3)$ (c) $f(t-1)$

5. $g(t) = \dfrac{t+8}{t^2 - 16}$

(a) $g(0)$ (b) $g(4)$ (c) $g(2a)$

---

218 ■ CHAPTER 2 Basic Functions and Graphs

6. $f(x) = \begin{cases} 2 - |x| & x \le 4 \\ 3x - 10 & x > 4 \end{cases}$

(a) $f(-5)$ (b) $f(4)$ (c) $f(6)$

In Exercises 7 and 8, produce a graph of the given function. Use the graph to estimate the domain and range of the function.

7. $h(x) = |x - 8| - |x|$ 8. $g(x) = \sqrt[3]{x^2 - 8}$

In Exercises 9 and 10, determine the domain of the given function.

9. $r(x) = \dfrac{x+2}{5x - x^2}$ 10. $s(t) = \sqrt{t^2 + 1}$

11. Which point(s) of the graph of f corresponds to the value(s) of x for which $f(x) = 0$?

12. Use a graph of $f(x) = \dfrac{x}{x-2}$ to estimate the value of x so that (a) $f(x) = -4$ and (b) $f(x) = 2.25$.

Section 2.2

13. Determine the slope of the line containing $(7, -2)$ and $(5, 1)$.

14. Write the equation $2y + x = 14$ in slope-intercept form. What are the slope and y-intercept of the graph of the equation?

In Exercises 15 and 16, write an equation of the line that satisfies the given conditions.

15. Contains $A(1, -3)$, $m = -4$

16. Contains $P(-4, 2)$ and $Q(-4, 0)$

In Exercises 17 and 18, write a linear function satisfying the given conditions.

17. $f(0) = 0$, $m = -2$

18. $f(-3) = 6$, $f(-1) = 6$

19. Determine the intercepts of the line $3y - x = 15$.

20. Write a linear function that intersects the graph of $f(x) = \sqrt{x - 3}$ where $x = 4$ and $x = 12$.

21. Use graphs of linear functions to estimate the solutions of each equation or inequality. Then solve algebraically.

(a) $1 - 3x = x + 13$ (b) $1 - 3x < x + 3$

22. **Military Aircraft Losses** The number of military aircraft lost in crashes decreased from 132 in 1989 to 54 in 1997. (*Source*: Department of Defense.) Let t represent the number of years since 1985 and write a linear function $A(t)$ to model the data. According to the model, what was the number of aircraft losses in 1987?

Section 2.3

23. **Sport Utility Vehicles** The table shows the percentage of vehicles sold for over $28,000 that were sport utility vehicles. (*Source*: Mercedes-Benz.)

Year	Percentage
1993	33%
1995	38%
1997	46%

Use a calculator to determine a linear regression equation to model the data. (Let t represent the number of years since 1990.) According to the model, in what year will the percentage exceed 60%?

24. Suppose that a table of values for a function is given. Explain how to decide if the function could be linear.

25. **Women's Earnings** The accompanying table shows women's annual earnings as a percentage of men's annual earnings for the period 1980–1992. (*Source*: Department of Labor, Women's Bureau.)

Year	Percentage
1980	60.2
1982	61.7
1984	63.7
1986	64.3
1988	66.0
1990	71.6
1992	70.6

(a) Letting x represent the number of years since 1980, produce a linear regression equation to model the data.

(b) Produce a scatterplot of the data and the graph of the equation in part (a). How well does the graph appear to fit the data?

(c) Use the model to estimate the year in which women's and men's annual earnings are projected to be equal.

26. In Exercise 25, what is the correlation coefficient? What does this value indicate about the quality of the linear regression equation as a model of the data?

INDEX of Real-Life Applications

INDEX of Real-Data Applications

College Algebra

Visualizing and Determining Solutions

Preamble

Getting Started with a Graphing Calculator

In the 1998 baseball season, two players, Mark McGwire and Sammy Sosa, broke a home run record that had stood for 37 years.

Many of their home runs traveled unusually long distances. At most stadiums, one person had the job of estimating those distances, which were then displayed on the scoreboard and reported in the news media. Those estimates were calculated with the aid of a mathematical model of the flight path of the ball. By taking into account such variables as the height of the ball and the bat speed, the computer operator could evaluate an algebraic expression and arrive at an estimate of the distance that the ball traveled.

The common and often repetitive tasks of evaluating expressions and producing graphs are made easier by the technological assistance that a computer or graphing calculator can provide. In this Preamble, we introduce the graphing calculator and some of its basic features. As you become more and more proficient with the use of your graphing calculator, you will begin to appreciate its value as a learning tool and as a time- and labor-saving device.

1

Preamble: Getting Started with a Graphing Calculator

Numerical Expressions   ■   *Variables and Algebraic Expressions*

We live in an age of information. New technologies allow us to collect and transmit great amounts of data at high speeds. But information alone is not useful unless it can be organized, analyzed, and interpreted in such a way that meaningful conclusions can be reached.

No matter what your eventual career may be, you will frequently be called upon to be a problem solver or to recommend a course of action. Mathematics gives you the tools needed to represent (or *model*) the conditions of a problem and to arrive at solutions. To use mathematics effectively, you will need to be well grounded in fundamental algebraic concepts and skills, in the effective use of technology, and in general problem-solving strategies.

Our focus in this Preamble is on the basic features of a graphing calculator. Your calculator will be an invaluable aid in performing operations quickly and accurately and will help you to visualize important algebraic concepts. Being familiar with these calculator techniques (and others that will be introduced later) will be important throughout this text.

Numerical Expressions

A **numerical expression** is any combination of numbers, grouping symbols, and arithmetic operations. Each of the following is a numerical expression.

$$2 \cdot 9 - 7 \qquad 5 + 3^2 \qquad \sqrt{16 - 12} + 1 \qquad \frac{3 + 11}{12 - 3(2)}$$

Evaluating a numerical expression means determining the value of the expression by performing all the indicated operations.

You can evaluate a numerical expression by hand, or you can use a scientific calculator to perform the arithmetic. To use a graphing calculator, you should be aware of several basic menu selections.

Home   The home screen is the calculator's primary screen for entering expressions to be evaluated or instructions to be carried out.

Quit   Pressing this key will return you to the home screen.

Clear   One use of the *Clear* key is to erase the home screen.

Mode   You can choose to report your numerical results to as many significant digits as the screen can display (*Float* mode), or you can choose the number of decimal places to which you want the result rounded off.

Fraction   You can choose to report your numerical results in simplified fraction form.

EXAMPLE **1**   **Evaluating a Numerical Expression on the Home Screen**

Evaluate $\dfrac{3+11}{12-3(2)}$.

(a) in floating decimal form.
(b) rounded to two decimal places.
(c) in simplified fraction form.

SOLUTION

(a) After we select *Float* from the mode menu (see the second line in Figure P.1), we enter the expression on the home screen and press *Enter*. (See the first two lines of Figure P.2.) Note the use of parentheses around the numerator and the denominator.

FIGURE P.1   FIGURE P.2

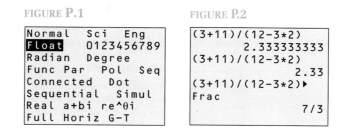

(b) We select *2* (for 2 decimal places) from the second line of the mode menu and repeat the entry of the expression. (See the third and fourth lines of Figure P.2.)
(c) This time, we follow the entry of the expression with *Frac* (typically from the math menu) to obtain the result in simplified fraction form. (See the last three lines of Figure P.2.)

Sometimes we need to compare the values of two numerical expressions. Your calculator's *Test* feature can help you do that.

> Test   The test menu contains the symbols $=$, $<$, $\geq$, and so on.

EXAMPLE **2**   **Using the Test Feature with Numerical Expressions**

Test the truth of each of the following.

(a) $2(5+3) = 2 \cdot 5 + 2 \cdot 3$
(b) $\dfrac{12}{13} > \dfrac{13}{14}$

SOLUTION

Figure P.3 shows the entry for each part. The symbols $=$ and $>$ were selected from the test menu.

FIGURE P.3

```
2(5+3)=2*5+2*3
                    1
12/13>13/14
                    0
```

(a) The returned value of 1 indicates that $2(5 + 3) = 2 \cdot 5 + 2 \cdot 3$ is true.

(b) The returned value of 0 indicates that $\dfrac{12}{13} > \dfrac{13}{14}$ is false.

In addition to the $+$, $-$, $\times$, and $\div$ keys, special calculator keys or menu selections are needed to perform certain arithmetic operations.

Negative   Some calculators distinguish between the *Minus* key for subtraction and the *Negative* key for entering negative numbers. Check your calculator to see whether a special *Negative* key is used.

Exponent   Typically the exponent key is ^. For squaring a number, you may have a special x² key.

Square Root   Look for the $\sqrt{\ }$ key on your calculator.

Absolute Value   Your calculator may have a special key for absolute value, or you may need to find the option in a menu. Typically the symbol is ABS.

▪ *NOTE   You will need to use parentheses to enclose the quantity whose square root or absolute value you are calculating.*

EXAMPLE 3   **Evaluating Numerical Expressions**

Evaluate each of the following. In part (b), round to the nearest thousandth.

(a) $5^2 + (-3)^4$ (b) $\sqrt{59}$ (c) $|2 - 6|$

SOLUTION

Figure P.4 shows the entries for the three numerical expressions and their values.

FIGURE P.4

```
5²+(-3)^4
                 106
√(59)
               7.681
abs(2-6)
                   4
```

Variables and Algebraic Expressions

A **variable** is a symbol (usually a letter) that represents some unknown number. A variable has no known value until we replace the variable with some specific number.

> **Store** With a graphing calculator, assigning a specific value to a variable is called storing the value in the variable.

Figure P.5 shows that we have substituted 5 for the variable x; that is, 5 has been *stored* in the calculator variable X. Note that if you then press the X key, the calculator reports the current value that is stored in X.

FIGURE P.5

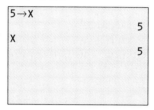

An **algebraic expression** is any combination of numbers, variables, grouping symbols, and operations symbols. Each of the following is an algebraic expression.

$$4x^3 - 5 \qquad -2(a + 2b) - x \qquad \frac{x^2 + 1}{x - 1} \qquad 2z - 3\sqrt{x}$$

We **evaluate** an algebraic expression by replacing the variable(s) with specific values and then evaluating the resulting numerical expression.

Several methods for evaluating an algebraic expression with a calculator are available. One method involves entering the expression to be evaluated on the home screen or on the Y screen.

> **Y Screen** An expression to be evaluated is entered on the Y screen. The expression is assigned to a Y-variable that holds the value of the expression.
>
> **Y-Var** A list of Y-variables can be displayed by pressing the Y-VAR (or simply VAR) key.

FIGURE P.6

```
Plot1  Plot2  Plot3
\Y₁=
\Y₂=
\Y₃=
\Y₄=
\Y₅=
\Y₆=
\Y₇=
```

The *Clear* key, used to erase the home screen, can also be used to erase individual entries on the Y screen. Figure P.6 shows the Y screen with no expressions entered.

Another method for evaluating an expression is to create a *table of values*. From previous experience, you may be familiar with x-y tables (sometimes called *t-tables*). Your calculator can produce such tables very rapidly.

> **Table Set** Usually you can specify how you want a table to be displayed by setting the initial *x*-value (*Table start*) and the increments of *x*-values (Δ *Table*). With the *Auto* option, you can display a full table of values of an expression that you have entered on the Y screen. With the *Ask* option, you can evaluate the expression for any *x*-value you enter.
>
> **Table** Pressing the *Table* key displays a table of values as specified in your table settings. You can use an arrow key to scroll through the table to see other entries.

EXAMPLE 4 **Evaluating an Algebraic Expression**

Evaluate $5x - (3 - 2x)$ for $x = -4$ by

(a) storing -4 in X and evaluating on the home screen.
(b) entering the expression on the Y screen.
(c) creating a table of values.

SOLUTION

(a) In Figure P.7, we have stored -4 in X and then entered the expression. The calculator reports the value of the expression for the stored value of X.

FIGURE P.7   FIGURE P.8

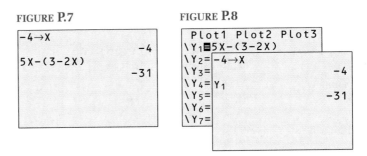

(b) We begin by entering the expression as Y_1 on the Y screen. Then we return to the home screen, store -4 in X, and select Y_1 from the list of Y-variables. (See Figure P.8.) The calculator reports the value of Y_1 (that is, the value of the expression) for the stored value of X.
(c) With the expression entered as Y_1 on the Y screen, we display a table of values and scroll to the line for which $x = -4$. (See Figure P.9.) The value in the Y_1 column is the value of the expression.

FIGURE P.9

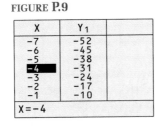

Evaluating an expression is a basic algebraic skill. Knowing how to do this with your calculator will save you time, improve your accuracy, and allow you to focus on expanding the scope of your knowledge.

Preamble Exercises

1. What is the difference between a numerical expression and an algebraic expression?

2. In the expression $3x + 2\pi$, explain why x is a variable but π is not a variable.

In Exercises 3–8, match each task given in Column A with a calculator key or menu item from Column B.

Column A	Column B
3. Select the number of decimal places.	(a) store
	(b) mode
4. Determine whether $\dfrac{3}{8} < 0.37$.	(c) clear
	(d) test
5. From the Y screen, go to the home screen.	(e) Y-var
	(f) quit
6. Enter Y_1 on the home screen.	
7. Replace x with 9.	
8. Erase the home screen.	

In Exercises 9–14, use your calculator to evaluate the given numerical expression.

9. $9 - 5^2$

10. $4(11) + 5 \cdot 3^2$

11. $\sqrt{16 - 12} + 1$

12. $3\sqrt{36} - 36 \div 2$

13. $|4^2 - 17|$

14. $10 - |2 - 3^2|$

15. (a) To the nearest hundredth, determine the decimal representation of each number.

    (i) $\dfrac{\sqrt{27}}{\sqrt{147}}$ (ii) $\dfrac{\pi}{3}$ (iii) $\dfrac{25}{\sqrt{10}}$

    (b) For which number(s) will the fraction option display a fraction? What fraction is displayed?

16. Mentally estimate the value of $1.9 \cdot 3.01 - \sqrt{50}$. Then check your estimate by determining the value of the expression to the nearest tenth.

In Exercises 17–20, use your calculator to evaluate the given numerical expression. Obtain the answer in simplified fraction form, if possible. Otherwise, round the answer to the nearest hundredth.

17. $\dfrac{9}{20} - \dfrac{15}{14} + \dfrac{5}{6}$

18. $\dfrac{3 \cdot 5 - 6}{4 - 3(-1)}$

19. $\dfrac{1}{2\pi} + \dfrac{\pi}{5}$

20. $\dfrac{\sqrt{24}}{\sqrt{15}}$

In Exercises 21–24, use your calculator's test feature to determine whether the given statement is true. If the statement is false, replace the symbol with $<$, $>$, or $=$ to make the statement true.

21. $-\dfrac{21}{20} < -\dfrac{20}{21}$

22. $\sqrt{5} + \sqrt{3} = \sqrt{8}$

23. $|-10| < 10$

24. $\sqrt{90} > 3\pi$

In Exercises 25–28, suppose that you have made the entries shown in the figure.

```
Plot1  Plot2  Plot3
\Y1■X+3
\Y2=-5→X
\Y3=                    -5
\Y4=
\Y5=
\Y6=
\Y7=
```

What value will your calculator return if you make the given entry on your home screen?

25. X   **26.** Y_1 **27.** $X = Y_1$ **28.** $X + Y_1$

29. If you need to evaluate $(x - 2)^2 + 7$ for six different values of x, why would entering the expression on the Y screen be more efficient than using only the home screen?

30. Suppose that you want to evaluate $x^2 + 2x$ for $x = 3$ by storing 3 in X. What additional step, if any, is needed if you enter the expression on the

    (a) home screen?   (b) Y screen?

In Exercises 31–34, use the Y-screen method to evaluate the algebraic expression for the given values of the variable. Round the results to the nearest hundredth.

31. $3x + 7$; $-4, \dfrac{9}{4}, 8.65, \sqrt{10}$

32. $|7 - x|$; $13, 3\dfrac{5}{7}, -0.68, 2\pi$

33. πr^3; $4, 3\dfrac{2}{9}, 3.7, 2\sqrt{7}$

34. $-x^2 + 3x + 10$; $-15, \dfrac{15}{7}, 1.48, \dfrac{\sqrt{2}}{2}$

35. Suppose that you want to use a table of values to evaluate an expression for the values 0.2, 0.5, 0.8, ..., 2.0. What table settings should be used so that the table displays only the given x-values?

36. Suppose that you need to evaluate $\sqrt{x^2 - 1}$ for all integer x-values between 5 and 12. What is the advantage of the table method over the Y-screen method?

In Exercises 37 and 38, use an appropriate calculator table to help you complete the given table of values.

37.

x	−0.5	0	0.5	1
$2x - x^2$	(a)	(b)	(c)	(d)

38.

x	0.6	0.8	1	1.2
$\dfrac{3}{x}$	(a)	(b)	(c)	(d)

In Exercises 39 and 40, complete the given table by using your calculator's table feature with the *Ask* option.

39.

x	−7	−2.6	5	9.4
$x + \lvert x \rvert$	(a)	(b)	(c)	(d)

40.

x	5.24	−0.59	−4	46.84
$\sqrt{x + 5}$	(a)	(b)	(c)	(d)

Basic Concepts Revisited

Zeno, an early Greek philosopher, argued that the speedy Achilles could never catch up to a crawling tortoise as long as the tortoise had any head start. By the time Achilles reaches the tortoise's starting point A, the tortoise will have moved ahead to point B. When Achilles reaches point B, the tortoise will have moved forward to point C, and so on. Thus, according to Zeno's paradox, Achilles will always lag behind the tortoise.

Motion, which involves distance, speed, and time, can be modeled by an algebraic expression. In general, expressions have a variety of forms, including polynomial expressions, rational expressions, and radical expressions. Although a graphing calculator can be used to evaluate an algebraic expression, we often need to simplify the expression or rewrite the expression in some other more convenient form.

In this chapter, we review the fundamental concepts, skills, and language of basic algebra. Although you may already be familiar with much of this material, conscientiously reviewing it will provide you with the essential foundation for all the topics in this course.

R.1 *The Real Numbers*

The Real Number System ■ *Order and Absolute Value* ■
Properties of the Real Numbers

Because much of algebra consists of generalizations and rules about numbers, you need to be familiar with the various sets of numbers that, taken together, form the set of real numbers. In this section, we review the structure and properties of the real number system.

The Real Number System

The basic sets of numbers in the real number system are the **natural numbers,** the **whole numbers,** and the **integers.**

$$N = \{1, 2, 3, 4, \ldots\}$$ The set of natural numbers
$$W = \{0, 1, 2, 3, \ldots\}$$ The set of whole numbers
$$J = \{\ldots, -3, -2, -1, 0, 1, 2, 3, \ldots\}$$ The set of integers

The **rational numbers** are numbers that can be written as quotients of integers. The set of rational numbers can be represented with **set-builder notation** as follows:

$$Q = \left\{\frac{p}{q} \mid p \text{ and } q \text{ are integers, } q \neq 0\right\}$$

Set Q is read "the set of all numbers of the form $\frac{p}{q}$ such that p and q are integers and q is not 0."

Rational numbers have decimal names that either terminate or have repeating blocks.

Rational number	Decimal name	Type of decimal
$\frac{2}{5}$	0.4	Terminating
$\frac{7}{3}$	$2.333333\ldots = 2.\overline{3}$	Repeating
$\frac{9}{11}$	$0.818181\ldots = 0.\overline{81}$	Repeating

Numbers that do not have terminating or repeating decimal names are called **irrational numbers.** For example, the numbers $\sqrt{5} = 2.23606797\ldots$ and $\pi = 3.14159265\ldots$ are irrational numbers.

The symbol $\approx$ (read as "approximately equal to") is used when decimal numbers are rounded off.

$$\frac{2}{3} \approx 0.67 \qquad \sqrt{5} \approx 2.24 \qquad \pi \approx 3.142$$

The set of **real numbers,** denoted by **R,** is the set of all numbers with decimal names. Figure R.1 shows the structure of the real numbers.

FIGURE **R.1**

THE REAL NUMBERS

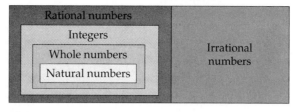

EXAMPLE **1**   **Identifying Numbers**

Identify the numbers in the following set that are

(a) whole numbers.　　　　(b) integers.
(c) rational numbers.　　　(d) irrational numbers.

$$\left\{ \frac{5}{8}, \sqrt{7}, -\frac{\pi}{2}, 3, 1.28, 0, -\frac{11}{9}, -693 \right\}$$

SOLUTION

(a) Whole numbers: $3, 0$

(b) Integers: $3, 0, -693$

(c) Rational numbers: $\dfrac{5}{8}, 3, 1.28, 0, -\dfrac{11}{9}, -693$

(d) Irrational numbers: $\sqrt{7}, -\dfrac{\pi}{2}$

Order and Absolute Value

We often use a **number line** to visualize real numbers and their relation to one another. Figure R.2 shows a number line with the numbers $-5, -3.5, 0, \sqrt{3}, 2, \frac{10}{3}, 4.12,$ and 2π graphed on it.

FIGURE **R.2**

Symbols such as $=, <,$ and $>$ indicate the order of two real numbers. For example, because $\frac{3}{2}$ and 1.5 are associated with the same point of the number line, we say that $\frac{3}{2} = 1.5$. In Figure R.2, we see that -3.5 is to the right of -5 on the number line, so we write $-3.5 > -5$ (read as "-3.5 is greater than -5"). Similarly, Figure R.2 shows that $\sqrt{3}$ lies to the left of 2 on the number line, so we write $\sqrt{3} < 2$ (read as "$\sqrt{3}$ is less than 2").

On a number line, the distance between any number and 0 is called the **absolute value** of the number. We indicate the absolute value of a number a by writing $|a|$.

> **Definition of Absolute Value**
>
> For any real number a, $|a| = \begin{cases} a & \text{if } a \geq 0 \\ -a & \text{if } a < 0 \end{cases}$

For example, if $a = 7$, then a ≥ 0 and we use the first part of the definition: $|7| = 7$. However, if $a = -2$, then $a < 0$ and we use the second part of the definition: $|-2| = -(-2) = 2$.

To determine the distance between any two points of the number line, we simply subtract the coordinates of the points. We can subtract in either order provided that we take the absolute value of the result in order to guarantee that the distance is nonnegative.

> **Distance Between Two Points of a Number Line**
>
> Let a and b be the coordinates of two points of a number line. The distance between the points is $d = |a - b| = |b - a|$.

EXAMPLE 2   **The Distance Between Two Points of a Number Line**

Show two ways to determine the distance d between the points whose coordinates are -2 and 6.

SOLUTION

(i) $d = |-2 - 6| = |-8| = 8$
(ii) $d = |6 - (-2)| = |8| = 8$

Properties of the Real Numbers

In this section, we summarize the fundamental properties of the real numbers. These properties are the basic rules of algebra from which other rules can be derived. When we simplify expressions or solve equations, for example, these and other properties are needed in order to justify the steps that we take.

Property
Example

> **The Fundamental Properties of Algebra**
>
> If a, b, and c are algebraic expressions, then
>
> | *Property* | *Example* | |
> |---|---|---|
> | $a + b = b + a$ | $2x + 5 = 5 + 2x$ | **Commutative Property of Addition** |

$ab = ba$	$(x - 2) \cdot 4 = 4(x - 2)$	**Commutative Property of Multiplication**
$a + (b + c) = (a + b) + c$	$3x + (2x + 1) = (3x + 2x) + 1$	**Associative Property of Addition**
$a(bc) = (ab)c$	$\dfrac{3}{5}(5z) = \left(\dfrac{3}{5} \cdot 5\right)z$	**Associative Property of Multiplication**
$a + 0 = 0 + a = a$	$-4x^3 + 0 = -4x^3$	**Additive Identity Property**
$a \cdot 1 = 1 \cdot a = a$	$x^2 + 3 = 1(x^2 + 3)$	**Multiplicative Identity Property**
$a + (-a) = 0$	$7x + (-7x) = 0$	**Additive Inverse Property**
$a \cdot \dfrac{1}{a} = 1, a \neq 0$	$(x - 1) \cdot \dfrac{1}{x - 1} = 1, x \neq 1$	**Multiplicative Inverse Property**
$a(b + c) = ab + ac$	$x(2y + 5) = 2xy + 5x$	**Distributive Property**

On the number line, 5 and -5 are on opposite sides of 0 and are the same distance from 0. We call these numbers *opposites*. In general, the numbers represented by x and $-x$ are opposites. If x represents a positive number, then the symbol $-x$ represents a negative number. However, if x represents a negative number, then the symbol $-x$ represents a positive number. For this reason, we read $-x$ as "the opposite of x," not "negative x."

The following properties of opposites are often used to simplify expressions.

The Properties of Opposites

If $a, b,$ and c are algebraic expressions, $b \neq 0$, then

Property	*Example*	
$-a = -1 \cdot a$	$-(x + 2) = -1(x + 2)$	**Multiplication Property of -1**
$-(-a) = a$	$-(-3x) = 3x$	**Property of the Opposite of an Opposite**
$-(a - c) = c - a$	$-(x - 5) = 5 - x$	**Property of the Opposite of a Difference**
$-\dfrac{a}{b} = \dfrac{-a}{b} = \dfrac{a}{-b}$	$-\dfrac{2x}{y} = \dfrac{-2x}{y} = \dfrac{2x}{-y}$	**Property I of Signs in Quotients**
$-\dfrac{-a}{-b} = -\dfrac{a}{b}$	$-\dfrac{-(2 - x)}{-3c} = -\dfrac{2 - x}{3c}$	**Property II of Signs in Quotients**

The number 0 plays an important role in multiplication and division as well as in addition.

Properties of Zero

$a \cdot 0 = 0 \cdot a = 0$ **Multiplication Property of 0**

For $a \neq 0$, $\dfrac{0}{a} = 0$.

For any real number a, $\dfrac{a}{0}$ is undefined.

EXAMPLE 3   **Applying the Properties of Real Numbers**

Use the indicated property to supply the missing expression.

(a) Associative Property of Multiplication: $\dfrac{2}{3}\left[\dfrac{3}{2}(x + 1)\right] =$ ▨▨▨

(b) Multiplication Property of -1: $5x + (-x) =$ ▨▨▨

(c) Commutative Property of Addition: $3(5 + y) =$ ▨▨▨

(d) Property of the Opposite of a Difference: $-(4 - 3t) =$ ▨▨▨

(e) Distributive Property: $8(x + z) =$ ▨▨▨

(f) Distributive Property: $5x + 3x =$ ▨▨▨

(g) Additive Inverse Property: $-(a + c) + (a + c) =$ ▨▨▨

SOLUTION

(a) $\dfrac{2}{3}\left[\dfrac{3}{2}(x + 1)\right] = \left(\dfrac{2}{3} \cdot \dfrac{3}{2}\right)(x + 1)$

(b) $5x + (-x) = 5x + (-1x)$

(c) $3(5 + y) = 3(y + 5)$

(d) $-(4 - 3t) = 3t - 4$

(e) $8(x + z) = 8x + 8z$

(f) $5x + 3x = x(5 + 3)$

(g) $-(a + c) + (a + c) = 0$

Speaking the Language

1. If c and d are integers and $d \neq 0$, then the number $\dfrac{c}{d}$ is called a(n) ▨▨▨ number.

2. The set containing 0, the natural numbers, and their additive inverses is called the set of ▨▨▨ .

3. If a number x is to the right of a number y on the number line, then we say that y is ▨▨▨ x.

4. The distance between a number and 0 on the number line is called the ▨▨▨ of the number.

R.1   Exercises

1. There is an observable pattern in the decimal number 0.12112111211112 . . . . Does this mean that the number is a rational number? Why?

2. Repeating decimals are often written with a bar over the repeating block: 1.828282 . . . = $1.\overline{82}$. Use this notation to write the decimal name of $\frac{1}{7}$.

In Exercises 3 and 4, identify the numbers in the given set that are

(a) natural numbers   (b) whole numbers
(c) integers   (d) irrational numbers
(e) rational numbers

3. $\left\{ -15, 4.\overline{52}, \frac{\pi}{2}, \sqrt[3]{36}, -\frac{9}{5}, -8.2, \sqrt{10}, \sqrt{64}, 6\frac{3}{7} \right\}$

4. $\left\{ 31, -\pi, 0, -\sqrt{7}, \frac{12}{4}, -4.375, -\frac{9}{4}, \sqrt{\frac{25}{9}}, 67.\overline{43} \right\}$

In Exercises 5–8, write the number as a terminating or repeating decimal.

5. $\dfrac{41}{33}$

6. $\dfrac{29}{11}$

7. $6\dfrac{7}{8}$

8. $\dfrac{3}{16}$

9. We say that A is a *subset* of B if every element of A is also an element of B. Explain why **R** is not a subset of the set Q of rational numbers.

10. Explain how to enter a mixed number such as $5\frac{3}{7}$ into a calculator.

11. If $\dfrac{1}{3} = 0.33333$ . . . and $\dfrac{2}{3} = 0.66666$ . . . , to what number is 0.99999 . . . equivalent?

12. If $c < a$ and $b > a$, what is the order (from left to right) of a, b, and c on the number line?

In Exercises 13 and 14, write the list of numbers in order beginning with the smallest number.

13. $\dfrac{1319}{500}, 2.\overline{23}, \dfrac{3\pi}{4}, \sqrt{5}, \dfrac{26}{11}, \sqrt{15} - \sqrt{2}$

14. $\sqrt{10} + \sqrt{17}, 7.8\overline{53}, \dfrac{987}{132}, \sqrt{60}, \dfrac{89}{12}, \dfrac{5\pi}{2}$

In Exercises 15–18, a number n is described in reference to a number line. State all possible values of n.

15. The number n is 5 units from -3.

16. The number n is $|-4|$ units from 0.

17. When decreased by 3, the number n is 2 units from 0.

18. The number n has an absolute value of 17.

In Exercises 19–24, determine the distance on a number line between the two numbers. Round answers to the nearest hundredth.

19. $4, -7$

20. $-5, -16$

21. $\dfrac{11}{5}, 1.3$

22. $-12.2, 3.6$

23. $\pi, 4$

24. $-5, -\sqrt{20}$

25. Suppose that a number line "ruler" is used to measure the distance between points P and Q and that the coordinates of points P and Q are -9 and 3, respectively. If the number line is shifted 5 units to the left, how would the coordinates of P and Q be affected? How would the distance between P and Q be affected?

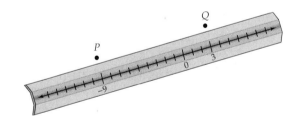

26. In the definition of absolute value, we say that $|a|$ is either a or $-a$. Does this mean that $|a|$ could be negative? Explain.

In Exercises 27–30, evaluate the given expression.

27. $|-4| - |-10|$

28. $-|-6| + |5|$

29. $|-2 - |-7||$

30. $||-5| - |-9||$

In Exercises 31–34, fill in the blank with $<$, $>$, or $=$.

31. If $x > 0$, then $|x|$ ▮▮▮ $-x$.

32. If $x < 0$, then $|x|$ ▮▮▮ x.

33. If $x < 0$, then $-x$ ▮▮▮ 0.

34. If $x < 0$, then $|x|$ ▮▮▮ $|-x|$.

35. If either a or b is 0, then the rule $|ab| = |a| \, |b|$ is true. Use the definition of absolute value to show that the rule also holds when
(a) a and b are both positive.
(b) a and b are both negative.
(c) a and b have opposite signs.

36. For real numbers a, b, and c, the Transitive Property of Equality states that if $a = b$ and $b = c$, then $a = c$. For lines L, M, and N, show whether the relations *parallel* and *perpendicular* are transitive.

37. Give examples to show that subtraction is neither commutative nor associative.

38. Give examples to show that division is neither commutative nor associative.

In Exercises 39–46, identify the property that justifies the given statement.

39. $3(x - y) = 3x - 3y$ **40.** $-4x + 4x = 0$

41. $6\left[\dfrac{2}{3}(x + 1)\right] = \left(6 \cdot \dfrac{2}{3}\right)(x + 1)$

42. $(x^2 + 1) \cdot \dfrac{1}{x^2 + 1} = 1$

43. $3y - 7y = (3 - 7)y$ **44.** $-8a + 0 = -8a$

45. $3x + yz = yz + 3x$ **46.** $1(y + 7) = y + 7$

In Exercises 47–50, fill in the blank to make the statement true. State the property that justifies your answer.

47. $3x \cdot \text{\_\_\_\_} = 1$ **48.** $-y + \text{\_\_\_\_} = 0$

49. $-\dfrac{5}{8} \cdot \text{\_\_\_\_} = -\dfrac{5}{8}$ **50.** $\text{\_\_\_\_}(3 - 5y) = -3 + 5y$

In Exercises 51–54, use the indicated property to supply the missing expression.

51. Commutative Property of Addition;
$4 - x = \text{\_\_\_\_}$

52. Associative Property of Multiplication;
$-5(3y) = \text{\_\_\_\_}$

53. Associative Property of Addition;
$(x + 5) + 6 = \text{\_\_\_\_}$

54. Commutative Property of Multiplication;
$(x + 3)(x - 5) = \text{\_\_\_\_}$

In Exercises 55–62, use the Distributive Property to rewrite the expression.

55. $5(3a + b)$ **56.** $-x(3 - y)$

57. $-4(2x - 5y)$ **58.** $2s(3t + 4)$

59. $15x + 20y$ **60.** $-8a - 4b$

61. $3ab + a$ **62.** $-2ac + 6bc$

$R.2$ *Expressions and Exponents*

Expressions and the Order of Operations ■
Properties of Exponents ■ *Scientific Notation*

We have seen that a calculator can be a useful aid in evaluating numerical and algebraic expressions. However, your understanding of the standard order in which operations are performed is essential to your success in mathematics.

In this section, we revisit algebraic expressions, placing particular emphasis on expressions that involve exponents. You will see how the properties of exponents can be used to evaluate and simplify expressions. Finally, we use exponents to represent numbers in a form called *scientific notation*, which you are likely to encounter even in nonscience courses.

Expressions and the Order of Operations

In the Preamble, we described an algebraic expression as any combination of variables, numbers, grouping symbols, and operations symbols.

By grouping symbols, we usually mean parentheses, brackets, and braces. However, fraction bars, radical symbols, and absolute value symbols are also examples of grouping symbols.

The four basic operations that can be indicated in an algebraic expression are addition, subtraction, multiplication, and division. In a sum $a + b$, the numbers a and b are called **addends.** In a product ab, the numbers a and b are called **factors.**

A positive integer exponent can be used to indicate repeated multiplication by the same factor.

Exponential form	*Expanded form*
$(-3)^2$	$(-3)(-3)$
x^5	$x \cdot x \cdot x \cdot x \cdot x$
$(x + 1)^3$	$(x + 1)(x + 1)(x + 1)$

> **Definition of a Positive Integer Exponent**
>
> If a is a real number and n is a positive integer, then
>
> $$a^n = \underbrace{a \cdot a \cdot a \cdot \ldots \cdot a}_{n \text{ factors of } a}$$

The expression a^n is called an **exponential expression.** The number a is the **base,** and the number n is the **exponent.**

In general, a^n is read as "a to the nth power." However, a^2 and a^3 are usually read as "a squared" and "a cubed," respectively. When an exponent is not written, it is understood to be 1: $x = x^1$.

The addends of an expression are called **terms**. You can begin to identify the terms of an expression by drawing a line under each product in the expression. (Remember that an exponential expression is a product.) Then convert each indicated subtraction that is not underlined into an addition. The resulting products are the terms of the expression.

$$\underline{2(x + 3)} - \underline{5x} = \underline{2(x + 3)} + \underline{(-5x)}$$

The terms are $2(x + 3)$ and $-5x$.

Variable terms contain at least one variable, whereas **constant terms** are simply numbers. The **numerical coefficient** (or simply **coefficient**) of a term is the numerical factor.

Expression	*Variable terms*	*Constant term*	*Coefficients*
$x^3 + 9x^2 - 6x + 4$	$x^3, 9x^2, -6x$	4	$1, 9, -6, 4$
$8ab + 3(y - 2)$	$8ab, 3(y - 2)$	0	$8, 3$

Algebraic expressions can be evaluated only after the variables have been replaced with specific values. Algebraic expressions are **equivalent** if they have the same value for all permissible replacements of the variable(s). For example, the Distributive Property guarantees that $2(x + y)$ and $2x + 2y$ are equivalent expressions because they will have the same value for any replacements of x and y.

In the Preamble, we discussed a variety of ways to use a calculator to evaluate expressions. All calculators are designed to perform operations in a standard order, called the **Order of Operations.**

The Order of Operations

1. If grouping symbols are present, perform all operations inside of them according to the following order. For grouping symbols within grouping symbols, start with the innermost group and work outward.
2. Evaluate exponents and radicals.
3. Perform multiplication and division from left to right.
4. Perform addition and subtraction from left to right.

Example 1 illustrates how we use the Order of Operations to evaluate a numerical expression.

EXAMPLE **1** **Evaluating Numerical Expressions**

Evaluate the following numerical expressions.

(a) $-3(2 \cdot 6 - 10)$ (b) $\dfrac{4^2 - 2 \cdot 5}{(3 - 1)^2}$

SOLUTION

(a) $-3(2 \cdot 6 - 10) = -3(12 - 10)$ **Begin with the multiplication inside the grouping symbols.**

$= -3(2)$ **Still inside the grouping symbols, perform the subtraction.**

$= -6$

(b) The fraction bar serves as a grouping symbol. Therefore, we evaluate the numerator and the denominator separately and perform the division last. If you use a calculator to evaluate the expression, you may need to insert parentheses.

$\dfrac{4^2 - 2 \cdot 5}{(3 - 1)^2} = \dfrac{16 - 2 \cdot 5}{(2)^2}$ **In the numerator, the exponent has the highest rank. In the denominator, work inside the parentheses.**

$= \dfrac{16 - 10}{4}$ **Multiplication ranks higher than subtraction.**

$= \dfrac{6}{4} = \dfrac{3}{2}$ **Subtract and simplify the fraction.**

When we evaluate an algebraic expression, we replace each occurrence of a variable with the given value of that variable. The result is a numerical expression that is evaluated with the Order of Operations.

EXAMPLE 2   **Evaluating Algebraic Expressions**

Evaluate the given algebraic expressions as indicated.

(a) $x^3 + 9x^2 - 6x + 4$ for $x = -3$
(b) $8ab + 3(b - 2)$ for $a = 2$ and $b = -1$

SOLUTION

(a) $x^3 + 9x^2 - 6x + 4 = (-3)^3 + 9(-3)^2 - 6(-3) + 4$
$= -27 + 9(9) - 6(-3) + 4$
$= -27 + 81 + 18 + 4$
$= 76$

(b) $8ab + 3(b - 2) = 8(2)(-1) + 3(-1 - 2)$
$= -16 + 3(-3)$
$= -25$

Alpha   A special key is used when a variable other than x is entered in a calculator.

Figure R.3 shows the home-screen method for evaluating the expression in Example 2(b).

FIGURE **R.3**

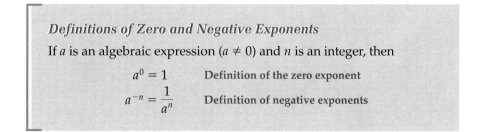
```
2→A
                              2
-1→B
                             -1
8AB+3(B-2)
                            -25
```

Properties of Exponents

At the beginning of this section, we defined a positive integer exponent. The following definitions allow exponents to be *any* integer.

> ### *Definitions of Zero and Negative Exponents*
>
> If a is an algebraic expression ($a \neq 0$) and n is an integer, then
>
> $$a^0 = 1$$ **Definition of the zero exponent**
>
> $$a^{-n} = \frac{1}{a^n}$$ **Definition of negative exponents**

$\overline{\text{EXAMPLE}}\ 3$ **Zero and Negative Exponents**

(a) Compare the values of $3x^0$ and $(3x)^0$.
(b) Write $5x^{-2}$ and $(5x)^{-2}$ with a positive exponent.
(c) Evaluate $3^{-2} \cdot 27$.
(d) Evaluate $5^{-1} - 5^0$.
(e) Evaluate $4^2 \cdot 4^{-2} \div \dfrac{1}{4^0}$.

SOLUTION

(a) $3x^0 = 3 \cdot 1 = 3$ The base is x.
 $(3x)^0 = 1$ The base is $3x$.

(b) $5x^{-2} = 5 \cdot x^{-2} = 5 \cdot \dfrac{1}{x^2} = \dfrac{5}{x^2}$ The base is x.

 $(5x)^{-2} = \dfrac{1}{(5x)^2}$ The base is $5x$.

(c) $3^{-2} \cdot 27 = \dfrac{1}{3^2} \cdot 27 = \dfrac{1}{9} \cdot 27 = 3$

(d) $5^{-1} - 5^0 = \dfrac{1}{5} - 1 = \dfrac{1}{5} - \dfrac{5}{5} = -\dfrac{4}{5}$

(e) $4^2 \cdot 4^{-2} \div \dfrac{1}{4^0} = 4^2 \cdot \dfrac{1}{4^2} \div \dfrac{1}{1} = 16 \cdot \dfrac{1}{16} \div 1 = 1$

The following properties apply to all exponential expressions with integer exponents. In all cases, assume that the bases and denominators are not 0.

Properties of Exponents

If a and b are algebraic expressions, and if m and n are integers, then

Property	*Example*	
$a^m \cdot a^n = a^{m+n}$	$x^2 \cdot x^5 = x^7$	**Product Rule for Exponents**
$\dfrac{a^m}{a^n} = a^{m-n}$	$\dfrac{(3y)^5}{(3y)^3} = (3y)^2$	**Quotient Rule for Exponents**
$(a^m)^n = a^{mn}$	$(z^{-3})^2 = z^{-6} = \dfrac{1}{z^6}$	**Power to a Power Rule**
$(ab)^m = a^m b^m$	$(-2x^2)^3 = (-2)^3(x^2)^3 = -8x^6$	**Product to a Power Rule**
$\left(\dfrac{a}{b}\right)^m = \dfrac{a^m}{b^m}$	$\left(\dfrac{3}{x^{-2}}\right)^3 = \dfrac{3^3}{(x^{-2})^3} = \dfrac{3^3}{x^{-6}} = 27x^6$	**Quotient to a Power Rule**
$\left(\dfrac{a}{b}\right)^{-n} = \left(\dfrac{b}{a}\right)^n$	$\left(\dfrac{2x}{y}\right)^{-3} = \left(\dfrac{y}{2x}\right)^3$	**Quotient to a Negative Power Rule**

Even when all the exponents in an algebraic expression are positive, applying the properties of exponents to simplify the expression may result in negative exponents. We usually write the simplified expression with positive exponents.

EXAMPLE **4**   **Simplifying Exponential Expressions**

Simplify the given exponential expressions.

(a) $(2a^3x^2)(-3a^2x)$ (b) $(-x^3y)^2$

(c) $\dfrac{15c^3}{9c^7}$ (d) $(2xy^2)^3(3x^2y)^2$

SOLUTION

(a) $(2a^3x^2)(-3a^2x^1) = 2(-3)a^{3+2}x^{2+1}$ **Product Rule for Exponents**
$$= -6a^5x^3$$

(b) $(-x^3y)^2 = (-x^3)^2(y)^2$ **Product to a Power Rule**
$$= x^6y^2$$

(c) $\dfrac{15c^3}{9c^7} = \dfrac{3}{3} \cdot \dfrac{5}{3} \cdot c^{3-7}$ **Quotient Rule for Exponents**

$$= 1 \cdot \dfrac{5}{3} \cdot c^{-4}$$

$$= \dfrac{5}{3} \cdot \dfrac{1}{c^4}$$

$$= \dfrac{5}{3c^4}$$

(d) $(2xy^2)^3(3x^2y)^2 = (2)^3(x)^3(y^2)^3 \cdot (3)^2(x^2)^2(y)^2$
$$= 8x^3y^6 \cdot 9x^4y^2$$
$$= 8 \cdot 9 \cdot x^{3+4} \cdot y^{6+2}$$
$$= 72x^7y^8$$

When operations are to be performed with exponential expressions, you can begin by changing all negative exponents to positive exponents. Usually, however, a more efficient approach is to begin by performing the operations and then writing the result with positive exponents.

EXAMPLE **5**   **Simplifying Exponential Expressions**

■ *NOTE   In all parts of Example 5, moving a factor from the numerator to the denominator or from the denominator to the numerator changed the sign of the exponent. However, keep in mind that this technique applies only to factors, not to terms.*

Simplify the given exponential expressions and write the result with positive exponents. Assume that all variables represent positive numbers.

(a) $(-2x^{-3}y)(x^5y^{-2}) = (-2 \cdot 1)(x^{-3} \cdot x^5)(y^1 \cdot y^{-2})$
$$= -2x^2y^{-1} = \dfrac{-2x^2}{y}$$

(b) $(-x^2y^{-3}z)^{-1} = -x^{-2}y^3z^{-1} = -\dfrac{y^3}{x^2z}$

(c) $\dfrac{12x^2y^{-3}}{4x^{-3}y} = \dfrac{12}{4} \cdot \dfrac{x^2}{x^{-3}} \cdot \dfrac{y^{-3}}{y} = 3x^{2-(-3)}y^{-3-1} = 3x^5y^{-4} = \dfrac{3x^5}{y^4}$

(d) $\left(\dfrac{-2x^{-2}}{z}\right)^{-3} = \dfrac{(-2x^{-2})^{-3}}{z^{-3}} = \dfrac{(-2)^{-3}(x^{-2})^{-3}}{z^{-3}} = \dfrac{x^6z^3}{(-2)^3} = -\dfrac{x^6z^3}{8}$

Scientific Notation

The distance from the Earth to the sun is approximately 93,000,000 miles. An example of a small number is the mass of one hydrogen atom, which is approximately 0.00000000000000000000000017 gram. Writing and calculating with very large or very small numbers such as these can be facilitated by the use of exponents to write the numbers in **scientific notation.**

$$93,000,000 = 9.3 \cdot 10^7$$

$$0.00000000000000000000000017 = 1.7 \cdot 10^{-24}$$

A number in scientific notation has the form $n \cdot 10^p$, where $1 \le n < 10$ and p is an integer. A positive exponent on 10 indicates the number of decimal places to move the decimal point to the right. A negative exponent on 10 indicates the number of decimal places to move the decimal point to the left.

EXAMPLE **6** **Changing from Scientific Notation to Decimal Form**

The given numbers are in scientific notation. Write the numbers in decimal form.

(a) $6.23 \cdot 10^5$
(b) $4.0 \cdot 10^{-6}$

SOLUTION

(a) $6.23 \cdot 10^5$ The exponent is 5, so move the decimal point
6.23000.0   5 places to the right.

$6.23 \cdot 10^5 = 623{,}000$

(b) $4.0 \cdot 10^{-6}$ The exponent is -6, so move the decimal
0.000004.0   point 6 places to the left.

$4.0 \cdot 10^{-6} = 0.000004$

EXAMPLE **7** **Changing from Decimal Form to Scientific Notation**

The given numbers are in decimal form. Write the numbers in scientific notation $n \cdot 10^p$.

(a) 125,000,000,000
(b) 0.000000000734

SOLUTION

(a) 125,000,000,000.0

1.25000000000.0   Move the decimal point so that $n = 1.25$. The original
decimal point was 11 places to the right, so $p = 11$.

$125,000,000,000 = 1.25 \cdot 10^{11}$

(b) 0.000000000734

0.0000000007.34   Move the decimal point so that $n = 7.34$. The original
decimal point was 10 places to the left, so $p = -10$.

$0.000000000734 = 7.34 \cdot 10^{-10}$

Scientific   Most calculators allow you to enter numbers in scientific notation.
Typically the **EE** key is used. For example, 5.3E6 is interpreted as $5.3 \cdot 10^6$.

Mode   Typical options in the mode menu are *Normal*, *Sci*, and *Eng*. Select *Sci* to
report results in scientific notation.

EXAMPLE **8**   **Calculating with Scientific Notation**

Use your calculator to evaluate $(7.4 \cdot 10^8)(2.5 \cdot 10^{-3})$. Enter the numbers
in scientific notation and report the results in scientific notation.

SOLUTION

After selecting *Sci* from the mode menu, we use the **EE** key to enter the
expression. (See Figure R.4.) We interpret the result as $1.85 \cdot 10^6$.

FIGURE **R.4**

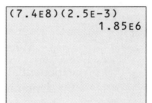

```
(7.4E8)(2.5E-3)
             1.85E6
```

Speaking the Language

1. The numbers x and y are called ▨▨▨▨▨ in the sum $x + y$, whereas they
   are called ▨▨▨▨▨ in the product xy.
2. An expression that can be evaluated is called a(n) ▨▨▨▨▨ expression,
   whereas an expression such as $2x + 3y - 5$ is called a(n) ▨▨▨▨▨
   expression.
3. Parentheses and fraction bars are examples of ▨▨▨▨▨.
4. The number 2300 is said to be in ▨▨▨▨▨ form, whereas the same
   number in ▨▨▨▨▨ is $2.3 \cdot 10^3$.

R.2 Exercises

1. What are equivalent algebraic expressions?

2. To evaluate $\dfrac{x^2 - 3}{\sqrt{x + 5}}$ with a calculator, you must insert parentheses. Why?

In Exercises 3–8, perform the indicated operations.

3. $-3^2 + 2(-1)^5 + 6$

4. $7 - 3 \cdot 5 + 24 \div (-8)$

5. $9 - 3[2 - (5 - 1)]$

6. $5 - [2(4 - 1)^2 - 3]/3$

7. $\dfrac{-5 + \sqrt{(-5)^2 - 4(2)(-7)}}{2(2)}$

8. $\sqrt{[3 - (-2)]^2 + (-7 - 5)^2}$

In Exercises 9–14, evaluate the expression for the given value of the variable.

9. $5x^2 + 7x - 6; \quad -3$

10. $4x - 3(1 - 2x); \quad 6$

11. $\dfrac{3}{x} + \dfrac{5}{x - 2}; \quad 4$

12. $\dfrac{x^2 + 3}{x + 2}; \quad -1$

13. $\sqrt{15 - 2x}; \quad -5$

14. $8 - |x - 2|; \quad 5$

In Exercises 15–18, evaluate the expression for $a = -3$, $b = -2$, and $c = 4$.

15. $-a^2 - 5b - c$

16. $(a + 2c)^b$

17. ab^c

18. $\sqrt{a^2 + c^2}$

19. Try to evaluate the following expressions with your calculator.

   (i) $-2^4, (-2)^4$ (ii) $0^0, 0^{-3}$ (iii) $-2^3, (-2)^3$

   (a) If $a \neq 0$ and n is a positive integer, under what conditions is $-a^n = (-a)^n$?

   (b) Why do you receive an error message for the expressions in (ii)?

20. Evaluate the expression 2^{2^3}. If you attempt to evaluate the expression by entering 2^2^3 in a calculator, the result may not be correct. Why?

In Exercises 21–30, evaluate the expression.

21. $(-2)^3(-2)^4$

22. $\dfrac{5^8}{5^5}$

23. -4^0

24. 7^0

25. 6^{-2}

26. $(-5)^{-3}$

27. $\left(\dfrac{2}{3}\right)^{-1}\left(\dfrac{3}{2}\right)^3$

28. $3^4 \cdot 3^{-5} \div 3^{-7} \div 3^4$

29. $\dfrac{-12 \cdot 3^{-2}}{4^{-2}}$

30. $\dfrac{5^0 - 2^{-1}}{4^0 + 3^{-2}}$

In Exercises 31–42, simplify the expression.

31. $(-6x^4)(3x^2)$

32. $(a^2b^3)(3ab^4)$

33. $(4y)^3$

34. $(-3x^2)^4$

35. $(-2a^6b^3)^5$

36. $(4r^4s)^3$

37. $\dfrac{z^8}{z^{10}}$

38. $\dfrac{5w^3}{w^7}$

39. $\dfrac{-15a^6b}{20ab^6}$

40. $\dfrac{24x^4y^7}{40x^2y^3}$

41. $(3x^4y)^2(xy^3)^4$

42. $(-3xy^4)^2(-2x^3y^2)^3$

43. A common error is to write $2x^{-1}$ as $\dfrac{1}{2x}$. Produce a table of values for each expression and observe that the two expressions never have the same value. Use the tables to write an inequality that relates the two expressions for $x > 0$.

44. A common error is to write $\dfrac{x^{-2} + 3}{y}$ as $\dfrac{3}{x^2y}$ or $\dfrac{3}{x^2 + y}$. What must be true in order to move an exponential expression from the numerator to the denominator?

In Exercises 45–56, simplify the expression and write the result with positive exponents.

45. $\dfrac{x^{-2}}{x^3}$

46. $y^{-3} \cdot y^5$

47. $\dfrac{3x^{-5}}{x^4}$

48. $\dfrac{2a^{-4}}{3b^{-2}}$

49. $\dfrac{1}{5y^{-6}}$

50. $3w^{-4}z^3$

51. $\left(\dfrac{2b^{-2}}{c^{-3}}\right)^{-3}$

52. $(-3xy^{-2})^{-3}(2x^2y^{-1})^3$

53. $\dfrac{8x^{-5}y^{-2}}{24x^3y^{-4}}$

54. $\dfrac{x^{-2}y^3}{x^{-7}y^{-2}}$

55. $\left(\dfrac{3x^2}{y^{-1}}\right)^{-2}\left(\dfrac{y}{x}\right)^{-3}$

56. $\dfrac{(a^2b^{-2})^3}{(3a^{-2}b)^{-1}}$

In Exercises 57–60, you will need the formula for computing the value A of an investment of P dollars at an annual interest rate r compounded n times per year for t years. The formula is

$$A = P\left(1 + \frac{r}{n}\right)^{nt}$$

57. Suppose that you borrow \$3000 at 9% compounded monthly. How much interest do you owe if you pay off the loan at the end of (a) 8 months? (b) 15 months?

58. Suppose that you invest \$2500 for 18 months at 6.5%. What is the value of the investment if the interest is compounded (a) monthly? (b) daily?

59. In 1998 the average cost of a compact car was approximately \$15,000. Assuming an annual inflation rate of 4.7%, what is the predicted cost of a compact car in 2004?

60. In 1998, the average annual cost of food for a family of four was \$7500. With an annual inflation rate of 5.2%, what is the expected annual cost of food in 2008?

In Exercises 61–68, write the given number in decimal form.

61. $3 \cdot 10^7$

62. $5.37 \cdot 10^5$

63. $4.36 \cdot 10^3$

64. $7.4 \cdot 10^6$

65. $3.34 \cdot 10^{-6}$

66. $5 \cdot 10^{-3}$

67. $4.2 \cdot 10^{-4}$

68. $7.36 \cdot 10^{-8}$

In Exercises 69–76, write the given number in scientific notation.

69. 7,030,000

70. 97,200,000

71. 532,000

72. 8000

73. 0.000437

74. 0.0000072

75. 0.002

76. 0.00065

In Exercises 77–82, use your calculator to perform the indicated operations. Enter numbers in scientific notation and express the result in scientific notation.

77. $(4.6 \cdot 10^5)(3.1 \cdot 10^{-2})$

78. $(6.13 \cdot 10^{-12})(5.2 \cdot 10^4)$

79. $(4.6 \cdot 10^7)^5$

80. $(2.6 \cdot 10^{-3})^2$

81. $\dfrac{4.2 \cdot 10^{-6}}{7 \cdot 10^{-2}}$

82. $\dfrac{6.5 \cdot 10^3}{9.1 \cdot 10^{-6}}$

In Exercises 83 and 84, express the result in scientific notation.

83. The average weekly food cost is \$36 per person. What is the average amount spent on food per year in the United States? (Assume that the population is 267 million.)

84. The $1 \cdot 10^5$ building material and garden supply stores in the United States have total annual sales of $\$1.25 \cdot 10^{11}$. What is the average annual sales per store?

In Exercises 85 and 86, perform the operations with scientific notation, but express the results in decimal form.

85. Building contractors employ $3.6 \cdot 10^6$ construction workers and have a payroll of $\$8.3 \cdot 10^{10}$. What is the average pay of a construction worker?

86. In 1998, the $1.46 \cdot 10^8$ automobiles in the United States were driven $1.62 \cdot 10^{12}$ miles. What was the average number of miles driven per automobile?

87. The largest hydroelectric facility in the United States is at the Grand Coulee Dam on the Columbia River in Washington. Its power capacity is rated at 6180 megawatts per hour. (*Source:* U.S. Army Corps of Engineers.)

(a) What do the prefixes "kilo" and "mega" mean?

(b) Use scientific notation to write Grand Coulee's power rating in watts.

(c) Approximately 0.3 pound of coal is required to generate the equivalent of 1 kilowatt for one hour. How many tons of coal would be required to match Grand Coulee's power capacity for one hour?

0.3 pound   1 kilowatt hour
coal   electricity

R.3 *Polynomials*

Basic Forms ■ *Operations with Polynomials* ■
Special Products

A *polynomial* is a specific kind of algebraic expression that is often used to represent the conditions of an applied problem or to model real-world data. In this section, we review the vocabulary associated with polynomials and the basic operations that we perform with them.

Basic Forms

A **monomial** is a number or the product of a number and one or more variables with nonnegative exponents. The **degree** of a nonzero monomial is the sum of the exponents on the variables. The degree of a nonzero constant is 0, but the monomial 0 has no degree.

Each of the following is a monomial and its degree.

Monomial	Degree	
7	0	The degree of a nonzero constant is 0.
$2x^3$	3	The degree is the exponent on the variable.
$-a^2b$	3	The degree is the sum of the exponents on the variables: $2 + 1 = 3$.
xy	2	The exponent on each variable is 1, so the degree of the monomial is $1 + 1 = 2$.
$\frac{2}{3}xy^2z$	4	The degree is the sum of the exponents on the variables: $1 + 2 + 1 = 4$.
0	None	The monomial 0 has no degree.

A **polynomial** is a finite sum of monomials.

Definition of a Polynomial in x

If $a_0, a_1, a_2, \ldots, a_n$ are real numbers and n is a nonnegative integer, then the standard form of a **polynomial in x** is

$$a_n x^n + a_{n-1} x^{n-1} + \ldots + a_1 x + a_0$$

where the **leading coefficient** $a_n \neq 0$.
The number a_0 is the **constant term.** The **degree** of the polynomial is n.

Polynomials of two and three nonzero terms are **binomials** and **trinomials,** respectively. A polynomial in one variable is in **descending order** if it is written with decreasing powers of the variable. The degree of a polynomial in more than one variable is the highest degree of its terms.

EXAMPLE **1**   **Features of Polynomials**

For the following polynomials, write the coefficients, the constant term, and the degree.

(a) $3x - 5x^3 + 6 - x^2$ (b) $8xy^2 + x^3y - 2xy$

SOLUTION

Polynomial	Coefficients	Constant term	Degree
(a) $-5x^3 - x^2 + 3x + 6$	$-5, -1, 3, 6$	6	3
(b) $8xy^2 + x^3y - 2xy$	$8, 1, -2$	0	4

Operations with Polynomials

Like terms are terms that have the same variables with the same exponents. The Distributive Property is used to combine like terms.

$$5y + 8y = (5 + 8)y = 13y$$

$$-4a^2b + 2ab^2 + 7a^2b = (-4 + 7)a^2b + 2ab^2 = 3a^2b + 2ab^2$$

To add or subtract polynomials, simply remove parentheses and combine like terms. Recall that when grouping symbols are preceded by a minus symbol or an opposite symbol, the sign of each term inside the group must be changed.

$$-(x - 5) + 2x = -x + 5 + 2x = x + 5$$

$$7y - (2y + 3) = 7y - 2y - 3 = 5y - 3$$

EXAMPLE **2**   **Adding and Subtracting Polynomials**

(a) $(8x^3 - 5x^2 + 4) + (2x^3 + 6x^2 - 7x - 1)$

$= 8x^3 - 5x^2 + 4 + 2x^3 + 6x^2 - 7x - 1$ Remove parentheses.

$= (8x^3 + 2x^3) + (-5x^2 + 6x^2) - 7x + (4 - 1)$ Group like terms.

$= 10x^3 + x^2 - 7x + 3$ Combine like terms.

(b) $(3x^4 - 2x^3 + x - 1) - (x^4 - 5x^3 - 4x^2 + 9)$

$= 3x^4 - 2x^3 + x - 1 - x^4 + 5x^3 + 4x^2 - 9$ Change signs in the second polynomial.

$= 2x^4 + 3x^3 + 4x^2 + x - 10$ Combine like terms.

The Distributive Property in the form $a(b + c) = ab + ac$ is used to multiply a polynomial by a monomial. By repeatedly applying the Distributive Property, we can multiply two polynomials by multiplying each term of one polynomial by each term of the other.

EXAMPLE **3**   **Multiplying Polynomials**

(a) $3x^2y(2x - y^2 + 4) = 3x^2y(2x) - 3x^2y(y^2) + 3x^2y(4)$

$= 6x^3y - 3x^2y^3 + 12x^2y$

(b) $(2x + 5)(3x^2 - 4x + 7)$

$= 2x(3x^2 - 4x + 7) + 5(3x^2 - 4x + 7)$

$= 6x^3 - 8x^2 + 14x + 15x^2 - 20x + 35$

$= 6x^3 + 7x^2 - 6x + 35$

When we multiply polynomials, a vertical format has the advantage of aligning like terms.

EXAMPLE 4   **Multiplying Polynomials**

$$
\begin{array}{l}
2x^2 + 3x - 5 \\
\underline{3x^2 - 4x + 7} \\
6x^4 + 9x^3 - 15x^2 \\
\quad\quad\; -8x^3 - 12x^2 + 20x \\
\quad\quad\quad\quad\quad\quad 14x^2 + 21x - 35 \\
\underline{\hspace{5cm}} \\
6x^4 + \;\; x^3 - 13x^2 + 41x - 35
\end{array}
$$

$3x^2(2x^2 + 3x - 5)$

$-4x(2x^2 + 3x - 5)$

$7(2x^2 + 3x - 5)$

Combine like terms.

As with any polynomials, we can apply the Distributive Property to the product of two binomials.

$$
\begin{aligned}
(2x + 5)(3x - 4) &= 2x(3x) + 2x(-4) + 5(3x) + 5(-4) \\
&= 6x^2 - 8x + 15x - 20 \\
&= 6x^2 + 7x - 20
\end{aligned}
$$

The **First-Outside-Inside-Last (FOIL)** pattern can be used as an efficient method for multiplying two binomials, especially when the binomials have the same form.

First terms

Last terms   **F   O   I   L**

$$(2x + 5)(3x - 4) = 2x(3x) + 2x(-4) + 5(3x) + 5(-4)$$

Inner terms

Outer terms

If the outside and inside products are like terms, we can combine them mentally and thereby write the result directly.

$$
\begin{aligned}
(2x + 5)(3x - 4) &= 6x^2 + (-8x + 15x) - 20 \\
&= 6x^2 + 7x - 20
\end{aligned}
$$

Special Products

Knowing certain special patterns for the products of binomials makes multiplying more efficient and assists in factoring, which is the topic of the next section.

Special Products of Binomials

If A and B represent algebraic expressions, then the following special patterns can be used for multiplying binomials.

Pattern	*Example*
Product of a Sum and Difference of Two Terms	
$(A + B)(A - B) = A^2 - B^2$	$(2x + 3)(2x - 3) = (2x)^2 - 3^2$
	$\quad\quad\quad\quad\quad\quad\quad = 4x^2 - 9$

Pattern · *Example*

Square of a Binomial

$$(A + B)^2 = A^2 + 2AB + B^2$$

$$(x + 5)^2 = x^2 + 2(x)(5) + 5^2$$
$$= x^2 + 10x + 25$$

$$(A - B)^2 = A^2 - 2AB + B^2$$

$$(x - 2y)^2 = x^2 - 2(x)(2y) + (2y)^2$$
$$= x^2 - 4xy + 4y^2$$

Cube of a Binomial

$$(A + B)^3$$
$$= A^3 + 3A^2B + 3AB^2 + B^3$$

$$(z + 3)^3 = z^3 + 3z^2(3) + 3z(3)^2 + 3^3$$
$$= z^3 + 9z^2 + 27z + 27$$

$$(A - B)^3$$
$$= A^3 - 3A^2B + 3AB^2 - B^3$$

$$(x - 4)^3 = x^3 - 3x^2(4) + 3x(4)^2 - 4^3$$
$$= x^3 - 12x^2 + 48x - 64$$

These patterns can be used as templates for special products of binomials and other polynomials that fit the patterns.

EXAMPLE 5 **Special Products**

Determine each product.

(a) $(2a^2 - 5b)^2$ (b) $(3x + 2y)^3$ (c) $(yx - 2z)^3$
(d) $(x - 2 + y)(x - 2 - y)$

SOLUTION

(a) $(A - B)^2 = A^2 - 2\ A\ B + B^2$ Square of a binomial

$$(2a^2 - 5b)^2 = (2a^2)^2 - 2\,(2a^2)\,(5b) + (5b)^2$$
$$= 4a^4 - 20a^2b + 25b^2$$

(b) $(A + B)^3 = A^3 + 3\ A^2\ B + 3\ A\ B^2 + B^3$ Cube of a binomial

$$(3x + 2y)^3 = (3x)^3 + 3\,(3x)^2(2y) + 3\,(3x)\,(2y)^2 + (2y)^3$$
$$= 27x^3 + 3(9x^2)\,(2y) + 3\,(3x)\,(4y^2) + 8y^3$$
$$= 27x^3 + 54x^2y + 36xy^2 + 8y^3$$

(c) $(A - B)^3 = A^3 - 3\ A^2\ B + 3\ A\ B^2 - B^3$ Cube of a binomial

$$(yx - 2z)^3 = (yx)^3 - 3\,(yx)^2\,(2z) + 3\,(yx)\,(2z)^2 - (2z)^3$$
$$= y^3x^3 - 3(y^2x^2)\,(2z) + 3\,(yx)\,(4z^2) - 8z^3$$
$$= x^3y^3 - 6x^2y^2z + 12xyz^2 - 8z^3$$

(d) $(A + B)\ (A - B) = A^2 - B^2$

$$[(x - 2) + y][(x - 2) - y] = (x - 2)^2 - y^2$$
$$= x^2 - 4x + 4 - y^2$$

Simplifying an expression may involve a combination of operations, as we see in the next example.

EXAMPLE **6** **Combined Operations**

Simplify each of the following.

(a) $(2x - 5)(x + 3) - (x - 4)^2$
(b) $4x(x + 4)(x - 7)$
(c) $(y + 4)(y + 1)(y - 4)$

SOLUTION

(a) Using the Order of Operations, we multiply first and subtract last.

$(2x - 5)(x + 3) - (x - 4)^2$ Use FOIL and the square of a binomial.

$$= (2x^2 + x - 15) - (x^2 - 8x + 16)$$
$$= 2x^2 + x - 15 - x^2 + 8x - 16 \qquad \text{Remove parentheses.}$$
$$= x^2 + 9x - 31 \qquad \text{Combine like terms.}$$

(b) Usually, the easiest approach is to multiply the binomials and then distribute the monomial.

$$4x(x + 4)(x - 7) = 4x(x^2 - 3x - 28) \qquad \text{Use FOIL.}$$
$$= 4x^3 - 12x^2 - 112x \qquad \text{Distribute } 4x.$$

(c) Detecting special products makes simplifying easier.

$$(y + 4)(y + 1)(y - 4) = [(y + 4)(y - 4)](y + 1)$$
$$= (y^2 - 16)(y + 1)$$
$$= y^3 + y^2 - 16y - 16$$

Speaking the Language

1. The expressions $3a$, -5, and xyz^2 are all called _____.
2. The _____ of $-3xy^2z^3$ is 6.
3. We say that $5x^2y$ and $-3x^2y$ are _____ terms.
4. The FOIL pattern is used for multiplying two _____.

R.3 Exercises

1. In the following list, if any statement is false, identify it and explain why it is false.

    (i) All monomials are also polynomials.
    (ii) The sum of two polynomials of degree 2 is also a polynomial of degree 2.
    (iii) The polynomial $x^2 + 0x - 5$ is a trinomial.

2. Explain why the monomials 1 and 0 do not have the same degree.

In Exercises 3–8, determine whether the expression is a polynomial.

3. $2x - x^4 + \sqrt{3}$

4. $\dfrac{y^5}{2} + y^2 - \dfrac{y}{3} + \dfrac{2}{5}$

5. $x^{-2} + 3x^{-1} - 7$

6. $\sqrt{x} + 7$

7. 15

8. $\dfrac{5x}{\pi}$

In Exercises 9–14, determine the degree of each polynomial.

9. $7x - 3x^5 + x^3$

10. $2^4 - 4x$

11. -5^4

12. xy^3

13. $x^3y - xy^4 + 3x^2$

14. $n^4 + n^6m + n^4m^5 + m^3n^2$

In Exercises 15–22, perform the indicated operations.

15. $(1 - 3x) + (x^2 + 2x - 4)$

16. $(xy + xy^2 + 3) + (2xy - xy^2 - 4)$

17. $(3x + 2y - 6) - (x + 3y - 3)$

18. $(3x^4 - 7x + 3x^3 - 2) - (-3x^3 + x^4 - 2x - x^2)$

19. $(4x^3 - 2x - x^2 + 3) - (5 - 2x) + (x^2 + 2x^3)$

20. $(x^5 + 2x - x^4 - 2) + (x^3 - 3x + 2) - (x^4 - x^3 + x)$

21. $(2b - 3c + 7) - (3c - 4a) + (a - b - 6)$

22. $(x + 3y - z) + (5y - 7z + 3) - (4x - 9z + 7)$

23. In the following list, if any statement is false, identify it and explain why it is false.

    (i) $(x + y)^2 = x^2 + y^2$

    (ii) The product of two first-degree polynomials is a second-degree polynomial.

    (iii) The product of two binomials is a trinomial.

24. (a) In the following list, determine the number of terms in each expression.

    (b) Identify any expressions that are equivalent and explain why they are.

        (i) $(x + 3)(x - 2)$

        (ii) $x^2 + x - 6$

        (iii) $x(x - 2) + 3(x - 2)$

In Exercises 25–32, multiply.

25. $-2xy^3(3y^2 - 5y + 6)$

26. $ab^2(a^4b^2 + a^3b - ab + 5)$

27. $(a - 3)(a^2 + 3a - 5)$

28. $(x + 5)(2x^2 - x - 2)$

29. $(2x + 1)(x^2 - x - 4)$

30. $(x + y)(x^2 + 2xy - y^2)$

31. $(x + 3y - 2)(4x + y - 1)$

32. $(x^2 - x - 4)(2x^2 + x + 3)$

33. Suppose that $(a + b)^2 = 7$ and $ab = -5$. What is the value of $a^2 + b^2$? Explain how you know.

34. Suppose that you produce a table of values for each of the following expressions.

    (i) $(x + 3)(x - 1)$

    (ii) $x^2 - 3$

    (iii) $x^2 + 2x - 3$

Which tables would be exactly the same? Why?

In Exercises 35–42, multiply the binomials.

35. $(y + 7)(y - 5)$

36. $(x - 6)(x - 2)$

37. $(3x + 2)(2x + 3)$

38. $(7 - 2y)(4 - 3y)$

39. $(y - 3x)(2y + 5x)$

40. $(5b - 2)(4b - 1)$

41. $(5 - 3z)(2 + 3z)$

42. $(xy + 2z)(3xy + z)$

43. Based on tables of values of the expressions, which expression appears to be the result of expanding $(x - 3)^2$?

    (i) $x^2 + 9$ (ii) $x^2 - 9$ (iii) $x^2 - 6x + 9$

44. Based on tables of values of the expressions, which expression appears to be the result of expanding $(x + 1)^3$?

    (i) $x^3 + 1$ (ii) $x^3 + 3x^2 + 3x + 1$ (iii) $x^3 - 1$

In Exercises 45–56, use special products to determine the product.

45. $(7x - 2)(7x + 2)$

46. $(4 + 3b)(4 - 3b)$

47. $(y^2 + 4z)(y^2 - 4z)$

48. $(x + 5y^2)(x - 5y^2)$

49. $(5 - y)^2$

50. $(3x + 4y)^2$

51. $(a + 3b)^2$

52. $(2x - y)^2$

53. $(y - 2)^3$

54. $(3x + 1)^3$

55. $(2y + 3z)^3$

56. $(ab - 2c)^3$

In Exercises 57–64, perform the indicated operations.

57. $3x^2(2x - 1)(3x + 2)$

58. $-a(a^2 + 3)(a^2 - 3)$

59. $(x + 3)(x - 2)(x - 1)$

60. $(2y - 1)(2y + 3)(y + 1)$

61. $(x + 2)(2x - 1) - (x - 1)(x + 3)$

62. $2x(5x + 7) - (2x + 5)(x - 3)$

63. $[(2y + 4) - (3 - y)] + y(y - 4)$

64. $x(x + 3) - [(3x - 4) - (x^2 - 3)]$

In Exercises 65–68, use special products to perform the indicated operations.

65. $(3x + 1)(3x - 1) - (3x - 1)^2$

66. $(3x + 2)(9x^2 + 4)(3x - 2)$

67. $(x + 6)(x - 6) - (x + 1)(x - 1)$

68. $(x - 2)^2 - (x + 2)^2$

69. Show two ways to group the factors of the following expression so that a special product can be used to determine the product.

$$(t + 3)(3 - t)(t + 3)$$

70. Show how to use special products to determine the product.

$$(x^2 - 4x + 1)(x^2 + 4x + 1)$$

In Exercises 71–76, use special products to determine the product.

71. $(2x + y + 5)(2x + y - 5)$

72. $(a + b + c)(a - b - c)$

73. $(x + 2)(x - 2)(x^2 + 4)$

74. $(x^3 + 1)(x^3 - 1)(x^6 + 1)$

75. $(a + b + 3)^2$

76. $(3 - 2x - y)^2$

R.4 Factoring

Common Factors ■ *Special Factoring Forms* ■ *Quadratic Trinomials* ■ *The Grouping Method*

For such tasks as simplifying expressions and solving equations, we often need to write an algebraic expression in the form of a product. This section provides a review of the basic techniques for *factoring* a polynomial.

Common Factors

Writing a polynomial as a product of polynomials is called **factoring** the polynomial. A polynomial is **prime over the integers** (or simply **prime**) if it cannot be factored into polynomials with integer coefficients. A polynomial is **completely factored** if each of its factors (other than a common factor) is prime.

The recommended first step in factoring is to determine whether the expression has a **common factor,** a factor that appears in each term of the expression. If the expression has a common factor, then we can use the Distributive Property to "remove" it.

To determine the greatest common factor (GCF) of two expressions, we look for the greatest number of times that any factor appears in both expressions.

$$12x^3y^2 = 2^2 \cdot 3 \cdot x^3 \cdot y^2$$

$$16xy^4 = 2^4 \cdot x^1 \cdot y^4$$

$$\text{GCF: } 2^2 \cdot x^1 \cdot y^2 = 4xy^2$$

Note that we can construct the GCF simply by selecting the smallest exponent on each common factor.

EXAMPLE 1 **Removing Common Factors**

(a) $15y^3 + 10y = (5y)(3y^2) + (5y)(2) = 5y(3y^2 + 2)$

(b) $4a(x - 5) - 3b(x - 5) = (x - 5)(4a - 3b)$

(c) $-20x^3y^3 - 12x^2y^5 = -4x^2y^3(5x + 3y^2)$

Note that the GCF could be $-4x^2y^3$ or $4x^2y^3$. When the leading term of the expression has a negative coefficient, we often use a GCF with a negative coefficient.

Special Factoring Forms

Each of the following special factoring patterns can be verified by multiplying the factors to obtain the original expression.

Special Factoring Patterns

If A and B represent algebraic expressions, then the following special patterns can be used for factoring polynomials.

Pattern

Difference of Two Squares

$$A^2 - B^2 = (A + B)(A - B)$$

Example

$$16x^2 - 9y^2 = (4x)^2 - (3y)^2$$
$$= (2x + 3y)(2x - 3y)$$

Perfect Square Trinomial

$$A^2 + 2AB + B^2 = (A + B)^2$$

$$u^2 + 10u + 25 = u^2 + 2u(5) + 5^2$$
$$= (u + 5)^2$$

$$A^2 - 2AB + B^2 = (A - B)^2$$

$$x^2 - 4x + 4 = x^2 - 2x(2) + 2^2$$
$$= (x - 2)^2$$

Sum and Difference of Two Cubes

$$A^3 + B^3 = (A + B)(A^2 - AB + B^2)$$

$$y^3 + 27 = y^3 + 3^3$$
$$= (y + 3)(y^2 - 3y + 9)$$

$$A^3 - B^3 = (A - B)(A^2 + AB + B^2)$$

$$8x^3 - 1 = (2x)^3 - 1^3$$
$$= (2x - 1)(4x^2 + 2x + 1)$$

An important step in the factoring process is to recognize clues that point to the possibility of a special pattern. Sometimes the removal of a common factor will reveal a special pattern.

EXAMPLE **2** **Factoring a Difference of Two Squares**

(a) $12x^2 - 27 = 3(4x^2 - 9)$ Remove the common factor.
$\qquad\quad = 3[(2x)^2 - 3^2]$ $4x^2 - 9$ is a difference of two squares.
$\qquad\quad = 3(2x + 3)(2x - 3)$

(b) $x^4 - 16 = (x^2)^2 - 4^2$ $x^4 - 16$ is a difference of two squares.
$\qquad\quad = (x^2 + 4)(x^2 - 4)$ The expression is factored but not completely factored because $x^2 - 4$ is a difference of two squares.
$\qquad\quad = (x^2 + 4)(x + 2)(x - 2)$

(c) $(2x - 5)^2 - z^2$ A difference of two squares.
$\qquad = [(2x - 5 + z][(2x - 5) - z]$
$\qquad = (2x - 5 + z)(2x - 5 - z)$

An expression of the form $A^2 + $ � $+ B^2$ may be a perfect square trinomial, but the middle term must be checked. Note the sign patterns for the two forms.

$$A^2 + 2AB + B^2 = (A + B)^2 \qquad\qquad A^2 - 2AB + B^2 = (A - B)^2$$
$$\uparrow\!\rule{0pt}{0pt}\text{— Same signs —}\!\uparrow \qquad\qquad \uparrow\!\rule{0pt}{0pt}\text{— Same signs —}\!\uparrow$$

EXAMPLE 3   **Factoring a Perfect Square Trinomial**

(a) $49x^2 - 70x + 25 = (7x)^2 - 2(7x)(5) + 5^2$ **Perfect square trinomial**
$$= (7x - 5)^2$$

(b) $(x + 2)^2 + 6(x + 2) + 9$
$$= (x + 2)^2 + 2(x + 2)(3) + 3^2$$ **Perfect square trinomial**
$$= [(x + 2) + 3]^2$$
$$= (x + 5)^2$$

(c) $a^6 + 12a^3b^2 + 36b^4 = (a^3)^2 + 2(a^3)(6b^2) + (6b^2)^2$
$$= (a^3 + 6b^2)^2$$

If the two terms of a binomial are perfect cubes, we can apply one of the special factoring patterns. Compare the sign patterns in the original expression and the factors.

Same signs   Same signs

$$A^3 + B^3 = (A + B)(A^2 - AB + B^2) \qquad A^3 - B^3 = (A - B)(A^2 + AB + B^2)$$

Different signs   Different signs

EXAMPLE 4   **Factoring Sums and Differences of Cubes**

(a) $32 - 4x^3 = 4(8 - x^3)$ **Remove the common factor.**
$$= 4(2^3 - x^3)$$ **A difference of two cubes**
$$= 4(2 - x)(4 + 2x + x^2)$$

(b) $x^6 + 8 = (x^2)^3 + 2^3$ **A sum of two cubes**
$$= (x^2 + 2)(x^4 - 2x^2 + 4)$$

Quadratic Trinomials

A trinomial of the form $ax^2 + bx + c, a \neq 0$, is called a **quadratic trinomial.**

In the preceding section, we saw that a quadratic trinomial is frequently the result of multiplying two binomials. Thus one method for factoring a quadratic trinomial is to test feasible products of binomials until the correct combination is found or until it is determined that the expression is prime. When we use this *trial-and-check* method, we must multiply the factors to verify that the original expression has been obtained.

The easiest quadratic trinomial to factor is one in which $a = 1$ — that is, a trinomial of the form $x^2 + bx + c$. Suppose that the factorization has the following form:

$$x^2 + bx + c = (x + m)(x + n)$$

Then factoring the trinomial involves finding numbers m and n such that $mn = c$ and $m + n = b$.

EXAMPLE **5**   **Factoring Quadratic Trinomials, *a* = 1**

Factor the given trinomials.

(a) $x^2 + 7x + 12$ (b) $x^2 - 4x - 21$ (c) $x^2 + 3x + 6$
(d) $-x^2 - 2x + 35$ (e) $x^2 - 6xy + 5y^2$

SOLUTION

(a) $x^2 + 7x + 12$
To list the feasible factorizations, we seek numbers m and n such that $mn = 12$ and $m + n = 7$.

Feasible factorizations	Middle term	
$(x + 1)(x + 12)$	$13x$	
$(x + 2)(x + 6)$	$8x$	
$(x + 3)(x + 4)$	$7x$	Correct middle term

$$x^2 + 7x + 12 = (x + 3)(x + 4)$$

Note that if b and c are both positive, then the required numbers m and n are both positive.

(b) $x^2 - 4x - 21$

Feasible factorizations	Middle term	
$(x - 1)(x + 21)$	$20x$	
$(x + 1)(x - 21)$	$-20x$	
$(x - 3)(x + 7)$	$4x$	
$(x + 3)(x - 7)$	$-4x$	Correct middle term

$$x^2 - 4x - 21 = (x + 3)(x - 7)$$

Note that if c is negative, then the required numbers m and n have opposite signs.

(c) $x^2 + 3x + 6$

Feasible factorizations	Middle term
$(x + 1)(x + 6)$	$7x$
$(x + 2)(x + 3)$	$5x$

Because neither combination produces the correct middle term, $x^2 + 3x + 6$ is prime.

(d) $-x^2 - 2x + 35 = -1(x^2 + 2x - 35)$ We factor out -1 to make $a = 1$.

Feasible factorizations	Middle term	
$(x + 35)(x - 1)$	$34x$	
$(x - 35)(x + 1)$	$-34x$	
$(x - 7)(x + 5)$	$-2x$	
$(x + 7)(x - 5)$	$2x$	Correct middle term

$$-x^2 - 2x + 35 = -1(x + 7)(x - 5)$$

(e) $x^2 - 6xy + 5y^2$

Because $c = 5$ is positive and $b = -6$ is negative, the required numbers m and n must both be negative. Thus the only feasible factorization is $(x - 5y)(x - y)$, which can be verified by multiplying the factors.

$$x^2 - 6xy + 5y^2 = (x - 5y)(x - y)$$

Quadratic trinomials for which $a \neq 1$ can still be factored with the trial-and-check method. However, the number of feasible factorizations can become large. Remember to remove the GCF first. Then you can minimize the number of possible combinations by disregarding any binomials that contain a common factor.

Suppose that you want to factor $15x^2 + x - 2$. The product of the first terms of the binomial factors must be $15x^2$, and the product of the last terms must be -2.

$$15x^2 + x - 2 = (\underline{\qquad} \ \underline{\qquad})(\underline{\qquad} \ \underline{\qquad})$$

As before, you must check the middle term to learn whether you have found the correct combination. Verify that the factorization of $15x^2 + x - 2$ is $(5x + 2)(3x - 1)$.

EXAMPLE **6** **Factoring Quadratic Trinomials, $a \neq 1$**

Factor the given trinomials.

(a) $6x^2 - 11x - 10$ (b) $16x^3 + 52x^2 + 30x$

SOLUTION

(a) $6x^2 - 11x - 10$

Feasible factorizations	Middle term	
$(6x + 1)(x - 10)$	$-59x$	
$(6x - 1)(x + 10)$	$59x$	
$(6x + 5)(x - 2)$	$-7x$	
$(6x - 5)(x + 2)$	$7x$	
$(2x + 1)(3x - 10)$	$-17x$	
$(2x - 1)(3x + 10)$	$17x$	
$(2x + 5)(3x - 2)$	$11x$	
$(2x - 5)(3x + 2)$	$-11x$	**Correct middle term**

Note that factors such as $6x + 2$ and $2x - 10$ are not feasible because they contain common factors.

$$6x^2 - 11x - 10 = (2x - 5)(3x + 2)$$

(b) $16x^3 + 52x^2 + 30x$

$$16x^3 + 52x^2 + 30x = 2x(8x^2 + 26x + 15)$$ **Remove the common factor.**

Feasible factorizations of $8x^2 + 26x + 15$	*Middle term*
$(4x + 15)(2x + 1)$	$34x$
$(4x + 1)(2x + 15)$	$62x$
$(4x + 5)(2x + 3)$	$22x$
$(4x + 3)(2x + 5)$	$26x$
.	.
.	.
.	.

Although there are other feasible factorizations that are not shown in the list, we obtain the correct middle term with $(4x + 3)(2x + 5)$.

$$16x^3 + 52x^2 + 30x = 2x(4x + 3)(2x + 5)$$

The Grouping Method

When a polynomial has more than three terms, you may be able to factor it with the *grouping method*.

As the name suggests, the grouping method involves strategically grouping the terms of the expression so that each group can be factored. Using previously discussed techniques, we may then be able to factor the resulting expression.

EXAMPLE 7   **Factoring with the Grouping Method**

Factor the given expressions.

(a) $x^3 + 3x^2 - 5x - 15$ (b) $x^2 + 6x + 9 - y^2$

SOLUTION

(a) $x^3 + 3x^2 - 5x - 15$ **Group by pairs.**
$= (x^3 + 3x^2) - (5x + 15)$ **Note the sign changes in the second pair.**
$= x^2(x + 3) - 5(x + 3)$ **Remove common factors from each group.**
$= (x + 3)(x^2 - 5)$ **The binomial $x + 3$ is a common factor.**

Observe that we began with four terms. After factoring each group, we had two terms, which means that the expression was still not factored. Indeed, we would not have been able to complete the factorization had the binomial not been the same in each term.

(b) $x^2 + 6x + 9 - y^2$ **Group the first three terms.**
$= (x^2 + 6x + 9) - y^2$ **The first group is a perfect square trinomial.**
$= (x + 3)^2 - y^2$ **The resulting expression is a difference of two squares.**
$= (x + 3 + y)(x + 3 - y)$

To factor an expression, we might need to use a combination of two or more of the factoring techniques described in this section.

EXAMPLE 8   **Combining Factoring Methods**

Factor the given expressions.

(a) $t^4 - 10t^2 + 9$ (b) $4x^3 + x^2 - 16x - 4$
(c) $-12t^5 - 30t^4 + 18t^3$

SOLUTION

(a) $t^4 - 10t^2 + 9$

Although the first and last terms are perfect squares, we can verify that the expression is not a perfect square trinomial. Therefore, we use the trial-and-check method.

$$t^4 - 10t^2 + 9 = (t^2 - 9)(t^2 - 1)$$
$$= (t + 3)(t - 3)(t + 1)(t - 1)$$

Both binomials are differences of two squares.

(b) $4x^3 + x^2 - 16x - 4 = (4x^3 + x^2) - (16x + 4)$ Group by pairs.
$$= x^2(4x + 1) - 4(4x + 1)$$ Factor each group.
$$= (4x + 1)(x^2 - 4)$$ The second factor is a difference of two squares.
$$= (4x + 1)(x + 2)(x - 2)$$

(c) $-12t^5 - 30t^4 + 18t^3 = -6t^3(2t^2 + 5t - 3)$ Remove the GCF.
$$= -6t^3(2t - 1)(t + 3)$$ Trial-and-check

Speaking the Language

1. Because $x^2 + x + 1$ cannot be factored into polynomials with integer coefficients, we say that the expression is ▨▨▨▨▨ .

2. The first step in factoring is to determine whether the expression has a(n) ▨▨▨▨▨ .

3. The expression $x^2 - 4y^2$ is an example of a special pattern known as a(n) ▨▨▨▨▨ .

4. The ▨▨▨▨▨ method is a factoring method to try when the expression has four or more terms.

R.4 Exercises

1. Explain how to determine the GCF of the two monomials a^4b^2 and a^3b^7.

2. Each of the following binomials is the sum of two squares. Identify any binomial that is prime and factor the other(s).

(i) $9x^2 + 9$ (ii) $x^6 + 1$ (iii) $9x^2 + 1$

In Exercises 3–8, determine the GCF and factor completely.

3. $6a + 9$ 4. $8y - 4z + 12$
5. $y^3 - y^2$ 6. $2x^3 + 2x^2 - 6x$
7. $3x(x - 2) + y(x - 2)$
8. $2a(x^2 + 5) + (x^2 + 5)$

In Exercises 9–12, factor the difference of two squares.

9. $36x^2 - 1$ **10.** $81 - t^2$

11. $49 - 25c^2$ **12.** $9x^2y^2 - 64$

In Exercises 13–16, factor the perfect square trinomial.

13. $x^2 - 6x + 9$ **14.** $4y^2 + 20y + 25$

15. $25 + 30z + 9z^2$ **16.** $16t^2 - 8t + 1$

17. Explain how to use a table of values to check the result of factoring the polynomial in Exercise 9.

18. Explain how to use a table of values to check the result of factoring the polynomial in Exercise 13.

In Exercises 19–22, factor the sum or difference of cubes.

19. $64 - w^3$ **20.** $y^3 + 64z^3$

21. $27x^3 + 8$ **22.** $y^3 - 1$

23. Which of the following is a correct factorization of $12 - x - x^2$? Explain how you know.

(i) $-(x - 3)(x + 4)$ (ii) $(3 - x)(4 + x)$

24. Which of the following expressions would have the same table of values as $x^2 - 5x + 6$? Explain how you know.

(i) $(x - 2)(x - 3)$
(ii) $(2 - x)(3 - x)$
(iii) $(x - 6)(x + 1)$

In Exercises 25–34, factor the trinomial.

25. $y^2 - y - 6$ **26.** $28 - 3x - x^2$

27. $12y^2 + 8y - 15$ **28.** $4a^2 + 5ab - 9b^2$

29. $4 + 3x - x^2$ **30.** $x^2 + 13x + 12$

31. $8x^2 - 11x + 3$ **32.** $6x^2 + 17x + 10$

33. $x^4 + 5x^2 + 6$ **34.** $y^4 + 4y^2 - 12$

35. In the following list, if any statement is false, identify it and explain why it is false.

(i) $x^2 + 4x + 16$ is a perfect square trinomial.
(ii) When $x^4 - 16$ is factored completely, one factor is $x^2 - 4$.
(iii) $x^6 - 64$ is both a difference of two squares and a difference of two cubes.

36. Factor $a^6 - 1$ as (a) the difference of two squares and (b) the difference of two cubes. Which is a complete factorization?

In Exercises 37–40, use the grouping method to factor the given polynomial.

37. $ax + 2bx - ay - 2by$ **38.** $yz - 3x - 3z + xy$

39. $x^3 - 5x^2 + 2x - 10$ **40.** $9x^3 + 3x^2 - 6x - 2$

In Exercises 41–82, factor completely.

41. $b^2 + 13b + 42$ **42.** $2c^2 - 24c + 64$

43. $121m^2 - 4n^2$ **44.** $1 - 25t^2$

45. $5x^3 + 25x^2 + x + 5$

46. $3x^3 + 21x^2 + 7x + 49$

47. $40y^{40} - 30y^{30}$ **48.** $a^4b^6 + a^3b^7 - a^5b^3$

49. $20a^4 + a^2b - 12b^2$ **50.** $24x^4 - 25x^2y + 6y^2$

51. $4x^2 - 29xy + 42y^2$ **52.** $18y^2 + 27yz + 10z^2$

53. $81 - 36y + 4y^2$ **54.** $64t^2 + 16t + 1$

55. $4x^4 + 4x$ **56.** $-2y^3 - 16z^3$

57. $8x - 2x^2 - 3x^3$ **58.** $27y^2 - 12y + 1$

59. $81a^4 - 16$ **60.** $25a^4 - 4b^2$

61. $x^2 - 7xy - 30y^2$ **62.** $a^2b^2 + 10ab - 24$

63. $1 + 8x^6$ **64.** $125 - x^3y^6$

65. $4a^2 + 28a + 49$ **66.** $a^2b^2 - 12abc + 36c^2$

67. $x^2 - 4x + 6$ **68.** $y^2 - 3y + 4$

69. $x^3 + 3x^2 - 4x - 12$

70. $2x^3 + 5x^2 - 18x - 45$

71. $y(x - 3) - 4z(3 - x)$

72. $4(m + n) + n^2(m + n)$

73. $2t^6 + 11t^3 + 12$ **74.** $x^4 - 7x^2 - 8$

75. $-3x^3 + 18x^2y - 27xy^2$

76. $9t^7 + 24t^6 + 16t^5$

77. $98 - 8y^2$ **78.** $8x^3 - 8x$

79. $2x^5y^2 + 5x^4y^3 - 3x^3y^4$

80. $-5x^6 + 10x^5 + 15x^4$

81. $16y^2 - 54y^5$ **82.** $2w^3 - 128$

In Exercises 83–92, factor completely.

83. $9(4 - x^2)^2 + 24(4 - x^2) + 16$

84. $(y - 3)^2 - 16(y - 3) + 63$

85. $(2y + z)^2 - 100$

86. $1 - (a + b)^3$

87. $(3y - 2)(y + 2) + (4 - 5y)(2 - 3y)$

88. $(x^2 - 9)(x - 7) + (1 - x)(9 - x^2)$

89. $(x + 2)^5 - (x + 2)^3$

90. $y^6 - 6y^5 + 9y^4 - y^2 + 6y - 9$

91. $x^2 + 6x + 9 - y^2$

92. $z^2 - x^2 + 2x - 1$

93. Show how you can obtain the rule for factoring the difference of two cubes from the rule for factoring the sum of two cubes.

94. Show how you can factor $x^4 + 4y^4$ by adding and subtracting $4x^2y^2$ to and from the expression.

95. Although we cannot factor $x^2 - 5$ over the integers, we can factor it over the real numbers as $(x + \sqrt{5})(x - \sqrt{5})$. Use this method to factor each of the following binomials.

(a) $x^2 - 10$ (b) $15 - y^2$ (c) $6x^2 - 1$

96. Use factoring to evaluate each of the following expressions with a minimal amount of arithmetic.

(a) $200^2 - 199^2$ (b) $505^2 - 495^2$

R.5 Rational Expressions

Restricted Values and Properties ■ *Simplifying Rational Expressions* ■ *Operations with Rational Expressions* ■ *Complex Fractions*

Just as fractions are an essential part of arithmetic, *rational expressions* play an important role in algebra. Applications that involve comparisons or ratios often require rational expressions to model the conditions of the problem. In this section, we review the methods for simplifying and performing operations with rational expressions.

Restricted Values and Properties

A **fractional expression** is a quotient of algebraic expressions. If the expressions are polynomials, then the quotient is called a **rational expression.**

Unlike polynomials, for which the variable(s) can be replaced by any real number, rational expressions are limited to replacements that do not cause the denominator to be 0. Any replacement for which the denominator of a rational expression is 0 is called a **restricted value.**

Factoring the denominator of a rational expression is a good first step in determining restricted values. The rational expression is not defined if any one of the factors has a value of 0.

EXAMPLE 1   **Restricted Values**

Determine the restricted values for each rational expression.

(a) $\dfrac{x - 5}{x^2 - 1}$ (b) $\dfrac{2x + 1}{x^2 + 5x - 14}$ (c) $\dfrac{x - 5}{x^2 + 9}$

SOLUTION

(a) $x^2 - 1 = (x + 1)(x - 1)$ **Factor the denominator.**

Note that if $x = -1$ or $x = 1$, the denominator of $\dfrac{x - 5}{x^2 - 1}$ is 0. Thus the restricted values are -1 and 1.

(b) $x^2 + 5x - 14 = (x + 7)(x - 2)$ **Factor the denominator.**

Replacing x with either -7 or 2 would cause the denominator of $\dfrac{2x + 1}{x^2 + 5x - 14}$ to be 0. Thus the restricted values are -7 and 2.

(c) The denominator is $x^2 + 9$. Because no replacement for x would make the denominator 0, the expression has no restricted values.

In the examples and exercises that follow, we will not specifically list all the restricted values. We will assume that all equalities involving rational expressions hold for those values for which the expressions are defined.

Operations with fractions are based on the following properties.

Properties of Fractions

If $a, b, c,$ and d are expressions, where $b \neq 0$ and $d \neq 0$, then

Property	*Example*	
$\dfrac{a}{b} = \dfrac{ac}{bc}, c \neq 0$	$\dfrac{2}{3} = \dfrac{2 \cdot 5}{3 \cdot 5} = \dfrac{10}{15}$	**Renaming fractions**
$\dfrac{ac}{bc} = \dfrac{a}{b}, c \neq 0$	$\dfrac{4(x + 1)}{5(x + 1)} = \dfrac{4}{5}$	**Dividing out common factors**
$\dfrac{a}{b} \cdot \dfrac{c}{d} = \dfrac{ac}{bd}$	$\dfrac{2}{y} \cdot \dfrac{x - 3}{a} = \dfrac{2(x - 3)}{ay}$	**Multiplication**
$\dfrac{a}{b} \div \dfrac{c}{d} = \dfrac{a}{b} \cdot \dfrac{d}{c}, c \neq 0$	$\dfrac{2y}{x} \div \dfrac{z}{5} = \dfrac{2y}{x} \cdot \dfrac{5}{z} = \dfrac{10y}{xz}$	**Division**
$\dfrac{a}{b} + \dfrac{c}{b} = \dfrac{a + c}{b}$	$\dfrac{x}{x - 1} + \dfrac{2y}{x - 1} = \dfrac{x + 2y}{x - 1}$	**Addition, like denominators**

Simplifying Rational Expressions

The property $\dfrac{ac}{bc} = \dfrac{a}{b}$ is the basis for simplifying a rational expression. This property allows us to divide out common factors from the numerator and the denominator. Factoring the polynomials is usually the first step in simplifying a rational expression.

EXAMPLE 2   **Simplifying Rational Expressions**

Simplify the given rational expressions and state the restricted values.

(a) $\dfrac{x^2 - 2x - 15}{2x^2 - 18}$ (b) $\dfrac{a^2 - b^2}{3b - 3a}$

SOLUTION

(a) $\dfrac{x^2 - 2x - 15}{2x^2 - 18} = \dfrac{(x + 3)(x - 5)}{2(x^2 - 9)}$ **Factor the numerator and the denominator.**

$\qquad\qquad\quad = \dfrac{(x + 3)(x - 5)}{2(x + 3)(x - 3)}$ **Divide out common factors.**

$\qquad\qquad\quad = \dfrac{x - 5}{2(x - 3)}$

Figure R.5 shows a table of values for the original expression (Y_1) and the simplified expression (Y_2).

FIGURE R.5

X	Y_1	Y_2
-3	ERROR	.66667
-2	.7	.7
-1	.75	.75
0	.83333	.83333
1	1	1
2	1.5	1.5
3	ERROR	ERROR

X=-3

Note that the table for the simplified expression displays an error message only for $x = 3$, which indicates that the only restricted value is 3. However, the table for the original expression indicates two restricted values, -3 and 3.

When we divided out the common factor $x + 3$, we assumed that $x \neq -3$. Always determine the restricted values of the original expression. Expressions are equivalent if they have the same values for all permissible replacements of *both* expressions.

(b) $\dfrac{a^2 - b^2}{3b - 3a} = \dfrac{(a + b)(a - b)}{-3(a - b)}$ **Factor the numerator and the denominator.**

$\qquad\qquad\quad = \dfrac{a + b}{-3}$ **Divide out common factors.**

Note that factoring out -3 rather than 3 resulted in common factors that could be divided out.

The denominator of the expression is 0 if $a = b$. Thus a and b are restricted to values for which $a \neq b$.

Operations with Rational Expressions

To multiply rational expressions, we use the fraction rule $\dfrac{a}{b} \cdot \dfrac{c}{d} = \dfrac{ac}{bd}$.

Factoring and dividing out common factors before multiplying is the recommended way to begin.

To divide rational expressions, we multiply by the reciprocal of the divisor.

EXAMPLE **3**   **Multiplying and Dividing Rational Expressions**

Perform the indicated operations and list all restricted values.

(a) $\dfrac{x^2 - x - 12}{x^2 - 16} \cdot \dfrac{3x + 12}{x^2 - 4x - 21} = \dfrac{(x+3)(x-4)}{(x+4)(x-4)} \cdot \dfrac{3(x+4)}{(x+3)(x-7)}$

$$= \dfrac{3}{x-7}$$

The restricted values are $-4, 4, -3,$ and 7.

(b) $\dfrac{x^2 - 4}{x^3} \div \dfrac{2 - x}{x^2} = \dfrac{x^2 - 4}{x^3} \cdot \dfrac{x^2}{2 - x}$

$$= \dfrac{(x+2)(x-2)}{x^2 \cdot x} \cdot \dfrac{x^2}{-1(x-2)}$$

$$= -\dfrac{x+2}{x}$$

Here the denominators cannot be 0, but neither can the numerator of the divisor. The restricted values are 0 and 2.

To add or subtract rational expressions with like denominators, combine the numerators and retain the common denominator.

EXAMPLE **4**   **Subtracting Rational Expressions with Like Denominators**

$\dfrac{5x + 7}{x + 1} - \dfrac{x + 3}{x + 1} = \dfrac{(5x + 7) - (x + 3)}{x + 1}$ Treat the numerators as grouped quantities.

$$= \dfrac{5x + 7 - x - 3}{x + 1}$$ Remove grouping symbols and change signs.

$$= \dfrac{4x + 4}{x + 1}$$ Combine like terms.

$$= \dfrac{4(x + 1)}{x + 1}$$ Factor and simplify.

$$= 4$$

To add or subtract rational expressions with unlike denominators, the fractions must be renamed with like denominators. The procedure involves determining the **least common multiple (LCM)** of the denominators.

EXAMPLE **5**   **Determining Least Common Multiples**

Determine the least common multiple (LCM) of the given expressions.

(a) $x + 7, x - 4$
(b) $x^2 - 9, 2x - 6$
(c) $x^2 + 3x - 10, x^2 - 6x + 8, x^2 + 10x + 25$

SOLUTION

When possible, we factor the given expressions. Then the LCM consists of each factor that appears in any factorization, and the exponent on each factor of the LCM is the largest exponent that appears on that factor in any factorization.

(a) LCM: $(x + 7)(x - 4)$ **The LCM is simply the product of the factors.**

(b) $x^2 - 9 = (x - 3)(x + 3)$ **Factor the expressions.**
    $2x - 6 = 2(x - 3)$
    LCM: $2(x - 3)(x + 3)$

(c) $x^2 + 3x - 10 = (x + 5)(x - 2)$
    $x^2 - 6x + 8 = (x - 2)(x - 4)$
    $x^2 + 10x + 25 = (x + 5)^2$
    LCM: $(x - 2)(x - 4)(x + 5)^2$ **Note that two factors of $x + 5$ are needed.**

To add or subtract rational expressions with unlike denominators, we first determine the LCM of the denominators. Then we rename the fractions so that each denominator is the LCM. We call the resulting denominator the **least common denominator (LCD).**

EXAMPLE **6**   **Adding Rational Expressions with Unlike Denominators**

Add $\dfrac{2x}{x^2 + 6x + 9}$ and $\dfrac{5}{x^2 + 3x}$.

SOLUTION

$$\frac{2x}{x^2 + 6x + 9} + \frac{5}{x^2 + 3x} = \frac{2x}{(x + 3)(x + 3)} + \frac{5}{x(x + 3)}$$

From the factored denominators, we see that the LCD is $x(x + 3)(x + 3)$. Multiply the numerator and denominator of each fraction by the factors in the LCD that are not factors of the original denominator.

$$\frac{2x}{x^2 + 6x + 9} + \frac{5}{x^2 + 3x} = \frac{2x \cdot x}{x(x + 3)(x + 3)} + \frac{5(x + 3)}{x(x + 3)(x + 3)}$$

■ *NOTE Avoid the temptation to simplify the rational expressions after you have renamed them. Wait until after you have performed the operation before simplifying.*

$$= \frac{2x^2}{x(x + 3)^2} + \frac{5x + 15}{x(x + 3)^2}$$

$$= \frac{2x^2 + 5x + 15}{x(x + 3)^2} \qquad \text{The result cannot be simplified.}$$

EXAMPLE **7**   **Subtracting Rational Expressions with Unlike Denominators**

Subtract: $\dfrac{y}{y^2 + 4y + 3} - \dfrac{2}{y^2 - 2y - 3}$

SOLUTION

$$\frac{y}{y^2 + 4y + 3} - \frac{2}{y^2 - 2y - 3}$$

$$= \frac{y}{(y+3)(y+1)} - \frac{2}{(y-3)(y+1)}$$ Factor the denominators.
The LCD is $(y-3)(y+3)(y+1)$.

$$= \frac{y(y-3)}{(y-3)(y+3)(y+1)} - \frac{2(y+3)}{(y-3)(y+3)(y+1)}$$

$$= \frac{y^2 - 3y}{(y-3)(y+3)(y+1)} - \frac{2y+6}{(y-3)(y+3)(y+1)}$$

$$= \frac{(y^2 - 3y) - (2y + 6)}{(y-3)(y+3)(y+1)}$$ Subtract the numerators; retain the denominator.

$$= \frac{y^2 - 3y - 2y - 6}{(y-3)(y+3)(y+1)}$$ Remove parentheses and change signs.

$$= \frac{y^2 - 5y - 6}{(y-3)(y+3)(y+1)}$$

$$= \frac{(y-6)(y+1)}{(y-3)(y+3)(y+1)}$$ Factor the numerator.

$$= \frac{y-6}{(y-3)(y+3)}$$ Divide out the common factor $y + 1$.

Complex Fractions

A **complex fraction** is a fractional expression that has fractions in the numerator, the denominator, or both.

One method for simplifying a complex fraction is to follow the Order of Operations. We begin by performing the operations in the numerator and in the denominator. Then we perform the division.

EXAMPLE **8** **Simplifying a Complex Fraction**

$$\frac{1 + \dfrac{y}{x}}{\dfrac{2}{x} + \dfrac{2}{y}} = \frac{\dfrac{x+y}{x}}{\dfrac{2y+2x}{xy}}$$ Combine the fractions in the numerator and the denominator.

$$= \frac{x+y}{x} \cdot \frac{xy}{2y+2x}$$ Multiply by the reciprocal of the divisor.

$$= \frac{x+y}{x} \cdot \frac{xy}{2(y+x)} = \frac{y}{2}$$ Factor and divide out common factors.

An alternative method is to multiply each term in the numerator and the denominator by the LCD of all the fractions in both the numerator and the denominator.

<u>EXAMPLE</u> **9** **Simplifying a Complex Fraction**

The LCD of all the fractions in Example 8 is xy.

$$\frac{1 + \dfrac{y}{x}}{\dfrac{2}{x} + \dfrac{2}{y}} = \frac{xy \cdot \left(1 + \dfrac{y}{x}\right)}{xy \cdot \left(\dfrac{2}{x} + \dfrac{2}{y}\right)}$$ Multiply the numerator and the denominator by the LCD.

$$= \frac{xy \cdot 1 + xy \cdot \dfrac{y}{x}}{xy \cdot \dfrac{2}{x} + xy \cdot \dfrac{2}{y}}$$ Distributive Property

$$= \frac{xy + y^2}{2y + 2x}$$

$$= \frac{y(x + y)}{2(y + x)} = \frac{y}{2}$$ Factor and divide out common factors.

Speaking the Language

1. If A and B are polynomials, $B \neq 0$, then $\dfrac{A}{B}$ is called a(n) ▨▨▨▨ expression.

2. For the expression $\dfrac{x - 5}{x + 5}$, we call -5 a(n) ▨▨▨▨ .

3. To perform $\dfrac{1}{x + 2} + \dfrac{1}{x - 3}$, we must first determine the ▨▨▨▨ , which in this case is $(x + 2)(x - 3)$.

4. A fractional expression with a fraction in the numerator or in the denominator or in both is called a(n) ▨▨▨▨ .

R.5 Exercises

1. The simplified form of $\dfrac{x^2 - 1}{x - 1}$ is $x + 1$. How would the tables of values for these two expressions differ? Why?

2. For the quotient $\dfrac{x + 1}{x} \div \dfrac{x + 3}{x - 1}$, why would stating that 0 and 1 are the only restricted values be incorrect?

In Exercises 3–8, determine the restricted values.

3. $\dfrac{x - 1}{(x + 5)(x - 4)}$

4. $\dfrac{2x + 3}{(x^2 + 1)(x - 1)}$

5. $\dfrac{x^2 + 2x + 1}{x^2 + 3x - 10}$

6. $\dfrac{x^2 - 5}{x^3 + 1}$

7. $\dfrac{-8}{y^3 + 3y^2 - y - 3}$

8. $\dfrac{a}{-4a^2 + 24a - 36}$

In Exercises 9–18, simplify.

9. $\dfrac{4c - 8}{c^2 - 4c + 4}$

10. $\dfrac{y^2 - 10y + 25}{y^2 - 25}$

11. $\dfrac{x^2 - 2x - 15}{5 - x}$

12. $\dfrac{n^3 - n}{1 - n}$

13. $\dfrac{x^3 - 8}{x^2 + 2x - 8}$

14. $\dfrac{x^2 + 5x + 4}{x^2 - 4x - 5}$

15. $\dfrac{m^3 - 2m^2 - 8m}{2m^3 - m^2 - 10m}$

16. $\dfrac{x^4 - 9x^2}{x^3 + 5x^2 + 6x}$

17. $\dfrac{x^2 - 4}{x(x + 5) + 6}$

18. $\dfrac{x(x + 7) - 18}{2 - 3x + x^2}$

In Exercises 19–34, perform the indicated multiplications and divisions.

19. $\dfrac{a^2 - 1}{4a - 2} \cdot \dfrac{12}{a^2 - a}$

20. $\dfrac{5x}{5x + 5} \cdot \dfrac{5x^2 - 5}{15x^2}$

21. $\dfrac{x^2 - x - 6}{x^2 - 9} \cdot \dfrac{x^2 + x - 6}{x^2 + 5x + 6}$

22. $\dfrac{y^2 - 25}{2y - 2} \cdot \dfrac{y^2 + 4y - 5}{y^2 + 10y + 25}$

23. $\dfrac{x^2 + 4x}{x + 4} \div (x^3 - 16x)$

24. $\dfrac{9n - 15}{2n - 2} \div \dfrac{6n - 10}{n^2 - 1}$

25. $\dfrac{4x - 8}{-3x} \div \dfrac{12 - 6x}{15}$

26. $\dfrac{1 - x}{x^2 - 3x} \div \dfrac{1}{x^3 - 9x}$

27. $\dfrac{x^2 - x - 6}{x^2 + x - 2} \cdot \dfrac{x^2 + 3x - 4}{x^2 - 2x - 3}$

28. $\dfrac{y^2 + y + 1}{y + 1} \cdot \dfrac{(y - 1)^2}{y^3 - 1}$

29. $\dfrac{a^4 - 1}{a^2 - 1} \div \dfrac{a^3 + a^2 + a + 1}{4}$

30. $\dfrac{-2x^2 + 8x + 10}{5 - x} \div (-x - 1)$

31. $\dfrac{2x - 6}{5} \cdot \dfrac{x^2 - 1}{x^2 + 2x - 3} \div \dfrac{x^2 - 2x - 3}{x + 3}$

32. $\dfrac{y - 1}{y} \div \dfrac{1 - y}{y^2 - 3} \cdot \dfrac{4 - y}{y^3 - 4y^2 - 3y + 12}$

33. $\dfrac{x^2 - 9}{x + 1} \cdot (x^2 - 1) \div (x - 3)$

34. $\dfrac{x(x + 1) - 2}{x^2 - 1} \div \dfrac{x - 3}{x(x + 4) + 3} \cdot \dfrac{5x - 15}{x + 2}$

35. Suppose that a student performs the following operation:

$$\dfrac{2}{x + 1} - \dfrac{x + 3}{x + 1} = \dfrac{2 - x + 3}{x + 1} = \dfrac{-x + 5}{x + 1}$$

What would the tables of values for the first and last expressions reveal? How would you correct the student's work?

36. Verify that the following is true.

$$\dfrac{x^2}{x + 1} + \dfrac{2 + 3x}{x + 1} = x + 2$$

Are the two expressions equivalent? Why? Are the tables of values for the two expressions identical? Why?

In Exercises 37–44, perform the indicated additions and subtractions.

37. $\dfrac{2x + y}{x - y} + \dfrac{x - 2y}{x - y}$

38. $\dfrac{3x - 5}{xy} + \dfrac{3y + 5}{xy}$

39. $\dfrac{4x + 1}{x} - \dfrac{3x - 1}{x}$

40. $-\dfrac{2 - y}{y - x} - \dfrac{3 - x}{y - x}$

41. $\dfrac{2x + 9}{x - 4} + \dfrac{4x - 33}{x - 4}$

42. $\dfrac{5x - y}{2x + y} + \dfrac{x + 4y}{2x + y}$

43. $\dfrac{n^2 - 4n}{n^2 - n - 6} - \dfrac{n - 6}{n^2 - n - 6}$

44. $\dfrac{3a + 2b - c}{2a + 2c} - \dfrac{2(a + b - c)}{2a + 2c}$

In Exercises 45–50, determine the least common multiple (LCM) of the given expressions.

45. $x^2 - 9, x^2 - x - 6$

46. $x^3 + 3x^2 + 2x, x^2 + 2x + 1$

47. $x^3 - 1, x^2 - 1$ **48.** $3x^2 - 27, 6x - 18$

49. $(x + 1)^3, x^2 + 2x + 1, x^2 - 1$

50. $x^2 + x - 12, x^2 + 2x - 8, x^2 - 5x + 6$

In Exercises 51–66, perform the indicated additions and subtractions.

51. $\dfrac{3x}{x - 5} + \dfrac{2}{5 - x}$

52. $\dfrac{t^2}{1 - t} + \dfrac{t + 1}{t - 1}$

53. $\dfrac{1}{a - b} - \dfrac{1 - b}{b - a}$

54. $\dfrac{x + y}{x - y} - \dfrac{x + 2y}{y - x}$

55. $\dfrac{x}{x + 3} + \dfrac{2}{x + 5}$

56. $\dfrac{s}{2s + 1} - \dfrac{1}{s + 1}$

57. $\dfrac{m + n}{m - n} - \dfrac{m - n}{m + n}$

58. $\dfrac{x + 2}{x^2} + \dfrac{x}{x - 1}$

59. $\dfrac{x}{x-2} + \dfrac{x-16}{x^2+3x-10}$

60. $\dfrac{x-18}{x^2-x-12} + \dfrac{x}{x+3}$

61. $\dfrac{3x}{x^2-7x+10} - \dfrac{2x}{x^2-8x+15}$

62. $\dfrac{x-2}{4x+8} - \dfrac{x-3}{5x+10}$

63. $\dfrac{a^2+2}{a^2-a-2} + \dfrac{1}{a+1} - \dfrac{a}{a-2}$

64. $\dfrac{y-6}{y^2-4} - \dfrac{y+1}{y+2} + \dfrac{y-1}{y-2}$

65. $\dfrac{5}{2-n} - \dfrac{1}{2+n} + \dfrac{2n}{n^2-4}$

66. $\dfrac{x+4y}{2x-y} + \dfrac{(x-2y)^2}{2x^2-3xy+y^2}$

In Exercises 67–74, simplify the complex fraction.

67. $\dfrac{\dfrac{x-5}{y}}{\dfrac{x^2-25}{5y}}$

68. $\dfrac{\dfrac{x^2-y^2}{xy}}{\dfrac{x-y}{y}}$

69. $\dfrac{\dfrac{1}{a^3}-a}{\dfrac{1}{a^2}-1}$

70. $\dfrac{\dfrac{1}{x}+\dfrac{1}{y}}{\dfrac{1}{x^3}+\dfrac{1}{y^3}}$

71. $\dfrac{\dfrac{1}{x}-\dfrac{1}{x+h}}{h}$

72. $\dfrac{\dfrac{1}{x-h}-\dfrac{1}{x}}{h}$

73. $\dfrac{\dfrac{x}{1-x}+\dfrac{1+x}{x}}{\dfrac{1-x}{x}+\dfrac{x}{1+x}}$

74. $\dfrac{\dfrac{3}{x}+\dfrac{3}{y}-\dfrac{6}{xy}}{\dfrac{4}{x}+\dfrac{4}{y}-\dfrac{8}{xy}}$

R.6   *Radicals and Rational Exponents*

Properties of Radicals ■ *Simplifying Radicals* ■
Operations with Radicals ■ *Rationalizing Denominators* ■
Rational Exponents ■ *The Pythagorean Theorem*

Properties of Radicals

Because $6^2 = 36$ and $(-6)^2 = 36$, the two **square roots** of 36 are 6 and -6. Similarly, $2^3 = 8$, so 2 is the **cube root** of 8. In general, the **nth root** of a number is one of its n equal factors.

> ### Definition of nth Root of a Number
> Suppose that a and b are real numbers and n is an integer, where $n \geq 2$.
>
> If $b^n = a$, then b is an **nth root** of a.
>
> If n is even, then a has two nth roots; if n is odd, then a has one nth root.

When a number has two nth roots, they are opposites. For example, 16 has two fourth roots, 2 and -2, because $2^4 = 16$ and $(-2)^4 = 16$. The **principal** fourth root is the positive number of the pair.

> **Definition of Principal *n*th Root**
>
> Suppose that a is a real number with at least one nth root. Then the **principal *n*th root** of a, denoted by the **radical symbol** $\sqrt[n]{a}$, is the nth root that has the same sign as a.

The number a is the **radicand**, and the positive integer n is the **index** of the radical. For $n = 2$, the index is omitted: $\sqrt[2]{a} = \sqrt{a}$. Also, $\sqrt[n]{0} = 0$.

EXAMPLE **1**   **Even and Odd Roots**

(a) $\sqrt{100} = 10$ and $-\sqrt{100} = -10$

   Generalization: If a is positive and n is even, then a has two real number nth roots, $\sqrt[n]{a}$ and $-\sqrt[n]{a}$, but only one principal nth root, $\sqrt[n]{a}$.

(b) $\sqrt[3]{8} = 2$ and $\sqrt[3]{-8} = -2$

   Generalization: If a is any real number and n is odd, then a has exactly one real number nth root, $\sqrt[n]{a}$.

(c) $\sqrt{-25}$ and $\sqrt[6]{-1}$ are not real numbers because there is no real number b for which $b^2 = -25$ or for which $b^6 = -1$.

   Generalization: If a is negative and n is even, then a has no real number nth root.

In the following summary of properties of radicals, we assume that all nth roots are real numbers.

> **Properties of Radicals**
>
> If a and b represent algebraic expressions, and if m and n are integers greater than 1, then
>
> | *Property* | *Example* | |
> |---|---|---|
> | $\sqrt[n]{a^n} = \begin{cases} \lvert a \rvert & \text{for even } n \\ a & \text{for odd } n \end{cases}$ | $\sqrt{(-3)^2} = \lvert -3 \rvert = 3$
 $\sqrt[3]{(-3)^3} = -3$ | |
> | $\sqrt[n]{a} \cdot \sqrt[n]{b} = \sqrt[n]{ab}$ | $\sqrt[3]{3} \cdot \sqrt[3]{9} = \sqrt[3]{27} = 3$ | **Product Rule for Radicals** |
> | $\dfrac{\sqrt[n]{a}}{\sqrt[n]{b}} = \sqrt[n]{\dfrac{a}{b}}, b \neq 0$ | $\dfrac{\sqrt[4]{32}}{\sqrt[4]{16}} = \sqrt[4]{\dfrac{32}{16}} = \sqrt[4]{2}$ | **Quotient Rule for Radicals** |
> | $\left(\sqrt[n]{a}\right)^n = a$ | $\left(\sqrt[3]{7}\right)^3 = 7$ | |
> | $\sqrt[n]{a^m} = \left(\sqrt[n]{a}\right)^m$ | $\sqrt[4]{16^3} = \left(\sqrt[4]{16}\right)^3 = 2^3 = 8$ | |

Simplifying Radicals

If $\sqrt[n]{a}$ is a rational number, then a is a **perfect nth power.** For example, 49 is a **perfect square** because $\sqrt{49}$ is the rational number 7, and $\frac{8}{27}$ is a **perfect cube** because $\sqrt[3]{\frac{8}{27}}$ is the rational number $\frac{2}{3}$.

A radical $\sqrt[n]{a}$ is considered simplified if certain conditions are met.

1. The radicand contains no perfect nth power factors.
2. The radicand is not a fraction.
3. There is no radical in the denominator of a fraction.
4. The index cannot be reduced.

The Product Rule for Radicals is often used to satisfy the first of these conditions.

EXAMPLE **2**   **Simplifying Radical Expressions**

(a) $\sqrt[3]{40} = \sqrt[3]{8 \cdot 5}$ 8 is the largest perfect cube factor of 40.

$\qquad\quad = \sqrt[3]{8} \cdot \sqrt[3]{5}$ **Product Rule for Radicals**

$\qquad\quad = 2\sqrt[3]{5}$

(b) $\sqrt{45x^3} = \sqrt{9x^2 \cdot 5x}$ $9x^2$ is the largest perfect square factor of $45x^3$.

$\qquad\quad\; = \sqrt{(3x)^2} \cdot \sqrt{5x}$ **Product Rule for Radicals**

$\qquad\quad\; = |3x|\sqrt{5x}$ $\sqrt{a^2} = |a|$

(c) $\sqrt{x^2 - 8x + 16} = \sqrt{(x-4)^2}$ **Factor the radicand.**

$\qquad\qquad\qquad\;\; = |x - 4|$

(d) $\sqrt[3]{a^5 b^7} = \sqrt[3]{a^3 \cdot a^2 \cdot b^6 \cdot b}$ **The largest perfect cube factors are a^3 and b^6.**

$\qquad\quad\; = ab^2 \sqrt[3]{a^2 b}$ **Absolute value is not needed for odd roots.**

EXAMPLE **3**   **Simplifying Radical Expressions**

(a) $\sqrt{6xy^3} \cdot \sqrt{12x^8 y^9} = \sqrt{72x^9 y^{12}}$ **Multiply first, then simplify.**

$\qquad\qquad\qquad\quad = \sqrt{36 \cdot 2 \cdot x^8 \cdot x \cdot y^{12}}$ **The largest perfect square factors are 36, x^8, and y^{12}.**

$\qquad\qquad\qquad\quad = 6x^4 y^6 \sqrt{2x}$

(b) $\dfrac{\sqrt[3]{54x^2 y^{10}}}{\sqrt[3]{2x^8 y}} = \sqrt[3]{\dfrac{27y^9}{x^6}} = \dfrac{3y^3}{x^2}$ **Divide first, then simplify.**

In the remainder of this section, we assume that all variables in radicands represent positive numbers and that no denominator is 0.

Operations with Radicals

Like radicals are radicals with the same index and radicand. Using the Distributive Property, we can combine like radical expressions into a single term.

EXAMPLE **4**   **Combining Radicals**

(a) $2\sqrt{75} - 3\sqrt{12} = 2\sqrt{25}\,\sqrt{3} - 3\sqrt{4}\,\sqrt{3}$ Product Rule for Radicals

$= 2 \cdot 5\sqrt{3} - 3 \cdot 2\sqrt{3}$

$= 10\sqrt{3} - 6\sqrt{3}$

$= (10 - 6)\sqrt{3} = 4\sqrt{3}$ Distributive Property

(b) $\sqrt[3]{24x^4} + 5\,\sqrt[3]{81x} = \sqrt[3]{8x^3}\,\sqrt[3]{3x} + 5 \cdot \sqrt[3]{27}\,\sqrt[3]{3x}$ Product Rule for Radicals

$= 2x\,\sqrt[3]{3x} + 15\,\sqrt[3]{3x}$

$= (2x + 15)\,\sqrt[3]{3x}$ Distributive Property

The Distributive Property, the FOIL pattern, and special product patterns all apply to products of radical expressions.

EXAMPLE **5**   **Multiplying Radicals**

Perform the indicated operations.

(a) $\left(\sqrt{x} - 2\right)\left(3\sqrt{x} + 1\right)$ (b) $\left(\sqrt{3} - \sqrt{x}\right)^2$

SOLUTION

(a) $\left(\sqrt{x} - 2\right)\left(3\sqrt{x} + 1\right)$

$= \sqrt{x} \cdot 3\sqrt{x} + \sqrt{x} \cdot 1 - 2 \cdot 3\sqrt{x} - 2 \cdot 1$ FOIL

$= 3x + \sqrt{x} - 6\sqrt{x} - 2$

$= 3x - 5\sqrt{x} - 2$ Combine like radicals.

(b) $\left(\sqrt{3} - \sqrt{x}\right)^2 = \left(\sqrt{3}\right)^2 - 2\left(\sqrt{3}\right)\left(\sqrt{x}\right) + \left(\sqrt{x}\right)^2$ $(A - B)^2 = A^2 - 2AB + B^2$

$= 3 - 2\sqrt{3x} + x$

Rationalizing Denominators

Condition 2 for a simplified radical is easily met by applying the Quotient Rule for Radicals. For example, $\sqrt{\dfrac{x}{3}}$ violates condition 2 because the radicand is a fraction. However, we can use the Quotient Rule for Radicals to write the expression as $\dfrac{\sqrt{x}}{\sqrt{3}}$. This result satisfies condition 2, but it now violates condition 3 because there is a radical in the denominator. If we multiply the numerator and denominator by $\sqrt{3}$, the resulting denominator will be a rational number, and the expression will be simplified.

$$\frac{\sqrt{x}}{\sqrt{3}} = \frac{\sqrt{x}}{\sqrt{3}} \cdot \frac{\sqrt{3}}{\sqrt{3}} = \frac{\sqrt{3x}}{\sqrt{9}} = \frac{\sqrt{3x}}{3}$$

This process is called **rationalizing the denominator.**

$\overline{\text{EXAMPLE}}$ **6** **Rationalizing Denominators**

(a) $\dfrac{x}{3\sqrt{5}} = \dfrac{x}{3\sqrt{5}} \cdot \dfrac{\sqrt{5}}{\sqrt{5}} = \dfrac{x\sqrt{5}}{3(5)} = \dfrac{x\sqrt{5}}{15}$

(b) $\dfrac{1}{\sqrt[3]{x}} = \dfrac{1}{\sqrt[3]{x}} \cdot \dfrac{\sqrt[3]{x^2}}{\sqrt[3]{x^2}}$ Produce a perfect cube radicand in the denominator.

$ = \dfrac{\sqrt[3]{x^2}}{\sqrt[3]{x^3}}$

$ = \dfrac{\sqrt[3]{x^2}}{x}$

The binomials $3 + \sqrt{2}$ and $3 - \sqrt{2}$ are examples of **square root conjugates,** which are expressions of the form $a + b$ and $a - b$, where at least one of a and b is a square root expression. The product of square root conjugates is $a^2 - b^2$, an expression that does not contain a square root radical. This fact can be used to rationalize a binomial denominator.

$\overline{\text{EXAMPLE}}$ **7** **Rationalizing a Denominator**

$\dfrac{4}{x - \sqrt{3}} = \dfrac{4}{x - \sqrt{3}} \cdot \dfrac{x + \sqrt{3}}{x + \sqrt{3}}$ Multiply the numerator and the denominator by the conjugate of the denominator.

$\phantom{\dfrac{4}{x - \sqrt{3}}} = \dfrac{4(x + \sqrt{3})}{(x)^2 - (\sqrt{3})^2}$

$\phantom{\dfrac{4}{x - \sqrt{3}}} = \dfrac{4(x + \sqrt{3})}{x^2 - 3}$

Rational Exponents

Rational exponents are defined in terms of radicals.

Definition of Rational Exponents

If a is a real number and $\sqrt[n]{a}$ exists, then

$$a^{1/n} = \sqrt[n]{a}$$

If m is a positive integer such that m and n have no common factor, and if $\sqrt[n]{a}$ exists, then

$$a^{m/n} = (a^{1/n})^m = \left(\sqrt[n]{a}\right)^m$$

$$a^{m/n} = (a^m)^{1/n} = \sqrt[n]{a^m}$$

In words, to convert a rational exponential expression $a^{m/n}$ into a radical, take the nth root of a (if it exists) and then raise the result to the mth power.

EXAMPLE **8**   **Evaluating Expressions with Rational Exponents**

(a) $(-27)^{2/3} = \left(\sqrt[3]{-27}\right)^2 = (-3)^2 = 9$

(b) $36^{-1/2} = \dfrac{1}{36^{1/2}} = \dfrac{1}{\sqrt{36}} = \dfrac{1}{6}$

(c) $(-32)^{-4/5} = \dfrac{1}{(-32)^{4/5}} = \dfrac{1}{\left(\sqrt[5]{-32}\right)^4} = \dfrac{1}{(-2)^4} = \dfrac{1}{16}$

EXAMPLE **9**   **Writing Exponential Expressions as Radicals**

(a) $x^{1/5} = \sqrt[5]{x}$

(b) $7y^{3/5} = 7\sqrt[5]{y^3}$

(c) $x^{2/3}y^{1/3} = (x^2y)^{1/3} = \sqrt[3]{x^2y}$

(d) $(x^2 + a^2)^{3/4} = \left(\sqrt[4]{x^2 + a^2}\right)^3 = \sqrt[4]{(x^2 + a^2)^3}$

All the previously stated properties of exponents hold for rational exponents.

EXAMPLE **10**   **Operations with Rational Exponents**

(a) $(2y^{3/4})(3y^{-1/3}) = 6y^{3/4\, -\, 1/3} = 6y^{9/12\, -\, 4/12} = 6y^{5/12}$

(b) $\left(\dfrac{8x^{1/2}}{x^{-3/2}}\right)^{1/3} = (8x^{1/2\, -\, (-3/2)})^{1/3} = (8x^2)^{1/3} = 8^{1/3}(x^2)^{1/3} = 2x^{2/3}$

Rational exponents can be used to satisfy condition 4 (reducing the index) for a simplified radical.

EXAMPLE **11**   **Reducing the Index**

(a) $\sqrt[8]{x^2} = x^{2/8} = x^{1/4} = \sqrt[4]{x}$

(b) $\sqrt[3]{\sqrt{27}} = (27^{1/2})^{1/3} = 27^{1/6} = (3^3)^{1/6} = 3^{1/2} = \sqrt{3}$

Although most calculators have square root and cube root keys, rational exponents can be used to compute higher roots.

—————
EXAMPLE 12   **Using a Calculator to Compute Roots**

Use your calculator to perform the indicated computations. Round results to the nearest hundredth.

(a) $\sqrt[5]{-58}$ 　　　　　(b) $\sqrt{\sqrt{13} - 2}$

(c) $\left(1 + \sqrt{3}\right)^{1.62}$ 　　(d) $\sqrt[7]{6} - \sqrt[6]{7}$

SOLUTION

(a) $\sqrt[5]{-58} = (-58)^{1/5} \approx -2.25$ (See Figure R.6.)

(b) $\sqrt{\sqrt{13} - 2} \approx 1.27$ (See Figure R.6.)

(c) $\left(1 + \sqrt{3}\right)^{1.62} \approx 5.09$ (See Figure R.7.)

(d) $\sqrt[7]{6} - \sqrt[6]{7} = 6^{1/7} - 7^{1/6} \approx -0.09$ (See Figure R.7.)

FIGURE **R.6**

```
(-58)^(1/5)
              -2.25
√(√(13)-2)
               1.27


```

FIGURE **R.7**

```
(1+√(3))^1.62
               5.09
6^(1/7)-7^(1/6)
               -.09


```

The Pythagorean Theorem

Certain formulas may involve a radical or may lead to a radical when the formula is applied. An example is the formula that is given in the Pythagorean Theorem.

The **Pythagorean Theorem** relates the lengths of the sides of a right triangle.

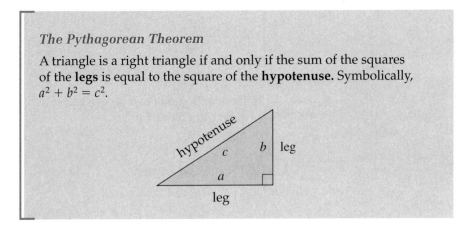

The Pythagorean Theorem

A triangle is a right triangle if and only if the sum of the squares of the **legs** is equal to the square of the **hypotenuse.** Symbolically, $a^2 + b^2 = c^2$.

—————
EXAMPLE 13   **Surveying**

Working near a small lake, a surveyor places poles at A, B, and C, and measures the distances from the pole at C to the poles at A and B. (See Figure R.8.) What is the distance AB across the lake?

FIGURE **R.8**

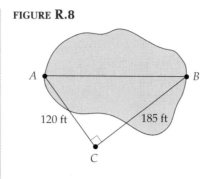

SOLUTION

Because $\triangle ABC$ is a right triangle, we can determine the distance AB by applying the Pythagorean Theorem.

$$(AB)^2 = 185^2 + 120^2 = 48{,}625$$

$$AB = \sqrt{48{,}625} \approx 220.5 \text{ feet}$$

Speaking the Language

1. Although 9 has two square roots, $\sqrt{9}$ represents the _____ square root of 9.

2. In the expression $\sqrt[3]{2x}$, the _____ is $2x$ and the _____ is 3.

3. Rewriting $\dfrac{x}{2\sqrt{x}}$ as $\dfrac{\sqrt{x}}{2}$ is called _____ the denominator.

4. The expressions $x + \sqrt{y}$ and $x - \sqrt{y}$ are examples of square root _____ .

R.6 Exercises

1. If $b^2 = 16$, then we say that b is *a* square root of 16. However if $b^3 = 64$, then we say that b is *the* cube root of 64. Explain the difference.

2. Determine whether the tables of values for the given pair of expressions are the same. Explain why or why not.

   (a) $y_1 = \sqrt{x^2}, y_2 = x$ (b) $y_1 = \sqrt[3]{x^3}, y_2 = x$

In Exercises 3–12, determine the indicated root. Assume that all variables represent any real number.

3. $\sqrt{(-7)^2}$

4. $\sqrt{(-15)^2}$

5. $\sqrt{25y^6}$

6. $\sqrt{36t^{12}}$

7. $\sqrt[5]{-32}$

8. $\sqrt[4]{(-3)^4}$

9. $\sqrt[3]{-8x^{15}}$

10. $\sqrt[5]{y^{25}}$

11. $\sqrt{(x-5)^2}$

12. $\sqrt{x^2 + 14x + 49}$

In Exercises 13–24, use the Product Rule for Radicals to simplify the given expression. Assume that all variables represent positive numbers.

13. $\sqrt{75}$

14. $\sqrt{72}$

15. $\sqrt[3]{40}$

16. $\sqrt[4]{48}$

17. $\sqrt{x^5 y^6}$

18. $\sqrt{20a^8 b^9}$

19. $\sqrt{28x^{11}}$

20. $-\sqrt{36a^7 b^{15}}$

21. $\sqrt[3]{-24x^4}$

22. $\sqrt[4]{y^{11}}$

23. $\sqrt{9x^2 + 36}$

24. $\sqrt{b^{10} + b^8}$

25. Consider a radical that has the form $\sqrt[n]{x^m}$. Write an inequality involving m and n that must be true in order for the radical to be considered simplified.

26. Why is $\dfrac{\sqrt{x}}{\sqrt{3}}$ not considered simplified? If we write the expression as $\sqrt{\dfrac{x}{3}}$, is the radical now simplified? Why?

In Exercises 27–30, multiply and simplify. Assume that all variables represent positive numbers.

27. $3\sqrt{6a} \cdot \left(-2\sqrt{2a}\right)$

28. $\sqrt{15y^3} \cdot \sqrt{3y^6}$

29. $\left(\sqrt{7ab^3}\right)\left(\sqrt{14a^3b^5}\right)$

30. $\left(\sqrt[3]{4x^7}\right)\left(\sqrt[3]{16x^5}\right)$

31. Explain how to use a table of values to check your result in Exercise 27.

32. Suppose that you use the Product Rule for Radicals to write $\sqrt{(x+3)}\,\sqrt{(x-3)} = \sqrt{x^2 - 9}$. Are the tables of values for the two expressions the same? Why?

In Exercises 33–40, use the Quotient Rule for Radicals to simplify the given expression. Assume that all variables represent positive numbers.

33. $\sqrt{\dfrac{24x^{11}y^3}{2xy^9}}$

34. $\sqrt{\dfrac{72ab^7}{3a^5b^4}}$

35. $\dfrac{\sqrt{40x^{15}y^4}}{\sqrt{5x^{10}y^6}}$

36. $\dfrac{\sqrt{90y^9z^8}}{\sqrt{2y^2z^3}}$

37. $\dfrac{\sqrt{18t^3}}{\sqrt{50t^{13}}}$

38. $\dfrac{\sqrt{48xy^{10}}}{\sqrt{6x^7y^3}}$

39. $\sqrt[3]{\dfrac{5t^5}{40t^2}}$

40. $\dfrac{\sqrt[3]{54yz^8}}{\sqrt[3]{250y^4z^2}}$

In Exercises 41–46, perform the indicated operations. Assume that all variables represent positive numbers.

41. $3x\sqrt{7} + x\sqrt{7}$

42. $5\sqrt{6x} - 4\sqrt{6x}$

43. $\sqrt{45} + \sqrt{125}$

44. $7\sqrt{24} - 5\sqrt{54}$

45. $2t\sqrt{48} - 4\sqrt{27t^2}$

46. $a\sqrt{12a} - \sqrt{27a^3} + \sqrt{9a^4}$

In Exercises 47–50, multiply and simplify the result. Assume that all variables represent positive numbers.

47. $3\sqrt{2}\left(5\sqrt{2} - \sqrt{3}\right)$

48. $\left(\sqrt{2x} - 3\right)\left(\sqrt{2x} + 3\right)$

49. $\left(\sqrt{5a} - 2\right)\left(\sqrt{2} + \sqrt{5a}\right)$

50. $\left(\sqrt{x} + \sqrt{7}\right)^2$

51. Rationalize the denominator of $\dfrac{1}{\sqrt{x} + 1}$. Then display a table of values for each expression for nonnegative values of x. Why is one expression defined for all nonnegative values of x, but the other is not?

52. Rationalize the denominator of $\dfrac{1}{\sqrt{\sqrt[3]{x}}}$.

In Exercises 53–58, rationalize the denominator. Assume that all variables represent positive numbers.

53. $\dfrac{12}{\sqrt{3}}$

54. $\dfrac{-4}{\sqrt{y}}$

55. $\dfrac{5}{2\sqrt{3}}$

56. $\dfrac{15}{2\sqrt{12}}$

57. $\dfrac{7}{\sqrt[3]{x}}$

58. $\dfrac{10}{\sqrt[3]{4}}$

In Exercises 59–64, rationalize the denominator. Assume that all variables represent positive numbers.

59. $\dfrac{6}{\sqrt{6} + 3}$

60. $\dfrac{8}{\sqrt{7} - \sqrt{3}}$

61. $\dfrac{\sqrt{5} + \sqrt{x}}{\sqrt{5} - \sqrt{x}}$

62. $\dfrac{\sqrt{t}}{4 + \sqrt{t}}$

63. $\dfrac{7a}{\sqrt{a} + a}$

64. $\dfrac{1}{2 - \sqrt{x + 1}}$

65. Evaluate the expressions that represent real numbers. Explain why the others do not represent real numbers.

  (i) $\sqrt[3]{-\sqrt[4]{1}}$

  (ii) $\sqrt[4]{-\sqrt[3]{1}}$

  (iii) $(-1)^{-1/2}$

  (iv) $-1^{-1/2}$

66. The definition of $b^{m/n}$ requires that m and n have no common factors. Use $[(-1)^6]^{1/2}$ to explain why this requirement is necessary.

In Exercises 67–74, evaluate the given expression.

67. $25^{-1/2}$

68. $-64^{1/3}$

69. $-27^{-1/3}$

70. $81^{-1/4}$

71. $\left(\dfrac{27}{8}\right)^{-4/3}$

72. $\left(\dfrac{1}{16}\right)^{-3/2}$

73. $64^{5/6}$

74. $(-8)^{5/3}$

In Exercises 75 and 76, fill in the blank with the equivalent exponential or radical form of the given expression. Assume that all variables represent positive numbers.

75. *Radical form*   *Exponential form*

$\sqrt[3]{y}$ (a)

(b) $(2a)^{1/4}$

(c) $2a^{3/4}$

$-\sqrt[3]{x^2}$ (d)

76. *Radical form*   *Exponential form*

(a) $7n^{1/5}$

(b) $(x+3)^{1/2}$

$\sqrt[3]{a^2}$ (c)

$\sqrt{3b}$ (d)

In Exercises 77–84, use the properties of exponents to simplify the given expression. Assume that all variables represent positive numbers.

77. $y^{-2/3} \cdot y^{5/3}$

78. $(a^8 b^4)^{3/2}$

79. $\left(\dfrac{m^6}{n^{12}}\right)^{-2/3}$

80. $\dfrac{x^{1/2}}{x^{1/3}}$

81. $(x^{-6})^{-1/3}$

82. $(-27t^{12})^{-1/3}$

83. $\dfrac{x^{1/3} y^{2/3}}{y^{-1/3}}$

84. $\left(\dfrac{x^{-2/3}}{y^{-3/2}}\right)^{-6}$

85. We defined $b^{1/n}$ for integers $n > 1$. Try to evaluate $\sqrt[0.5]{16}$ with your calculator. Then write the radical as an expression with a rational exponent and explain why the calculator result is reasonable.

86. Explain why the Product Rule for Radicals cannot be used to write $\sqrt[3]{5}\,\sqrt{5}$ as a single radical. Show how to use another property to rewrite the expression as a single radical.

In Exercises 87–96, write the expression with rational exponents and simplify. Express the result as a single radical. Assume that all variables represent positive numbers.

87. $\sqrt[6]{y^2}$

88. $\sqrt[9]{27t^3 w^6}$

89. $\left(\sqrt[3]{20}\right)^2$

90. $\left(\sqrt[6]{b}\right)^3$

91. $\sqrt[3]{\sqrt[4]{6}}$

92. $\sqrt[3]{\sqrt{x}}$

93. $\sqrt[3]{a} \cdot \sqrt{a}$

94. $\sqrt{2} \cdot \sqrt[4]{8}$

95. $\sqrt{27 \sqrt[3]{9}}$

96. $\sqrt[3]{x^4 \sqrt{x^3}}$

In Exercises 97–102, use a calculator to perform the indicated operations. Round results to the nearest hundredth.

97. $\left(\sqrt{7} - 1\right)^{4.37}$

98. $\sqrt[7]{-150}$

99. $\dfrac{\sqrt{6}+1}{\sqrt[3]{20}-5}$

100. $\left(1 + \dfrac{2}{\sqrt{3}}\right)^{1.85}$

101. $\sqrt[4]{100} + \sqrt[5]{-50}$

102. $\sqrt{\sqrt[3]{20}+1}$

In Exercises 103–106, perform the indicated operations and simplify the result. Assume that all variables represent positive numbers.

103. $\dfrac{\sqrt{y}+3}{\sqrt{y}}$

104. $\dfrac{\sqrt{3}}{\sqrt{2}} + \dfrac{1}{\sqrt{6}}$

105. $\sqrt{8} - \dfrac{1}{\sqrt{2}}$

106. $\dfrac{2}{\sqrt{5}+2} - \dfrac{1}{\sqrt{5}-2}$

In Exercises 107–110, the lengths of the legs of a right triangle are a and b, and the length of the hypotenuse is c. For the given two lengths, determine the length of the third side. Round to the nearest tenth.

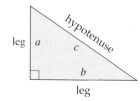

107. $a = 5, b = 11$

108. $b = 6, c = 20$

109. $a = \sqrt{5}, c = \sqrt{15}$

110. $a = 3\sqrt{2}, b = 2\sqrt{3}$

111. Two ships are located at points M and N, and a beacon is located at point B. The ship at M is 4.2 miles from the beacon, and the ship at N is 8.3 miles from the beacon. If $\overline{MB}$ and $\overline{MN}$ form a right angle, how far apart are the ships?

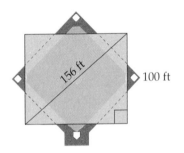

112. A rectangular tarp is used to cover the infield of a baseball diamond. One side of the tarp is 100 feet long, and the diagonal distance across the tarp is 156 feet. To the nearest square foot, determine the area of the tarp.

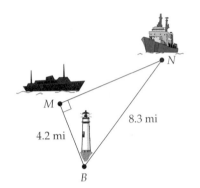

113. Referring to the figure, determine the radius of the circle.

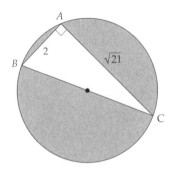

114. A tent pole 30 feet in length is supported by two guy wires on each side of the pole. The wires run from the top of the pole to the ground, where they are fastened 10 feet and 20 feet, respectively, from the base of the pole. How long are the guy wires?

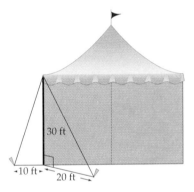

Chapter Review Exercises

Section R.1

1. Identify the numbers in the following list that are
  (a) natural numbers.
  (b) whole numbers.
  (c) integers.
  (d) rational numbers.
  (e) irrational numbers.

$$\frac{2}{3}, 0, -6, -\sqrt{25}, 0.\overline{6}, \frac{\pi}{3}, -0.29, \sqrt{9}$$

2. Determine whether the statement is true or false.
  (a) Every integer is a whole number.
  (b) There are infinitely many real numbers that are not irrational numbers.
  (c) Irrational numbers have nonterminating, repeating decimal names.

3. Calculate $0.\overline{3} - 0.3$.

4. Explain why 8 is a rational number.

5. Give an example of two irrational numbers whose sum is a rational number.

6. Evaluate.

   (a) the distance on a number line between -6 and -7

   (b) $|-6| - |-7|$

7. Insert $<$, $>$, or $=$ to make each relation true.

   (a) $|-3|$ _____ $-|3|$

   (b) 12.7 _____ $12.\overline{7}$

   (c) 1.6 _____ $\dfrac{8}{5}$

8. Identify the property that justifies each statement.

   (a) $5(y - 3) = 5y - 15$

   (b) $\dfrac{3}{4}[4(x + 2)] = \left(\dfrac{3}{4} \cdot 4\right)(x + 2)$

9. Fill in the blank to make each statement true. State the property that justifies your answer.

   (a) $\dfrac{1}{y^2 + 4} \cdot$ _____ $= 1$

   (b) _____ $(3 - z) = z - 3$

10. Use the indicated property to rewrite each expression.

    (a) Additive Identity: $-5x + 5x =$ _____

    (b) Commutative Property of Multiplication: $(x - 6)(x + 3) =$ _____

Section R.2

11. Evaluate.

    (a) $-2 - (-3)(-6) + 2^5$ (b) $\dfrac{-3^2 + (-5)}{\sqrt{36} + 2^3}$

12. Evaluate the given expressions for $x = -5$ and $y = 2$.

    (a) $x^2 - 2x + 3$ (b) $xy^3 + y - x$

13. Evaluate the given expressions for $a = 3$ and $b = -1$.

    (a) $(2a - 1)(3a + 5)$ (b) $\dfrac{a^2 - b^2}{a^2 + b^2}$

14. Evaluate $\dfrac{8^{-1} + 4^{-2}}{2^{-3}}$.

15. Explain the difference between -5^{-2} and $(-5)^{-2}$.

16. Write $\left(\dfrac{a^2}{b^3}\right)^5$ without parentheses.

In Exercises 17 and 18, simplify and write the result with positive exponents. Assume that all variables represent positive numbers.

17. (a) $x^{-3} \cdot x^2$ (b) $(3xy^{-4})^{-2}$

18. (a) $\dfrac{12x^{-2}y}{3x^{-4}y^2}$ (b) $\left(\dfrac{2x^{-2}}{y^3}\right)^{-1}$

19. Write 12 billion in scientific notation.

20. Write $7.6 \cdot 10^{-6}$ in decimal form.

Section R.3

21. Which of the following is not a monomial? Why?

    (i) 4 (ii) $\dfrac{3x}{2}$ (iii) $-5x^2y^3$ (iv) $3x^{-1}$ (v) 0

22. Determine the degree of $3x - 5x^2y + y^2 - 6$.

23. Combine like terms: $2xy^2 + 3x^2y - 5x^2y + xy^2$.

24. Add $x^2 + 3x - 4$ and $x^3 - x^2 - 5x + 6$.

25. Find the difference of $a^2 - ab + b^2$ and $3ab - 4b^2$.

In Exercises 26–30, determine the product.

26. $7a(2a - 3b + 5)$

27. $(x - 2y)(x^2 + 2xy + 3y^2)$

28. $(3c + 8)(2c - 5)$

29. (a) $(xy + 3)(xy - 3)$ (b) $(2x - 5)^2$

30. (a) $(x^2 - 1)^3$ (b) $(a + 2b)^3$

Section R.4

31. Explain why $(2x + 4)(2x - 4)$ is not a complete factorization of $4x^2 - 16$.

32. Factor $8x^2y^3 + 12x^2y - 16y$.

33. Remove the common factor -2 from the expression $-4x^2 - 2x + 8$.

In Exercises 34–40, factor completely.

34. (a) $9x^2 - c^2$ (b) $(x + 3)^2 - y^2$

35. (a) $4x^2 - 4xy + y^2$ (b) $2a^2c + 20ac + 50c$

36. (a) $x^3 - 8$ (b) $x^6 - 27$

37. (a) $a^3 + 1000$ (b) $x^3y^3 + 1$

38. (a) $x^2 + 3x - 28$ (b) $x^2 - 8x + 15$

39. (a) $4x^2 - x - 5$ (b) $12x^2 + 7x - 10$

40. (a) $x^3 + 3x^2 + 3x + 9$ (b) $2x^2 + 2ax - bx - ab$

Section R.5

41. Determine the restricted values of the given expressions.

(a) $\dfrac{1}{3x - 3}$ (b) $\dfrac{x - 3}{x^2}$ (c) $\dfrac{x + 5}{x^2 + 3x + 2}$

42. What is the difference between a fractional expression and a rational expression?

43. Simplify $\dfrac{2x + 6}{4x^2 + 8x - 12}$.

44. Simplify $\dfrac{x^2 + 4x + 4}{3x^2 - 12}$.

45. Multiply.

(a) $\dfrac{x + 1}{x^2 + x - 6} \cdot \dfrac{x^2 - 4}{x^2 + 3x + 2}$

(b) $\dfrac{x - 3}{x + 1} \cdot \dfrac{x^2 + 2x + 1}{6 - 2x}$

46. Divide.

(a) $\dfrac{x^2 y}{x + 1} \div \dfrac{xy}{(x + 1)^2}$

(b) $\dfrac{a^2 + ab - a - b}{a - 1} \div (a + b)$

47. Perform the indicated operations.

(a) $\dfrac{x}{2x + 1} - \dfrac{x - 1}{2x + 1}$

(b) $\dfrac{2x - y}{x + y} + \dfrac{2y - x}{x + y}$

48. Perform the indicated operations.

(a) $\dfrac{x}{x^2 - 9} + \dfrac{5}{x + 3}$

(b) $\dfrac{5x}{x^2 - x - 2} - \dfrac{2x}{x^2 + x - 6}$

49. Use the Order of Operations to simplify the complex fraction.

$$\dfrac{\dfrac{1}{x} + 1}{1 - \dfrac{1}{x^2}}$$

50. Use the LCD method to simplify the complex fraction.

$$\dfrac{\dfrac{1}{x + 1} + \dfrac{1}{x}}{\dfrac{1}{x^2 - 1}}$$

Section R.6

51. (a) What are the square roots of 49?

(b) What is $\sqrt{49}$?

52. Simplify. Assume that $x > 0$ and that c and y are any real numbers.

(a) $\sqrt{48x^3}$ (b) $\sqrt{c^2}$ (c) $\sqrt[3]{24y^5}$

53. Perform the indicated operations and simplify the result.

(a) $\sqrt{5x^5 y} \cdot \sqrt{10x^6 y^3}$ (b) $\dfrac{\sqrt{28x^{12} y^5}}{\sqrt{7x^3 y^9}}$

54. Rationalize the denominators.

(a) $\sqrt{\dfrac{1}{x^3}}$ (b) $\dfrac{2}{\sqrt[3]{x^2}}$

55. Rationalize the denominator of $\dfrac{2}{3 + \sqrt{x}}$.

56. Write each expression with rational exponents and simplify. Then express the result as a single radical. Assume that all variables represent positive numbers.

(a) $\sqrt[6]{x^4}$ (b) $\sqrt{\sqrt[3]{x^2}}$

57. Evaluate.

(a) $-8^{2/3}$ (b) $8^{-2/3}$ (c) $(-8)^{-2/3}$

58. Perform the indicated operations and simplify the result.

(a) $\sqrt{80a} + \sqrt{20a}$ (b) $3\sqrt{2}\left(2\sqrt{8} + \sqrt{5}\right)$

59. Evaluate $\sqrt[3]{15} + 6^{2.4}$ to the nearest hundredth.

60. A 20-foot ladder rests against a wall with the bottom of the ladder 5 feet from the base of the wall. To the nearest hundredth, how far up the wall does the top of the ladder rest?

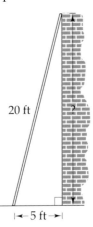

20 ft

|← 5 ft →|

Chapter **1**

Equations and Inequalities

When a mother and her daughter sit at the ends of a teeter-totter, the mother's greater weight will cause her end of the board to rest on the ground, and the daughter will remain in the air. However, the mother can move forward to a point that will allow her and her daughter to be in perfect balance.

The mother can find the balance point by experimenting with various locations. However, the point can be determined mathematically with an equation that describes the conditions of the problem. If the weights of the daughter and the mother are W_D and W_M, respectively, then the equation $W_D \cdot d_1 = W_M \cdot d_2$ describes the balanced condition shown in the figure below. Solving this equation for d_2 gives us the exact distance that is needed to achieve perfect balance.

The focus of this chapter is on a variety of types of equations and inequalities and on graphing and algebraic methods for solving them. Solving equations and inequalities is a fundamental skill in algebra and is basic to your ability to solve application problems in real life.

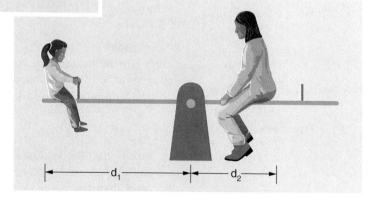

1.1 *Coordinate Systems and the Graphing Calculator*

The Cartesian Coordinate System ■ *The Viewing Window of a Calculator* ■ *Graphs of Algebraic Expressions*

In this chapter, we discuss ways in which you can merge technology with familiar algebraic methods for solving equations and inequalities. You will find that a graphing calculator can be a powerful ally in visualizing the concept of *solving,* and that it can assist you in the solving process.

We begin in this section with a brief review of the Cartesian coordinate system. Then we discuss the various ways in which a graphing calculator can be used to display a coordinate system. Finally, we consider how we can use a graph to depict the values of an expression.

The Cartesian Coordinate System

FIGURE **1.1**

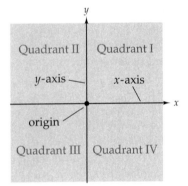

The graphing system for associating the corresponding values of two variables is called the **Cartesian (or rectangular) coordinate system.** (It is called *Cartesian* in honor of the French mathematician René Descartes, who is credited with creating the system.)

Figure 1.1 shows that the Cartesian coordinate system consists of two perpendicular number lines, which are called the **x-axis** and the **y-axis,** intersecting at a point called the **origin.** The two axes partition the **coordinate plane** into four **quadrants.**

If a number x corresponds to some number y, the two numbers can be written as an **ordered pair** (x, y). This pair of numbers is uniquely associated with a point in the coordinate plane. The first number, or **x-coordinate,** indicates how far to the right or left the point is from the vertical axis; the second number, or **y-coordinate,** indicates how far above or below the point is from the horizontal axis. Highlighting a point that represents an ordered pair is called **plotting** the point. Figure 1.2 shows the point $P(-4, 3)$ plotted in the coordinate system.

FIGURE **1.2**

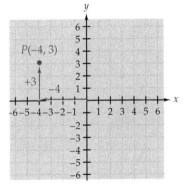

Because there is a **one-to-one correspondence** between ordered pairs and points, we use "plotting points" and "plotting ordered pairs" interchangeably. To identify a particular point, we give the point a letter name. For example, point P in Figure 1.2 is the point that represents the ordered pair $(-4, 3)$.

Before graphing technology, drawing coordinate systems and plotting points had to be done by hand. Now computers and graphing calculators can be used to display a coordinate system, which can be customized to any specifications, and to rapidly plot points in that system.

The Viewing Window of a Calculator

The feature that distinguishes a graphing calculator from a scientific calculator is the viewing *window* for displaying the Cartesian coordinate system. We can adjust this window to display any portion of the coordinate plane, and we can set the distance (*scale*) between the *tick marks* on either axis.

Graph   The *Graph* key is used to display a coordinate system in the viewing window.

Cursor   A general cursor can be moved to any point in the viewing window, and the coordinates of that point are displayed.

Most graphing calculators have several options for automatically sizing the viewing window.

Standard   On both axes, the tick marks are from −10 (*Xmin* and *Ymin*) to 10 (*Xmax* and *Ymax*). The scale is 1 (*Xscl* and *Yscl*) on both axes. (See Figure 1.3.)

FIGURE 1.3   **FIGURE 1.4**   **FIGURE 1.5**

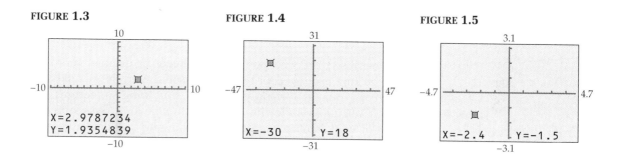

In Figure 1.3, we have moved the general cursor as close as we can to the point (3, 2). The standard window is usually not the best setting when we wish to locate exact points with the cursor.

Integer   The advantage of this setting is that displayed points have at least one coordinate that is an integer. In Figure 1.4, we were able to locate (−30, 18) exactly.

Decimal   This setting is useful for locating exact points close to the origin. Coordinates are reported in tenths. (See Figure 1.5.)

You can customize the viewing window settings by entering your own minimum and maximum values for the axes and by selecting the scale.

Window   The *Window* key displays the current viewing window settings. You can change any or all of these to settings of your own choice.

EXAMPLE **1**   **Customizing the Viewing Window**

Display Quadrant I in the viewing window as follows:

(i) On the *x*-axis, the maximum value is 12 and the tick marks represent 3 units.

(ii) On the *y*-axis, the maximum value is 10 and the tick marks represent 2 units.

SOLUTION

Figure 1.6 shows the desired specifications entered on the window screen. The resulting display of Quadrant I of the coordinate system is shown in Figure 1.7.

FIGURE 1.6

FIGURE 1.7

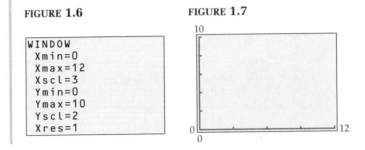

```
WINDOW
 Xmin=0
 Xmax=12
 Xscl=3
 Ymin=0
 Ymax=10
 Yscl=2
 Xres=1
```

The integer and decimal settings are examples of "friendly" window settings because displayed points have integer or decimal coordinates rather than the long decimal coordinates associated with the standard setting. You can create your own friendly windows, but doing so requires a knowledge of your calculator's viewing screen that is beyond the scope of this book. For your convenience, we offer the following calculator *program* that will allow you to create friendly windows automatically.

FRIENDLY WINDOW

From time to time, we will refer to a program that you can enter into your calculator (see Appendix C). A program is a list of instructions that you can store in your calculator's memory. When you execute the program, those instructions will be carried out in the specified order. Programs allow you to perform certain repetitive tasks without having to enter all the instructions each time.

Graphs of Algebraic Expressions

Consider the algebraic expression $\frac{1}{2}x + 3$. For any selected value of x, the expression has a corresponding value. We can show the correspondence with ordered pairs of the following form:

(value of the variable, value of the expression)

In the Preamble, we saw that if $\frac{1}{2}x + 3$ is entered as Y_1 on the Y screen, then Y_1 holds the value of the expression for the currently stored value of x. Thus another way to write the ordered pairs is (X, Y_1).

Figure 1.8 shows part of a table of values for the expression $\frac{1}{2}x + 3$, with the x-values in increments of 4. In Figure 1.9, we have used the information in the table to form ordered pairs.

FIGURE **1.8**

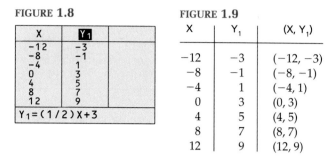

FIGURE **1.9**

X	Y_1	(X, Y_1)
−12	−3	$(−12, −3)$
−8	−1	$(−8, −1)$
−4	1	$(−4, 1)$
0	3	$(0, 3)$
4	5	$(4, 5)$
8	7	$(8, 7)$
12	9	$(12, 9)$

We can plot these ordered pairs (and many others) in a coordinate system (see Figure 1.10). We call this process **graphing the expression.**

FIGURE **1.10**

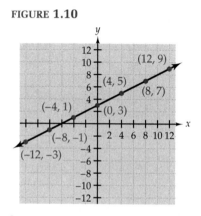

Because there are infinitely many possible replacements for x, we cannot plot every possible point to create a complete graph. However, we can plot enough points to see the general appearance and important features of the graph. For example, Figure 1.10 suggests that a complete graph of $\frac{1}{2}x + 3$ would be a straight line.

With the expression $\frac{1}{2}x + 3$ entered on the Y screen, and having selected the integer setting for the viewing window, we use the *Graph* key to produce the graph of the expression. (See Figure 1.11.)

FIGURE **1.11**

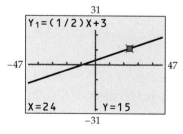

■ *NOTE   In this text, "sketching" a graph refers to drawing a graph by hand (as in Figure 1.10), whereas "producing" a graph refers to displaying the graph with a calculator (as in Figure 1.11).*

The essential concept that you must understand about the graph in Figure 1.11 is this: For every point of the graph of $\frac{1}{2}x + 3$, the first coordinate is a value of x, and the second coordinate is the corresponding value of $\frac{1}{2}x + 3$. In general, when we graph an expression, the first coordinate of each point is a value of the variable, and the second coordinate is the corresponding value of the expression.

The points of a graph can be highlighted with a second kind of cursor.

> **Trace** A tracing cursor can be activated with the *Trace* key. The tracing cursor can be moved along a graph and the coordinates of the highlighted points will be displayed.

In Figure 1.11, the tracing cursor is on the point (24, 15). This means that when the value of x is 24, the value of $\frac{1}{2}x + 3$ is 15. In general, graphing an expression gives us another way of evaluating the expression. The following summarizes the suggested procedure.

■ *NOTE Keep in mind that the success of the graphing method for evaluating an expression depends on your being able to trace to exact points. If you are unable to find a proper window setting to do that, then your result is only an estimate and other evaluation methods are needed to determine the exact value.*

> ### Evaluating an Expression with the Graphing Method
> 1. Enter the expression on the Y screen.
> 2. Select a window setting that is likely to display the points that you want to see.
> 3. Produce the graph of the expression.
> 4. To evaluate the expression for a particular value of x, trace the graph to the point whose first coordinate is x. The second coordinate of that point is the value of the expression.

The points at which a graph crosses the x-axis and the y-axis are called **intercepts**.

> ### Definition of Intercepts
> An x-**intercept** of a graph is a point at which the graph intersects the x-axis. The second coordinate of an x-intercept is 0.
>
> A y-**intercept** of a graph is a point at which the graph intersects the y-axis. The first coordinate of a y-intercept is 0.

By tracing, we can determine (or estimate) the coordinates of the intercepts of a graph.

EXAMPLE 2   **Estimating the Intercepts of a Graph**

Use the integer setting to produce the graph of the expression $2x - 9$. Trace the graph to estimate the intercepts.

SOLUTION

Figure 1.12 shows the graph of the expression $2x - 9$. To determine the y-intercept, we trace to the point whose x-coordinate is 0, as shown in Figure 1.12. We see that the y-intercept is $(0, -9)$.

FIGURE 1.12

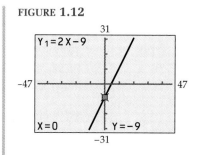

If we attempt to trace the graph in Figure 1.12 to the x-intercept, we find that we can locate $(4, -1)$ and $(5, 1)$, but we cannot highlight the point whose second coordinate is 0. The x-intercept appears to lie between $(4, 0)$ and $(5, 0)$.

One method for obtaining an improved estimate of the x-intercept in Example 2 is to magnify (*zoom in* on) the graph near that point.

> **Zoom In**   The *Zoom* key allows you to zoom in (or zoom out) at the current cursor location.

■ *NOTE   Even with many zoom-and-trace cycles, you will not necessarily ever arrive at the exact point that you seek. In such cases, you must regard your result as an estimate. Other methods will be needed to determine the exact coordinates.*

Figure 1.13 shows the graph of $2x - 9$ with the tracing cursor near the x-intercept. In Figure 1.14, we have zoomed in to obtain a magnified portion of the graph. Tracing again, we find that the x-intercept is $(4.5, 0)$.

FIGURE 1.13   **FIGURE 1.14**

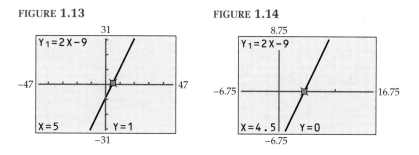

Graphing methods can be used to determine (or at least estimate) the value(s) of a variable for which two expressions have the same value.

EXAMPLE 3   **Two Expressions with the Same Value**

Use the integer setting to produce the graphs of the expressions $0.5x + 15$ and $-(x + 12)$. Then trace to the point of intersection of the two graphs.

(a) What ordered pair is represented by the point of intersection?
(b) Compare the values of the two expressions at the point of intersection.

SOLUTION

On the Y screen, we enter $0.5x + 15$ as Y_1, and $-(x + 12)$ as Y_2. Figure 1.15 shows the graphs of the expressions with the tracing cursor at the point of intersection.

FIGURE 1.15

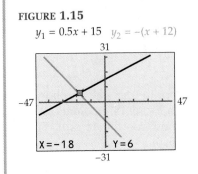

$y_1 = 0.5x + 15 \quad y_2 = -(x + 12)$

(a) The point of intersection represents the ordered pair $(-18, 6)$.
(b) When $x = -18$, the value of both expressions is 6. (You should verify this by substitution.)

As was the case with intercepts, tracing may not take you to an exact point of intersection.

EXAMPLE 4 Estimating a Point of Intersection

Use the graphing method to estimate the value of x for which the expressions $x + 5$ and $-0.4x + 1$ have the same value. After two zoom-and-trace cycles, round your estimate of the x-value to the nearest hundredth.

■ *NOTE Depending on your initial window setting and other considerations, your estimates can vary slightly from those given in the text. You can be satisfied with an estimate as long as it is reasonably close to the one that we give.*

SOLUTION

Figure 1.16 shows the graphs of the two expressions in a standard window setting. The tracing cursor is near the point of intersection.

FIGURE 1.16

FIGURE 1.17

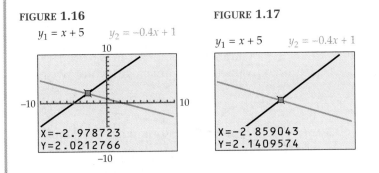

$y_1 = x + 5 \quad y_2 = -0.4x + 1$

$y_1 = x + 5 \quad y_2 = -0.4x + 1$

Figure 1.17 shows the tracing cursor near the point of intersection after two zoom-and-trace cycles. The x-value for which the expressions have the same value is about -2.86.

In this section, we have introduced you to many of the basic features of a graphing calculator. They will serve you well throughout this text. Try to become comfortable and skillful with your calculator. It will help you to visualize many important algebraic concepts, and it will be a time-saving device for carrying out calculations and operations.

1.1 Quick Reference

The Cartesian Coordinate System

■ The **Cartesian coordinate system** consists of two perpendicular number lines, called the *x*-**axis** and the *y*-**axis,** that intersect at the **origin.**

■ The axes divide the coordinate plane into four **quadrants.**

■ If a number *x* corresponds to a number *y*, the numbers can be written as an **ordered pair** (x, y). When the ordered pair is **plotted** in a coordinate system, we call *x* the *x*-**coordinate** and *y* the *y*-**coordinate.**

■ There is a **one-to-one correspondence** between ordered pairs and the points that represent them.

The Viewing Window of a Calculator

■ Any portion of a coordinate system can be displayed in the viewing window of a calculator. The viewing window can be displayed with built-in specifications, or it can be customized.

■ A *general cursor* is used to highlight any point in the coordinate system and to display the coordinates of that point.

Graphs of Algebraic Expressions

■ For an algebraic expression in one variable, ordered pairs can be written in the form (value of the variable, value of the expression).

■ Plotting all such ordered pairs is called **graphing the expression.**

■ For any point of the graph of an expression, the first coordinate is the value of the variable and the second coordinate is the value of the expression.

■ A *tracing cursor* can be used to highlight any point of a graph and to display the coordinates of that point.

■ Two special points of a graph are the *x*-**intercept,** whose *y*-coordinate is 0, and the *y*-**intercept,** whose *x*-coordinate is 0.

■ When two expressions are graphed in the same coordinate system, the point of intersection, if any, represents the value of the variable for which the expressions have the same value.

Speaking the Language

1. In a Cartesian coordinate system, the *x*-axis and the *y*-axis intersect at the ▩▩▩▩▩, and they partition the coordinate plane into ▩▩▩▩▩ .

2. When $Q(5, -7)$ is plotted in a coordinate system, we call 5 the ▩▩▩▩▩ .

3. For any point of the graph of an algebraic expression, the first coordinate is the value of the ▩▩▩▩▩ , and the second coordinate is the value of the ▩▩▩▩▩ .

4. A point of a graph whose *y*-coordinate is 0 is called a(n) ▩▩▩▩▩ .

1.1 Exercises

1. Plot the points $A(2, 5)$, $B(2, 3)$, and $C(2, -4)$. Based on the pattern that you observe, describe the location of any point of the form $P(2, y)$.

2. Which calculator menu would you select to enter the viewing windows settings for Quadrant IV? What would be the signs of the settings for $Xmax$ and $Ymin$?

In Exercises 3 and 4, plot the points associated with the given ordered pairs.

3. $(4, 2), (3, -5), (0, 2), (-2, 1), (3, 0)$
4. $(-3, -4), (-4, 0), (0, 5), (-1, -3), (3, -1)$

In Exercises 5–10, use the given values of x and your calculator to produce a table of values for the expression. Use the points associated with entries in the table to sketch the graph of the expression. Then use a calculator to produce the graph of the expression and compare the graph to your sketch.

5. $3 - \dfrac{1}{2}x$; $-6, -2, 0, 4, 6$
6. $2x + 5$; $-3, -1, 1, 3, 6$
7. $x^2 - 4x$; $-1, 0, 1, 2, 5$
8. $x^2 + 1$; $-2, -1, 0, 1, 2$
9. $3 - |x|$; $-6, -3, 0, 2, 5$
10. $\sqrt{x + 6}$; $-6, -5, -2, 3, 10$

In Exercises 11–14, select several values for x and use your calculator to determine the corresponding values of the given expression. Use those values to sketch the graph of the expression. Then use a calculator to produce the graph of the expression and compare the graph to your sketch.

11. $2x - 6$
12. $5 - x$
13. $x^2 + 3x$
14. $|x - 5|$

In Exercises 15–20, use the integer, standard, and decimal settings to produce the graph of the given expression. Which window setting appears to be the best for displaying the graph?

15. $x^3 - x$
16. $-\dfrac{1}{4}x^2 + 3x + 16$
17. $\dfrac{1}{2}x - 12$
18. $x + \dfrac{2}{x}$
19. $\sqrt{25 - x^2}$
20. $\dfrac{1}{x}$

21. What do we mean when we say that $(5, 3)$ belongs to the graph of the expression $\sqrt{x^2 - 16}$?

22. Explain why the graph of the expression $x^3 + x$ has infinitely many points.

In Exercises 23–26, trace the graph of the given expression to determine the missing coordinates so that the ordered pairs belong to the graph. (*Hint:* Use the integer and decimal settings to produce the graph and choose the one that appears to be more appropriate for the expression and the given values.)

23. $2x + |x|$;
    $(\ (a)\ , 3), (5, \ (b)\ ), (-6, \ (c)\ ), (\ (d)\ , -1)$

24. $\dfrac{x - 1}{x - 2}$;
    $(0.8, \ (a)\ ), (2.6, \ (b)\ ), (\ (c)\ , 2),\quad (\ (d)\ , 0.375)$

25. $\dfrac{1}{16}x^2 - \dfrac{1}{4}x$;
    $(-2, \ (a)\ ), (\ (b)\ , -0.25), (1.6, \ (c)\ ), (3.6, \ (d)\ )$

26. $3\sqrt{36 - x}$;
    $(\ (a)\ , 15), (\ (b)\ , 18), (-13, \ (c)\ ), (32, \ (d)\ )$

In Exercises 27–32, produce the graph of the given expression and match your graph with one of the given graphs. What window settings would you use so that your graph and the matching graph are identical in appearance? (*Hint:* All window settings are integers.)

(A)

(B)

(C)

(D)

(E)   (F)

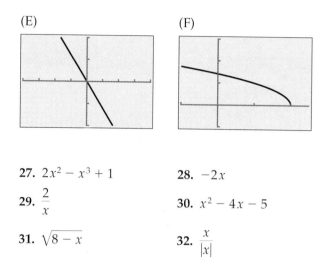

27. $2x^2 - x^3 + 1$ **28.** $-2x$

29. $\dfrac{2}{x}$ **30.** $x^2 - 4x - 5$

31. $\sqrt{8 - x}$ **32.** $\dfrac{x}{|x|}$

In Exercises 33–38, perform the following steps.

(a) Create a table of values for the given expression. Scroll through the table to find the value(s) of x for which the value of the expression is 0.

(b) Use the integer or decimal setting to produce the graph of the expression and trace to the point(s) where the graph intersects the x-axis. Compare the x-coordinates of those points to your results in part (a).

33. $2x - 7$ **34.** $5 - |x|$

35. $x^2 - 3x - 4$ **36.** $x^3 - 9x$

37. $\sqrt{x + 2}$ **38.** $\sqrt{x^2 - 1}$

39. What is the significance of the point of intersection of the graphs of two expressions?

40. Suppose that you are tracing to determine the x-intercept of the graph of an expression. How do you know that you have traced to the exact point?

In Exercises 41–48, use the integer or decimal setting to produce the graph of the given expression. Then trace to determine the x- and y-intercepts of the graph of the expression.

41. $0.5x - 1$ **42.** $x - 9$

43. $\dfrac{1}{4}x^2 - 16$ **44.** $3x - x^2$

45. $12 - |x|$ **46.** $1 - \sqrt{x + 4}$

47. $\dfrac{1}{4}x^3 - x$ **48.** $x - \dfrac{x^3}{100}$

In Exercises 49–52, use the standard setting to produce the graph of the given expression. If necessary, adjust the window setting for one axis so that you can see the intercepts. Then trace to estimate the x- and y-intercepts. If you cannot trace exactly to an intercept, use two zoom-and-trace cycles and estimate your answer to the nearest tenth.

49. $5 - \dfrac{x}{3}$

50. $2x + 15$

51. $x^2 - x - 14$

52. $-\dfrac{1}{5}x^2 + 2x + 3$

In Exercises 53–58, use the integer or decimal setting to produce the graphs of the two given expressions. Trace to estimate the coordinates of the point(s) of intersection.

53. $\dfrac{1}{3}x;\quad 2 - \dfrac{2}{3}x$

54. $3 - x;\quad 2x - 6$

55. $x^2 + 2x - 4;\quad x + 8$

56. $\sqrt{24 - 2x};\quad x$

57. $|x - 3|;\quad \dfrac{1}{2}x + 9$

58. $x;\quad \dfrac{1}{x}$

In Exercises 59–62, use the standard setting to produce the graphs of the two given expressions. If necessary, adjust the window setting for one axis so that you can see the point(s) of intersection of the graphs. Then trace to estimate the coordinates of the point(s) of intersection. Use two zoom-and-trace cycles and estimate your answer to the nearest tenth.

59. $9 - 2x;\quad x + 11$

60. $x - 9;\quad 8 - \dfrac{x}{2}$

61. $4 + 2x - \dfrac{x^2}{4};\quad 5 - x$

62. $x^2 - 4x - 12;\quad 2x - 1$

1.2 **Linear Equations**

Equations and Solutions ■ *Calculator Methods* ■
Algebraic Methods

Your success as a problem solver will depend on your ability to describe the conditions of a given problem with an equation or an inequality. Then you need to be able to solve the equation or the inequality and to interpret the result.

Although you are probably already familiar with the basic algebraic methods for solving, this section reviews those methods and gives special emphasis to the visual assistance that a graphing calculator can provide.

Equations and Solutions

An **equation** is a statement that two algebraic expressions have the same value. The following are some examples of equations.

$$x^2 + 3x + 2 = 0 \qquad \frac{1}{x-1} = \frac{4}{x+1}$$

$$y = 2x - 7 \qquad x^2 + y^2 = 4$$

For the equation $4x + 7 = 3 - 2x$, the expression $4x + 7$ is called the *left side of the equation*, and the expression $3 - 2x$ is called the *right side of the equation*.

An equation may contain one or more variables. For an equation with only one variable, a **solution** of that equation is any replacement for the variable that makes the equation true. Determining all the solutions of an equation is called *solving the equation*. The set of all solutions of an equation is the **solution set.**

An equation that is true for every permissible replacement of the variable is called an **identity,** and its solution set is the set **R** of all real numbers. An equation with no solution is called a **contradiction,** and its solution set is the empty set, denoted by Ø. A **conditional equation** is an equation that has at least one solution but is not an identity.

$2(x + 3) = 2x + 6$	Identity	The equation is true for all values of x.
$-(x - 3) = -x - 3$	Contradiction	The equation is false for all values of x.
$3x + 1 = 2x + 5$	Conditional	The equation is true for $x = 4$ but false for all other values of x.

In this section, our focus is on a specific type of conditional equation called a **linear equation in one variable.**

A **linear equation in one variable** is an equation that can be written in the form $Ax + B = 0$, where A and B are real numbers and $A \neq 0$.

Calculator Methods

To solve a linear equation in one variable, we must determine the value of the variable for which the expressions on the left and right sides of the equation have the same value.

EXPLORATION

Solving Equations Graphically

Consider the equation $x - 1 = -0.5x + 2$. Use the decimal setting to produce the graphs of the left and right sides of the equation.

(a) What does the y-coordinate represent for
  (i) every point of the graph of $x - 1$?
  (ii) every point of the graph of $-0.5x + 2$?
  (iii) the point of intersection of the graphs?
(b) For what x-coordinate do the two expressions have the same value?
(c) What is the solution of the equation?

We can use calculator tables and graphs for visual assistance in solving an equation.

DEVELOPING THE CONCEPT

Solving Equations Graphically

Consider the equation $\frac{2}{3}x + 11 = 2x - 9$.

The Table Method

To solve the equation, we can use the Y screen to enter $\frac{2}{3}x + 11$ as Y_1 and $2x - 9$ as Y_2. Then we produce the table of values shown in Figure 1.18.

FIGURE **1.18**   FIGURE **1.19**

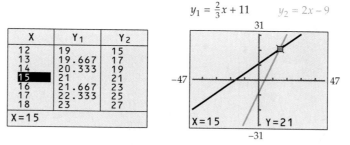

From the table, we see that $Y_1 = Y_2$ (that is, $\frac{2}{3}x + 11 = 2x - 9$) when $x = 15$. Thus 15 is the solution of the equation.

The Graphing Method

With the left and right sides of the equation entered on the Y screen, we produce the graphs of the two expressions (see Figure 1.19). Tracing to

FIGURE **1.20**

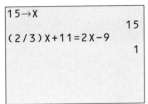

the point of intersection of the graphs, we find that $Y_1 = Y_2$ (that is, $\frac{2}{3}x + 11 = 2x - 9$) when $x = 15$. Thus 15 is the solution of the equation.

Although a table of values may reveal the exact solution of an equation, we should regard graphing-method solutions as estimates until the solution is verified. With the left and right sides already entered on the Y screen, we can evaluate each expression for $x = 15$ to verify that they are equal. Alternatively, we can use the test feature, as shown in Figure 1.20. Recall that the returned value of 1 indicates that the equation is true when $x = 15$.

The graphing method is limited by your ability to trace to the exact point of intersection. If you cannot, then performing zoom-and-trace cycles will usually improve your estimate of the solution.

Your calculator may have other features that will allow you to obtain the exact solution or a very accurate estimate of it.

Intersect   Begin by moving the tracing cursor as close as possible to the point of intersection. Then select the *Intersect* feature. Your calculator will compute the coordinates of the point of intersection.

We will illustrate the *intersect* feature in Example 1.

Algebraic Methods

Equivalent equations are equations with the same solution set. The equations $2x + 5 = 13$, $2x = 8$, and $x = 4$ are equivalent equations because the solution set of each is {4}. On the other hand, $x^2 = 25$ and $x = 5$ are not equivalent equations because the solution sets are {−5, 5} and {5}, respectively.

Simple equations, such as $x = 5$ and $x - 2 = 0$, can be solved *by inspection.* For some equations, the following properties of equality are used to produce successively simpler equivalent equations until the solution is obvious.

■ *NOTE   The last two properties are stated with the phrase "if and only if," which indicates that the statement and its converse are both true. For example, "If $a = b$, then $ac = bc$" and "If $ac = bc$, then $a = b$."*

Properties of Equality

Suppose that a, b, and c are algebraic expressions.

If $a = b$, then $b = a$.	**Symmetric Property of Equality**
If $a = b$ and $b = c$, then $a = c$.	**Transitive Property of Equality**
If $a = b$, then a can be replaced by b in any expression.	**Substitution Property**
$a = b$ if and only if $a + c = b + c$.	**Addition Property of Equality**
$a = b$ if and only if $ac = bc$, $c \neq 0$.	**Multiplication Property of Equality**

The Addition Property of Equality allows us to add (or subtract) any number to (or from) both sides of an equation. This property is usually used to eliminate terms from one or both sides of an equation and to isolate the variable term.

The Multiplication Property of Equality allows us to multiply (or divide) both sides of an equation by any nonzero number. In the solving process, this property is often used as the final step of isolating the variable.

The following routine is applicable to solving linear equations in one variable.

Solving a Linear Equation in One Variable

1. Clear fractions, if any, by multiplying both sides of the equation by the LCD of all the fractions.
2. Simplify both sides of the equation by removing grouping symbols and combining like terms.
3. Isolate the variable term.
4. Isolate the variable itself.
5. Solve the resulting equivalent equation by inspection.
6. Verify the solution.

Although checking solutions is highly recommended, we will not always include that step in our examples.

EXAMPLE 1   Solving Equations

Solve the following equations.

(a) $2x - 6 = \dfrac{3}{7}x + 16$ (b) $-(x - 8) = 3[2x - 3(x + 1)]$

SOLUTION

(a) VISUALIZING THE SOLUTION

FIGURE 1.21

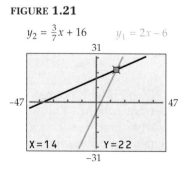

$y_2 = \frac{3}{7}x + 16$ $y_1 = 2x - 6$

X = 14   Y = 22

Figure 1.21 shows the graphs of $2x - 6$ and $\dfrac{3}{7}x + 16$ produced with the integer setting. With the tracing cursor on the point of intersection, we estimate the solution to be 14.

DETERMINING THE SOLUTION

$$2x - 6 = \frac{3}{7}x + 16 \qquad \text{Multiply both sides by 7 to clear the fraction.}$$

$$7(2x - 6) = 7\left(\frac{3}{7}x + 16\right) \qquad \text{Multiplication Property of Equality}$$

$$14x - 42 = 3x + 112 \qquad \text{Distributive Property}$$

$$14x - 3x - 42 = 3x - 3x + 112 \qquad \text{Addition Property of Equality}$$

$$11x - 42 = 112 \qquad \text{Combine like terms.}$$

$$11x - 42 + 42 = 112 + 42 \qquad \text{Addition Property of Equality}$$

$$11x = 154$$

$$\frac{11x}{11} = \frac{154}{11} \qquad \text{Multiplication Property of Equality}$$

$$x = 14$$

As predicted graphically, the solution is 14.

(b) VISUALIZING THE SOLUTION

After we use the integer setting to produce the graphs of $-(x - 8)$ and $3[2x - 3(x + 1)]$, we trace as close as possible to the point of intersection (see Figure 1.22). The estimated solution is -8.

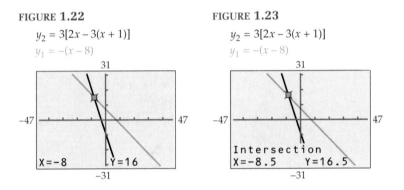

FIGURE **1.22**

$y_2 = 3[2x - 3(x + 1)]$
$y_1 = -(x - 8)$

X=-8   Y=16

FIGURE **1.23**

$y_2 = 3[2x - 3(x + 1)]$
$y_1 = -(x - 8)$

Intersection
X=-8.5   Y=16.5

Suspecting that the tracing cursor is not on the exact point of intersection, we use the *Intersect* feature to improve the estimates of the coordinates (see Figure 1.23). The revised approximate solution is -8.5.

DETERMINING THE SOLUTION

$$-(x - 8) = 3[2x - 3(x + 1)] \qquad \text{Begin with the innermost grouping symbols.}$$

$$-(x - 8) = 3[2x - 3x - 3] \qquad \text{Distributive Property}$$

$$-(x - 8) = 3[-x - 3] \qquad \text{Combine like terms.}$$

$$-x + 8 = -3x - 9 \qquad \text{Distributive Property}$$

$$2x = -17 \qquad \text{Addition Property of Equality}$$

$$x = -\frac{17}{2} \qquad \text{Multiplication Property of Equality}$$

The solution is -8.5.

If solving an equation leads to a true equivalent equation with no variable (such as $0 = 0$), then the original equation is an identity, and the solution set is **R**. If a false equivalent equation with no variable (such as $0 = 1$) is obtained, then the original equation is a contradiction, and the solution set is Ø.

EXAMPLE **2**   **Identities and Contradictions**

Write the solution set of each of the given equations.

(a) $2(x - 3) + 1 = 5(x - 1) - 3x$
(b) $4n + 3 = 2 - (1 - 3n) + n$

SOLUTION

(a) $2(x - 3) + 1 = 5(x - 1) - 3x$
$2x - 6 + 1 = 5x - 5 - 3x$
$2x - 5 = 2x - 5$
$-5 = -5$ True.

The equation is an identity and the solution set is **R**.

(b) $4n + 3 = 2 - (1 - 3n) + n$
$4n + 3 = 2 - 1 + 3n + n$
$4n + 3 = 4n + 1$
$3 = 1$ False.

The equation is a contradiction and the solution set is Ø.

Some calculators have very sophisticated features for computing the solution(s) of an equation without any reference to graphs. The following feature is presented in Appendix B, and you may want to try the method. However, because the menus and keystrokes vary so widely among calculator models, we will not rely on this method in our future topics.

> **Solve**   With Y_1 and Y_2 representing the left and right sides of the equation, respectively, this method begins with the entering of $Y_1 - Y_2$ in an equation editor. Pressing the *Solve* key then produces the solution of the equation.

Equation solving is a fundamental task in algebra because it is the process by which application problems are solved. Your ability to merge technology with algebraic skills and procedures will be important factors in your success with mathematics.

1.2   Quick Reference

Equations and Solutions   ■   An **equation** is a statement that two algebraic expressions have the same value.

■   For an equation in one variable, a **solution** is any replacement for the variable that makes the equation true. The set of all solutions of an equation is called the **solution set.**

■   An equation that is true for every permissible replacement of the variable is an **identity,** and its solution set is **R.**

■ An equation with no solution is called a **contradiction,** and its solution set is Ø.

■ A **conditional equation** is an equation that has at least one solution but is not an identity.

■ A **linear equation in one variable** is an equation that can be written in the form $Ax + B = 0$, where A and B are real numbers and $A \neq 0$.

Calculator Methods ■ By entering the left and right sides of an equation as Y_1 and Y_2, we can produce a table of values to determine the value of x for which $Y_1 = Y_2$. This value of x is the solution of the equation.

■ We can also graph the two expressions and trace to any point of intersection. The x-coordinate of that point is a solution of the equation.

■ Zoom-and-trace cycles or the calculator's *Intersect* feature can be used to improve the estimates of the point of intersection.

Algebraic Methods ■ **Equivalent equations** are equations with the same solution set.

■ The properties of equality can be used to write successively simpler equivalent equations until the solution can be determined *by inspection.*

Speaking the Language

1. Because $x + 3 = 3 + x$ is true for all values of x, we call the equation a(n) ▨▨▨.

2. An equation of the form $Ax + B = 0$, $A \neq 0$, is called a(n) ▨▨▨ equation in one variable.

3. When we graph the left and right sides of an equation, the ▨▨▨ of a point of intersection represents a solution of the equation.

4. Equations with the same solution set are called ▨▨▨ equations.

1.2 Exercises

1. Explain why a linear equation in one variable is a conditional equation.

2. Compare the solution sets of an identity and a contradiction.

In Exercises 3–10, use the decimal or integer setting and the graphing method to solve the given equation. Use your calculator to verify the solutions.

3. $2x - 7 = 5 - x$

4. $-3(x - 4) = 2(x - 16) - 1$

5. $\dfrac{1}{6}x^2 - \dfrac{1}{2}x = 18$

6. $x^2 - 2 = x$

7. $x^3 + x^2 = -2(x - 2)$

8. $0.1x^3 - 2x = 20.3$

9. $\sqrt{x - 1} = 5 - 2x$

10. $15 - \sqrt{2x + 1} = x - 2$

In Exercises 11–14, use the standard setting and the graphing method to estimate the solution(s) of the given equation. Perform zoom-and-trace cycles until your estimate is accurate to the nearest tenth.

11. $4x - 3 = x + 1$

12. $6x - 17 = 19 - 2x$

13. $\dfrac{1}{x} = x + 1$

14. $\sqrt{2x + 7} = x - 3$

15. From the following list, identify the two equations that are equivalent and explain why the other equation is not equivalent.
   (i) $5x - 7 = 2x + 1$
   (ii) $5x + 7 = 2x - 1$
   (iii) $3x + 8 = 0$

16. To solve $\frac{3}{7}x + 2 = \frac{1}{3}x + \frac{11}{2}$, you might clear the fractions by multiplying by the LCD, 42. Why must the term 2 be multiplied by 42 even though it is not a fraction?

In Exercises 17–30, solve the given equation algebraically.

17. $3 - 2(2x - 3) = -5(x + 1)$

18. $6y - 3(y + 3) = 7 - 4(y - 3)$

19. $16t + 8(1 - t) = 9t + 7$

20. $2(t + 1) = 5(t - 1) - 2t$

21. $3x + 5(x - 2) = 2 - 4(3 - 5x)$

22. $6x - (x + 2) = 13 - (3 - x) + x$

23. $\frac{2}{3}x + 3 = \frac{1}{2}x - 1$

24. $\frac{3}{4}(y + 2) - \frac{1}{5}(5 - y) = 0$

25. $x - 2[x - 2(x + 1)] = 5x$

26. $3[5(x + 1) - (2x + 1)] = 5 - x$

27. $-\{x + 2[x - (2x + 1)]\} = 3(x + 1)$

28. $-(x - 2) - \{3 - [x - (x + 1)]\} = x$

29. $\frac{5}{6}\left[\frac{2}{3}(x - 1) - 6(x + 1)\right] = \frac{1}{6}(x + 3)$

30. $\frac{2}{5}[10x - (x + 1)] = \frac{1}{3}x + 2$

In Exercises 31–36, use your calculator's *Intersect* feature to estimate the solution of the given equation. Round your estimate to the nearest hundredth. Then solve the equation algebraically to determine the exact solution.

31. $3x - (x + 1) = 2(1 - x) + x + 2$

32. $\frac{3}{4}(x - 3) - 2(3 - x) = \frac{1}{2}(3x - 5)$

33. $\frac{4x}{9} - \frac{1}{6} = \frac{-x}{3} - 5$

34. $\frac{3x - 8}{2} = -8 - \frac{x}{3} - x$

35. $-2[x - 3(x + 1)] = -\frac{2x}{5}$

36. $0.3x - 1.8(x - 2) = 0.5(x - 7)$

37. Describe the graphs of the left and right sides of an equation that is (a) an identity and (b) a contradiction.

38. The equations $5x + 2 = 3x - 8$, $2x = -10$, and $x = -5$ are all equivalent because the solution set of each is $\{-5\}$. If the graphs of the left and right sides of these equations were produced in the same coordinate system, would all six graphs have the same point of intersection? Explain how you know.

In Exercises 39–44, use the graphing method to form a conclusion as to whether the given equation is a conditional equation, an identity, or a contradiction.

39. $\frac{1}{2}x - 1 = 2x + 1 - \frac{3x}{2}$

40. $-(2 - x) = 2x - (x + 2)$

41. $3 - x(3 - x) = x + 3$

42. $x^2 + 3 = x - 6$

43. $\sqrt{x^2} = |x|$

44. $\sqrt{x^2} = x$

In Exercises 45–50, solve the given equation algebraically. For an identity or a contradiction, write the solution set.

45. $2(x + 3) + 3x + 7 = 2(x - 2) - 3(1 - x)$

46. $-[2 - (3 - x)] = 3 - (2 - x)$

47. $\frac{1}{2}(x - 3) + 3 = \frac{1}{2}(1 - x)$

48. $\frac{5x}{3} - (x - 1) = 2\left(3 + \frac{x}{3}\right)$

49. $7x - 2(3x - 1) = 2 + x$

50. $\frac{1}{2}x(2x - 8) = -x(4 - x)$

1.3 Inequalities

Linear Inequalities and Interval Notation ■ *Graphing Methods*
■ *Algebraic Methods* ■ *Compound Inequalities* ■
Absolute Value Equations and Inequalities

The conditions of a real-world application problem are often described with such words as *less than* or *at least*. To solve this type of problem, you need to know how to solve an *inequality* or a combination of inequalities.

In this section, we introduce graphing methods for visualizing the solving concept, and we review the algebraic procedures for determining the exact solutions of inequalities.

Linear Inequalities and Interval Notation

The left and right sides of an inequality can be any algebraic expression. A particular type of inequality is a **linear inequality in one variable.**

> A **linear inequality in one variable** is an inequality that can be written in the form $Ax + B < 0$, where A and B are real numbers and $A \neq 0$. This definition can also be stated for the inequality symbols $>$, $\leq$, and $\geq$.

Solution sets of inequalities are often written with **interval notation,** as illustrated in the following list.

■ *NOTE A bracket indicates that the end of the interval is included, whereas a parenthesis indicates that the end of the interval is not included. A parenthesis is used for ∞ or $-\infty$ to indicate that the interval extends without end.*

Inequality	Number line graph	Interval notation
$x > 2$		$(2, \infty)$
$x \geq 2$		$[2, \infty)$
$x < 5$		$(-\infty, 5)$
$x \leq 5$		$(-\infty, 5]$
$1 < x \leq 6$		$(1, 6]$
$1 \leq x < 6$		$[1, 6)$

Graphing Methods

As was true for equations, the graphing method for solving an inequality begins with entering the expressions on the left and right sides of the inequality on the Y screen and producing their graphs.

EXPLORATION

Solving a Linear Inequality

Consider the inequality $x + 5 \leq -(x - 21)$. Use the integer setting to produce the graphs of the left and right sides of the inequality.

(a) Trace to the point of intersection. Is the x-coordinate of that point a solution of the inequality? Why?

(b) Trace to points of the two graphs that are to the right of the point of intersection. For these points, are the values of $x + 5$ greater than or less than the values of $-(x - 21)$? Are the x-coordinates of these points solutions of the inequality?

(c) Repeat part (b) for points of the two graphs that are to the left of the point of intersection.

(d) From what you learned in parts (a), (b), and (c), use interval notation to write the solution set of the inequality.

_____

DEVELOPING THE CONCEPT

Solving a Linear Inequality

FIGURE **1.24**

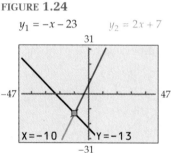

$y_1 = -x - 23$ $y_2 = 2x + 7$

The inequality $2x + 7 > -x - 23$ states that the value of $2x + 7$ is greater than the value of $-x - 23$. Figure 1.24 shows the graphs of the two expressions with the tracing cursor on the point of intersection, $(-10, -13)$.

Note that the point of intersection does *not* represent a solution because the inequality symbol is $>$ rather than $\geq$.

By tracing the graphs of the two expressions to the right of the point of intersection, we see that for any x-coordinate greater than -10, the value of $2x + 7$ is greater than the value of $-x - 23$. In effect, we only need to observe where the graph of $2x + 7$ is *above* the graph of $-x - 23$.

Because the inequality is true for all x-values greater than -10, the solution set is $(-10, \infty)$. Figure 1.25 shows a number line graph of the solution set.

FIGURE **1.25**

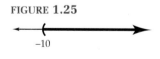

-10

_____

FIGURE **1.26**

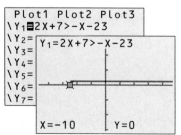

```
Plot1 Plot2 Plot3
\Y₁◼2X+7>-X-23
\Y₂=
\Y₃=
\Y₄=
\Y₅=
\Y₆=
\Y₇=
```

$Y_1 = 2X + 7 > -X - 23$

$X = -10$ $Y = 0$

Another interesting calculator method for estimating solutions of an inequality makes use of the calculator's *Test* feature. Figure 1.26 shows the entry of the preceding inequality on the Y screen.

When we produce the graph of Y_1, the calculator returns a value of 1 for all values of x for which the inequality is true. (See Figure 1.26.) In effect, we obtain a simulated number line graph of the solution set of the inequality. Note that the tracing cursor is on the end of the interval. The y-value of 0 indicates that -10 is not a solution. However, the y-value is 1 for all points that are to the right of -10. Thus the solution set is $(-10, \infty)$.

Algebraic Methods

The algebraic routine for solving a linear *equation* in one variable is also applicable to solving a linear *inequality* in one variable. With one important exception, the addition and multiplication properties of inequality are similar to the corresponding properties of equality.

> ### Properties of Inequality
>
> If $a, b, c,$ and d are real numbers, $d \neq 0$, then
> 1. $a < b$ if and only if $a + c < b + c$.   **Addition Property of Inequality**
> 2. if $d > 0, a < b$ if and only if $ad < bd$;   **Multiplication Property of**
>   if $d < 0, a < b$ if and only if $ad > bd$.   **Inequality**

Note that we must reverse the inequality symbol when we multiply or divide both sides of an inequality by a negative number. These properties of inequalities are used to produce simpler equivalent inequalities until the solution set is obvious.

Example 1 shows how graphing and algebraic methods can be combined in solving an inequality.

EXAMPLE 1   **Solving a Linear Inequality**

Solve the given inequalities.

(a) $3x - 11 \geq 5x + 7$ (b) $\frac{3}{4}(x + 5) < -(1 - x) + 3$

SOLUTION

(a) VISUALIZING THE SOLUTIONS

Figure 1.27 shows the graphs of the left and right sides of the inequality. Using the *Intersect* feature, we see that the x-coordinate of the point of intersection is -9, which is a solution because the inequality symbol is $\geq$.

We also see that the graph of $3x - 11$ is *above* the graph of $5x + 7$ to the left of the point of intersection. Thus $3x - 11 \geq 5x + 7$ for $x = -9$ and for all values of x that are less than -9. The estimated solution set is $(-\infty, -9]$.

DETERMINING THE SOLUTIONS

We obtain the exact solution set by solving algebraically.

$$3x - 11 \geq 5x + 7$$
$$3x - 5x - 11 \geq 5x - 5x + 7 \qquad \text{\textbf{Addition Property of Inequality}}$$

FIGURE 1.27

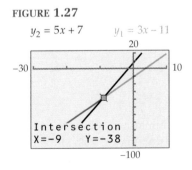

$y_2 = 5x + 7 \qquad y_1 = 3x - 11$

Intersection
X=-9   Y=-38

$$-2x - 11 \geq 7 \qquad \text{Combine like terms.}$$
$$-2x - 11 + 11 \geq 7 + 11 \qquad \text{Addition Property of Inequality}$$
$$-2x \geq 18$$
$$\frac{-2x}{-2} \leq \frac{18}{-2} \qquad \text{Dividing by } -2 \text{ reverses the inequality symbol.}$$
$$x \leq -9$$

The solution set is $(-\infty, -9]$.

(b) $\frac{3}{4}(x + 5) < -(1 - x) + 3$

■ *NOTE   If you try the graphing method in Example 1(b), you will find that the lines are close together. In such cases, zooming in near the point of intersection will help you to see which line is above or below the other.*

$$3(x + 5) < -4(1 - x) + 12 \qquad \text{Multiply each term by 4 to clear the fraction.}$$
$$3x + 15 < -4 + 4x + 12 \qquad \text{Distributive Property}$$
$$3x + 15 < 4x + 8$$
$$7 < x \qquad \text{Subtract } 3x \text{ and 8 from both sides.}$$

The solution set is $(7, \infty)$.

Compound Inequalities

A **compound inequality** consists of two inequalities connected by the word *and* (the compound inequality is called a **conjunction**) or the word *or* (the compound inequality is called a **disjunction**).

Compound inequality	Connective	Name
$x > -3$ and $x \leq 5$	and	Conjunction
$x \leq 2$ or $x > 7$	or	Disjunction

Suppose that A and B are the respective solution sets of the two components of a compound inequality. If the compound inequality is a conjunction, then the solution set is the **intersection** of A and B (written $A \cap B$). If the compound inequality is a disjunction, then the solution set is the **union** of A and B (written $A \cup B$).

EXAMPLE 2   **Solution Sets of Compound Inequalities**

Use number line graphs and interval notation to write the solution sets of the given compound inequalities.

(a) $x > -3$ and $x \leq 5$ (b) $x \leq 2$ or $x > 7$

SOLUTION

Number line graphs   *Interval notation*

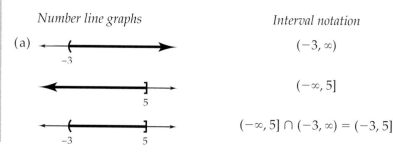

(a) $(-3, \infty)$

$(-\infty, 5]$

$(-\infty, 5] \cap (-3, \infty) = (-3, 5]$

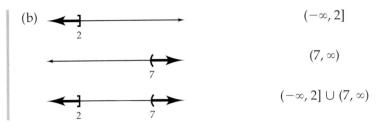

(b) $(-\infty, 2]$

$(7, \infty)$

$(-\infty, 2] \cup (7, \infty)$

The following procedure is generally applicable to solving compound inequalities.

Solving a Compound Inequality

1. Solve each component of the compound inequality.
2. Draw a number line graph of the solution set of each component.
3. Draw a third number line graph that shows the union or intersection of the graphs in step 2.
4. Use the third number line graph to write the solution set in interval notation.

EXAMPLE 3 **Solving Compound Inequalities**

Solve the given compound inequalities.

(a) $4x - 5 \geq 7$ or $3x + 8 \leq 2$
(b) $2x - 5 \leq 5$ and $3 - x < 0$
(c) $-15 < 2x - 7 \leq 11$

SOLUTION

(a) $4x - 5 \geq 7 \quad$ or $\quad 3x + 8 \leq 2 \qquad$ Solve each component separately.

$\qquad 4x \geq 12 \quad$ or $\quad 3x \leq -6$

$\qquad x \geq 3 \quad$ or $\quad x \leq -2$

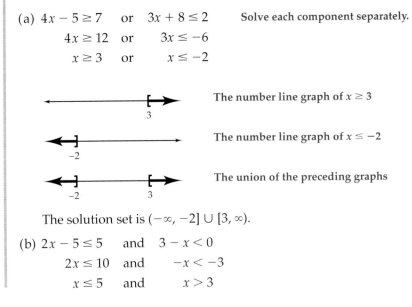

The number line graph of $x \geq 3$

The number line graph of $x \leq -2$

The union of the preceding graphs

The solution set is $(-\infty, -2] \cup [3, \infty)$.

(b) $2x - 5 \leq 5 \quad$ and $\quad 3 - x < 0$

$\qquad 2x \leq 10 \quad$ and $\quad -x < -3$

$\qquad x \leq 5 \quad$ and $\quad x > 3$

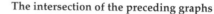

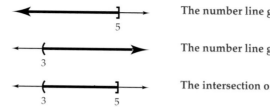

The number line graph of $x \le 5$

The number line graph of $x > 3$

The intersection of the preceding graphs

The solution set is $(3, 5]$.

(c) We call $-15 < 2x - 7 \le 11$ a **double inequality**, which is a special kind of conjunction that means $-15 < 2x - 7$ and $2x - 7 \le 11$.

VISUALIZING THE SOLUTIONS

FIGURE 1.28

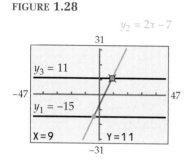

Figure 1.28 shows the graphs of all three components of the double inequality. The points of intersection are $(-4, -15)$ and $(9, 11)$, but only $(9, 11)$ represents a solution. The graph of $y_2 = 2x - 7$ is *between* the graphs of $y_1 = -15$ and $y_3 = 11$ for all x-values between -4 and 9. Thus the solution set is $(-4, 9]$.

DETERMINING THE SOLUTIONS

The properties of inequality can be applied to all three components at once.

$$-15 < 2x - 7 \le 11$$
$$-8 < 2x \quad\;\; \le 18 \qquad \text{Add 7 to all three components.}$$
$$-4 < \;\; x \quad\;\;\; \le \;\; 9 \qquad \text{Divide all three components by 2.}$$

The solution set is $(-4, 9]$.

Absolute Value Equations and Inequalities

The following are examples of simple absolute value equations and inequalities.

$$|2x - 3| = 11 \qquad |4 - x| < 3 \qquad |x + 7| \ge 5$$

The graphing method provides visual assistance in solving such equations and inequalities.

EXAMPLE 4   **Solving Absolute Value Equations and Inequalities**

In each part, use the graphing method to solve.

FIGURE 1.29

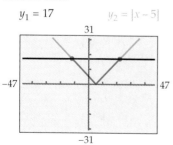

(a) $|x - 5| = 17$ (b) $|x - 5| \le 17$ (c) $|x - 5| > 17$

SOLUTION

Figure 1.29 shows the graphs of $y_2 = |x - 5|$ and $y_1 = 17$.

(a) In Figure 1.29, the points of intersection are $(-12, 17)$ and $(22, 17)$, so the solutions of the equation $|x - 5| = 17$ are -12 and 22.

(b) Figure 1.29 shows that the graph of $y_2 = |x - 5|$ intersects or is below the graph of $y_1 = 17$ for x-values between -12 and 22, inclusive. Thus the solution set of $|x - 5| \le 17$ is $[-12, 22]$.

(c) Finally, Figure 1.29 shows that the graph of $y_2 = |x - 5|$ is above the graph of $y_1 = 17$ for x-values that are less than -12 or greater than 22. Thus the solution set of $|x - 5| > 17$ is $(-\infty, -12) \cup (22, \infty)$.

Example 4 suggests the following algebraic rules for absolute value equations and inequalities.

■ *NOTE Numerous special cases arise when C = 0 or C < 0. Keep in mind that |A| is never negative.*

> ### Absolute Value Equations and Inequalities
>
> If A is an algebraic expression and C is a positive real number, and
>
> 1. if $|A| = C$, then $A = C$ or $A = -C$.
> 2. if $|A| < C$, then $-C < A < C$.
> 3. if $|A| > C$, then $A > C$ or $A < -C$.

EXAMPLE 5   **Solving Absolute Value Equations and Inequalities**

Solve each given equation or inequality algebraically.

(a) $|2x + 3| = 7$ (b) $|-(x - 2)| < 5$ (c) $|3x - 4| \geq 1$

SOLUTION

(a) $|2x + 3| = 7$

$$2x + 3 = 7 \quad \text{or} \quad 2x + 3 = -7 \qquad \text{If } |A| = C, \text{ then } A = C \text{ or } A = -C.$$
$$2x = 4 \quad \text{or} \qquad 2x = -10$$
$$x = 2 \quad \text{or} \qquad x = -5$$

The solutions are -5 and 2.

(b) $|-(x - 2)| < 5$

$$|2 - x| < 5 \qquad -(a - b) = b - a$$
$$-5 < 2 - x < 5 \qquad \text{If } |A| < C, \text{ then } -C < A < C.$$
$$-7 < \quad -x < 3$$
$$7 > \quad x > -3 \qquad \text{Multiplying by } -1 \text{ reverses the inequalities.}$$

The solution set is $(-3, 7)$.

(c) $|3x - 4| \geq 1$

$$3x - 4 \geq 1 \quad \text{or} \quad 3x - 4 \leq -1 \qquad \text{If } |A| > C, \text{ then } A > C \text{ or } A < -C.$$
$$3x \geq 5 \quad \text{or} \qquad 3x \leq 3$$
$$x \geq \frac{5}{3} \quad \text{or} \qquad x \leq 1$$

The solution set is $(-\infty, 1] \cup [\frac{5}{3}, \infty)$.

1.3 Quick Reference

Linear Inequalities and Interval Notation

■ A **linear inequality in one variable** is an inequality that can be written in the form $Ax + B < 0$, where A and B are real numbers and $A \neq 0$. This definition can also be stated for the inequality symbols $>$, $\leq$, and $\geq$.

■ Solution sets of inequalities are often written with **interval notation.**

Graphing Methods

■ When the left and right sides of an inequality are graphed, the point of intersection represents a solution for $\leq$ and $\geq$ inequalities, but not for $<$ and $>$ inequalities.

■ If an inequality has the form $Y_1 < Y_2$, the solution set consists of all x-values for which the graph of Y_1 is below the graph of Y_2.

Algebraic Methods

■ Properties of Inequality

If a, b, c, and d are real numbers, $d \neq 0$,

1. $a < b$ if and only if $a + c < b + c$. (Addition Property of Inequality)

2. if $d > 0$, $a < b$ if and only if $ad < bd$; if $d < 0$, $a < b$ if and only if $ad > bd$. (Multiplication Property of Inequality)

■ The inequality symbol must be reversed when both sides of an inequality are multiplied or divided by a negative number.

Compound Inequalities

■ A **compound inequality** consists of two inequalities connected by the word *and* (**conjunction**) or the word *or* (**disjunction**).

■ For a conjunction, the solution set is the **intersection** of the solution sets of the components. For a disjunction, the solution set is the **union** of the solution sets of the components.

■ The following procedure can be used to solve a compound inequality algebraically.

1. Solve each component of the compound inequality.

2. Draw a number line graph of the solution set of each component.

3. Draw a third number line graph that shows the union or intersection of the graphs in step 2.

4. Use the third number line graph to write the solution set in interval notation.

■ A **double inequality** is a compound inequality of the form $A < B < C$, where $<$ can be replaced with $\leq$, $>$, or $\geq$.

■ A double inequality is solved by applying the properties of inequality to all three components at once.

Absolute Value Equations and Inequalities

■ Absolute value equations and inequalities can be solved graphically in the same manner as linear equations and inequalities.

■ The following rules can be used to solve an absolute value equation or inequality: If A is an algebraic expression and C is a positive real number, and

1. if $|A| = C$, then $A = C$ or $A = -C$.

2. if $|A| < C$, then $-C < A < C$.

3. if $|A| > C$, then $A > C$ or $A < -C$.

Speaking the Language

1. When we write the solution set of $x \leq 3$ as $(-\infty, 3]$, we are using ▨▨▨▨▨▨ notation.

2. We call two inequalities joined by the word *or* a(n) ▨▨▨▨▨, whereas the connective for a(n) ▨▨▨▨▨ is *and*.

3. For the compound inequality $x < 3$ and $x > 1$, we write the solution set as the ▨▨▨▨▨ of $(-\infty, 3)$ and $(1, \infty)$. For the compound inequality $x \geq 5$ or $x < 0$, we write the solution set as the ▨▨▨▨▨ of $[5, \infty)$ and $(-\infty, 0)$.

4. The conjunction $x < 0$ and $x > -3$ can be written as $-3 < x < 0$, which is called a(n) ▨▨▨▨▨.

1.3 Exercises

Concepts and Skills

1. Suppose that $a > b$ and $c < 0$. Identify the statements that are false and explain why they are false.
   (i) $ac > bc$
   (ii) $a - c > b - c$
   (iii) $\dfrac{a}{c} < \dfrac{b}{c}$
   (iv) $-ac > -bc$

2. Why would the notation $[3, \infty]$ be inappropriate and misleading?

In Exercises 3–14, use interval notation to write the solutions of the given inequality or write an inequality whose solutions are represented by the given interval.

Inequality	Interval
3. $x < 5$	▨▨▨
4. $x > -3$	▨▨▨
5. ▨▨▨	$(1, \infty)$
6. ▨▨▨	$(-\infty, 2]$
7. $x \geq -8$	▨▨▨
8. $x \leq 2$	▨▨▨
9. $-5 < x < -2$	▨▨▨
10. $0 < x \leq 4$	▨▨▨
11. ▨▨▨	$[-5, 3)$
12. ▨▨▨	$[2, 7]$
13. $0 \leq x < 1$	▨▨▨
14. $8 > x > -3$	▨▨▨

In Exercises 15–22, solve the linear inequality.

15. $9 - 5t \leq -11$

16. $2(1 - 3y) > 20$

17. $6 - (1 - 3x) < x + 5$

18. $4(2y + 3) - 3(y - 1) < 0$

19. $t - \dfrac{7}{4} \geq \dfrac{5}{4}t - 1$

20. $\dfrac{1}{2}y + \dfrac{7}{12} < \dfrac{1}{4}(y + 3)$

21. $3(2x + 5) > 6(x - 1)$

22. $4x + 9 < x + 3(x - 4)$

23. Describe the difference between the solution sets for the given pairs of compound inequalities.
   (a) $x < -5$ and $x > 3$ $x < -5$ or $x > 3$
   (b) $x > -5$ and $x < 3$ $x > -5$ or $x < 3$

24. Identify the pairs that are not equivalent and explain why.
   (i) $1 < x$ and $x > 7$ $1 < x > 7$
   (ii) $x > -3$ and $x < -1$ $-3 < x < -1$
   (iii) $2 < x$ or $x < -4$ $2 < x < -4$

In Exercises 25–32, solve the compound inequality.

25. $-3x \leq 12$ and $-2x \geq -10$

26. $\dfrac{1}{4}x - \dfrac{1}{2} \leq 0$ and $\dfrac{x}{5} + \dfrac{1}{4} > \dfrac{1}{4}$

27. $x - 4 > 1$ or $x + 4 < 3$

28. $8 \leq 5 - x$ or $3x - 2 > 10$

29. $2x - 5 < 3$ and $3x + 1 < -5$

30. $4x + 5 \leq 0$ and $2 \leq -1 + 3x$

31. $3(x + 2) \leq -3$ or $4x - 2 \geq 0$

32. $7 - 2x \leq -7$ or $6x > 5x + 2$

In Exercises 33–36, solve the double inequality.

33. $-10 < -5t \leq 20$ **34.** $1 \leq x - 4 \leq 10$

35. $\dfrac{2}{3} < \dfrac{1}{3} - \dfrac{1}{2}x < \dfrac{10}{3}$ **36.** $0.4 \leq 2 - 0.2x < 1.8$

37. Describe the difference between the solutions of the following equations.

  (i) $|x - 3| = x - 3$ (ii) $|x - 3| = 3 - x$

38. Describe the difference between the solutions of the following inequalities.

  (i) $|x + 2| > 0$ (ii) $|x + 2| \geq 0$

In Exercises 39–46, solve the absolute value equation.

39. $|1 - 2x| = 7$ **40.** $|x + 11| = 5$

41. $|3x - 11| - 6 = 4$ **42.** $5 - |x - 9| = 1$

43. $|x + 1| + 9 = 2$ **44.** $4 - |x - 1| = 7$

45. $|3x + 7| - 4 = -4$ **46.** $3 + |5x + 1| = 12$

In Exercises 47–54, solve the given absolute value inequality.

47. $|2t - 3| > 8$ **48.** $5 - 3|x - 5| \leq -13$

49. $7 + 3|2x + 1| \geq 22$ **50.** $|y + 1| + 7 > 3$

51. $|4 - 3x| \leq 15$ **52.** $7 - |x + 1| \geq -2$

53. $|x - 23| + 4 < -1$ **54.** $|1 - t| - 7 < -12$

Concept Extension

In Exercises 55 and 56, solve the given compound inequality.

55. $4x \leq 2(x + 3) \leq 3x + 13$

56. $x - 1 > x - 2 > x - 3$

57. Describe the solutions of $|A| < C$ where $C \leq 0$.

58. Describe the solutions of $|A| \geq C$ where $C \leq 0$.

In Exercises 59–62, use the graphing method to estimate the solution set of the given equation or inequality.

59. $|x| - |x - 2| = 2$

60. $|x + 1| + |x - 1| = 2$

61. $|x - 2| + |x + 3| < 8$

62. $|x + 1| - |x| \geq 0$

63. Use the graphing method to describe the difference between the solutions of the given inequalities.

  (i) $|x + 3| > \dfrac{1}{2}x - 1$

  (ii) $|x + 3| < \dfrac{1}{2}x - 1$

1.4   *Quadratic Equations and Inequalities*

The Graphing Method ▪ *Factoring Methods* ▪ *The Even Root Property* ▪ *Completing the Square* ▪ *The Quadratic Formula* ▪ *Higher-Degree Equations and Inequalities*

In most career fields and even in our personal lives, we often need to manage affairs in a way that creates the most favorable (*optimum*) conditions. For example, you might want to minimize your mortgage payments or, in business, maximize your profits.

Mathematics gives us the tools to *optimize* certain quantities, and *quadratic equations* often serve as the models for doing so. In this section, we consider graphic and algebraic methods for solving quadratic and higher-degree equations and inequalities.

(b)
$$12x^2 - 11x = 5$$
$$12x^2 - 11x - 5 = 0 \qquad \text{Subtract 5 from both sides so that one side is 0.}$$
$$(3x + 1)(4x - 5) = 0 \qquad \text{Factor so that one side is a product.}$$
$$3x + 1 = 0 \quad \text{or} \quad 4x - 5 = 0 \qquad \text{Zero Factor Property}$$
$$3x = -1 \quad \text{or} \quad 4x = 5$$
$$x = -\frac{1}{3} \quad \text{or} \quad x = \frac{5}{4}$$

The solutions are $-\dfrac{1}{3}$ and $\dfrac{5}{4}$.

(c)
$$x^3 = 9x \qquad \text{Note that this is not a quadratic equation.}$$
$$x^3 - 9x = 0 \qquad \text{Subtract 9x from both sides so that one side is 0.}$$
$$x(x^2 - 9) = 0 \qquad \text{Factor so that one side is a product.}$$
$$x(x + 3)(x - 3) = 0$$
$$x = 0 \quad \text{or} \quad x + 3 = 0 \quad \text{or} \quad x - 3 = 0 \qquad \text{Zero Factor Property}$$
$$x = 0 \quad \text{or} \quad x = -3 \quad \text{or} \quad x = 3$$

The solutions are 0, -3, and 3.

The Even Root Property

Another algebraic method for solving a quadratic equation involves the **Even Root Property.**

The Even Root Property

For real numbers A and B and a positive even integer n, if $A^n = B$, then
1. for $B > 0$, $A = \pm\sqrt[n]{B}$;
2. for $B = 0$, $A = 0$;
3. for $B < 0$, $A^n = B$ has no real number solution.

EXAMPLE 2 **Solving with the Even Root Property**

Solve the equation $3x^2 - 15 = 0$.

SOLUTION
$$3x^2 - 15 = 0$$
$$3x^2 = 15 \qquad \text{Write the equation in the form } A^n = B.$$
$$x^2 = 5$$
$$x = \pm\sqrt{5} \approx \pm 2.24 \qquad \text{Even Root Property with } B > 0.$$

Completing the Square

We can use the Even Root Property to solve a quadratic equation only if the equation can be written in the form $A^2 = B$. To write the equation in this form, we can use the method of **completing the square.**

> ### Completing a Perfect Square Trinomial
>
> If the first part of a perfect square trinomial is $x^2 + bx$, then the entire perfect square trinomial is
>
> $$x^2 + bx + \left(\frac{b}{2}\right)^2$$
>
> In words, the constant term is the square of half the coefficient of x.

The following is a summary of the method of completing the square.

> ### Solving a Quadratic Equation by Completing the Square
>
> 1. If necessary, divide both sides by the leading coefficient to obtain a coefficient of 1 for the quadratic term.
> 2. Write the equation in the form $x^2 + bx = k$.
> 3. Add $\left(\frac{b}{2}\right)^2$ to both sides to complete the square.
> 4. Factor the resulting perfect square trinomial.
> 5. Solve with the Even Root Property.

EXAMPLE **3**   **Completing the Square**

Solve $2x^2 - 8x - 9 = 0$.

SOLUTION

$$2x^2 - 8x - 9 = 0$$

Divide both sides by 2 so that the coefficient of the quadratic term is 1.

$$x^2 - 4x - \frac{9}{2} = 0$$

$$x^2 - 4x = \frac{9}{2}$$

Move the constant term to the right side.

$$x^2 - 4x + \left(\frac{-4}{2}\right)^2 = \frac{9}{2} + \left(\frac{-4}{2}\right)^2$$

Complete the square on the left side by adding the square of half the coefficient of x.

$$x^2 - 4x + 4 = \frac{9}{2} + 4$$

$$(x - 2)^2 = \frac{17}{2}$$

Factor the perfect square trinomial. Now the equation is in the form $A^2 = B$ and the Even Root Property can be applied.

$$x - 2 = \pm\sqrt{\frac{17}{2}}$$

$$x = 2 \pm\sqrt{\frac{17}{2}}$$

$$\approx -0.92, 4.92$$

The Quadratic Formula

We can generalize the method of completing the square to derive a formula for the solutions of any quadratic equation.

$$ax^2 + bx + c = 0$$

$$ax^2 + bx = -c$$

Move the constant term to the right side.

$$x^2 + \frac{b}{a}x = -\frac{c}{a}$$

The leading coefficient must be 1, so divide both sides by a.

$$x^2 + \frac{b}{a}x + \frac{b^2}{4a^2} = -\frac{c}{a} + \frac{b^2}{4a^2}$$

Complete the square by adding $\left(\frac{1}{2}\cdot\frac{b}{a}\right)^2 = \frac{b^2}{4a^2}$ to both sides.

$$x^2 + \frac{b}{a}x + \frac{b^2}{4a^2} = \frac{-4ac}{4a^2} + \frac{b^2}{4a^2}$$

The LCD on the right side is $4a^2$.

$$x^2 + \frac{b}{a}x + \frac{b^2}{4a^2} = \frac{b^2 - 4ac}{4a^2}$$

Combine the fractions on the right side.

$$\left(x + \frac{b}{2a}\right)^2 = \frac{b^2 - 4ac}{4a^2}$$

Factor the perfect square trinomial.

$$x + \frac{b}{2a} = \frac{\pm\sqrt{b^2 - 4ac}}{2a}$$

Even Root Property

$$x = \frac{-b}{2a} \pm \frac{\sqrt{b^2 - 4ac}}{2a}$$

Subtract $\frac{b}{2a}$ to solve for x.

$$x = \frac{-b \pm \sqrt{b^2 - 4ac}}{2a}$$

Combine the fractions.

We call this formula the **Quadratic Formula.**

> *The Quadratic Formula*
>
> If a quadratic equation is written in the standard form $ax^2 + bx + c = 0$, then the solutions of the equation are given by the **Quadratic Formula,**
>
> $$x = \frac{-b \pm \sqrt{b^2 - 4ac}}{2a}$$

QUADRATIC FORMULA

By entering this program in your calculator, you can determine the real number solutions of $ax^2 + bx + c = 0$ simply by entering the numbers a, b, and c.

If applicable, the factoring method and the Even Root Property are the most efficient methods for solving a quadratic equation. Alternatively, the Quadratic Formula can be used to solve *any* quadratic equation.

In the Quadratic Formula, the expression $b^2 - 4ac$ is called the **discriminant.** If the coefficients of a quadratic equation are rational numbers, then we can use the discriminant to predict the number and type of real number solutions of the equation.

■ *NOTE   This table indicates the number and type of real number solutions only. If the discriminant is negative, the equation has two complex number solutions. Complex numbers will be discussed in the chapter on Polynomial and Rational Functions.*

Description of Discriminant	Description of Solutions	
	Number of Solutions	Type of Solutions
Positive (perfect square)	2	Rational
Positive (not a perfect square)	2	Irrational
Zero	1	Rational
Negative	0	—

The graphing method can be used to anticipate the number of solutions of a quadratic equation and to estimate the solutions. Then we can determine the exact solutions algebraically and observe whether the solutions are consistent with those suggested by the graph. If you also want to solve related inequalities, then returning to the graph can be very helpful in seeing the solutions and writing the solution sets. Example 4 illustrates the process.

EXAMPLE **4**   **Solving Quadratic Equations and Inequalities**

In each part, solve the given equation and the corresponding inequalities.

(a) $x^2 + 7 = 6x$ $x^2 + 7 < 6x$ $x^2 + 7 > 6x$
(b) $25 = 20x - 4x^2$ $25 < 20x - 4x^2$ $25 > 20x - 4x^2$
(c) $x(3x + 2) = -1$ $x(3x + 2) \le -1$ $x(3x + 2) > -1$

FIGURE 1.32

$y_1 = x^2 - 6x + 7$

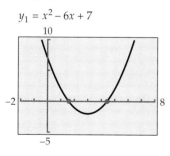

SOLUTION

(a) VISUALIZING THE SOLUTIONS

$$x^2 + 7 = 6x$$
$$x^2 - 6x + 7 = 0 \qquad \text{Write the equation in standard form.}$$

Figure 1.32 shows the graph of $y_1 = x^2 - 6x + 7$. The value of the expression is 0 at the *x*-intercepts, whose *x*-coordinates appear to be about 1.5 and 4.5. Thus the estimated solutions are 1.5 and 4.5.

DETERMINING THE SOLUTIONS

To solve the equation algebraically, we use the Quadratic Formula with $a = 1$, $b = -6$, and $c = 7$.

$$x = \frac{-b \pm \sqrt{b^2 - 4ac}}{2a}$$

$$= \frac{-(-6) \pm \sqrt{(-6)^2 - 4(1)(7)}}{2(1)}$$

$$= \frac{6 \pm \sqrt{8}}{2} \qquad \text{The discriminant is positive, so there are two solutions.}$$

$$= \frac{6 + 2\sqrt{2}}{2} = 3 \pm \sqrt{2}$$

$$\approx 4.41 \text{ or } 1.59$$

These solutions are consistent with our estimates in Figure 1.32.

VISUALIZING THE SOLUTIONS

Figure 1.32 also shows that the graph of $y_1 = x^2 - 6x + 7$ is *below* the x-axis between the x-coordinates. Thus the solutions of $x^2 - 6x + 7 < 0$ are all values of x between 1.59 and 4.41. The approximate solution set is (1.59, 4.41).

Similarly, the graph of $y_1 = x^2 - 6x + 7$ is *above* the x-axis to the left of the x-intercept (1.59, 0) and to the right of the x-intercept (4.41, 0). Thus the solution set of $x^2 - 6x + 7 > 0$ is $(-\infty, 1.59) \cup (4.41, \infty)$.

(b) VISUALIZING THE SOLUTION

$$25 = 20x - 4x^2$$

$$4x^2 - 20x + 25 = 0 \qquad \text{Write the equation in standard form.}$$

Figure 1.33 shows the graph of $y_1 = 4x^2 - 20x + 25$. Because there appears to be only one x-intercept, we anticipate one solution, approximately 2.5.

DETERMINING THE SOLUTION

Although the equation could be solved by factoring, we apply the Quadratic Formula.

$$x = \frac{-b \pm \sqrt{b^2 - 4ac}}{2a}$$

$$= \frac{-(-20) \pm \sqrt{(-20)^2 - 4(4)(25)}}{2(4)} \qquad a = 4, b = -20, c = 25$$

$$= \frac{20 \pm \sqrt{0}}{8} \qquad \text{The discriminant is 0, so the equation has only one solution.}$$

$$= 2.5$$

We call the solution 2.5 a **double root.**

VISUALIZING THE SOLUTIONS

Figure 1.33 shows that no portion of the graph is below the x-axis. Thus $4x^2 - 20x + 25 < 0$ has no solution. The solution set is ∅.

FIGURE **1.33**

$y_1 = 4x^2 - 20x + 25$

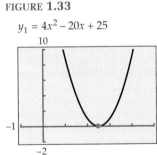

Except for the x-intercept, the entire graph is above the x-axis. Thus every real number except 2.5 is a solution of $4x^2 - 20x + 25 > 0$. The solution set is $(-\infty, 2.5) \cup (2.5, \infty)$.

(c) VISUALIZING THE SOLUTIONS

$$x(3x + 2) = -1$$
$$3x^2 + 2x = -1 \qquad \text{Distributive Property}$$
$$3x^2 + 2x + 1 = 0 \qquad \text{Write the equation in standard form.}$$

Figure 1.34 shows that the graph of $y_1 = 3x^2 + 2x + 1$ has no x-intercepts, which indicates that the equation has no real number solution. This can be confirmed by evaluating the discriminant.

$$b^2 - 4ac = (2)^2 - 4(3)(1) = -8$$

Because the discriminant is negative, the equation has no real number solution.

Because no portion of the graph is below the x-axis, the solution set of $3x^2 + 2x + 1 \le 0$ is Ø. However, the solution set of $3x^2 + 2x + 1 > 0$ is **R** because all of the graph is above the x-axis.

FIGURE **1.34**

$y_1 = 3x^2 + 2x + 1$

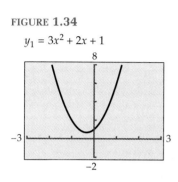

Higher-Degree Equations and Inequalities

A linear equation in one variable is a **first-degree equation,** and a quadratic equation is a **second-degree equation.** Equations of the form $a_n x^n + a_{n-1} x^{n-1} + a_{n-2} x^{n-2} + \cdots + a_1 x + a_0 = 0$, where $n \ge 3$, are called **higher-degree equations.**

The Even Root Property or the following Odd Root Property might apply to certain simple higher-degree equations.

> *The Odd Root Property*
>
> For real numbers A and B and an odd integer n, $n > 1$, if $A^n = B$, then $A = \sqrt[n]{B}$.

EXAMPLE **5**   **Solving Simple Higher-Degree Equations**

Solve. Round decimal answers to the nearest hundredth.

(a) $x^4 - 10 = 0$ (b) $x^3 + 28 = 4$

SOLUTION

(a) $x^4 - 10 = 0$

$$x^4 = 10 \qquad \text{Write the equation in the form } A^n = B.$$
$$x = \pm\sqrt[4]{10} \approx \pm 1.78 \qquad \text{Even Root Property}$$

(b) $x^3 + 28 = 4$

$$x^3 = -24 \qquad \text{Write the equation in the form } A^n = B.$$
$$x = \sqrt[3]{-24} = \sqrt[3]{-8}\,\sqrt[3]{3} = -2\sqrt[3]{3} \approx -2.88 \qquad \text{Odd Root Property}$$

If the polynomial expression can be factored, then the Zero Factor Property can be used to solve the equation or the related inequality.

EXAMPLE 6   **Solving a Higher-Degree Equation and Inequality by Factoring**

Solve.

(a) $x^3 + 5x^2 - 9x - 45 = 0$ (b) $x^3 + 5x^2 - 9x - 45 > 0$

SOLUTION

(a) VISUALIZING THE SOLUTIONS

FIGURE 1.35

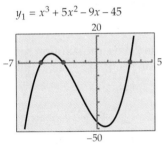

Figure 1.35 shows the graph of $y_1 = x^3 + 5x^2 - 9x - 45$. The three x-intercepts represent the solutions of the equation. We estimate the solutions to be -5, -3, and 3.

DETERMINING THE SOLUTIONS

$$x^3 + 5x^2 - 9x - 45 = 0$$
$$(x^3 + 5x^2) - (9x + 45) = 0 \qquad \textbf{Factor with the grouping method.}$$
$$x^2(x + 5) - 9(x + 5) = 0$$
$$(x + 5)(x^2 - 9) = 0$$
$$(x + 5)(x + 3)(x - 3) = 0$$
$$x + 5 = 0 \quad \text{or} \quad x + 3 = 0 \quad \text{or} \quad x - 3 = 0 \qquad \textbf{Zero Factor Property}$$
$$x = -5 \quad \text{or} \qquad x = -3 \quad \text{or} \qquad x = 3$$

As suggested by Figure 1.35, the solutions are -5, -3, and 3.

(b) VISUALIZING THE SOLUTIONS

To solve $x^3 + 5x^2 - 9x - 45 > 0$, we look for those portions of the graph that are above the x-axis. From Figure 1.35, we see that the solution set is $(-5, -3) \cup (3, \infty)$.

In Example 6, we were able to determine the exact solutions of both the equation and the inequality because the equation can be solved with a known factoring method. Even if the x-intercepts of a graph cannot be determined algebraically, we can still estimate them graphically and thereby obtain estimated solution sets of higher-degree equations and inequalities.

1.4 Quick Reference

The Graphing Method   ■ A **quadratic equation** is an equation that can be written in the standard form $Ax^2 + Bx + C = 0$, where A, B, and C are real numbers and $A \neq 0$.

■ A **quadratic inequality** is an inequality that can be written in the standard form $Ax^2 + Bx + C < 0$, where A, B, and C are real numbers and $A \neq 0$. The definition also applies to the inequality symbols $>$, $\leq$, and $\geq$.

■ When an equation is written in standard form, we call Ax^2 the **quadratic term,** Bx the **linear term,** and C the **constant term.**

■ When a quadratic equation is written in standard form, the solutions of the equation are represented by the x-intercepts of the graph of the left side.

Factoring Methods ■ The Zero Factor Property

If a and b are algebraic expressions and $ab = 0$, then $a = 0$ or $b = 0$.

■ If the left side of a quadratic equation in standard form can be factored, then the Zero Factor Property can be applied to determine the solutions algebraically.

The Even Root Property ■ The Even Root Property

For real numbers A and B and a positive even integer n, if $A^n = B$, then
1. for $B > 0$, $A = \pm\sqrt[n]{B}$;
2. for $B = 0$, $A = 0$;
3. for $B < 0$, $A^n = B$ has no real number solution.

Completing the Square ■ If the first part of a perfect square trinomial is $x^2 + bx$, then the entire perfect square trinomial is $x^2 + bx + \left(\dfrac{b}{2}\right)^2$.

In words, the constant term is the square of half the coefficient of x.

■ Use the following procedure to solve a quadratic equation by completing the square:
1. If necessary, divide both sides by the leading coefficient to obtain a coefficient of 1 for the quadratic term.
2. Write the equation in the form $x^2 + bx = k$.
3. Add $\left(\dfrac{b}{2}\right)^2$ to both sides to complete the square.
4. Factor the resulting perfect square trinomial.
5. Solve with the Even Root Property.

The Quadratic Formula ■ If a quadratic equation is written in the standard form $ax^2 + bx + c = 0$, then the solutions of the equation are given by the **Quadratic Formula,**

$$x = \frac{-b \pm \sqrt{b^2 - 4ac}}{2a}$$

■ In the Quadratic Formula, the expression $b^2 - 4ac$ is called the **discriminant.** If the coefficients of a quadratic equation are rational numbers, then we can use the discriminant to predict the number and type of real number solutions of the equation.

Higher-Degree Equations and Inequalities ■ A linear equation in one variable is a **first-degree equation,** and a quadratic equation is a **second-degree equation.** Equations of the form $a_n x^n + a_{n-1} x^{n-1} + a_{n-2} x^{n-2} + \cdots + a_1 x + a_0 = 0$, where $n \geq 3$, are called **higher-degree equations.**

■ The Odd Root Property

For real numbers A and B and an odd integer n, $n > 1$, if $A^n = B$, then $A = \sqrt[n]{B}$.

■ The graphing methods for solving quadratic equations and inequalities can be extended to higher-degree equations and inequalities.

Speaking the Language

1. For the ▩▩▩ equation $3x^2 - 5x + 7 = 0$, we call $-5x$ the ▩▩▩ term and $3x^2$ the ▩▩▩ term.

2. If we can write an equation in the form $A^n = B$, where n is a positive even integer, then the ▩▩▩ Property is a convenient method for solving the equation.

3. Adding 9 to both sides of $x^2 - 6x = 1$ results in a perfect square trinomial on the left side. This step is known as ▩▩▩.

4. The formula that can be used to solve any quadratic equation is called the ▩▩▩.

1.4 Exercises

Concepts and Skills

1. Consider the equation $2x^2 + 5x = 3$. Which method (graphing, factoring, completing the square, quadratic formula) can be used to solve the equation?

2. Explain how to write a quadratic equation $x^2 + bx + c = 0$ so that the solutions are 4 and -5.

In Exercises 3–10, use the graphing method to estimate the solution(s) of the equation. Then determine the solution(s) with the factoring method.

3. $w^2 + 4w - 21 = 0$
4. $30 - 7x - x^2 = 0$
5. $14x - 2x^2 = 0$
6. $15x = 6x^2$
7. $x(4x - 11) = 20$
8. $6t^2 + 7t - 20 = 0$
9. $x^2 + 9 = 6x$
10. $5x(5x + 4) = -4$

In Exercises 11–18, use the Even Root Property to solve the equation.

11. $12 - 4x^2 = 0$
12. $x^2 - 18 = 3$
13. $x^6 - 12 = 30$
14. $x^4 - 20 = 0$
15. $x^2 + 15 = 3$
16. $1 - x^2 = 7$
17. $(x + 3)^2 = 10$
18. $(1 - 2x)^2 = 16$

19. For what values of b does the equation $x^2 + bx - 1 = 0$ have real number solutions? How do you know?

20. Explain how to determine k so that $x^2 + 5x + k$ is a perfect square trinomial.

In Exercises 21–28, solve the quadratic equation by completing the square.

21. $x^2 + 6x = -4$
22. $x^2 = 9 + 2x$
23. $x^2 + 32 = 10x$
24. $x^2 + 5x + 7 = 0$
25. $9x^2 + 6x - 1 = 0$
26. $2x^2 - 5x = 4$
27. $x + 5 = 3x^2$
28. $4x^2 + 3x - 2 = 0$

In Exercises 29–36, use the Quadratic Formula to solve the given equation.

29. $9x^2 + 12x - 1 = 0$
30. $4x^2 = 4x + 2$
31. $0 = 6x - 13 - x^2$
32. $2 + 6x + 9x^2 = 0$
33. $3x^2 + 2x = 2$
34. $x^2 - 7 = 5x$
35. $7 - 3x - 5x^2 = 0$
36. $4x = \frac{5}{2}x^2 - \frac{1}{5}$

37. Explain why $(-3, 5]$ cannot represent the solutions of a quadratic inequality.

38. Suppose that the solution of $ax^2 + bx + c \geq 0$ is $[m, n]$. What is the solution of $ax^2 + bx + c < 0$?

In Exercises 39–46, solve the quadratic inequality.

39. $x^2 + 2x - 15 \geq 0$
40. $2x^2 - 9x + 4 < 0$
41. $x(3x + 11) > -6$
42. $x^2 \leq 5x + 14$
43. $8(x + 2) \geq -x^2$
44. $(x + 3)(x + 1) < 3x$
45. $x^2 + 3x \leq 2$
46. $3 + x < x^2$

In Exercises 47–54, use the discriminant to determine the number and nature of the solutions of the given equation.

47. $9t^2 + 12t + 4 = 0$
48. $14x = x^2 + 49$
49. $2 + x^2 = 2x$
50. $2x^2 + 3x + 2 = 0$

51. $4y^2 + 5y = 0$

52. $63 + 29x - 24x^2 = 0$

53. $2 + t - 4t^2 = 0$ **54.** $3y^2 + y = 1$

In Exercises 55–60, solve the given equation.

55. $x^4 + 5x^2 - 36 = 0$ **56.** $x^4 - 4x^2 + 3 = 0$

57. $x^4 + 20 = 9x^2$ **58.** $14x^2 = 2x^4 + 20$

59. $x^4 - 25 = 0$ **60.** $16y^4 - 8y^2 + 1 = 0$

61. To solve the inequality $x^3 - 4x > 0$, you can begin by solving the equation $x^3 - 4x = 0$ to obtain the solutions -2, 0, and 2. Explain how to use a graph to complete the problem.

62. Describe how to determine the values of x for which $\sqrt{x^2 - 1}$ represents a real number.

In Exercises 63–66, use the Odd Root Property to solve the equation.

63. $20 - 5x^3 = 0$ **64.** $x^5 + 17 = 1$

65. $x^3 + 50 = 0$ **66.** $(5 - 2x)^3 = 30$

In Exercises 67–72, use the factoring method to solve the given equation.

67. $8x^3 - 50x = 0$ **68.** $6x^3 - 21x^2 + 9x = 0$

69. $x^3 + 3x^2 - 4x - 12 = 0$

70. $2x^3 - x^2 - 10x + 5 = 0$

71. $y^6 - 64 = 0$

72. $1 - x^6 = 0$

In Exercises 73–76, solve the inequality. (*Hint:* Use the results from Exercises 67–70.)

73. $8x^3 - 50x > 0$

74. $6x^3 - 21x^2 + 9x \le 0$

75. $x^3 + 3x^2 - 4x - 12 \le 0$

76. $2x^3 - x^2 - 10x + 5 > 0$

Concept Extension

In Exercises 77 and 78, use the Quadratic Formula to solve the equation.

77. $x^2 - 2\sqrt{2}x + 1 = 0$

78. $\sqrt{2}x^2 - \sqrt{3}x - \sqrt{8} = 0$

In Exercises 79 and 80, use the Quadratic Formula to solve the given equation for y.

79. $x = y^2 + 2y - 1$ **80.** $x = 2 - 2y - y^2$

1.5   *Rational and Radical Equations*

Rational Equations   ■   *Radical Equations*   ■   *Equations with*
Rational Exponents   ■   *Solving Formulas*

Linear and quadratic equations are the most common models for basic applications. However, for more sophisticated applications, other types of equations may be needed, and they play an important role in your study of science and mathematics.

We conclude this chapter with a discussion of graphing and algebraic methods for solving rational and radical equations and equations that involve rational exponents. Although the algebraic procedures that you use will depend on the type of equation that you are solving, the graphing methods discussed in preceding sections are applicable to all categories of equations.

Rational Equations

If the terms of an equation are rational expressions, we refer to the equation as a **rational equation.**

The graphing method can be used to estimate the solution(s) of a rational equation.

EXAMPLE **1**   **Solving a Rational Equation by Graphing**

Solve $x = \dfrac{12}{x} + 1$.

SOLUTION

One method is to graph the left and right sides of the equation and trace to the point(s) of intersection. In Figure 1.36, we have traced to the point $(-3, -3)$. By tracing, we find that the other point is $(4, 4)$. Thus the solutions of the equation are -3 and 4.

FIGURE **1.36**   FIGURE **1.37**

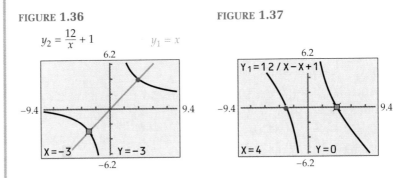

■ *NOTE Recall that solutions obtained by the graphing method must be regarded as estimates until they are checked. The solutions in Example 1 are easily verified by substitution.*

Alternatively, we can write the equation with 0 on one side, in which case the x-intercepts represent solutions. Figure 1.37 shows the graph of $y_1 = \dfrac{12}{x} - x + 1$ with the tracing cursor on $(4, 0)$. The other x-intercept is $(-3, 0)$. Again, the solutions are -3 and 4.

An important step in solving rational equations is to identify restricted values. Sometimes a number will be an apparent solution of an equation, but if the number is a restricted value, it cannot be a solution. We call such numbers **extraneous solutions.**

To solve a rational equation algebraically, the usual initial step is to *clear the fractions* by multiplying both sides of the equation by the LCD of all the denominators. The result may be a first-degree equation, a second-degree equation, or even a higher-degree equation.

EXAMPLE **2**   **Solving Rational Equations**

Solve the following rational equations.

(a) $\dfrac{3}{x} = \dfrac{2}{x + 4}$ (b) $x = \dfrac{12}{x} + 1$

SOLUTION

(a) Although we can clear the fractions by multiplying both sides by the LCD, which is $x(x + 4)$, we can use the **cross-multiplication** rule:

If $\dfrac{A}{B} = \dfrac{C}{D}$, then $AD = BC$.

$$\frac{3}{x} = \frac{2}{x+4}$$ **Note that 0 and −4 are restricted values.**

$$3 \cdot (x+4) = 2 \cdot x$$ **Cross-multiply.**

$$3x + 12 = 2x$$

$$x = -12$$

Because −12 is not a restricted value, the solution is −12.

(b) This is the equation that we solved with the graphing method in Example 1.

$$x = \frac{12}{x} + 1$$ **The restricted value is 0.**

$$x \cdot x = x \cdot \frac{12}{x} + x \cdot 1$$ **Multiply both sides by x to clear the fraction.**

$$x^2 = 12 + x$$ **The result is a quadratic equation.**

$$x^2 - x - 12 = 0$$ **Write the equation in standard form.**

$$(x+3)(x-4) = 0$$ **Factor.**

$$x + 3 = 0 \quad \text{or} \quad x - 4 = 0$$ **Zero Factor Property**

$$x = -3 \quad \text{or} \qquad x = 4$$

Neither number is a restricted value. As suggested by the graph in Example 1, the solutions are −3 and 4.

Combining the graphing method with the algebraic solving process can be very helpful in detecting extraneous solutions.

EXAMPLE **3**   **Combining Graphing and Algebraic Methods**

Solve $\dfrac{x+2}{x-1} = \dfrac{1}{x} + \dfrac{3}{x^2 - x}$.

SOLUTION

VISUALIZING THE SOLUTION

FIGURE 1.38

$$y_1 = \frac{x+2}{x-1} - \frac{1}{x} - \frac{3}{x^2 - x}$$

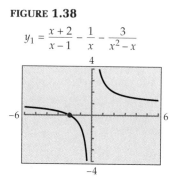

Figure 1.38 shows the graph of $y_1 = \dfrac{x+2}{x-1} - \dfrac{1}{x} - \dfrac{3}{x^2 - x}$. The estimated x-intercept is $(-2, 0)$, so the equation appears to have one solution, −2.

DETERMINING THE SOLUTION

To solve the equation algebraically, we clear fractions by multiplying both sides by the LCD.

$$\frac{x+2}{x-1} = \frac{1}{x} + \frac{3}{x^2 - x}$$

$$\frac{x+2}{x-1} = \frac{1}{x} + \frac{3}{x(x-1)}$$ **Factor to determine the LCD, $x(x-1)$.**

$$x(x-1) \cdot \frac{x+2}{x-1} = x(x-1) \cdot \frac{1}{x} + x(x-1) \cdot \frac{3}{x(x-1)}$$

$$x(x + 2) = x - 1 + 3$$
$$x^2 + 2x = x + 2$$
$$x^2 + x - 2 = 0$$
$$(x + 2)(x - 1) = 0$$
$$x + 2 = 0 \quad \text{or} \quad x - 1 = 0$$
$$x = -2 \quad \text{or} \quad x = 1$$

Observe that the graphing method revealed only one solution, whereas solving algebraically resulted in two apparent solutions. However, 1 is an extraneous solution because it is a restricted value. Thus, as indicated by the graph, the only solution is -2.

Radical Equations

A **radical equation** is an equation that contains at least one radical with a variable in the radicand.

■ *NOTE   To solve a radical equation, begin by isolating the radical.*

The general approach to solving a square root radical equation is to square both sides of the equation to eliminate the radical. For cube root equations, we cube both sides, and so on. The technique may need to be applied more than once until all radicals are eliminated.

Raising both sides of an equation to some power may introduce extraneous solutions. Therefore, checking all solutions in the original equation is essential.

EXAMPLE **4**   **Solving a Radical Equation**

Solve $\sqrt{2x + 9} = x - 3$.

SOLUTION

VISUALIZING THE SOLUTION

Figure 1.39 shows the graphs of the left and right sides of the equation. The one point of intersection represents the solution, which we estimate to be 8.

FIGURE **1.39**

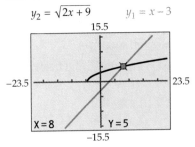

$y_2 = \sqrt{2x + 9}$ $y_1 = x - 3$

DETERMINING THE SOLUTION

We solve algebraically by squaring both sides of the equation to eliminate the radical.

$$\sqrt{2x+9} = x - 3$$

$$\left(\sqrt{2x+9}\right)^2 = (x-3)^2 \qquad \text{Square both sides of the equation.}$$

$$2x + 9 = x^2 - 6x + 9 \qquad \text{The result is a quadratic equation.}$$

$$x^2 - 8x = 0 \qquad \text{Write the equation in standard form.}$$

$$x(x - 8) = 0 \qquad \text{Factor.}$$

$$x = 0 \quad \text{or} \quad x - 8 = 0 \qquad \text{Zero Factor Property}$$

$$x = 0 \quad \text{or} \qquad x = 8$$

The algebraic solution resulted in two apparent solutions, whereas the graphing method shows only one solution. Replacing x with 0 in the original equation, we obtain $\sqrt{9} = -3$, which is false. Thus, 0 is an extraneous solution, and the only solution of the equation is 8.

When an equation contains two radicals, separating them usually makes the algebraic solving process easier.

EXAMPLE 5 **Solving a Radical Equation**

Solve $\sqrt{x-1} - \sqrt{2x-1} + 1 = 0$.

SOLUTION

$$\sqrt{x-1} - \sqrt{2x-1} + 1 = 0$$

$$\sqrt{x-1} + 1 = \sqrt{2x-1} \qquad \text{Separate the radicals.}$$

$$\left(\sqrt{x-1} + 1\right)^2 = \left(\sqrt{2x-1}\right)^2 \qquad \text{Square both sides.}$$

$$(x-1) + 2\sqrt{x-1} + 1 = 2x - 1$$

$$2\sqrt{x-1} = x - 1 \qquad \text{Isolate the remaining radical.}$$

$$\left(2\sqrt{x-1}\right)^2 = (x-1)^2 \qquad \text{Square both sides.}$$

$$4(x-1) = x^2 - 2x + 1 \qquad \text{The result is a quadratic equation.}$$

$$4x - 4 = x^2 - 2x + 1$$

$$x^2 - 6x + 5 = 0 \qquad \text{Write the equation in standard form.}$$

$$(x-1)(x-5) = 0 \qquad \text{Factor.}$$

$$x - 1 = 0 \quad \text{or} \quad x - 5 = 0 \qquad \text{Zero Factor Property}$$

$$x = 1 \quad \text{or} \qquad x = 5$$

A calculator or substitution can be used to verify that 1 and 5 are both valid solutions.

Equations with Rational Exponents

If an equation can be written in the form $A^{m/n} = B$, then we can raise both sides of the equation to the nth power in order to obtain an integer exponent on A.

EXAMPLE **6** **Equations with Rational Exponents**

Solve $(2x - 3)^{2/3} = 6$.

SOLUTION

VISUALIZING THE SOLUTIONS

FIGURE **1.40**

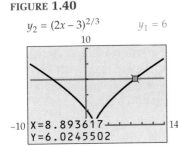

$y_2 = (2x - 3)^{2/3}$ $y_1 = 6$

From the graphs of the left and right sides of the equation, we can anticipate two solutions. (See Figure 1.40.) By tracing, we estimate one of the solutions to be about 8.9.

DETERMINING THE SOLUTIONS

To solve the equation algebraically, we raise both sides of the equation to the third power to eliminate the rational exponent.

$$(2x - 3)^{2/3} = 6$$
$$[(2x - 3)^{2/3}]^3 = 6^3$$
$$(2x - 3)^2 = 216$$
$$4x^2 - 12x + 9 = 216$$
$$4x^2 - 12x - 207 = 0$$

This quadratic equation can be solved with the Quadratic Formula. The resulting solutions are approximately -5.85 and 8.85, which are consistent with those suggested by the graphing method.

Solving Formulas

The algebraic methods for *solving a formula* for a specific variable are identical to those used for solving equations.

EXAMPLE **7** **Solving a Formula**

In a parallel circuit with two branches, the resistances R_1 and R_2 in the branches are related to the total resistance R of the circuit by the formula $\dfrac{1}{R} = \dfrac{1}{R_1} + \dfrac{1}{R_2}$. Solve this formula for R_1.

SOLUTION

$$\frac{1}{R} = \frac{1}{R_1} + \frac{1}{R_2} \qquad \textbf{Multiply both sides by } RR_1R_2 \textbf{ to clear the fractions.}$$

$$RR_1R_2 \cdot \frac{1}{R} = RR_1R_2 \cdot \frac{1}{R_1} + RR_1R_2 \cdot \frac{1}{R_2}$$

$$R_1R_2 = RR_2 + RR_1$$

$$R_1R_2 - RR_1 = RR_2 \qquad \textbf{Move the terms involving } R_1 \textbf{ to the left side.}$$

$$R_1(R_2 - R) = RR_2 \qquad \textbf{Factor out } R_1.$$

$$R_1 = \frac{RR_2}{R_2 - R} \qquad \textbf{Divide both sides by } R_2 - R.$$

Quick Reference

Rational Equations

■ If the terms of an equation are rational expressions, we refer to the equation as a **rational equation.**

■ To solve a rational equation algebraically, the usual initial step is to *clear the fractions* by multiplying both sides of the equation by the LCD of all the denominators.

■ When an apparent solution of a rational equation is a restricted value, we call the number an **extraneous solution.**

Radical Equations

■ A **radical equation** is an equation that contains at least one radical with a variable in the radicand.

■ The general approach to solving a square root radical equation is to square both sides of the equation to eliminate the radical. A similar method can be applied to higher roots.

■ Raising both sides of an equation to a power can introduce extraneous solutions.

Equations with Rational Exponents

■ If an equation can be written in the form $A^{m/n} = B$, then we can raise both sides of the equation to the nth power in order to obtain an integer exponent on A.

Solving Formulas

■ The algebraic methods for *solving a formula* for a specific variable are identical to those used for solving equations.

Speaking the Language

1. If the equation-solving process leads to an apparent solution that is a restricted value, the number is a(n) ▨▨▨▨▨ solution.

2. An efficient method for solving $\dfrac{x}{3} = \dfrac{5}{x}$ is to ▨▨▨▨▨ to obtain $x^2 = 15$.

3. The equation $5 - \sqrt{2x + 3} = 12$ is an example of a(n) ▨▨▨▨▨ equation.

4. Rewriting $A = \dfrac{1}{2}bh$ as $h = \dfrac{2A}{b}$ is called ▨▨▨▨▨ for h.

1.5 Exercises

Concepts and Skills

1. Explain how the LCD is used to perform the addition in (i) and to solve the equation in (ii).

(i) $\dfrac{x}{x+2} + \dfrac{x-1}{x+1} + 5$ (ii) $\dfrac{x}{x+2} + \dfrac{x-1}{x+1} = 5$

2. Consider the following two approaches to solving the equation $\dfrac{x^2 - 9}{x - 3} = 5$. Explain the apparent discrepancy.

(i) Simplify the fraction to obtain $x + 3 = 5$. Then the solution is 2.

(ii) Clear the fraction to obtain $x^2 - 5x + 6 = 0$. Then the solutions are 2 and 3.

In Exercises 3–16, solve the given rational equation.

3. $\dfrac{5}{y+4} = y$

4. $\dfrac{3}{x-1} = 3 - \dfrac{8}{x}$

5. $\dfrac{7}{12} + \dfrac{1}{2t} = \dfrac{5}{3t}$

6. $\dfrac{2}{y+2} = \dfrac{1}{y-1} + \dfrac{1}{y}$

7. $x = \dfrac{x + 10}{2x}$

8. $\dfrac{x}{6} + \dfrac{2}{3} = \dfrac{2}{x}$

9. $\dfrac{2}{x} + \dfrac{1}{x + 2} = -1$

10. $x + \dfrac{25}{x + 1} = 9$

11. $\dfrac{x}{x - 4} = 1 + \dfrac{1}{x^2 - 16}$

12. $\dfrac{2}{x - 1} - \dfrac{1}{x} = \dfrac{1}{x + 3}$

13. $\dfrac{t}{t - 3} = \dfrac{3}{t - 3} - \dfrac{2}{t}$

14. $1 + \dfrac{1}{x + 2} = \dfrac{6}{x} - \dfrac{2}{x(x + 2)}$

15. $\dfrac{x}{6} - \dfrac{2x - 3}{3x} = \dfrac{1}{x}$

16. $\dfrac{2x - 3}{x - 4} - \dfrac{9 - x}{x - 4} = 2$

17. Explain how you know that each equation has no solution.

   (i) $\sqrt{1 - x} + 3 = 0$ (ii) $\sqrt{1 - x} = \sqrt{x - 5}$

18. To solve $x^{2/3} = 5$, suppose that you raise both sides of the equation to the $\frac{3}{2}$ power to obtain $x = 5\sqrt{5}$. Explain why you have not found all solutions of the equation.

In Exercises 19–30, solve the radical equation.

19. $\sqrt{5 - 3x} = 5$

20. $\sqrt{x + 4} - 6 = -2$

21. $\sqrt[3]{1 - x} = -2$

22. $\sqrt[4]{2x + 7} = 1$

23. $\sqrt{4 - 3x} = 5$

24. $\sqrt[4]{x^2 + 12} = \sqrt[4]{7x}$

25. $x - 4 = \sqrt{4 - x}$

26. $x = -7 + \sqrt{1 - 2x}$

27. $x + 2 = \sqrt{12x - 11}$

28. $x = \sqrt{5x - 4}$

29. $\sqrt{x - 7} - \sqrt{x + 1} = -2$

30. $\sqrt{2x + 3} + \sqrt{x + 13} = 7$

In Exercises 31–36, solve.

31. $y^{1/4} = \dfrac{3}{2}$

32. $x^{1/3} = -5$

33. $x^{2/5} = 2$

34. $z^{3/2} = 9$

35. $(x + 5)^{2/3} = 4$

36. $(1 - x)^{3/2} = 8$

In Exercises 37–42, solve the given formula for the indicated variable.

37. $A = \dfrac{1}{2}h(b_1 + b_2)$ for b_2

38. $A = \pi r^2$ for r

39. $A = P + Prt$ for P

40. $P = \dfrac{nRT}{V}$ for V

41. $A = 2LW + 2LH + 2WH$ for H

42. $S = \dfrac{a}{1 - r}$ for r

Concept Extension

In Exercises 43 and 44, solve the given equation.

43. $x^{2/3} + 2x^{1/3} - 3 = 0$

44. $x^{1/3} - 5x^{1/6} + 4 = 0$

Chapter Review Exercises

Section 1.1

1. Which calculator menu would you select to enter the viewing window settings for Quadrant II? What would be the signs of the settings for *Xmin* and *Ymax*?

2. Plot the points associated with the following ordered pairs.

   $(-3, -5), (-1, -2), (1, 1), (3, 6), (5, 7), (7, 10)$

Then identify the point that does not appear to be a point of the line that contains all the other points.

3. Use the given values of x and your calculator to produce a table of values for the expression. Use the ordered pairs associated with the values in the table to sketch the graph of the expression. Then use a calculator to produce the graph of the expression and compare the graph to your sketch.

$$\dfrac{1}{2}x^2 + x; \quad -6, -4, -2, 0, 2, 4$$

4. Select six values of x to evaluate $|2x + 1|$. Use those values to sketch the graph of the expression. Then produce the graph with your calculator and compare it to your sketch.

5. Which of your built-in window settings (integer, decimal, or standard) appears to be the best for displaying the graph of the expression $\sqrt{x + 12}$?

6. Use the integer setting to produce the graph of the expression $0.1x^3 - 5$. Then trace the graph to determine the following missing coordinates.

$(\boxed{(a)}, -17.5), (0, \boxed{(b)}), (\boxed{(c)}, 1.4), (\boxed{(d)}, 16.6)$

7. Use a table of values to determine the x-intercepts of the graph of $x^2 + 3x - 10$.

8. What do you know about a graph whose x-intercept and y-intercept are the same point?

9. Use the integer setting to trace to the intercepts of the graph of $13 - |x - 7|$. Verify your results by substitution.

10. Use the standard setting to trace to the x-intercept of the graph of $0.9x - 6$. Perform two zoom-and-trace cycles and round the result to the nearest hundredth.

Section 1.2

11. Use the decimal setting and the graphing method to solve the equation $\frac{1}{2}x - 1 = -2x + \frac{3}{2}$. Use your calculator to verify the solution.

12. Use the standard setting and the graphing method to solve the equation $-2x + 5 = 4x + \frac{4}{3}$. After two zoom-and-trace cycles, round your answer to the nearest tenth.

13. Suppose that you produce the graphs of the left and right sides of the type of linear equation given in column A. Match the equation with the number of points of intersection given in column B.

Column A	*Column B*
(a) Contradiction	One
(b) Conditional	Infinitely many
(c) Identity	None

14. By what number would you multiply both sides of the equation $0.05x^2 - 0.4x = 0.007$ in order to clear the decimals?

In Exercises 15–18, solve the given equation.

15. $3(x - 4) = 2x - (9 - 5x)$

16. $15 - 3[x - (2x + 4)] = -[x - (x - 1)]$

17. $\frac{1}{2}x + \frac{1}{3} = 3 - \frac{5}{6}x$

18. $\frac{1}{2}\left[\frac{2}{3}(x + 1)\right] = \frac{1}{6}$

19. Use the standard setting and the graphing method to solve $3x - 2 = -0.2x + 5$. Use the calculator's *Intersect* feature to determine the exact solution.

20. Write the solution set of the given equations.

(a) $3x - 2(x + 5) = \frac{1}{2}(2x + 1)$

(b) $-x^2 + 3x = -6 - (x^2 - 3x - 6)$

Section 1.3

In Exercises 21 and 22, solve the given inequality.

21. $2x - 5 \le 4(x + 1)$

22. $\frac{x - 3}{4} > \frac{x + 3}{3}$

23. What name do we give to the inequality $-3 < 2x + 1 < 5$? To what compound inequality is this inequality equivalent?

24. Solve $1 > \frac{y}{2} - 1 \ge -2$.

In Exercises 25 and 26, solve the compound inequality.

25. $-8x \le 24$ and $3 - x \ge x - 5$

26. $-2x \le 3x + 10$ or $5 - 3x \ge -4x + 3$

27. Solve $|2x - 7| - 2 = 3$.

28. Use the graphing method to solve the inequality $|2x - 9| - 13 > 0$.

In Exercises 29 and 30, solve the given absolute value inequality.

29. $|x - 4| < 4$

30. $-2|5x + 1| \le -8$

Section 1.4

31. Why is it not possible to solve the equation $2x^2 + 3x = 0$ with the same solving routine that we use to solve $2x + 3 = 0$?

In Exercises 32 and 33, use the factoring method to solve the given quadratic equation.

32. $21x = -(10x^2 + 9)$

33. $x(x - 6) = -9$

34. Solve $-2x^3 + x^2 + 18x - 9 = 0$.

35. Use the Even Root or Odd Root Property to solve the equation.

(a) $x^4 - 5 = 11$

(b) $x^3 + 9 = 0$

36. How do you determine k so that $x^2 - 10x + k$ is a perfect square trinomial?

37. Solve $3x^2 + 2x - 6 = 0$.

38. Use the discriminant to determine the number and nature of the solutions of $x^2 - 4x + 8 = 0$.

39. Solve the inequalities.

(a) $3 - 2x - x^2 < 0$

(b) $3 - 2x - x^2 \ge 0$

40. Solve $9x^3 - x^5 \ge 0$.

Section 1.5

41. Even though you make no algebraic errors in solving a rational or radical equation, you should check the apparent solutions. Why?

In Exercises 42–45, solve the rational equation.

42. $\dfrac{x}{x-2} = \dfrac{x+1}{x}$

43. $\dfrac{1}{t+1} = 3 - \dfrac{5}{t+1}$

44. $\dfrac{3}{x-2} = 3 - \dfrac{8}{x-1}$

45. $\dfrac{x+5}{x+1} + \dfrac{4}{x^2+x} = \dfrac{1}{x}$

46. Solve the given equation.

    (a) $\sqrt{4-5x} = 7$ (b) $\sqrt[3]{1-x} = -3$

In Exercises 47 and 48, solve the radical equation.

47. $x - 1 = \sqrt{x+11}$

48. $\sqrt{x+2} - \sqrt{2x+6} = 1$

49. Solve $(t-4)^{-2/3} = 25$.

50. Solve $\dfrac{1}{f} = \dfrac{1}{p} + \dfrac{1}{q}$ for p.

Chapter 2

Basic Functions and Graphs

A politician's "approval rating" can change quickly and dramatically. The rating can be affected by such factors as the state of the economy, the conduct of foreign affairs, and the politician's stand on domestic issues.

By periodically polling small samples of people, political advisors can estimate the trend in a candidate's support among voters. Polling data can be modeled with a regression equation whose graph best fits the data, and this model can then be used to predict the level of support that a candidate can expect to receive.

In this chapter, we introduce the essential concept of a function, and we investigate the properties of the graphs of a variety of functions. In particular, you will learn how real data can be modeled with functions and how to judge the most appropriate model to use. Although many different models are available, our emphasis in this chapter is on linear and quadratic regression models.

This chapter provides you with a solid foundation for the important skills of analyzing, modeling, and interpreting real data.

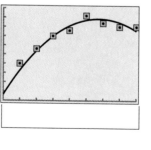

POLL RESULTS	
Week	Rating
1	20%
2	28%
3	35%
4	38%
5	46%
6	42%
7	40%
8	40%

2.1 Relations and Functions

Relations ■ Functions ■ Function Notation and Evaluation
■ Domain and Range ■ Piecewise Functions ■ Applications

In many real-world situations, the value of one quantity depends on the value of some other quantity. For example, the federal tax that you pay depends on your income, the distance that you can drive in a given time depends on your speed, and your success in a college course may depend on the number of hours that you study.

Being able to describe how one variable affects another variable is often an important step in analyzing a problem. In this section, we consider how to use a *function* to describe the association of two sets of numbers. The topic of functions is central to the study of mathematics.

Relations

The bar graph in Figure 2.1 shows how advertising affects the profits of a small business.

FIGURE 2.1

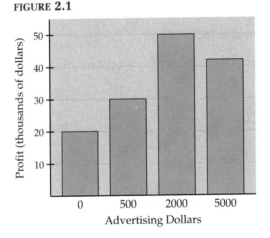

The graph shows that advertising does increase profits, but that beyond a certain level, the cost of too much advertising actually decreases profits. The advertising costs and corresponding profits shown in the bar graph can be written as a set of ordered pairs.

$$A = \{(0, 20{,}000), (500, 30{,}000), (2000, 50{,}000), (5000, 42{,}000)\}$$

Such a set is called a **relation.**

> ### Definition of a Relation
> A **relation** is a set of ordered pairs. The **domain** of a relation is the set of all first coordinates of the ordered pairs. The **range** is the set of all second coordinates.

In the preceding set A, the domain is the set of advertising costs, and the range is the set of profits.

$$\text{Domain: } \{0, 500, 2000, 5000\}$$

$$\text{Range: } \{20{,}000, 30{,}000, 50{,}000, 42{,}000\}$$

A relation can also be described with **set-builder notation.** For example, in the relation

$$B = \{(x, y) \mid y = x + 4\}$$

recall that the symbol | is read "such that," and we say that relation B is the set of all ordered pairs (x, y) such that the equation $y = x + 4$ is true. More often, we will simply say that the relation is $y = x + 4$.

Functions

The relation $y = x + 4$ consists of all ordered pairs of the form $(x, x + 4)$. For each value of x there is *exactly one* value of $x + 4$. For this reason, the set of all ordered pairs of the form $(x, x + 4)$ is an example of a **function.**

> ### Definition of a Function
>
> A **function** is a relation in which no two different ordered pairs have the same first coordinate.

To determine whether a set of ordered pairs is a function, we only need to apply the definition.

EXAMPLE **1**   **Testing for Functions**

Determine whether each given relation is a function.

(a) $D = \{(1, 2), (2, 3)\ (3, 4), (4, 5), \ldots\}$
(b) $K = \{(-3, 5), (-1, 2), (1, 4), (1, 5), (2, 7)\}$

SOLUTION

(a) Assuming that the indicated pattern continues, set D is a function because no two ordered pairs have the same first coordinate.
(b) Set K is not a function because the pairs $(1, 4)$ and $(1, 5)$ have the same first coordinate.

A **mapping diagram** is a visual way of showing the association of each member of a domain with the corresponding members of the range. In a mapping diagram, an arrow points from an element of the domain to the corresponding element(s) of the range. A mapping diagram can also be used to judge whether a relation is a function.

EXAMPLE 2   **Testing for Functions**

Figure 2.2 shows mapping diagrams for sets D and K in Example 1.

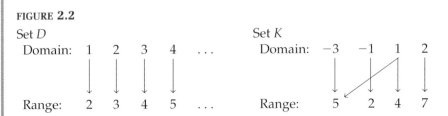

FIGURE 2.2

Use the mapping diagrams to determine whether the given sets are functions.

SOLUTION

For set D, the arrows indicate that each member of the domain corresponds to exactly one member of the range, so set D is a function.

However, for set K, the first coordinate 1 in the domain has arrows pointing to both 4 and 5 in the range, so set K is not a function.

Rather than list the ordered pairs of a function, we usually describe the function with a rule that indicates how the ordered pairs are generated.

> **Alternative Definition of a Function**
>
> A function is a rule or correspondence that assigns each element x in the domain to a unique element y in the range.

When a relation is expressed as a rule, we can determine whether the relation is a function by asking the question, "Is there any value of x for which there is more than one corresponding value of y?" If the answer is no, then the relation is a function.

EXAMPLE 3   **Testing for Functions**

For each equation, determine whether y is a function of x.

(a) $2x + 3y = 6$ (b) $x - y^2 - 5 = 0$

SOLUTION

(a) $2x + 3y = 6$

$$3y = -2x + 6$$

$$y = -\frac{2}{3}x + 2$$

For any value of x, there corresponds exactly one value of y. Thus y is a function of x.

(b) $x - y^2 - 5 = 0$

$$-y^2 = -x + 5$$
$$y^2 = x - 5$$
$$y = \pm\sqrt{x - 5}$$

If, for example, $x = 9$, then $y = 2$ or $y = -2$. Because this x-value corresponds to two different y-values, y is not a function of x.

We have seen that the ordered pairs generated by the function $y = x + 4$ have the form $(x, x + 4)$. These points can be plotted to create the **graph of the function.**

Note that the graph of the function $y = x + 4$ is the same as the graph of the expression $x + 4$. In general, if a function is written in the form $y =$ expression, then producing the graph of the function is the same as producing the graph of the expression.

A visual approach to determining whether a relation is a function is to inspect the graph of the relation.

DEVELOPING THE CONCEPT

The Vertical Line Test

Consider the graphs in Figure 2.3.

FIGURE **2.3**

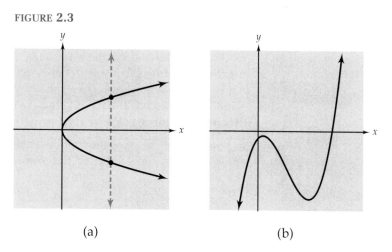

(a)   (b)

In Figure 2.3(a), a vertical line has been drawn that intersects the graph at two points. Because the points of a vertical line have the same first coordinate, this relation does not satisfy the definition of a function.

In Figure 2.3(b), it would be impossible to draw a vertical line that intersects the graph in more than one point. Thus the graph represents a function.

This visual method for determining whether a relation is a function is known as the **Vertical Line Test.**

The Vertical Line Test

If any vertical line can be drawn that intersects the graph of a relation at more than one point, then the graph does not represent a function. If no such line can be drawn, then the graph does represent a function.

Function Notation and Evaluation

Consider the function $f = \{(x, y) \mid y = 3x - 5\}$. For this function, x is called the **independent variable,** and y is called the **dependent variable** because the value of y depends on the value of x.

■ *NOTE   The functions $f(x) = 3x - 5$, $g(t) = 3t - 5$, and $h(z) = 3z - 5$ would all generate the same set of ordered pairs. Therefore, functions f, g, and h are regarded as equivalent functions.*

A more convenient way to describe function f is with **function notation:** $f(x) = 3x - 5$. The symbol $f(x)$ is read "*f* of *x*" (*not* "*f* times *x*"). The letter *f* is simply the name that we initially gave to the function. As is true for any set, a function can be named with any letter or symbol. The *x* in parentheses indicates that the independent variable in the expression is *x*.

Function notation is especially convenient for **evaluating a function.** If $f(x) = x^2 + 3x - 2$, then $f(5)$ is the value of the expression when x is replaced with 5.

$$f(x) = x^2 + 3x - 2$$
$$f(5) = (5)^2 + 3(5) - 2 = 25 + 15 - 2 = 38$$

Evaluating a function is the same as evaluating the expression by which the function is defined. Thus, previously discussed calculator methods for evaluating expressions apply to evaluating functions. We can use the home screen, the Y screen, a table of values (with either *Auto* or *Ask*), or a graph to evaluate a function.

In addition, some calculator models allow you to use function notation for entering a functional value on the home screen. [See Example 4(c).]

EXAMPLE **4**   **Evaluating Functions**

Use both substitution and a calculator method to evaluate the given functions as indicated.

(a) $f(x) = 3x + 5$; $f(4)$
(b) $g(x) = |2x - 1| + 5x$; $g(-3)$
(c) $h(t) = -t^2 + t$; $h(-2)$

SOLUTION

(a) $f(x) = 3x + 5$

   $f(4) = 3(4) + 5 = 17$ Replace x with 4.

   In Figure 2.4(a), we store 4 in X and enter the expression on the home screen. The returned value, 17, is $f(4)$.

(b) $g(x) = |2x - 1| + 5x$

   $g(-3) = |2(-3) - 1| + 5(-3)$ Replace x with -3.

   $= |-7| - 15$

   $= -8$

   After entering $|2x - 1| + 5x$ as Y_1 on the Y screen, we store -3 in X. [See Figure 2.4(b).] Then, by obtaining the value of Y_1, we learn that $g(-3) = -8$.

(c) $h(t) = -t^2 + t$

■ *NOTE   For the calculator evaluation, we use $h(x) = -x^2 + x$ because x is easier to enter than t. Remember that $h(t) = -t^2 + t$ and $h(x) = -x^2 + x$ are equivalent functions.*

   $h(-2) = -(-2)^2 + (-2) = -4 + (-2) = -6$ Replace t with -2.

   For convenience, we use x rather than t for the independent variable. We enter $-x^2 + x$ as Y_1 on the Y screen. Figure 2.4(c) shows that when $x = -2$, the value of the function is -6.

FIGURE 2.4

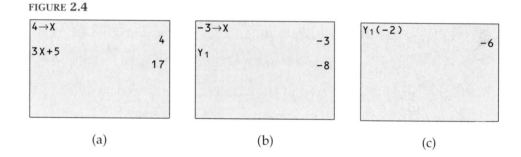

(a)   (b)   (c)

Each point of the graph of a function has a y-coordinate that is the value of the function for the corresponding x-value. Thus, by tracing the graph, we can estimate the value of a function for any displayed value of x, or we can estimate the value of x for which a function has a given value.

EXAMPLE 5   **Tracing the Graph of a Function**

Let $g(x) = 0.02x^3 + 5$. Use the integer setting to produce the graph of g. Then trace the graph to estimate each of the following.

(a) $g(10)$
(b) The value of x for which $g(x) = -15$

SOLUTION

(a) Figure 2.5(a) shows the graph of g with the tracing cursor on $(10, 25)$. The estimated value of $g(10)$ is 25.

FIGURE **2.5**

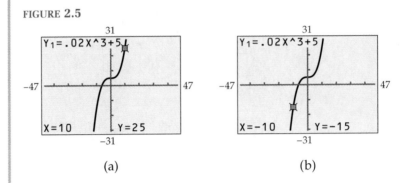

(a)   (b)

■ *NOTE   Recall that approximations of coordinates can be improved by zooming in on the graph at the point of interest.*

(b) Tracing to the point whose y-coordinate is -15 [see Figure 2.5(b)], we see that the value of x for which $g(x) = -15$ is -10.

We can also evaluate a function for a variable or an expression.

EXAMPLE **6**   **Evaluating Functions for Variable Expressions**

Let $f(x) = 4x - 7$ and $g(x) = x^2 + 3x$. Evaluate each of the following.

(a) $f(a + 2)$ (b) $g(t) - f(t)$ (c) $\dfrac{f(2 + h) - f(2)}{h}$

SOLUTION

(a) $f(a + 2) = 4(a + 2) - 7$ Replace x with $a + 2$.

$ = 4a + 8 - 7$ Distributive Property

$ = 4a + 1$

(b) $g(t) - f(t) = (t^2 + 3t) - (4t - 7)$ Replace x with t.

$ = t^2 + 3t - 4t + 7$ Simplify.

$ = t^2 - t + 7$

(c) $\dfrac{f(2 + h) - f(2)}{h} = \dfrac{[4(2 + h) - 7] - [4(2) - 7]}{h}$ To evaluate $f(2 + h)$, replace x with $2 + h$.

$\phantom{\dfrac{f(2 + h) - f(2)}{h}} = \dfrac{8 + 4h - 7 - 1}{h}$ Remove grouping symbols.

$\phantom{\dfrac{f(2 + h) - f(2)}{h}} = \dfrac{4h}{h}$ Simplify the numerator.

$\phantom{\dfrac{f(2 + h) - f(2)}{h}} = 4$ Divide out the common factor h.

Domain and Range

Because a function is a particular kind of relation, the definitions of domain and range of a function are the same as those for relations. An unspecified domain of a function is assumed to be the largest subset of real numbers for which the function is defined.

The graph of a function provides a visual way of estimating the domain and range. However, most graphs extend beyond the viewing window, so you must make sure that your window settings are such that you can see the important features of the graph.

EXAMPLE **7**   **Domains and Ranges**

Use graphing and algebraic methods to determine the domain and range of each of the given functions.

(a) $f(x) = 3\sqrt{12 - x}$ (b) $g(x) = x^4 - 3x^3 + 2x - 1$

SOLUTION

(a) VISUALIZING THE DOMAIN AND RANGE

Figure 2.6 shows the graph of $f(x) = 3\sqrt{12 - x}$ with the tracing cursor on the right endpoint of the graph.

The domain consists of the x-coordinates of all the points of the graph. The largest x-coordinate is 12 at the endpoint, and the graph appears to extend to the left forever. Thus the estimated domain is $(-\infty, 12]$.

The range of the function consists of the y-coordinates of the points of the graph. The smallest y-coordinate is 0 at the endpoint, and the graph appears to rise forever. The estimated range is $[0, \infty)$.

DETERMINING THE DOMAIN AND RANGE

The function $f(x) = 3\sqrt{12 - x}$ is defined only for those values of x for which the radicand $12 - x$ is nonnegative.

$$12 - x \geq 0$$
$$-x \geq -12$$
$$x \leq 12$$

The domain is $(-\infty, 12]$.

Because $\sqrt{12 - x}$ is nonnegative, $f(x) \geq 0$. Thus the range is $[0, \infty)$.

(b) VISUALIZING THE DOMAIN AND RANGE

Figure 2.7 shows the customized window for viewing the important features of the graph of function g. The graph extends to the left and right without end, so the domain appears to be **R**.

The tracing cursor is on the lowest point of the graph, where the y-coordinate is approximately -5.1, and the graph extends upward without end. The estimated range is $[-5.1, \infty)$.

FIGURE **2.6**

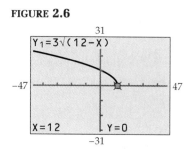

FIGURE **2.7**

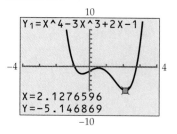

DETERMINING THE DOMAIN AND RANGE

For the function $g(x) = x^4 - 3x^3 + 2x - 1$, we can replace x with any real number. Thus the domain is **R**.

Determining the y-coordinate of the lowest point of the graph requires methods that are beyond the scope of this course. For now, we will be satisfied with our graphing estimate of the range.

When a function is described by an expression, the best method for determining the domain is to look for values of the independent variable that are *not* permissible.

EXAMPLE 8   Determining Domains

Determine the domain of each function.

(a) $g(x) = \dfrac{1}{x - 3}$ (b) $h(x) = \dfrac{x}{x^2 + 2}$

SOLUTION

■ *NOTE   In Example 8(a), we used set-builder notation to write the domain, whereas we used interval notation in Example 7(a). We will use the notation that seems most convenient for describing the set.*

(a) Replacing x with 3 would produce the undefined expression $\frac{1}{0}$. Any other replacement for x would be permissible. The domain of g is $\{x \mid x \neq 3\}$
(b) Because the denominator $x^2 + 2$ can never be 0, there are no restrictions on the value of x. The domain of h is **R**.

Piecewise Functions

Sometimes the rule that defines a function involves more than one expression. For example, recall that $|a| = a$ if $a \geq 0$, and $|a| = -a$ if $a < 0$. Therefore, if $f(x) = |x|$, we can write the following rule.

$$f(x) = \begin{cases} x & \text{if } x \geq 0 \\ -x & \text{if } x < 0 \end{cases}$$

Such functions are **piecewise-defined,** or simply **piecewise** functions.

> **Piecewise**   To graph a piecewise function with your calculator, you must enter each rule along with the domain for that rule. The *Dot* mode is often the best way to display the graph.

EXAMPLE 9   Evaluating a Piecewise Function

Suppose that $G(x)$ is a piecewise function defined as follows:

$$G(x) = \begin{cases} x + 1 & x \leq 0 \\ 2 - x & x > 0 \end{cases}$$

Evaluate function G as indicated.

(a) $G(-4)$ (b) $G(9)$

FIGURE 2.8

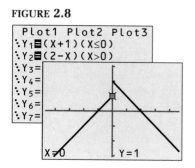

SOLUTION

VISUALIZING THE FUNCTION

Figure 2.8 shows how function G is entered on the Y screen and the graph of the function.

Observe that the tracing cursor is on $(0, 1)$. When $x = 0$, we use the rule $G(x) = x + 1$, and $G(0) = 1$.

DETERMINING VALUES OF THE FUNCTION

The value of the variable determines which rule is used to evaluate the function.

(a) For $x = -4$, we use the rule $G(x) = x + 1$. Then $G(-4) = -4 + 1 = -3$.
(b) For $x = 9$, we use the rule $G(x) = 2 - x$. Then $G(9) = 2 - 9 = -7$.

The values in parts (a) and (b) can also be found by tracing the graph in Figure 2.8 or by referring to a table of values.

Applications

Functions play an important role in modeling the conditions of a problem. To solve such problems, we often need to take the domain of the function into account. Evaluating the function is a typical step in generating the data to be analyzed.

EXAMPLE **10**   **The Length of a Meeting**

A time management consultant finds that the length L of a meeting (in minutes) can be modeled by the function $L(n) = 10(n^2 - n)$, where n is the number of people (up to 5) attending the meeting.

(a) For this situation, what is the domain of L?
(b) Create a table with the headings Number of People and Length of Meeting. Then complete the table by evaluating L.
(c) Although 1 is not a meaningful domain element in this situation, evaluate and interpret $L(1)$.

SOLUTION

(a) The domain consists of the permissible replacements for n, which represents the number of people attending the meeting. The domain is $\{2, 3, 4, 5\}$.

(b)

Number of People n	Length of Meeting (in minutes) $L(n) = 10(n^2 - n)$
2	20
3	60
4	120
5	200

(c) $L(1) = 10(1^2 - 1) = 0$. The length of a "meeting" with only 1 person attending is 0 minutes.

EXAMPLE 11   **A Savings Plan**

At the start of the year 2000, a person developed a long-range savings plan. For that year, she would save $1000, and for each of the next four years, she would increase her yearly savings by $500. Thereafter, her plan was to maintain the savings level reached in the year 2004.

(a) Letting x represent the number of years since 2000, write a function s that describes the amount saved during any year between 2000 and 2004.

(b) What is the savings level reached in 2004?

(c) Write a piecewise function f that describes the amount saved during any year between 2000 and 2010.

SOLUTION

(a) For any given year, the amount saved is $1000 plus $500 times the number of years since the year 2000.

$$s(x) = 1000 + 500x$$

(b) For 2004, $x = 4$.

$$s(4) = 1000 + 500(4) = 3000$$

(c) For $0 < x < 4$, the rule is function s. Between $x = 5$ (the year 2005) and $x = 10$ (the year 2010), the amount saved remains constant at $3000.

$$f(x) = \begin{cases} 1000 + 500x & 0 \le x \le 4 \\ 3000 & 5 \le x \le 10 \end{cases}$$

2.1 Quick Reference

Relations
■ A **relation** is a set of ordered pairs.

■ The **domain** of a relation is the set of all first coordinates of the ordered pairs. The **range** is the set of all second coordinates.

Functions
■ A **function** is a relation in which no two different ordered pairs have the same first coordinate.

■ A **mapping diagram** is a visual way of showing the association of each member of a domain with the corresponding members of the range. A mapping diagram can be used to judge whether a relation is a function.

■ Alternative Definition of a Function
A function is a rule or correspondence that assigns each element x in the domain to a unique element y in the range.

■ The graph of a function $f(x)$ consists of points whose first coordinates are values of x and whose second coordinates are the value of the function at x.

■ The Vertical Line Test
If any vertical line can be drawn that intersects the graph of a relation at more than one point, then the graph does not represent a function. If no such line can be drawn, then the graph does represent a function.

Function Notation and Evaluation

■ If a function is a set of ordered pairs (x, y), then x is the **independent variable** and y is the **dependent variable.**

■ In the **function notation** $f(x)$, f is the name of the function and x is the independent variable.

■ To **evaluate a function** $f(x)$ for some value a, calculate $f(a)$; that is, replace x with a in the expression that describes the function.

■ Evaluating a function is the same as evaluating the expression by which the function is defined. In addition to substitution, several calculator methods are available for evaluating a function.

■ A function can also be evaluated for a variable or an expression.

Domain and Range

■ By tracing a calculator graph of a function, we can estimate the domain and range.

■ When a function is described by an expression, the best method for determining the domain is to look for values of the independent variable that are *not* permissible.

Piecewise Functions

■ A **piecewise function** is a function that is described by different rules for different intervals of the domain.

2.1 Speaking the Language

1. A(n) ▨▨▨▨ is any set of ordered pairs, whereas a(n) ▨▨▨▨ is a set of ordered pairs in which no two different ordered pairs have the same first coordinate.

2. For the function $f(x) = 2x + 1$, we call x the ▨▨▨▨ variable.

3. For the function $g(x) = \sqrt{x}$, the variable x can be replaced with any non-negative real number. Thus we say that the ▨▨▨▨ of g is $\{x \mid x \geq 0\}$.

4. A function such as $f(x) = \begin{cases} x^2 & x < 0 \\ 2x & x \geq 0 \end{cases}$ is called a(n) ▨▨▨▨ function.

2.1 Exercises

Concepts and Skills

1. What is the difference between the domain and the range of a relation?

2. Why is it true that every function is a relation but not every relation is a function?

In Exercises 3–6, give the domain and range of the relation. Identify each relation that is a function.

3. $\{(-4, -1), (0, -1), (3, 5), (5, 6)\}$

4. $\{(0, 2), (2, 0), (0, -2)\}$

5. $\{(x, y) \mid y \text{ is a square root of } x, x = 0, 1, 4, 9\}$

6. $\{(x, y) \mid y \text{ is the square of } x, x = 0, 1, 2, 3\}$

7. Why is the Vertical Line Test a valid way to judge whether a graph represents a function?

8. How do the arrows in a mapping diagram indicate whether a relation is a function?

In Exercises 9–16, determine whether y is a function of x. (In some cases, you will need to begin by solving the equation for y.)

9. $x = y + 5$

10. $x^2 + 2 = y$

11. $y^2 = x - 1$

12. $5x - 3y = 15$

13. $y = \sqrt{x^2 - 9}$

14. $x^2 + y^2 = 49$

15. $xy = x^2 y + x + 1$

16. $xy - x = y + 2$

In Exercises 17–24, the graph of a relation is given. Estimate the domain and range. Is the relation a function?

17.

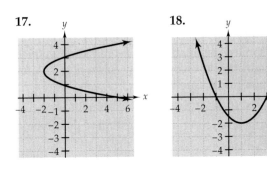

18.

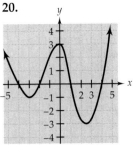

19.

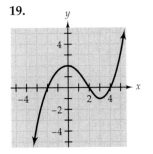

20.

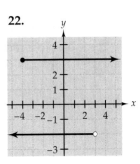

21.

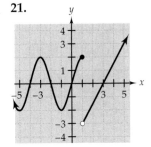

22.

23.

24.

25. Does $y > x$ represent a function? Why or why not?

26. For what integer values of n is $x = y^n$ a function?

In Exercises 27–32, evaluate the given function as indicated.

27. $f(t) = 4t - 7$
    (a) $f(3)$ (b) $f(0)$
    (c) $f\left(\frac{5}{2}\right)$ (d) $f(a + 1)$

28. $h(x) = 4 - 3x$
    (a) $h(6)$ (b) $h(-2)$
    (c) $h(-1)$ (d) $h(x - 2)$

29. $g(y) = -y^2 + 2y + 1$
    (a) $g(-1)$ (b) $g(2)$
    (c) $g(t + 3)$ (d) $g(1.5)$

30. $r(x) = \dfrac{x}{x^2 - 16}$
    (a) $r(0)$ (b) $r(-4)$
    (c) $r(4y)$ (d) $r(0.25)$

31. $h(t) = \sqrt{t^2 - 9}$
    (a) $h(0)$ (b) $h(3)$
    (c) $h(-5)$ (d) $h(3t)$

32. $f(x) = 3 - |x|$
    (a) $f(-3)$ (b) $f(5)$
    (c) $f(2.7)$ (d) $f(-x)$

33. Suppose that you have traced the graph of a function to the point $(2, -6)$. What is the significance of the second coordinate of that point?

34. Compare the graphs of the following functions:

$$f(x) = 2x - 3 \qquad g(x) = x - [2x + 3(1 - x)]$$

What is your conjecture? Show that your conjecture is correct by simplifying the expression for function g.

In Exercises 35–38, use a graph of the function to estimate the value of the function or the value of the variable so that the function has the indicated value. (*Hint*: Use the decimal setting.)

35. $f(x) = \dfrac{3}{4}x + 2$
    (a) $f(-4.3)$ (b) $f(x) = 3.05$
    (c) $f(x) = -0.775$

36. $g(x) = \dfrac{x + 1}{x}$
    (a) $g(-3.6)$ (b) $g(x) = 1.625$
    (c) $g(x) = -1.5$

37. $s(t) = 1 - \dfrac{1}{t}$

   (a) $s(0.4)$ (b) $s(t) = 0.375$
   (c) $s(t) = 2.25$

38. $A(w) = w^3 + w$

   (a) $A(1.1)$ (b) $A(w) = -2.928$
   (c) $A(w) = 0.101$

In Exercises 39–42, use a graph to estimate the value(s) of x so that the function has a value of 0.

39. $f(x) = -x^2 + 6x - 5$

40. $c(x) = \dfrac{1}{3}x^3 - x^2 - 6x$

41. $h(x) = \dfrac{x-3}{x+1}$ **42.** $f(x) = x^2 + 3x - 1$

In Exercises 43–46, evaluate and simplify.

43. $f(x) = 2x + x^2$ and $g(x) = x - 3$

   (a) $f(-4) + g(5)$ (b) $g(0) - f(-1)$
   (c) $f(t) + g(t + 3)$

44. $P(x) = |x + 2|$ and $Q(x) = -x^2$

   (a) $P(3) + Q(3)$ (b) $Q(-1) + P(-3)$
   (c) $P(t^2 - 2) + Q(-t)$

45. $s(t) = \sqrt{t + 3}$ and $r(t) = 3 - |t - 2|$

   (a) $s(6) - r(4)$ (b) $s(-3) + r(7)$
   (c) $s(t^4 - 3) - r(t^2 + 2)$

46. $h(y) = y^3 - y + 1$ and $g(y) = \dfrac{12}{y+1}$

   (a) $g(2) + h(-2)$ (b) $g(4) - h(0)$
   (c) $h(x + 1) + g\left(\dfrac{1}{x} - 1\right)$

In Exercises 47 and 48, $f(x) = 1 - x$ and $g(x) = x^2$. Evaluate and simplify the given expression.

47. $\dfrac{f(-2 + h) - f(-2)}{h}$ **48.** $\dfrac{g(3 + h) - g(3)}{h}$

In Exercises 49–52, evaluate and simplify the expression $\dfrac{f(x + h) - f(x)}{h}$.

49. $f(x) = 4x + 1$ **50.** $f(x) = 5 - 2x$

51. $f(x) = 3x^2 - x$ **52.** $f(x) = \dfrac{4}{x+1}$

In Exercises 53–58, produce the graph of the function. Then use the graph to estimate the domain and range of the function.

53. $f(x) = 2 - \dfrac{3}{2}x$ **54.** $h(x) = 3 - 2x - x^2$

55. $g(x) = 5 - |x + 2|$ **56.** $A(x) = |x + 6| - |x - 5|$

57. $f(x) = -\sqrt{16 - x^2}$ **58.** $P(x) = 3x^2 - x^3 - 5$

In Exercises 59–64, determine the domain of the function.

59. $r(x) = x^2 + x + 3$ **60.** $f(t) = 3 - |t|$

61. $g(x) = \sqrt{x + 4}$ **62.** $h(x) = \sqrt[4]{5 - 2x}$

63. $g(t) = \dfrac{1}{t} - \dfrac{3}{t-1}$ **64.** $r(x) = \dfrac{x+1}{x-5}$

65. Explain why the following is improperly defined.

$$f(x) = \begin{cases} x + 3 & x \le 5 \\ x^2 & x \ge 5 \end{cases}$$

66. (a) Compare the displays of the graph of $f(x) = x^2 - 14x + 24$ with the standard and integer settings. Which is better?
   (b) Which is the better setting for the graph of $f(x) = x^3 - x^2 - 2x$?
   (c) Why is neither of these settings good for displaying the graph of $f(x) = \dfrac{x}{x^2 + 1}$?

In Exercises 67–70, evaluate the given function as indicated.

67. $f(x) = \begin{cases} x + 1 & x < 0 \\ x^2 & x \ge 0 \end{cases}$

   (a) $f(0)$ (b) $f(-1)$
   (c) $f(3)$ (d) $f(-6)$

68. $p(x) = \begin{cases} |x| & x \le 2 \\ 5 & x > 2 \end{cases}$

   (a) $p(2)$ (b) $p(-6)$
   (c) $p(0)$ (d) $p(7)$

69. $g(x) = \begin{cases} x & x < 1 \\ x + 1 & 1 \le x < 4 \\ x + 2 & x \ge 4 \end{cases}$

   (a) $g(-2)$ (b) $g(3)$
   (c) $g(4)$ (d) $g(6)$

70. $h(x) = \begin{cases} x^2 & x \le -1 \\ 4 & -1 < x < 2 \\ x^2 & x \ge 2 \end{cases}$

   (a) $h(-2)$ (b) $h(-1)$
   (c) $h(1)$ (d) $h(3)$

Applications

71. Test Scores   The accompanying graph shows three test scores for each of four students. The data in the graph can be written as a relation R with 12 ordered pairs of the form (student, score). For example, one ordered pair in R is $(1, 75)$.

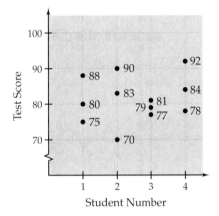

(a) Is R a function? Why or why not?

(b) The *inverse* of R is the set of 12 ordered pairs of the form (score, student). For example, one ordered pair is $(83, 2)$. Is this set a function? Why or why not?

72. Principal Religions   The following table shows the principal religions of six countries. *(Source: Information Please Almanac.)*

Country	Religion
India	Hindu
Brazil	Christian
Iran	Moslem
Mexico	Christian
Pakistan	Moslem
Japan	Shintoist

(a) If you were to draw a mapping diagram of countries to religions, would the diagram indicate a function?

(b) If you were to draw a mapping diagram of religions to countries, would the diagram indicate a function?

When yearly data are given, a convenient technique for modeling the data is to let the variable represent the number of years since some base year. For example, in Exercise 73, we let the variable x represent the number of years since 1976. This means that $x = 0$ for 1976, $x = 1$ for 1977, $x = 2$ for 1978, and so on.

73. Bottled Water   Bottled water has become a popular beverage in the United States, where the average consumption in 1996 was 11 gallons per person. *(Source: Beverage Digest.)* The function $B(x) = \frac{1}{2}x + 1$ models the average annual consumption B (in gallons), where x is the number of years since 1976.

(a) According to the model function, what is the projected average annual consumption of bottled water in the year 2002?

(b) Trace the graph of B to estimate the year when the average annual consumption was 8 gallons.

74. Taxi Fares   For a single-passenger fare, a taxi company charges $0.50 for the first one-tenth of a mile and $0.35 for each additional tenth of a mile. Thus the cost C_1 for one passenger can be modeled by $C_1(m) = 0.50 + 0.35m$, where m represents the number of *tenths* of a mile after the first tenth of a mile. However, the total charge is increased by 10% for each additional passenger.

(a) Write a simplified function C_2 for the cost of a trip for *two* passengers.

(b) Produce the graph of C_2 and trace it to estimate the cost of a 4-mile trip with two passengers.

(c) Determine the cost of a 4-mile trip with two passengers by evaluating $C_2(39)$. Compare the result to your estimate.

75. Custom Windows   A manufacturer of custom windows uses rows and columns of one-unit panes, where the number of rows is always 1 greater than the number of columns. The cost of a window is $45 per unit pane.

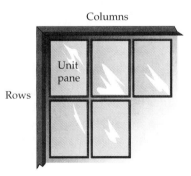

(a) If a window has x columns, how is the number of rows represented? Then what expression represents the total number of one-unit panes?

(b) Write a function C for the cost of a window with x columns.

(c) If a window can have no more than 6 rows, what is the domain of function C?

(d) Evaluate C to determine the total cost of a window with 4 columns.

76. Estate Tax An "estate" consists of all the net assets of a deceased person. The value of an estate includes cash, plus the total value of investments, personal property, life insurance, and so on.

In 1997, if the value of an estate was less than $600,000, the federal government imposed no tax on the estate. For an estate with a value of at least $600,000, the government levied a 50% tax on the difference between the value of the estate and $600,000. For example, if the value of the estate was $900,000, the government collected 50% of ($900,000 − $600,000), or $150,000.

In this example, the $600,000 is called the "estate tax credit." The accompanying bar graph shows the projected increases in the estate tax credit through the year 2006. (*Source*: Internal Revenue Service.)

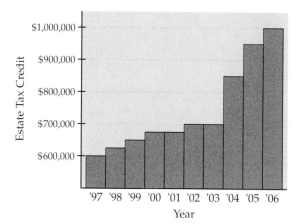

(a) Write a function f for the 1997 tax on an estate with a net worth of x dollars.

(b) Evaluate f to determine the 1997 tax owed on an estate worth $1.3 million.

(c) Referring to the bar graph, write a new function g for the 2006 tax on an estate with a net worth of x dollars.

(d) Write the function $h(x) = f(x) - g(x)$. What does this function represent?

77. Women's Health Research In 1992 approximately $1.1 billion was spent on medical research that was specifically targeted to women's health issues. That amount increased by $0.2 billion each year until 1995. (*Source*: National Institute of Health.)

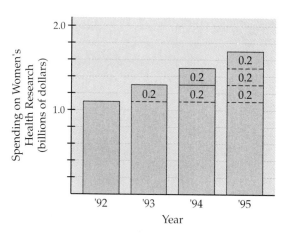

(a) Let n represent the number of years since 1992. Write a function R to describe the amount spent on women's health research during this period.

(b) Use function R to predict the total amount spent for the period 1996–1999.

78. Retired Population In 1994, the percentage of the United States population that was retired was 12%. For the indefinite future, an additional 0.2% of the population is projected to retire each year. (*Source*: TIAA/CREF.)

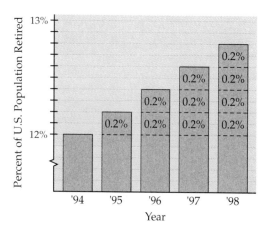

(a) Letting x represent the number of years since 1994, write a function r that describes the projected percentage of the population that is retired.

(b) Use function r to predict the percentage of the population that will be retired in the year 2030.

79. Exported Currency   The bar graph shows the trend in the percentage of U.S. currency in circulation that is held outside the country. (*Source*: Federal Reserve Board.)

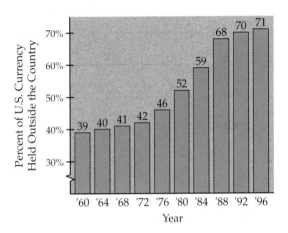

The data in the bar graph can be modeled by the following piecewise function, where x is the number of years since 1960:

$$f(x) = \begin{cases} 0.25x + 39 & 0 \le x \le 12 \\ -0.016x^2 + 2.25x + 14 & 12 < x \le 36 \end{cases}$$

(a) By evaluating f, show that the model function gives the exact percentages for the period 1960–1972.

(b) For which year in the period 1976–1996 does the model function give the most accurate percentage?

80. Steering-Wheel Dining   The accompanying graph depicts the trend in "steering-wheel dining." For the period 1984–1996, the graph shows the average number of meals (per person per year) that were purchased at a restaurant and eaten in a car. (*Source*: The NPD Group.)

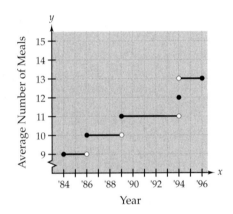

Letting x represent the number of years since 1984, write a piecewise function that models the data in the graph.

2.2   *Linear Functions*

Graphs of Linear Functions   ◼   *Slope of a Line*   ◼   *The Slope-Intercept Model*   ◼   *The Point-Slope Model*   ◼   *Rate of Change*

The earnings of an hourly employee depend on (are a *function* of) the number of hours the employee works. For each additional hour worked, the earnings increase by a constant amount, and the function that models the earnings is called a *linear* function.

Many other real-life situations can be modeled with linear functions. Because the graph of this most basic function is a straight line, we are particularly interested in the *slope* of the line as an indicator of how one variable is constantly changing with respect to the other variable.

In this section, we learn how to write, graph, and analyze linear functions, with particular emphasis on rate of change.

Graphs of Linear Functions

The following functions all have the form $f(x) = ax + b$.

$$f(x) = 3x - 7 \qquad a = 3, b = -7$$
$$g(x) = 2x \qquad a = 2, b = 0$$
$$h(x) = x \qquad a = 1, b = 0$$

Each of these is an example of a **linear function.**

> ### Definition of a Linear Function
> A **linear function** is a function that can be written in the form $f(x) = ax + b$, where a and b are real numbers and $a \neq 0$.

A **linear equation in two variables** is an equation that can be written in the form $Ax + By = C$, where A, B, and C are real numbers. If A and B are not 0, then the equation can be rewritten as follows:

$$Ax + By = C$$
$$By = -Ax + C$$
$$y = -\frac{A}{B}x + \frac{C}{B}$$
$$y = ax + b \qquad \text{Replace } -\frac{A}{B} \text{ with } a \text{ and } \frac{C}{B} \text{ with } b.$$

Because the linear *equation* $y = ax + b$ has the same form as the linear *function* $f(x) = ax + b$, we say that the equation defines a linear function.

Note that $Ax + By = C$ defines a linear function only if A and B are not 0. If $A = 0$, then the equation can be written in the form $y = $ constant, which is called a **constant function,** and its graph is a horizontal line. If $B = 0$, then the equation can be written in the form $x = $ constant. Its graph is a vertical line, which does not represent a function.

By producing the graphs of $y = ax + b$ for various values of a and b, we observe that the graphs are straight lines. (See Figure 2.9.)

FIGURE 2.9

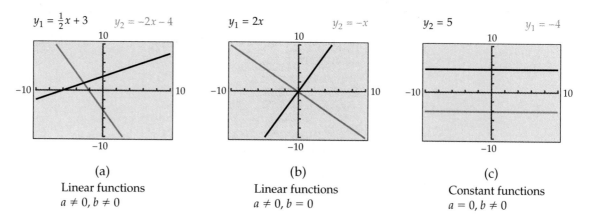

(a)
Linear functions
$a \neq 0, b \neq 0$

(b)
Linear functions
$a \neq 0, b = 0$

(c)
Constant functions
$a = 0, b \neq 0$

We can use Figure 2.9 to visualize the domain and range of the functions. In all cases, the graphs extend forever to the left and right, so the domain is **R.**

For the linear functions, the graphs rise and fall forever, so the range is also **R.** For the constant functions, the ranges are {−4} and {5}, respectively.

In general, for a linear function $f(x) = ax + b$, x can be any real number, so the domain is **R.** The value of the function can also be any real number, so the range is **R.**

For a constant function $f(x) = b$, x can be any real number, and the value of the function is always b. Thus the domain is **R,** and the range is {b}.

In the remainder of this section, we consider some basic properties of a straight-line graph.

Slope of a Line

One visually apparent property of a line is its steepness. The graph of $f(x) = ax + b$ may rise or fall rapidly or gradually. We now investigate how the steepness of a line can be measured more precisely.

DEVELOPING THE CONCEPT

Measuring the Steepness of a Line

Consider the function $y = \frac{3}{2}x$. Figure 2.10(a) shows a calculator table of selected x- and y-values that satisfy the equation. Observe that y increases by 3 units each time x increases by 2 units.

FIGURE 2.10

X	Y₁
−6	−9
−4	−6
−2	−3
0	0
2	3
4	6
6	9

$Y_1 = (3/2)X$

(a)

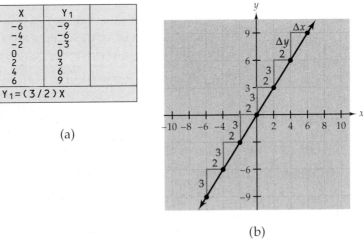

(b)

In Figure 2.10(b), we see these changes in x and y on the graph of the equation. In general, the change in x is called the *run* and is represented by Δx (read "delta x"); the change in y is called the *rise* and is represented by Δy (read "delta y").

A comparison (ratio) of the rise to the run is the basis for assigning a numerical measure to the steepness of a line. We call this number the **slope** of the line.

Figure 2.11 shows a line with two general points, $P(x_1, y_1)$ and $Q(x_2, y_2)$. Point $R(x_1, y_2)$ is plotted so that the *slope triangle PRQ* is a right triangle.

FIGURE 2.11

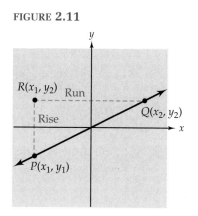

The rise PR is the difference of the y-coordinates, and the run QR is the difference of the x-coordinates.

$$\text{Rise} = y_2 - y_1 = \Delta y \qquad \text{Run} = x_2 - x_1 = \Delta x$$

We define the slope m of the line as the ratio of the rise to the run.

> ### *Definition of Slope*
> The **slope** m of a line containing $P(x_1, y_1)$ and $Q(x_2, y_2)$ is defined as follows:
> $$m = \frac{\Delta y}{\Delta x} = \frac{y_2 - y_1}{x_2 - x_1} \qquad x_1 \neq x_2$$

These important observations follow from the definition:

1. The definition does not apply if $x_1 = x_2$ — that is, if P and Q are points of a vertical line. Thus the slope of a vertical line is not defined.
2. The order in which Δy and Δx are calculated is unimportant because $\dfrac{y_2 - y_1}{x_2 - x_1} = \dfrac{y_1 - y_2}{x_1 - x_2}$. However, the order must be the same for both calculations.
3. The definition is applicable to any two points of the line.

EXAMPLE 1   Calculating Slope with the Slope Formula

Determine the slope of the line that contains each pair of given points.

(a) $P(-3, -5), Q(7, 2)$ (b) $A(-4, 1), B(2, -8)$
(c) $C(2, 4), D(-8, 4)$ (d) $R(3, -7), S(3, 5)$

SOLUTION

(a) $m = \dfrac{\Delta y}{\Delta x} = \dfrac{y_2 - y_1}{x_2 - x_1} = \dfrac{2 - (-5)}{7 - (-3)} = \dfrac{7}{10}$ [See Figure 2.12(a).]

(b) $m = \dfrac{1 - (-8)}{-4 - 2} = \dfrac{9}{-6} = -\dfrac{3}{2}$ [See Figure 2.12(a).]

(c) $m = \dfrac{4 - 4}{-8 - 2} = \dfrac{0}{-10} = 0$ [See Figure 2.12(b).]

(d) Because $x_1 = x_2$, the slope is undefined. [See Figure 2.12(b).]

FIGURE 2.12

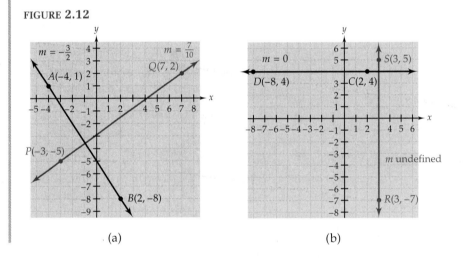

(a)   (b)

The results in Example 1 suggest the following general conclusions:

1. If $m > 0$, the line rises from left to right.
2. If $m < 0$, the line falls from left to right.
3. If $m = 0$, the line is horizontal.
4. If m is undefined, the line is vertical.

We now investigate the relationship between the slope of the line $y = ax + b$ and the number a.

EXPLORATION

The Role of a in y = ax + b

Consider several functions of the form $y = ax + b$.

(a) For each function that you select, find two solutions of the equation.
(b) Use the solutions to determine the slope of the line.
(c) Compare the slope to the value of a that you chose for your function.
(d) For $y = ax + b$, what general relationship between the slope of the line and the value of a do you suspect?

DEVELOPING THE CONCEPT

The Role of a in y = ax + b

The following linear equations are in the form $y = ax + b$:

$$y_1 = \frac{1}{2}x \qquad y_2 = 2x \qquad y_3 = -\frac{2}{3}x \qquad y_4 = -x$$

Note that $(0, 0)$ is a solution of each equation. We determine another solution and use it with $(0, 0)$ to calculate the slope of each line.

Equation	Point	Slope	a
$y_1 = \frac{1}{2}x$	$(8, 4)$	$m = \frac{4 - 0}{8 - 0} = \frac{1}{2}$	$\frac{1}{2}$
$y_2 = 2x$	$(3, 6)$	$m = \frac{6 - 0}{3 - 0} = 2$	2
$y_3 = -\frac{2}{3}x$	$(9, -6)$	$m = \frac{-6 - 0}{9 - 0} = -\frac{2}{3}$	$-\frac{2}{3}$
$y_4 = -1x$	$(10, -10)$	$m = \frac{-10 - 0}{10 - 0} = -1$	-1

In each case, the slope m is the same as the coefficient a.

The conclusion is true for any linear function. Because the coefficient of x is the same as the slope, we usually write a linear function as $f(x) = mx + b$ and a linear equation as $y = mx + b$.

The Slope-Intercept Model

In Section 1.1, we defined a point of intersection of a graph and an axis as an *intercept,* and we were able to trace a graph to estimate the coordinates of x- and y-intercepts.

Because x- and y-intercepts have the forms $(a, 0)$ and $(0, b)$, respectively, we can determine intercepts algebraically. The y-intercept is the point whose x-coordinate is 0, so we replace x with 0 and solve for y. Similarly, the x-intercept is the point whose y-coordinate is 0, so we replace y with 0 and solve for x.

EXAMPLE 2 **Determining Intercepts Algebraically**

Determine the intercepts of $3x - 4y = 12$.

SOLUTION

x-intercept

$$3x - 4y = 12$$
$$3x - 4(0) = 12 \quad \text{Replace } y \text{ with 0.}$$
$$3x = 12$$
$$x = 4$$

The *x*-intercept is (4, 0).

y-intercept

$$3x - 4y = 12$$
$$3(0) - 4y = 12 \quad \text{Replace } x \text{ with 0.}$$
$$-4y = 12$$
$$y = -3$$

The *y*-intercept is (0, −3).

For $f(x) = mx + b$, we have seen that the coefficient of x is the slope of the line. We now investigate how the constant term b affects the line.

EXPLORATION

The Role of b in y = mx + b

Consider functions of the form $y = -2x + b$.

(a) Choose several values of b, graph the functions, and trace to estimate the y-intercept.
(b) What relationship do you observe between the y-intercept and the value of b?
(c) Repeat the experiment for functions of the form $y = x + b$ and $y = -\dfrac{3}{2}x + b$. Does your conjecture in part (b) appear to be true for these functions?

DEVELOPING THE CONCEPT

The Role of b in y = mx + b

Figure 2.13 shows the graphs of $y = \frac{1}{2}x + b$ for $b = -5, -1, 2,$ and 6.

FIGURE 2.13

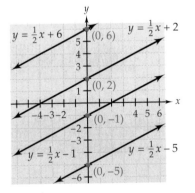

The figure suggests that the constant term b is the second coordinate of the y-intercept of the line. We can show that this is true by using the fact that the first coordinate of a y-intercept is 0.

$$f(x) = mx + b$$
$$f(0) = m(0) + b \qquad \text{For the } y\text{-intercept, } x = 0.$$
$$\qquad = b$$

Thus the y-intercept of the graph of $f(x) = mx + b$ is $(0, b)$.

Because the numbers m and b describe specific properties of a line, we give a special name to the equation $y = mx + b$.

> ### The Slope-Intercept Model of an Equation of a Line
>
> The **slope-intercept model** of an equation of a line is $y = mx + b$, where m is the slope of the line and $(0, b)$ is the y-intercept.

One advantage of the slope-intercept model is that the slope and y-intercept of a line can be determined directly from the equation.

EXAMPLE **3**   **Determining Slope and *y*-Intercept from an Equation**

Determine the slope and the y-intercept of each line whose equation is given.

(a) $y = \dfrac{3}{5}x + 4$

(b) $2x + 3y = 6$

SOLUTION

(a) The equation is in the slope-intercept form, where $m = \frac{3}{5}$ and $b = 4$. Thus the slope is $\frac{3}{5}$, and the y-intercept is $(0, 4)$.

(b) $2x + 3y = 6$ \qquad Write the equation in the slope-intercept form.

$\qquad 3y = -2x + 6$ \qquad Subtract $2x$ from both sides.

$\qquad\quad y = -\dfrac{2}{3}x + 2$ \qquad Divide both sides by 3.

The slope is $-\dfrac{2}{3}$, and the y-intercept is $(0, 2)$.

If we know the slope and one point of a line, we can sketch the graph of the line.

EXAMPLE 4   **Sketching Lines with Slopes and Points Given**

Draw the line that contains $Q(-4, 2)$ and whose slope is $-\frac{3}{2}$.

SOLUTION

We plot point $Q(-4, 2)$, and we start at this point to construct the slope triangle. (See Figure 2.14.) We can write $m = \frac{-3}{2}$ or $m = \frac{3}{-2}$. Therefore, we can move down 3 units and right 2 units, or we can move up 3 units and left 2 units. Both destination points are points of the line.

FIGURE 2.14

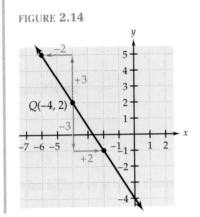

If we can write a given equation in the slope-intercept form $y = mx + b$, then we can use m and b to draw the line. We can also reverse the process. If we are given (or can determine) the slope and one point of a line, we can use the slope-intercept model for writing the equation of the line.

EXAMPLE 5   **Using the Slope-Intercept Model to Write an Equation**

Write an equation of the described line.

(a) The slope is $\frac{2}{3}$ and the y-intercept is $(0, 5)$.
(b) The slope is -1 and the line contains $Q(-4, 8)$.

SOLUTION

(a) $y = mx + b$ Use the slope-intercept model.

$y = \frac{2}{3}x + 5$ $m = \frac{2}{3}$ and $b = 5$.

(b) $y = mx + b$
$y = -1x + b$ $m = -1$
$8 = -1(-4) + b$ Because $Q(-4, 8)$ is a point of the line, the equation is true when $x = -4$ and $y = 8$.
$8 = 4 + b$ Solve for b.
$4 = b$
$y = -x + 4$ $m = -1$ and $b = 4$.

The Point-Slope Model

FIGURE **2.15**

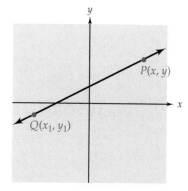

When the slope and y-intercept of a line are given, the slope-intercept model is ideal for writing the equation of the line. If a point other than the y-intercept is given, another approach can be used.

Figure 2.15 shows a line that has a known slope of m and that contains a known point $Q(x_1, y_1)$. The point $P(x, y)$ is some other point of the line.

We use points P and Q to calculate the slope of the line, and then we clear the fraction.

$$\frac{y - y_1}{x - x_1} = m \qquad \text{The LCD is } x - x_1.$$

$$y - y_1 = m(x - x_1) \qquad \text{Multiply both sides of the equation by } x - x_1.$$

The result is called the **point-slope model** of the equation of a line.

The Point-Slope Model of an Equation of a Line

The **point-slope model** of an equation of a line with slope m and containing the point (x_1, y_1) is

$$y - y_1 = m(x - x_1)$$

The point-slope model is especially convenient when the slope and a point other than the y-intercept are given. However, because the slope-intercept model is more useful for graphing, we usually write equations in that form.

EXAMPLE **6**   **Using the Point-Slope Model to Write an Equation**

Write an equation of the line that contains $P(-5, -2)$ and $Q(6, 0)$.

SOLUTION

$$m = \frac{-2 - 0}{-5 - 6} = \frac{-2}{-11} = \frac{2}{11} \qquad \text{Begin by calculating the slope.}$$

$$y - y_1 = m(x - x_1) \qquad \text{Use the point-slope form as a model.}$$

$$y - 0 = \frac{2}{11}(x - 6) \qquad m = \tfrac{2}{11}, x_1 = 6, y_1 = 0.$$

$$y = \frac{2}{11}x - \frac{12}{11} \qquad \text{The equation in slope-intercept form}$$

■ *NOTE   In Example 6, point Q(6, 0) was used as the known point (x_1, y_1). Verify that the same result would have been obtained if we had used P(−5, −2).*

To write the equation of a horizontal line, either the slope-intercept model or the point-slope model can be used. In either case, a model for the equation of a horizontal line is $y = b$, where the constant b is the second coordinate of the y-intercept $(0, b)$. [See Figure 2.16(a).]

FIGURE 2.16

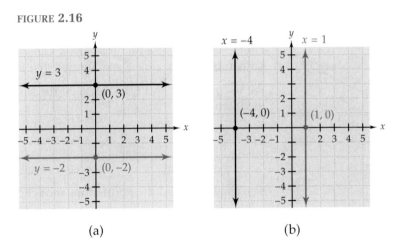

(a)   (b)

Because the slope of a vertical line is not defined, neither the slope-intercept model nor the point-slope model can be used. However, we know that the x-intercept of a vertical line has the form $(a, 0)$ and that every point of the line has a first coordinate a. [See Figure 2.16(b).] Therefore, a model for the equation of a vertical line is $x = a$.

Rate of Change

For a linear function, slope is a measure of the rate at which the value of the function changes for specified changes in the value of the independent variable.

EXAMPLE 7 **Slope as Rate of Change**

FIGURE 2.17

X	Y₁
0	12000
1	12600
2	13200
3	13800
4	14400
5	15000
6	15600

Y₁=15600

Figure 2.17 shows a table of x- and y-values for a certain linear function. Observe that each unit increase in x produces an increase of 600 in y. Also, when $x = 0$, $y = 12{,}000$. What linear function models these data?

SOLUTION

Because the rate of change is $\dfrac{\Delta y}{\Delta x} = \dfrac{600}{1} = 600$, $m = 600$.

The y-intercept of the graph is $(0, 12{,}000)$. Therefore, $f(x) = 600x + 12{,}000$.

The data in Example 7 might represent the population growth of a small town. The population is 12,000 in year $x = 0$, and it increases at an average rate of 600 in subsequent years. To project the population for any year, we simply evaluate the model function for the given value of x.

Real-data problems often involve the rate at which data are changing. If the data can be modeled with a linear function, then the slope of the graph represents the rate of change.

EXAMPLE 8   **Teachers' Salaries**

The average salary for classroom teachers rose from $31,200 in 1990 to $38,700 in 1996. (*Source*: Educational Research Service.)

(a) Letting x represent the number of years since 1990, write a linear function to model the average salary.
(b) What is the significance of the slope and y-intercept of the graph of the function?

SOLUTION

(a) If we write the ordered pairs in the form (x, y), where x is the number of years since 1990 and y is the average salary, two data points are $(0, 31{,}200)$ and $(6, 38{,}700)$. The slope is

$$m = \frac{38{,}700 - 31{,}200}{6 - 0} = \frac{7500}{6} = 1250$$

Because the first coordinate is 0, $(0, 31{,}200)$ is the y-intercept. We use the slope-intercept form to write the equation of the line.

$$y = mx + b \qquad \text{Slope-intercept form of a line.}$$
$$y = 1250x + 31{,}200 \qquad m = 1250 \text{ and } b = 31{,}200$$

(b) The y-intercept corresponds to the average salary ($31,200) in 1990, and the slope is the average yearly increase, or rate of change, in salary ($1250) from 1990 to 1996.

2.2 Quick Reference

Graphs of Linear Functions
■ A **linear function** is a function that can be written in the form $f(x) = ax + b$, where a and b are real numbers and $a \neq 0$. The domain and range are both **R**. The graph is a line.

■ A **constant function** has the form $f(x) = b$. The domain is **R,** and the range is $\{b\}$.

Slope of a Line
■ The **slope** m of a line containing $P(x_1, y_1)$ and $Q(x_2, y_2)$ is $m = \dfrac{y_2 - y_1}{x_2 - x_1}$, $x_1 \neq x_2$.

■ If $m > 0$, the line rises from left to right;
if $m < 0$, the line falls from left to right;
if $m = 0$, the line is horizontal;
if m is undefined, the line is vertical.

■ For a linear function $f(x) = ax + b$, the coefficient of x is the slope.

The Slope-Intercept Model
■ To determine an x-intercept of a graph, replace y with 0 and solve for x. To determine a y-intercept, replace x with 0 and solve for y.

■ For a linear function $f(x) = ax + b$, the constant term b is the second coordinate of the y-intercept.

■ The **slope-intercept model** of an equation of a line is $y = mx + b$, where m is the slope of the line and $(0, b)$ is the y-intercept.

The Point-Slope Model ■	The **point-slope model** of an equation of a line with slope m and containing the point (x_1, y_1) is $y - y_1 = m(x - x_1)$.
■	The model for the equation of a horizontal line is $y = b$; the model for the equation of a vertical line is $x = a$.
Rate of Change ■	Rate of change refers to the rate at which the value of a function changes for a specified change in the value of the independent variable.
■	For a linear function, the rate of change is the slope of the graph.

2.2 Speaking the Language

1. A function of the form $f(x) = ax + b$, where $a \neq 0$, is called a(n) ▨▨▨▨▨ function, whereas a function of the form $f(x) = b$ is called a(n) ▨▨▨▨▨ function.

2. A numerical measure of the steepness of a line is called the ▨▨▨▨▨ of the line.

3. The equation $y = mx + b$ is called the ▨▨▨▨▨ model of the equation of a line, whereas the equation $y = y_1 = m(x - x_1)$ is called the ▨▨▨▨▨ model.

4. For a linear function, the ▨▨▨▨▨ is constant and is equal to the slope of the graph of the function.

2.2 Exercises

Concepts and Skills

1. Suppose that P and Q are points of a line. What do you know about the slope of the line if
   (a) the first coordinates of P and Q are the same?
   (b) the second coordinates of P and Q are the same?

2. The graph of a constant function $f(x) = b$ is a line. What is the slope of the line? Why?

In Exercises 3–8, determine the slope of a line containing the given points.

3. $(4, -1), (-2, 2)$ 4. $(-3, -1), (-1, 3)$

5. $(-5, 3), (-2, 3)$ 6. $(1, 4), (1, -3)$

7. $(0, -6), (3, -1)$ 8. $(4, -1), (7, -6)$

In Exercises 9–12, determine the intercepts of the line.

9. $y + 9 - 2x = 0$ 10. $x = 6 - y$

11. $4x - y = 12$ 12. $2y - 5x - 10 = 0$

In Exercises 13–18, write a function $f(x) = mx + b$ that is defined by the equation. Determine the slope and y-intercept of the graph of the equation.

13. $0 = x - y - 5$ 14. $8x = 4y + 7$

15. $4x + 3y = 3$ 16. $10y + 25x = 2$

17. $1 - 2y = 0$ 18. $2y + 3 = -6$

In Exercises 19 and 20, use the same coordinate system to sketch the graphs of the lines containing the given point P with the given slope. Give the coordinates of two other points of each line.

19. $P(3, -2)$ (a) $m = 2$ (b) $m = -\dfrac{3}{2}$
    (c) $m = 0$ (d) Undefined slope

20. $P(-4, 3)$ (a) $m = 0$ (b) $m = \dfrac{5}{2}$
    (c) Undefined slope
    (d) $m = -1$

21. Compare the ranges of a linear function and a constant function.

22. Why is neither the slope-intercept model nor the point-slope model applicable in writing the equation of a vertical line? What model would you use instead?

In Exercises 23–26, write an equation of the line containing the given point and having the given slope.

23. $(0, -7)$, slope 3   **24.** $(6, -1)$, slope $-\dfrac{3}{4}$

25. $(-6, 0)$, slope undefined

26. $(-3, 7)$, slope 0

In Exercises 27–30, write a linear function f that satisfies the given conditions. [*Hint*: If $f(a) = b$, then (a, b) is a point of the line.]

27. $f(4) = 3$, slope $\dfrac{1}{2}$

28. $f(3) = -2$, slope -1

29. $f(0) = -3$, slope 0

30. $f(-2) = 1$, slope $-\dfrac{5}{3}$

In Exercises 31–34, write an equation of the line containing the given points.

31. $(-4, 3), (1, 0)$ **32.** $(5, 0), (0, -4)$

33. $(-2, 7), (6, 7)$ **34.** $(4, -2), (4, 1)$

In Exercises 35–38, write a linear function f that satisfies the given conditions. [*Hint*: If $f(a) = b$, then (a, b) is a point of the line.]

35. $f(2) = -5, f(-1) = 1$ **36.** $f(0) = 0, f(3) = 7$

37. $f(1.5) = -3, f(0.5) = -3$

38. $f(1.2) = 7, f(-3.5) = 7$

Concept Extension

39. Points are **collinear** if they are points of the same line.

     (a) Plot $P(-9, -5)$, $Q(0, 0)$, and $R(8, 4)$. Do the points appear to be collinear?

     (b) By calculating slopes, determine whether the points are collinear.

40. Another model for a linear equation in two variables is $\dfrac{x}{a} + \dfrac{y}{b} = 1$. What properties of the graph of the equation are represented by a and b?

In Exercises 41–44, determine c such that the line containing the points has the given slope.

41. $(3, -1), (7, c)$; $m = 1$

42. $(-4, 10), (c, 1)$; $m = -\dfrac{3}{2}$

43. $(c, 6), (-4, 1)$; m is undefined

44. $(2, c), (-6, 4)$; $m = 0$

45. If g is a linear function, what is the relationship between the rate of change of g and the slope of the graph of g?

46. If f is a constant function, what is the rate of change of f? Why?

In Exercises 47–50, suppose that $P(a, b)$ is a point of the line with the given slope. For the given increase/decrease of one variable, determine the corresponding increase/decrease of the other variable.

47. $m = -\dfrac{3}{5}$, b increased by 6

48. $m = \dfrac{7}{4}$, a increased by 4

49. $m = 2$, a decreased by 2

50. $m = -3$, b decreased by 12

In Exercises 51 and 52, determine the slope of the line containing the indicated points of the graph of f.

51. $f(x) = x^2$, where $x = 1$ and $x = -2$.

52. $f(x) = |x - 3|$, where $x = 2$ and $x = 6$.

In Exercises 53 and 54, produce the graphs of the given functions and use your calculator's *Intersect* feature to determine the points of intersection. Then determine the slope of the line that contains those points.

53. $f(x) = \dfrac{1}{2}x^2 - 4x$ and $g(x) = x - \dfrac{1}{2}x^2$

54. $f(x) = x^2 + x$ and $g(x) = 36 + x - 3x^2$

55. Suppose that $P(x_1, y_1)$ and $Q(x_2, y_2)$ are points of a line. Use the point-slope model and the slope formula to show that a two-point model of an equation of the line is

$$\frac{y - y_1}{x - x_1} = \frac{y_2 - y_1}{x_2 - x_1}$$

56. Suppose that $A(-4, 1)$ and $B(6, 3)$ are points of a line. Use the point-slope model and the two-point model in Exercise 55 to show that the same slope-intercept forms of the equations of the line are obtained with either model.

Applications

57. Senior Citizen Population In 1995 there were 4 million Americans who were 85 or older. That number is projected to reach 8.1 million by the year 2030. (*Source*: U.S. Bureau of Census.) Letting x represent the number of years since 1990, write a linear function to model the population of senior citizens.

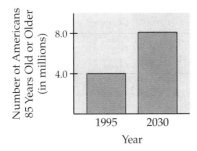

58. Senior Citizen Population In 1995 the percentage of the population who were 85 or older was 1.4%. That percentage is projected to reach 2.4% by the year 2030. (*Source*: U.S. Bureau of Census.) Letting x represent the number of years since 1990, write a linear function to model the percentage of the population who are senior citizens.

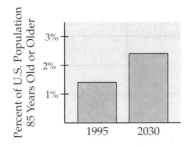

59. Beef Consumption In 1975 each American consumed an average of 83 pounds of beef annually. By 1990 the average had fallen to 64 pounds annually. (*Source*: U.S. Department of Agriculture.) Letting x represent the number of years since 1970, write a linear function to model the average annual beef consumption.

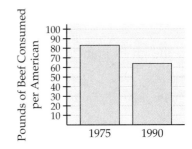

60. Coffee Consumption From 1970 to 1995, the average number of pounds of coffee used per person declined from 13.6 pounds to 8.2 pounds. (*Source*: U.S. Department of Agriculture.) Letting x represent the number of years since 1970, write a linear function to model the average annual coffee consumption.

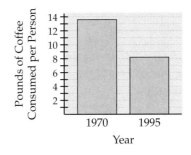

61. Maintenance Agreement For a fixed annual charge of $300, a heating and air conditioning company will provide service to residential customers with no hourly charge. A second plan has an annual charge of $100 and an hourly rate of $40. Let h represent the number of on-site hours.
(a) Write a function f for the annual cost under the first plan. What is such a function called?
(b) Write a function g for the annual cost under the second plan. What is the significance of the slope and y-intercept of the graph of g?

62. Depreciation An office copier had an original cost of $6000. Using the "straight-line" depreciation method, the accounting department valued the three-year-old copier at $2400. Let x represent the age (in years) of the copier.

(a) Write a function V for the value of the copier after x years.
(b) What is the significance of the slope and intercepts of the graph of V?

63. Union Workers A small construction company employs 25 workers, with x of the workers belonging to a union.

(a) Write a linear function N that describes the number of nonunion workers.
(b) Determine and interpret the intercepts of the graph of function N.

64. Veterinary Visits A veterinarian treats an average of 30 animals per day, of which $2x$ are cats and x are dogs.

(a) Write a linear function A that describes the number of animals other than dogs and cats that the veterinarian treats.
(b) Determine and interpret the intercepts of the graph of function A.

A manufacturer incurs a **fixed cost,** which is a cost that is the same regardless of the number of items produced, and a **variable cost,** which is a cost that depends on the number of items produced.

65. Greenhouse Plants The owner of a greenhouse incurs a cost of $5000 per year to maintain the greenhouse facilities. The cost of producing a tray of begonia plants is $2 per tray.

(a) Write a linear function C for the annual total cost of producing x trays of begonias.
(b) Determine and interpret the y-intercept of the graph of function C.
(c) Determine and interpret the slope of the graph of C.

66. Chocolate Truffles A confectioner has a candy shop specializing in hand-dipped chocolate truffles. The cost of maintaining the shop is $800 per month, and the cost of making the truffles is $6 per pound.

(a) Write a linear function C for the annual total cost of producing x pounds of truffles.
(b) Determine and interpret the y-intercept of the graph of function C.
(c) Determine and interpret the slope of the graph of C.

67. Clock Mechanism Suppose that the minute hand of a clock is 8 inches long and is pointing at 12.

(a) Write an expression for the distance that the tip of the minute hand travels in 1 hour. (*Hint*: The circumference C of a circle with radius r is $C = 2\pi r$.)
(b) Use the result in part (a) to write a function D for the distance (in inches) that the tip of the minute hand travels in t minutes, where $0 \le t \le 60$.

(c) How far has the tip of the minute hand traveled when the clock reads 12:35?

68. Temperature Conversion To convert a Fahrenheit temperature F to the corresponding Celsius temperature C, subtract 32 from the Fahrenheit temperature and multiply the result by $\frac{5}{9}$.

(a) Write C as a function of F in the slope-intercept form.
(b) Write F as a function of C in the slope-intercept form.

69. Value of a Dollar If your salary in 1985 was $20,000, you would need to have earned $28,400 in 1995 just to stay even with inflation. (*Source*: U.S. Chamber of Commerce.)

(a) Letting t represent the number of years since 1985, write a linear function f for the salary needed to maintain the purchasing power of $20,000 in 1985.

(b) If your salary were $30,000 in the year 2000, compare your purchasing power to that in 1985.

70. **Temperature Fall** On January 10, 1911, the temperature in Rapid City, South Dakota, fell from 55°F at 7:00 A.M. to 8°F at 7:15 A.M. (*Source:* National Climatic Center.) Letting x represent the number of minutes after 7:00 A.M., write a linear function T to model the temperature during this 15-minute period.

71. **Newspaper Subscribers and Profits** In 1998, a local newspaper with 20,000 subscribers earned a profit of $50,000. The publisher's goal is to add 4000 subscribers by the year 2001, and the projected profit is $60,000.

   (a) Produce a linear regression equation for the profit P as a function of the number x of subscribers.
   (b) Determine and interpret the slope of the graph.
   (c) Determine and interpret the y-intercept of the graph.

72. **Balanced Beam** The beam shown in the figure will be perfectly balanced if $\dfrac{W_1}{W_2} = \dfrac{L - x}{x}$. (Ignore the weight of the beam.)

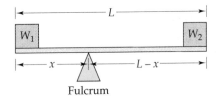

Fulcrum

Suppose that the weight W_1 is 5 feet from the fulcrum and that the beam is 15 feet long.

   (a) Write W_1 as a function of W_2. Is this a linear function?
   (b) If a 75-pound child stands at the right end of the beam, and a person stands at the left end, how much must the person weigh to balance the beam?
   (c) Suppose that L and W_2 are known, but the placement of the fulcrum is not known. Can W_1 be expressed as a linear function of x? Why?

A **direct variation** is described by a linear function of the form $f(x) = kx$, where $k \neq 0$. The associated equation is $y = kx$, and we say that y varies directly with x. Use this information for Exercises 73 and 74.

73. **Homeowners' Dues** A resident of a subdivision decides to form a homeowners' association with annual dues of $75 per residence.

   (a) Write a direct variation that describes the association's total annual income y from x members.
   (b) Produce the graph of y and trace it to estimate the number of members if the annual income is $2400.

74. **CD-ROM Games** The profit P (in dollars) from the sales of CD-ROM games varies directly with the number x of games sold.

   (a) If the direct variation is described by the function $P(x) = 30x$, what is the per-unit profit?
   (b) Compare the per-unit profit, the rate of change of P, and the slope of the graph of P.

2.3 *Linear Regression Models*

Scatterplots ■ *Linear Regression* ■ *The Correlation Coefficient*

In the real world, we do not always have access to all the data that we might require. For example, if the Doppler radar that we see on the evening weather forecast makes a full "sweep" every minute, we can observe the location of a storm cell from one minute to the next, but not in between those times.

One purpose of a mathematical model is to provide a means by which we can estimate missing data or predict possible future outcomes. The type of model that we select will depend on the nature of the data. In this section, we consider how to determine one kind of model, a linear function, and how to judge whether a linear model is appropriate for the given data.

Scatterplots

A **scatterplot** is simply a graph of plotted points.

EXAMPLE 1   **Creating a Scatterplot**

For the period 1977–1995, the table shows the percentage of high school athletes who were female. (*Source*: National High School Athletics Association.)

Year	Percentage
1977	32.3%
1982	34.7%
1987	35.1%
1992	36.9%
1997	40.2%

(a) Create a scatterplot of the data.
(b) For the given five-year periods, is the rate of change in the percentages constant?
(c) What trend is indicated by the scatterplot?

SOLUTION

(a) We let x represent the number of years since 1972 and y the corresponding percentage. Then we plot the points shown in Figure 2.18.

FIGURE 2.18

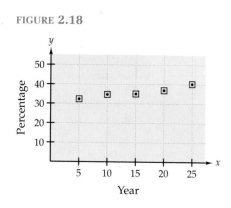

(b) The percentages do not increase by the same amount for each five-year period. Thus the rate of change is not constant, which means that no single line can be drawn through all the points of the scatterplot.
(c) The trend indicated by the scatterplot is that the percentage of female athletes rose at a *relatively* steady rate.

You should be able to use your calculator to produce a scatterplot.

Scatterplot   Producing a scatterplot begins with entering the ordered pairs in a statistics *list*. Then, after setting the window to display all the points, select the scatterplot option to produce the graph.

Figure 2.19(a) shows the entry of the data in Example 1 in the statistics list. To create this list, we again let x represent the number of years since 1972. Figure 2.19(b) shows the display of the scatterplot.

FIGURE **2.19**

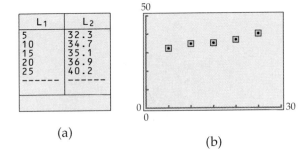

(a)

(b)

It is important to realize that the points in Figure 2.19(b) are not all points of one line. If they were, then, using the methods of the preceding section, we could write a linear function that would model the data exactly; that is, the graph of the function would contain all the points. Even so, we can envision a line that comes close to all the points, and the equation of that line would serve as a model of the data.

Deciding whether data appear to be linearly related is a matter of judgment. However, the lack of a linear relationship is often obvious. Figure 2.20 shows two scatterplots of data for which a linear function would not be a good model.

FIGURE **2.20**

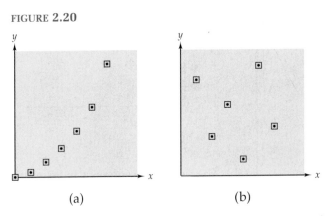

(a)

(b)

The graph in Figure 2.20(a) is rising at an increasingly rapid rate, and the data in Figure 2.20(b) show no apparent trend at all. In neither case would a straight line be representative of the data points.

For the remainder of this section, we will focus on the use of a linear equation as a model of data.

Linear Regression

If two people were each to draw a line through the scatterplot in Example 1, it is unlikely that they would draw exactly the same line, and there could be some dispute over which line is better.

Ideally, we want the line that is the "best fit" of the data. Determining such a line is called *curve fitting*. In general and roughly speaking, curve fitting involves finding a graph such that the distances from all the points to the graph is a minimum.

In applications, we say that data are approximately linear if we can plot the ordered pairs of data and then draw a line that comes close to containing all the points. The best such line is a line whose equation is called a **linear regression equation.** The methods for determining regression equations must be left to more advanced courses, but you can use your calculator to produce regression equations for you.

> **Regression**   A menu of regression equation forms is typically an option found with the **STAT** key. In this section, we will use *LinReg*, which means "linear regression."

EXAMPLE **2**   **A Linear Regression Equation for Real Data**

(a) Use your calculator to produce a linear regression equation to model the data in Example 1.
(b) Produce the graph of the regression equation in the same coordinate system as the scatterplot shown in Figure 2.19(b).

SOLUTION

(a) Because the data are already entered in the calculator, we only need to select the linear regression form from the menu of regression equations. Figure 2.21 shows a typical display of the results.

FIGURE **2.21**

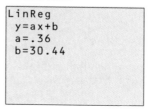

```
LinReg
 y=ax+b
 a=.36
 b=30.44
```

Using the displayed values of a and b, we write the linear regression equation: $y = 0.36x + 30.44$.

(b) We enter the regression equation on the Y screen and produce its graph on the same viewing screen as the scatterplot. Figure 2.22

shows that the graph of the regression equation comes very close to containing the data points.

FIGURE **2.22**

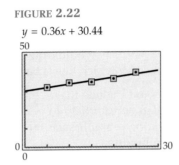

$y = 0.36x + 30.44$

Figure 2.22 provides convincing visual evidence that the regression equation $y = 0.36x + 30.44$ is a good model of the data. In fact, it can be shown that the graph of this equation is the best fit of any of the lines we can draw.

Once we have a model equation, we can use it to estimate missing data and to make reasonable predictions.

EXAMPLE **3**   **Using a Linear Regression Model**

Use the linear regression equation $y = 0.36x + 30.44$ for the female high school athlete data to answer the following questions:

(a) In 1994, what was the estimated percentage of high school athletes who were female?
(b) For the year 2003, what is the predicted percentage of high school athletes who are female?
(c) In what year are half of all high school athletes projected to be female?

SOLUTION

(a) For 1994, $x = 22$. Thus $y = 0.36(22) + 30.44 = 38.36\%$.
(b) For 2003, $x = 31$. Thus $y = 0.36(31) + 30.44 = 41.6\%$.
(c) We determine the year x when $y = 50\%$.

$$0.36x + 30.44 = 50$$
$$0.36x = 19.56$$
$$x \approx 54$$

The year when half of high school athletes will be female is predicted to be $1972 + 54 = 2026$.

Although simple models can be effective analytical tools in real-world applications, they cannot always be regarded as reliable predictors. For instance, we should be wary of the result in Example 3(c) because many social changes could occur that might dramatically alter the trend in female athletics. Modeling is a dynamic process, with new and more sophisticated models evolving as the influence of other variables is better understood.

The Correlation Coefficient

Your calculator uses accepted statistical methods when it computes a linear regression equation. However, the decision as to whether to use a linear model in the first place is yours. Even if the data do not exhibit a linear relationship, you can still instruct your calculator to produce a linear regression equation. However, the equation that you obtain may or may not be an acceptable model of the data. The question is, "What is acceptable?"

One measure of how well a regression equation models data is a number called the **correlation coefficient.** The conventional symbol for this number is r. When you use your calculator to produce a regression equation, the correlation coefficient is also computed and stored in the variable r.

> **Correlation**   You can instruct some calculator models to automatically display the value of *r* along with the regression equation. Usually, you can also find the value of *r* in a variables menu, such as **VARS**.

The formula for r produces a value between -1 and 1. When $r = 1$, the regression line has a positive slope, and every data point is a point of the line. [See Figure 2.23(a).] Conversely, when $r = -1$, the regression line has a negative slope, and every data point is a point of the line. [See Figure 2.23(b).]

FIGURE **2.23**

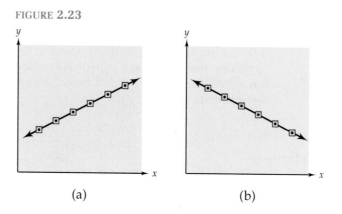

(a)   (b)

In most cases, the value of r falls between -1 and 1. Figure 2.24 shows examples of correlations that are less than perfect.

FIGURE **2.24**

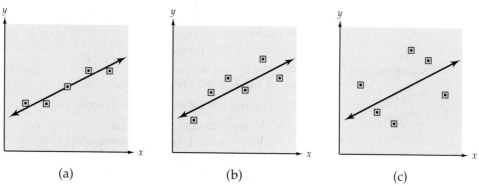

(a)   (b)   (c)

In Figure 2.24(a), we see a relatively high positive correlation, whereas Figure 2.24(b) shows a relatively low positive correlation. If we produce a linear regression equation for the data in Figure 2.24(c), the value of r is near 0, which indicates no linear correlation of the data.

EXAMPLE 4   **Calculating the Correlation Coefficient**

FIGURE 2.25

```
LinReg
 y=ax+b
 a=.36
 b=30.44
 r²=.9388038943
 r=.9689189307
```

Determine and interpret the correlation coefficient for the linear regression equation in Example 3.

SOLUTION
In Figure 2.25, we have computed the linear regression equation again, but this time we have included the value of r in the displayed result. The value of $r \approx 0.97$ indicates a very high positive correlation and suggests that the linear regression equation is a good model of the data.

Although the correlation coefficient is a helpful numerical measure of how well-associated data are, you must still make a judgment about the quality of your regression equation. In some instances, a value of $r = 0.80$ might be quite tolerable, but in other situations, the same value might be unacceptable.

Finally, as you read the statistical results of research studies, for example, be aware that the correlation coefficient is not an indication of cause and effect. If an educational journal reports a high negative correlation between hours of television viewing and grades, the value of r alone is not enough to conclude that long hours of television viewing result in poor grades. Other analytical factors are needed before a cause-and-effect relationship can be established.

2.3 Quick Reference

Scatterplots
■ A **scatterplot** is a graph of plotted points.
■ If a line can be drawn in a scatterplot so that all the points are close to the line, we regard the data as being linearly related.

Linear Regression
■ A **linear regression equation** is the equation of the line that is the best fit for linearly related data.
■ A calculator can be used to produce a linear regression equation.
■ A linear regression equation can be used to estimate missing data or to make predictions.

The Correlation Coefficient
■ Associated with a regression equation is a **correlation coefficient,** which is a measure of how well the equation models the given data.
■ The symbol for the correlation coefficient is r, and $-1 \le r \le 1$.
■ A calculator usually computes r along with regression equations. The closer r is to 1 (positive correlation) or -1 (negative correlation), the stronger is the association of the data.

2.3 Speaking the Language

1. A graph that consists of plotted points is called a(n) ▨▨▨▨▨ .
2. Determining the line or curve that is the best fit of given data is called ▨▨▨▨▨ .
3. The equation of a line that best fits linearly related data is called a linear ▨▨▨▨▨ equation.
4. A numerical measure of how well an equation fits given data is called the ▨▨▨▨▨ .

2.3 Exercises

1. Suppose that a scatterplot consists of three points that are not points of the same line. What is the maximum number of those points that the linear regression line could contain? Why?

2. What do we mean when we say that data are linearly related?

In Exercises 3–6, judge whether a linear model would be appropriate for the given scatterplot.

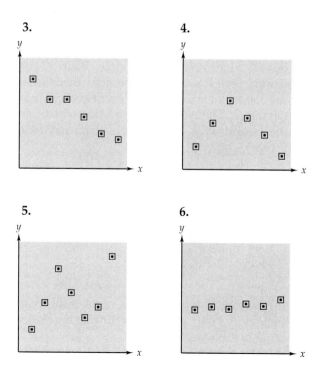

3.

4.

5.

6.

7. (a) Write the equation of the line that contains the points $P(-2, 5)$ and $Q(6, -3)$.
   (b) Use your calculator to produce a linear regression equation for the line in part (a).
   (c) Compare your results in parts (a) and (b).
   (d) Why is $r = -1$ rather than $+1$?

8. (a) Write the equation of the line that contains the points $A(0, 5)$ and $B(10, 20)$.
   (b) Use your calculator to produce a linear regression equation for the line in part (a).
   (c) Compare your results in parts (a) and (b).
   (d) Why is $r = +1$ rather than -1?

In Exercises 9 and 10, a table of values of a function is given.

   (a) By examining the changes in the x-values and the corresponding changes in the y-values, decide whether the function could be linear. If so, what is the function?
   (b) If you were to produce a linear regression equation, what would be the value of r? How do you know?
   (c) Verify your answers to parts (a) and (b) by producing the linear regression equation with your calculator.

9.

x	0	5	10	15	20
y	4	11	18	25	32

10.

x	-3.0	-1.0	3.0	11.0	27.0
y	3.6	-1.8	-12.6	-34.2	-77.4

11. Welfare Cases The bar graph shows the decrease in the number of welfare cases from 1994 to 1997. (*Source*: U.S. Health and Human Services Department.)

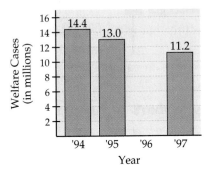

(a) Letting x represent the number of years since 1990, produce a linear regression equation to model the data.

(b) What is the significance of the coefficient of x in your equation?

(c) Use your equation to estimate the missing data for 1996.

12. Single Fathers Of the single-parent households headed by men in 1968, 5% of the fathers had never been married. By 1996, the percentage had increased to 30%. (*Source*: U.S. Census Bureau.)

(a) Letting x represent the number of years since 1965, produce a linear regression equation to model the data.

(b) What is the significance of the coefficient of x in your equation?

(c) Use your equation to project the percentage for the year 2004.

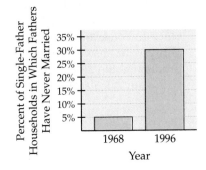

13. Use your calculator to produce a regression equation for the line that contains $A(5, 7)$ and $B(5, -1)$. Explain the result.

14. If you enter the data points $(2, 7)$, $(5, 7)$, and $(8, 7.1)$ and produce a linear regression equation $y = ax + b$, what would you expect the approximate value of a to be? Why?

In Exercises 15–18, produce a first-quadrant scatterplot of the data in the table and decide whether a linear model of the data is reasonable.

(a) If you think so, produce and graph a linear regression equation on the same screen as the scatterplot. Based on the resulting display, does a linear model still seem reasonable?

(b) How does the value of the correlation coefficient affect your judgment about the reasonableness of the model?

15.

x	10	20	30	40	50
y	95	42	20	32	87

16.

x	6	10	12	15	19
y	30	39	43	42	50

17.

x	4	6	8	10	12
y	16	18	23	35	38

18.

x	2	3	4	5	6
y	11	19	20	4	8

19. Speed Skating Times Bonnie Blair of the United States is a three-time gold medalist in the 500-meter speed skating event at the Winter Olympics. The table shows the U.S. winners in the period 1972–1994. (*Source*: *Encyclopedia of Sports*.)

Year	Gold medalist	Time (in seconds)
1972	Anne Henning	43.33
1976	Sheila Young	42.76
1988	Bonnie Blair	39.10
1992	Bonnie Blair	40.33
1994	Bonnie Blair	39.25

(a) Letting x represent the number of years since 1970, produce a linear regression equation for the data.

(b) Use the equation in part (a) to estimate the winning time in 1980.

(c) How does the value of the correlation coefficient affect your confidence in your answer to part (b)?

(d) Interpret the fact that the coefficient of x in your regression equation is negative.

20. Telephone Charges   Monthly telephone costs per household for long distance calls and other charges (excluding basic local service) have risen steadily. The bar graph shows the trend for the period 1980–1995. (*Source*: AT&T.)

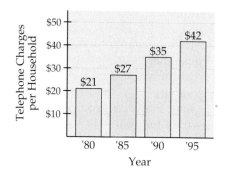

(a) Based on the bar graph, would you characterize the data as linear?

(b) Just by observing the data in the bar graph, estimate what you think the slope of a regression line would be.

(c) Letting x represent the number of years since 1980, produce a linear regression equation for the data and compare the slope with your estimate in part (b).

(d) Suppose that we let x represent the number of years since 1975 or 1962 rather than 1980. What effect would this have on the slope of the regression line?

21. Sports Radio Stations   The bar graph shows that the number of all-sports radio stations increased during the period 1994–1996. (*Source*: M Street Radio Directory.)

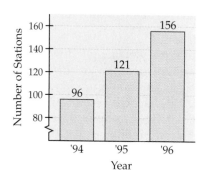

Produce a scatterplot of the points $A(1, 96)$, $B(2, 121)$, and $C(3, 156)$ and use it in the following parts.

(a) Write the equations of the lines that contain
  (i) A and B (ii) A and C (iii) B and C

(b) For each of the equations in part (a), produce the graph on the same screen as your scatterplot. Which of the three lines seems to model the data best?

(c) Now produce a linear regression equation and graph it with the scatterplot. Of the four graphs that you have produced, which do you think models the data best?

22. Recycled Appliances   The table shows the percentages of household appliances that were recycled for steel in the period 1993–1996. (*Source*: Steel Recycling Institute.)

Year	Percentage	Data point
1993	62%	$A(3, 62)$
1994	70%	$B(4, 70)$
1995	74%	$C(5, 74)$
1996	76%	$D(6, 76)$

To prepare for the following questions, produce a scatterplot of the data points.

(a) Write an equation of the line that contains
  (i) A and D (ii) B and D (iii) C and D

(b) One at a time, produce the graphs of the equations in part (a) on the same screen with the scatterplot. What do you observe about the slopes of the lines? How would you interpret this?

(c) If you were to produce a linear regression equation for all the data, what relationship would you anticipate among the slopes in equations (i) and (iii) and the regression equation?

(d) Produce the regression equation and verify your conjecture in part (c).

23. Long Distance Calls   The bar graph shows the number of long distance calls (in millions) on Mother's Day and on Father's Day for the years 1993–1996. (*Source*: AT&T.)

(a) Letting x represent the number of years since 1992, produce a scatterplot of the data for Mother's Day. Then produce and graph a linear regression equation for those data. How well does the graph appear to fit the data?

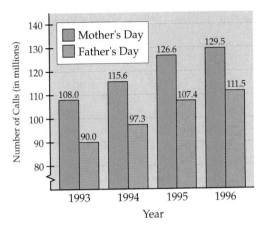

(b) Use the regression equation to determine and interpret y when $x = 0$.

(c) Suppose that the management of the telephone company predicts that approximately 161 million calls will be made on Mother's Day in the year 2000. In view of the value of r, how confident of this prediction can managers be?

24. Long Distance Calls   Refer to the bar graph in Exercise 23.

(a) Letting x represent the number of years since 1992, produce a scatterplot of the data for Father's Day. Then produce and graph a linear regression equation for those data. How well does the graph appear to fit the data?

(b) Trace the graph of the regression equation to estimate the year when the number of Father's Day calls is predicted to reach 150 million.

(c) In Exercise 23, the regression equation for Mother's Day is $y = 7.55x + 101.05$. Suppose that your manager wants you to predict the year when the number of calls for Mother's Day and Father's Day will be the same. How can you answer the question without graphing or solving an equation?

2.4 *Topics in Analytic Geometry*

Parallel and Perpendicular Lines ■ *The Distance Formula* ■ *The Midpoint Formula*

Analytic geometry is a fusion of algebra and geometry. By drawing points, lines, and figures in the coordinate plane, we can use algebraic methods to examine geometric relationships. In this section, we consider the slopes of parallel and perpendicular lines, determine distances between points, and determine midpoints of line segments.

Parallel and Perpendicular Lines

From geometry, two distinct lines L_1 and L_2 in a plane that do not intersect are **parallel.** Two lines L_1 and L_2 are **perpendicular** if they intersect to form a right angle.

The built-in settings for your calculator's viewing window are such that the distance between the tick marks on the x-axis are greater than the distance between the tick marks on the y-axis. This configuration can cause graphs to appear stretched or distorted. Usually, this is not a serious problem. However, if you want perpendicular lines to appear perpendicular, then you should consider a special window setting so that the tick marks are equally spaced along both axes.

> **Square**　The tick marks along the x- and y-axes should be equally spaced when the shape of a graph or the relationship between graphs is visually important.

We begin by investigating the relationship between the slopes of parallel lines.

EXPLORATION

Parallel Lines and Slope

Produce the graphs of the functions $y = -x + b$ for several values of b.

(a) What geometric relationship appears to exist between the lines?
(b) Repeat the experiment for functions of the form $y = mx + b$ by choosing a particular value of m and several different values of b.
(c) What general conjecture can you make regarding the geometric relationship between lines with the same slope?

DEVELOPING THE CONCEPT

Parallel and Perpendicular Lines and Slope

Figure 2.26 shows the graphs of the following groups of equations.

Column A	*Column B*	*Column C*
$y_1 = 2x + 9$	$y_1 = 2x + 1$	$y_1 = \frac{4}{5}x + 5$
$y_2 = 2x + 1$	$y_2 = -\frac{1}{2}x - 2$	$y_2 = -\frac{5}{4}x + 1$
$y_3 = 2x - 4$		
$y_4 = 2x - 10$		

FIGURE 2.26

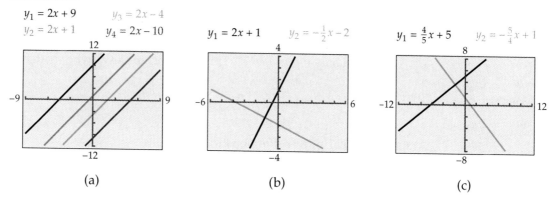

(a)　(b)　(c)

(a) In column A, the slopes are the same, and the lines appear to be parallel. [See Figure 2.26(a).] If we think of slope as rate of change, then we realize that two different lines with the same slope are rising or falling at the same rate. This means that the lines can never intersect and are therefore parallel.

(b) Look carefully at the slopes of the lines in columns B and C. In both cases, if m is the slope of one line, then the slope of the other line is $-\dfrac{1}{m}$. Another way to state this relationship is that the product of the slopes is -1.

For each pair of equations in columns B and C, the lines appear to be perpendicular. [See Figure 2.26(b) and (c).] Thus our conjecture is that two lines are perpendicular if the product of their slopes is -1.

These observations about the slopes of parallel and perpendicular lines can be proven to be true.

> *Parallel Lines and Slopes*
>
> Two distinct nonvertical lines in the same plane are **parallel** if and only if they have the same slope. Also, any two distinct vertical lines are parallel. The symbol for "is parallel to" is $\parallel$.

> *Perpendicular Lines and Slopes*
>
> If neither of two lines is vertical, then the lines are **perpendicular** if and only if the product of their slopes is -1. Also, a vertical line and a horizontal line are perpendicular. The symbol for "is perpendicular to" is $\perp$.

If the slopes of two lines are m_1 and m_2, then the lines are perpendicular if and only if $m_1 m_2 = -1$. This relationship can also be written $m_1 = -\dfrac{1}{m_2}$.

EXAMPLE 1   **Parallel and Perpendicular Lines**

Write an equation of each line L_1.

(a) L_1 contains $P(3, -4)$ and $L_1 \perp L_2$. The equation of L_2 is $5x - 2y = 8$.
(b) The y-intercept of L_1 is $(0, -5)$ and $L_1 \parallel L_2$. The equation of L_2 is $x = 3y + 12$.

SOLUTION

(a) $5x - 2y = 8$ The equation of L_2

$\quad\quad -2y = -5x + 8$ Write the equation in the slope-intercept form.

$$y = \frac{5}{2}x - 4 \quad\quad \text{The slope of } L_2 \text{ is } \tfrac{5}{2}.$$

Because $L_1 \perp L_2$, the slope of L_1 is $-\dfrac{2}{5}$.

$$y - y_1 = m(x - x_1) \quad\quad \text{Use the point-slope model for } L_1.$$

$$y - (-4) = -\frac{2}{5}(x - 3) \quad\quad m = -\tfrac{2}{5}, x_1 = 3, y_1 = -4$$

$$y + 4 = -\frac{2}{5}x + \frac{6}{5}$$

$$y = -\frac{2}{5}x - \frac{14}{5} \quad\quad \text{The equation of } L_1$$

(b) $\quad x = 3y + 12$ The equation of L_2

$\quad x - 12 = 3y$ Write the equation in the slope-intercept form.

$$y = \frac{1}{3}x - 4 \quad\quad \text{The slope of } L_2 \text{ is } \tfrac{1}{3}.$$

Because $L_1 \parallel L_2$, the slope of L_1 is $\dfrac{1}{3}$.

$$y = mx + b \quad\quad \text{Use the slope-intercept model for } L_1.$$

$$y = \left(\frac{1}{3}\right)x + (-5) \quad\quad m = \tfrac{1}{3}, b = -5$$

$$y = \frac{1}{3}x - 5 \quad\quad \text{The equation of } L_1$$

The Distance Formula

If x_1 and x_2 are numbers plotted on a number line, then the distance between the points is $|x_2 - x_1|$ or $|x_1 - x_2|$. Now we consider the distance between two points in a coordinate plane.

DEVELOPING THE CONCEPT

Distance Between Points

Figure 2.27 shows two general points, $P(x_1, y_1)$ and $Q(x_2, y_2)$, plotted in a coordinate plane. Let d represent the distance between these points.

FIGURE **2.27**

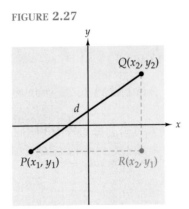

Point $R(x_2, y_1)$ is plotted so that the resulting triangle is a right triangle. The lengths of the legs of the right triangle are $PR = |x_2 - x_1|$ and $QR = |y_2 - y_1|$. Let d represent the length of the hypotenuse.

For a right triangle, the Pythagorean Theorem states that the sum of the squares of the legs is equal to the square of the hypotenuse.

$$d^2 = (PR)^2 + (QR)^2$$
$$= |x_2 - x_1|^2 + |y_2 - y_1|^2$$

Because, in general, $|a - b|^2$ and $(a - b)^2$ are equivalent, the equation can be written as

$$d^2 = (x_2 - x_1)^2 + (y_2 - y_1)^2$$

From the definition of square root, d must be a square root of the expression on the right. Because distance is positive, d is the principal square root.

$$d = \sqrt{(x_2 - x_1)^2 + (y_2 - y_1)^2}$$

We have derived the following formula for the distance between two points in a coordinate plane.

The Distance Formula

If (x_1, y_1) and (x_2, y_2) are points in a coordinate plane, then the distance d between the points is

$$d = \sqrt{(x_2 - x_1)^2 + (y_2 - y_1)^2}$$

We use different symbols to distinguish between a line segment and the length of that segment. In Example 2, $\overline{PQ}$ is the line segment whose endpoints are P and Q. The length of $\overline{PQ}$ is represented by PQ. Thus $d = PQ$.

EXAMPLE 2   **Using the Distance Formula**

Determine the length of the line segment $\overline{PQ}$ whose endpoints are $P(-3, 5)$ and $Q(4, 2)$.

SOLUTION

$$d = PQ = \sqrt{(-3-4)^2 + (5-2)^2} = \sqrt{49 + 9} = \sqrt{58} \approx 7.62$$

The Midpoint Formula

If M is the midpoint of the line segment $\overline{AB}$, then $AM = MB$. The midpoint coordinates are the averages of the corresponding endpoint coordinates. (See Figure 2.28.)

FIGURE 2.28

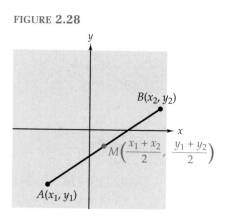

The Midpoint Formula

If $A(x_1, y_1)$ and $B(x_2, y_2)$ are the endpoints of a line segment, then the midpoint of the segment is

$$M\left(\frac{x_1 + x_2}{2}, \frac{y_1 + y_2}{2}\right)$$

EXAMPLE 3   **Using the Midpoint Formula**

Determine the midpoint M of the line segment whose endpoints are $A(-6, 1)$ and $B(4, -7)$.

SOLUTION

$$M\left(\frac{-6 + 4}{2}, \frac{1 + (-7)}{2}\right) = M\left(\frac{-2}{2}, \frac{-6}{2}\right) = M(-1, -3)$$

A **median** of a triangle is a line segment that connects a vertex to the mid-point of the opposite side.

EXAMPLE 4   A Median of a Triangle

In Figure 2.29, $\overline{AM}$ is a median of $\triangle ABC$. Determine the coordinates of point M. Then determine AM.

FIGURE 2.29

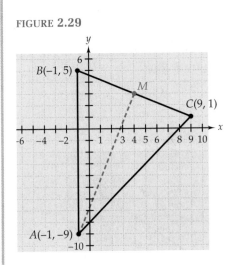

SOLUTION

$$M\left(\frac{-1+9}{2}, \frac{5+1}{2}\right) = M\left(\frac{8}{2}, \frac{6}{2}\right) = M(4, 3)$$

$$AM = \sqrt{(-1-4)^2 + (-9-3)^2} = \sqrt{(-5)^2 + (-12)^2}$$

$$= \sqrt{25 + 144}$$

$$= \sqrt{169} = 13$$

The **perpendicular bisector** of a line segment is a line that is perpendicular to the line segment at its midpoint.

EXAMPLE 5   The Equation of a Perpendicular Bisector

Determine an equation of the perpendicular bisector of the line segment whose endpoints are $A(-4, 2)$ and $B(8, -8)$.

SOLUTION

In Figure 2.30, line L is perpendicular to $\overline{AB}$ at its midpoint M.

FIGURE **2.30**

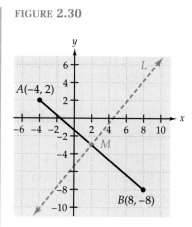

One point of L is

$$M\left(\frac{-4+8}{2}, \frac{2+(-8)}{2}\right) = M(2, -3)$$

The slope of $\overline{AB}$ is

$$m_{\overline{AB}} = \frac{-8-2}{8-(-4)} = \frac{-10}{12} = -\frac{5}{6}$$

Thus,

$$m_L = \frac{-1}{-\frac{5}{6}} = \frac{6}{5} \qquad\qquad m_L = \frac{-1}{m_{\overline{AB}}}$$

$$y - y_1 = m(x - x_1) \qquad\qquad \text{Use the point-slope model for } L.$$

$$y - (-3) = \frac{6}{5}(x - 2) \qquad\qquad x_1 = 2, y_1 = -3, m = \tfrac{6}{5}.$$

$$y = \frac{6}{5}x - \frac{12}{5} - 3$$

$$y = \frac{6}{5}x - \frac{27}{5} \qquad\qquad \text{The equation of } L.$$

2.4 Quick Reference

Parallel and Perpendicular Lines

■ Two distinct nonvertical lines in the same plane are **parallel** if and only if they have the same slope. Also, any two distinct vertical lines are parallel.

■ If neither of two lines is vertical, then the lines are **perpendicular** if and only if the product of their slopes is −1. Also, a vertical line and a horizontal line are perpendicular.

The Distance Formula

■ If (x_1, y_1) and (x_2, y_2) are points in a coordinate plane, then the distance d between the points is $d = \sqrt{(x_2 - x_1)^2 + (y_2 - y_1)^2}$.

The Midpoint Formula

■ If $A(x_1, y_1)$ and $B(x_2, y_2)$ are the endpoints of a line segment, then the midpoint of the segment is $M\left(\dfrac{x_1 + x_2}{2}, \dfrac{y_1 + y_2}{2}\right)$.

■ A **median** of a triangle is a line segment that connects a vertex to the midpoint of the opposite side.

■ The **perpendicular bisector** of a line segment is a line that is perpendicular to the line segment at its midpoint.

2.4 Speaking the Language

1. Two distinct lines in a plane are ▬▬▬▬ if they do not intersect and ▬▬▬▬ if they intersect to form a right angle.

2. Two lines are perpendicular if the product of their ▬▬▬▬ is -1.

3. The formula $d = \sqrt{(x_2 - x_1)^2 + (y_2 - y_1)^2}$ is used to find the ▬▬▬▬ between the points $P(x_1, y_1)$ and $Q(x_2, y_2)$.

4. If $P(x_1, y_1)$ and $Q(x_2, y_2)$ are the endpoints of a line segment, then $\left(\dfrac{x_1 + x_2}{2}, \dfrac{y_1 + y_2}{2}\right)$ are the coordinates of the ▬▬▬▬ of the line segment.

2.4 Exercises

Concepts and Skills

1. Why does the definition of parallel lines require that the lines be in the same plane?

2. Suppose that two nonvertical lines are perpendicular and the slope of one line is known. Describe an easy way to write the slope of the other line.

In Exercises 3–8, lines L_1 and L_2 are described. Determine whether the lines are parallel, perpendicular, or neither.

3. L_1: $3x - 2y - 2 = 0$
   L_2: $3y = -2(x - 6)$

4. L_1: $2x - 1 = 0$
   L_2: $y = 7$

5. L_1 contains $P(-5, 0)$ and $Q(5, 4)$
   L_2 contains $A(10, -4)$ and $B(5, -2)$

6. L_1 contains $M(0, -2)$ and $N(4, -3)$
   L_2 contains $C(-1, -5)$ and $D(3, -6)$

7. L_1 contains $R(4, 2)$ and $S(4, -5)$
   L_2: $3 - 2x = 0$

8. L_1: $3x - 7y + 14 = 0$
   L_2 contains $P(3, 6)$ and $Q(-6, -15)$

In Exercises 9–14, are the graphs of the linear functions parallel, perpendicular, or neither?

9. $f(x) = 2(1 - x) + 1, g(x) = 3 - 2(x + 3)$

10. $f(x) = \dfrac{x + 3}{2}, g(x) = -2x + 8$

11. $f(x) = 2x + 5, g(x) = -2x - 3$

12. $f(x) = 1.4x + 4, g(x) = -\dfrac{5}{7}x + 4$

13. f: $f(2) = -4, f(0) = -4$
    g: $g(5) = 1, g(-2) = 1$

14. f: $f(-3) = -8, f(9) = -4$
    g: $g(0) = 2, g(-5) = -13$

In Exercises 15–20, write the functions described in parts (a) and (b).
(a) Write a linear function f whose graph is perpendicular to line L_1 and satisfies the given condition.
(b) Write another linear function whose graph is parallel to line L_1 and satisfies the given condition.

15. L_1: $x - y = 6; f(2) = -5$

16. L_1: $2x + y + 3 = 0; f(0) = 8$

17. L_1 contains $A(6, -7)$ and $B(-3, 5)$; the graph of f has the same y-intercept as the line $y = 4 - 3x$.

18. L_1 contains $P(0, -6)$ and $Q(4, 4)$; the graph of f has the same x-intercept as the line $x + y = 2$.

19. L_1: $x + 5 = -1$; $f(4) = 6$

20. L_1: $3y = x$; $f(2) = 4$

In Exercises 21–26, write an equation of the described line L.

21. Line L has the same y-intercept as the line $y = 2 - 3x$ and L is perpendicular to the line $x - 2y = 4$.

22. The x-intercept of L is $A(-3, 0)$ and L is parallel to the line $3y - 2x = 0$.

23. Line L contains $R(5, -2)$ and is parallel to the line that contains $C(-1, 4)$ and $D(3, -8)$.

24. Line L contains $Q(3, -2)$ and is perpendicular to the x-axis.

25. Line L contains the points of the graph of $y = \dfrac{1}{x}$ where $x = 1$ and $x = -1$.

26. Line L contains the points of the graph of the equation $y = 5 - x^2$ where $x = 3$ and $x = -3$.

27. Determine the value of c such that the line $2x + cy = 12$

   (a) is perpendicular to the line $y = -3x + 1$.
   (b) has the same y-intercept as the line $x + 2y - 16 = 0$.

28. Determine the value of c such that the line $cx + 3y = 9$

   (a) is parallel to the x-axis.
   (b) has the same x-intercept as the line whose equation is $y = x + 1$.

29. If the endpoints of a line segment have integer coordinates, which statements are possible? Why?

   (i) The midpoint is $(1, -3)$.

   (ii) The midpoint is $\left(\dfrac{1}{2}, -\dfrac{3}{2}\right)$.

   (iii) The midpoint is $\left(\sqrt{3}, \sqrt{7}\right)$.

30. Explain how to determine the points that divide the line segment with endpoints $A(1, 4)$ and $B(15, 12)$ into 4 equal parts.

In Exercises 31–36, determine (a) PQ and (b) the midpoint of $\overline{PQ}$.

31. $P(3, -8)$, $Q(-2, 4)$

32. $P(5, 2)$, $Q(-3, -6)$

33. $P(-4, 0)$, $Q(5, -8)$

34. $P\left(1, 2\sqrt{5}\right)$, $Q\left(-3, 6\sqrt{5}\right)$

35. $P(a + b, b - a)$, $Q(a - b, a + b)$

36. $P(-3a, a)$, $Q(a, 4a)$

In Exercises 37 and 38, identify the point that is farther from the origin.

37. $(5, -12)$ or $(6, -11)$

38. $(5, 1)$ or $(3, 4)$

In Exercises 39–42, the endpoint A and the midpoint M of $\overline{AB}$ are given. Determine the coordinates of the endpoint B.

39. $A(0, 0)$, $M\left(-\dfrac{3}{2}, 2\right)$

40. $A(7, 4)$, $M\left(\dfrac{7}{2}, 3\right)$

41. $A(-3, 9)$, $M(-1, 6)$

42. $A(5, -3)$, $M\left(4, -\dfrac{9}{2}\right)$

43. Show that the distance from a point (a, b) to the origin is given by the formula $d = \sqrt{a^2 + b^2}$.

44. Show that $P(a, b)$ and $Q(b, a)$ are equal distances from the origin.

In Exercises 45–48, the vertices of a triangle are given. Use slopes or the Distance Formula to determine whether the triangle is a right triangle.

45. $(-6, 2)$, $(0, 5)$, $(-1, -8)$

46. $(10, -5)$, $(12.5, -4)$, $(7, 2.5)$

47. $(-1.5, 2)$, $(4.5, 6)$, $(6, -9)$

48. $(2, 11)$, $(0, 0)$, $(-4, 5)$

In Exercises 49–52, determine the area and perimeter (or circumference) of the given figure.[1]

49.

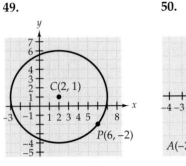

50.

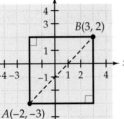

51.

52.

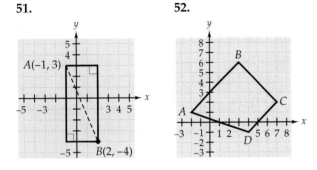

In Exercises 53–56, find the area and perimeter of the geometric figure whose vertices are given.

53. $A(-3, -3), B(1, -7), C(-8, -8)$

54. $P(-3.5, 5.5), Q(-8, 10), R(3, 12)$

55. $R(0, 8), S(-6, 6), T(-5, 13)$

56. $D(-8, -7), E(0, 9), F(-14, 6)$

In Exercises 57–60, write the equation of the perpendicular bisector of $\overline{PQ}$.

57. $P(-6, -3), Q(2, 5)$ **58.** $P(-4, 1), Q(3, 2)$

59. $P(2, -4), Q(7, 8)$ **60.** $P(-9, -4), Q(0, 0)$

61. The accompanying figure shows a quadrilateral *ABCD* drawn in a coordinate plane. Show that the midpoints of the sides, taken in order, form a parallelogram.

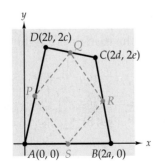

62. Draw a right triangle with vertices $A(0, 0)$, $B(0, 2b)$, and $C(2a, 0)$. Show that the midpoint of the hypotenuse is equidistant from all vertices.

63. Use the accompanying figure to show that the line segment joining the midpoints of two sides of a triangle is parallel to the third side and half the length of the third side.

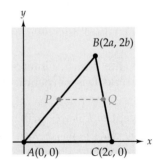

64. A **rhombus** is a parallelogram with equal sides. Use the accompanying figure to show that if the diagonals of a parallelogram are perpendicular, then the parallelogram is a rhombus. (*Hint*: Showing that $AB = AD$ is sufficient.)

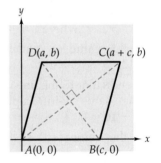

The distance from a point $A(x_1, y_1)$ to a line $y = mx + b$ is $d = \dfrac{|mx_1 + b - y_1|}{\sqrt{1 + m^2}}$.

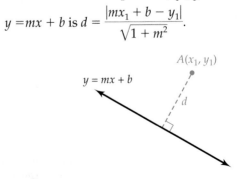

---

[1]Commonly used geometry formulas are listed on the inside cover of the book.

In Exercises 65–68, calculate the distance from the given point to the given line. Round results to the nearest hundredth.

65. $P(-4, -2); y = -2x + 5$

66. $Q(1, 3); x - y = -4$

67. $R(7, -3); x + 2y - 2 = 0$

68. $S(-3, 5); 3y - 2x = -27$

Applications

69. Median Ages of Cars In 1990 the median age of a car in the United States was 6.5 years. By 1996 the median age had increased to 7.9 years. (*Source*: The Polk Company.) Determine and interpret the midpoint of the line segment whose endpoints represent the given data.

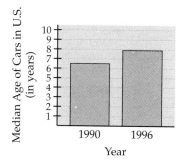

70. Grand Canyon Visitors In 1996 there were 2.65 million visitors to the Grand Canyon. This number was up from the 2.15 million visitors in 1984. (*Source*: National Park Service.) Determine and interpret the midpoint of the line segment whose endpoints represent the given data.

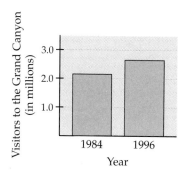

71. Surveyor's Map As shown in the figure, a small pond is drawn in the first quadrant of a coordinate map. A surveyor places pins at points *A*, *B*, and *C*. Points *A* and *C* belong to the line $y = -\frac{8}{9}x + 400$. If coordinates are given in feet, what is the perimeter (to the nearest foot) of triangle *ABC*?

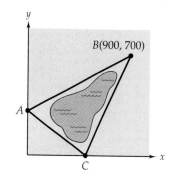

72. Distances Between Towns Three towns are plotted on a coordinate map, with coordinates given in miles. (See figure.) The slope of $\overline{AB}$ is $\frac{1}{2}$ and $\overline{AB} \perp \overline{BC}$. To the nearest tenth of a mile, what is the total distance from *A* to *B* to *C* and back to *A*?

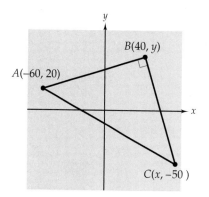

2.5 *Quadratic Functions*

A New Model ■ *The Intercepts of a Parabola* ■ *Graphs of Quadratic Functions* ■ *The Vertex of a Parabola* ■ *The Standard Form* ■ *Real-Life Applications*

A New Model

A linear function may be an appropriate model for data that are increasing or decreasing at an approximately constant rate. However, we often encounter other patterns of data for which a different model is needed.

DEVELOPING THE CONCEPT

Maximum Area of a Garden

Suppose that 200 feet of fencing is available for enclosing a rectangular garden. If we let x represent the width of the garden, then $100 - x$ represents the length. (See Figure 2.31.)

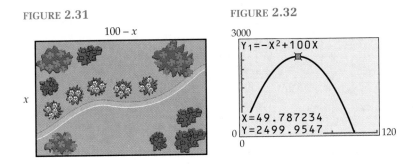

FIGURE **2.31**

FIGURE **2.32**

Recall that the area of a rectangle is the product of its length and its width. Thus, a function that models the area A of the garden is $A(x) = x(100 - x) = -x^2 + 100x$. Figure 2.32 shows a graph of the function $y = -x^2 + 100x$. The highest point of the curve appears to be about (50, 2500). We interpret these coordinates to mean that the maximum area of the garden is 2500 square feet, which is obtained when the width of the garden is 50 feet.

The function $A(x) = -x^2 + 100x$ is an example of a **quadratic function.** Its graph in Figure 2.32 is called a **parabola.**

> ### *Definition of a Quadratic Function*
> A **quadratic function** is a function of the form $P(x) = ax^2 + bx + c$, where $a \neq 0$. The domain of a quadratic function is **R.**

Although a parabola can have any orientation in a coordinate system, the graph of a quadratic function is a "vertical" parabola that opens upward or downward. The rest of our discussion in this section refers only to vertical parabolas.

The Intercepts of a Parabola

We can determine the y-intercept of a parabola algebraically.

$$f(x) = ax^2 + bx + c$$

$$f(0) = a(0)^2 + b(0) + c \qquad \text{The } x\text{-coordinate of a } y\text{-intercept is 0.}$$

$$f(0) = c$$

Thus $(0, c)$ is the y-intercept of any parabola.

An x-intercept of a parabola is a point whose y-coordinate is 0. Thus, to determine an x-intercept algebraically, we must find a value of x for which $f(x) = 0$. We call such an x-value a **zero** of function f. Finding a zero of $f(x) = ax^2 + bx + c$ is the same as finding the solution (or **root**) of the equation $ax^2 + bx + c = 0$.

> ### Zeros, Roots, and x-Intercepts
>
> A **zero** of a function f is a value of x for which $f(x) = 0$. For a quadratic function $f(x) = ax^2 + bx + c$, a zero of f is the same as a solution (or **root**) of the quadratic equation $ax^2 + bx + c = 0$. A zero of function f or a root of the corresponding equation is represented by the x-intercept of the graph of f.

EXAMPLE 1   **Intercepts and Zeros**

(a) Determine the intercepts of the graph of the function $f(x) = x^2 - 4x + 1$.
(b) What are the zeros of the function $f(x) = x^2 + 8x + 16$?

SOLUTION

(a) The function is written in the form $f(x) = ax^2 + bx + c$. Because $c = 1$, the y-intercept is $(0, 1)$.

For the x-intercepts, we determine the zeros of function f. In other words, we solve the equation $x^2 - 4x + 1 = 0$.

$$x^2 - 4x + 1 = 0 \qquad\qquad f(x) = 0$$

$$x = \frac{-b \pm \sqrt{b^2 - 4ac}}{2a} \qquad \begin{array}{l}\text{Use the Quadratic Formula}\\ \text{to determine } x.\end{array}$$

$$x = \frac{4 \pm \sqrt{16 - 4}}{2} \qquad\qquad a = 1, b = -4, c = 1$$

$$= 2 \pm \sqrt{3} \approx 3.73, 0.27$$

The (approximate) x-intercepts are $(0.27, 0)$ and $(3.73, 0)$.

■ *NOTE   In Section 1.4, we referred to a Quadratic Formula program that you can enter in your calculator. If you have done this, then you can use the program to determine zeros and x-intercepts for quadratic functions.*

(b) Because a zero is a value of x for which $f(x) = 0$, we solve the equation $x^2 + 8x + 16 = 0$.

$$x^2 + 8x + 16 = 0 \qquad f(x) = 0$$
$$(x + 4)^2 = 0 \qquad \text{Factor.}$$
$$x + 4 = 0 \qquad \text{Zero Factor Property}$$
$$x = -4$$

The only zero of the function is -4.

Recall that a quadratic equation may have zero, one, or two solutions. Therefore, a quadratic function may have zero, one, or two zeros, and its graph may have zero, one, or two x-intercepts.

Graphs of Quadratic Functions

The simplest quadratic function has the form $f(x) = ax^2$. We now consider the effect of the value of a on the shape and orientation of the graph of f.

EXPLORATION

The Effect of a on a Parabola

Choose several positive values of a and produce the graphs of the functions $y = ax^2$. Then repeat the experiment for negative values of a.

(a) What is the apparent effect of the sign of a on the graph?
(b) As the absolute value of a becomes larger, what is the apparent effect on the graph?

DEVELOPING THE CONCEPT

The Effect of a on a Parabola

Figure 2.33 shows the graph of $y = x^2$ along with graphs of $f(x) = ax^2$ for various positive values of a.

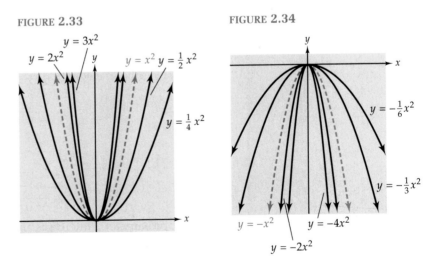

FIGURE 2.33

$y = 3x^2$
$y = 2x^2$
$y = x^2$ $y = \frac{1}{2}x^2$
$y = \frac{1}{4}x^2$

FIGURE 2.34

$y = -\frac{1}{6}x^2$
$y = -\frac{1}{3}x^2$
$y = -x^2$ $y = -4x^2$
$y = -2x^2$

Observe that the graph of $f(x) = ax^2$ is narrower ($a > 1$) or wider ($0 < a < 1$) than the graph of $y = x^2$. Also, note that the parabolas open upward when $a > 0$.

Figure 2.34 shows the graph of $y = -x^2$ along with graphs of $f(x) = ax^2$ for various negative values of a.

Observe that the graph of $f(x) = ax^2$ is narrower ($a < -1$) or wider ($-1 < a < 0$) than the graph of $y = -x^2$. Also, note that the parabolas open downward when $a < 0$.

These observations can be extended to all quadratic functions of the form $f(x) = ax^2 + bx + c$.

The Effect of a on a Parabola

The graphs of $y = x^2$ and $f(x) = ax^2 + bx + c$ are parabolas.

1. If $a > 0$, then the parabola opens upward; if $a < 0$, then the parabola opens downward.

2. If $|a| > 1$, then the graph of f is narrower than the graph of $y = x^2$; if $0 < |a| < 1$, then the graph of f is wider than the graph of $y = x^2$.

Sometimes the graph of a quadratic function is simply the graph of $y = x^2$ shifted to a new location.

EXPLORATION

Vertical and Horizontal Shifts

Produce the graph of $f(x) = x^2$.

(a) On the same coordinate system, produce the graphs of $y = x^2 + k$ for several values of k.
(b) How do the graphs appear to be related to the graph of f?
(c) Now repeat the experiment for functions $y = (x - h)^2$ for several values of h. Based on your observations, how would you describe the graph of $y = (x - h)^2 + k$ relative to the graph of $f(x) = x^2$?

DEVELOPING THE CONCEPT

Vertical and Horizontal Shifts

Figures 2.35 and 2.36 show the graphs of $v(x) = x^2 + k$ and $s(x) = (x - h)^2$ for various values of h and k.

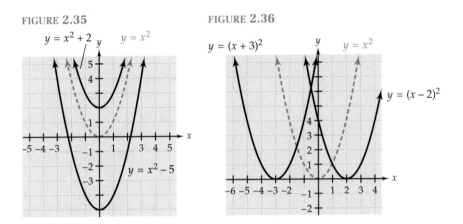

FIGURE **2.35**   FIGURE **2.36**

In Figure 2.35, observe that the graph of $v(x) = x^2 + k$ is the graph of $y = x^2$ shifted $|k|$ units vertically. The graph is shifted upward for $k > 0$ and downward for $k < 0$.

In Figure 2.36, the graph of $s(x) = (x - h)^2$ is the graph of $y = x^2$ shifted $|h|$ units horizontally. The graph is shifted to the right for $h > 0$ and to the left for $h < 0$.

The following is a summary of vertical and horizontal shifts of a parabola $y = x^2$.

Vertical and Horizontal Shifts

The graphs of $v(x) = x^2 + k$, and $s(x) = (x - h)^2$ are parabolas that have the same shape as the parabola $y = x^2$.

1. The graph of $v(x) = x^2 + k$ is the graph of $y = x^2$ shifted $|k|$ units vertically. The graph is shifted upward for $k > 0$ and downward for $k < 0$.

2. The graph of $s(x) = (x - h)^2$ is the graph of $y = x^2$ shifted $|h|$ units horizontally. The graph is shifted to the right for $h > 0$ and to the left for $h < 0$.

Vertical and horizontal shifts can be combined. For instance, to obtain the graph of $g(x) = (x - 4)^2 + 2$, shift the graph of $y = x^2$ to the right 4 units and upward 2 units. (See Figure 2.37.) The graph of $f(x) = -2(x + 1)^2 - 5$ is the graph of $y = -2x^2$ shifted to the left 1 unit and downward 5 units. (See Figure 2.38.)

FIGURE **2.37**

FIGURE **2.38**

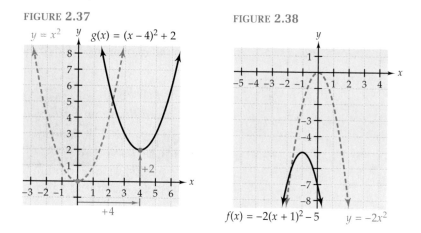

The Vertex of a Parabola

A key point of a parabola is its lowest point (if the parabola opens upward) or its highest point (if the parabola opens downward). This point of the graph is called the **vertex.**

When we shift a parabola horizontally and/or vertically, the vertex shifts accordingly. For example, Figure 2.39 shows the parabola $y = ax^2$ with the vertex $(0, 0)$. By shifting the graph to the right h units and upward k units, we obtain the graph of $f(x) = a(x - h)^2 + k$. Observe that the vertex of the shifted parabola is (h, k).

FIGURE **2.39**

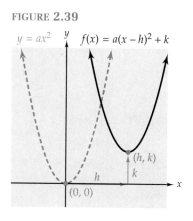

> ### Properties of the Graph of $f(x) = a(x - h)^2 + k$
>
> The graph of $f(x) = a(x - h)^2 + k$ is the graph of $y = ax^2$ with the vertex shifted to the point (h, k).

If a parabola opens upward ($a > 0$), the vertex corresponds to the minimum value of the function. If a parabola opens downward ($a < 0$), the vertex corresponds to the maximum value of the function.

Minimum and Maximum Values of a Function

The graph of the quadratic function $f(x) = a(x - h)^2 + k$ is a parabola with vertex $V(h, k)$.

1. For $a > 0$, $V(h, k)$ is a minimum point of the graph of f. The function f has a minimum value at $x = h$, and the minimum value is $f(h) = k$.

2. For $a < 0$, $V(h, k)$ is a maximum point of the graph of f. The function f has a maximum value at $x = h$, and the maximum value is $f(h) = k$.

EXAMPLE **2**   **Determining the Maximum or Minimum Value**

Determine the maximum or minimum of each of the given functions.

(a) $h(x) = -2(x - 2)^2 - 5$ (b) $g(x) = x^2 + 6x + 9$

SOLUTION

(a) $h(x) = -2(x - 2)^2 - 5$

Because the function is in the form $f(x) = a(x - h)^2 + k$, we see that $a = -2, h = 2$, and $k = -5$, so the vertex is $V(2, -5)$. Because a is negative, the parabola opens downward. Thus, the maximum value of the function is -5 when x is 2. (See Figure 2.40.)

FIGURE **2.40**   FIGURE **2.41**

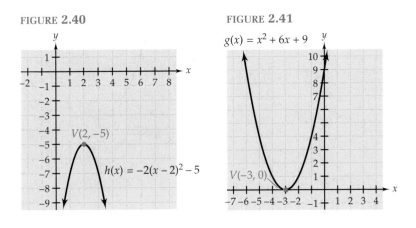

(b) $g(x) = x^2 + 6x + 9$

$g(x) = (x + 3)^2 + 0$ Write the function in the form $f(x) = a(x - h)^2 + k$.

Note that $a = 1, h = -3$, and $k = 0$. The vertex is $V(-3, 0)$, and the parabola opens upward. The minimum value of the function is 0 when x is -3. (See Figure 2.41.)

Maximum (Minimum)   You can move your tracing cursor to the vicinity of the vertex of a displayed parabola and have your calculator display the coordinates of the vertex.

Figure 2.42 shows a calculator's estimate of the maximum of the function in Example 2(a).

FIGURE **2.42**

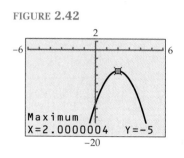

Because the second coordinate of a vertex represents the minimum or maximum value of a function, it can be used to determine the range of the function. For example, the vertex of the graph in Figure 2.42 is $V(2, -5)$. Because the graph opens downward, the maximum value of the function is -5. Thus the range of the function is $(-\infty, -5]$.

The second-quadrant portion of the graph of $y = x^2$ is a mirror image of the first-quadrant portion. (See Figure 2.43.) We say that the graph is *symmetric* with respect to the y-axis. The line $x = 0$ (y-axis) is the **axis of symmetry** for the function $y = x^2$.

FIGURE **2.43**

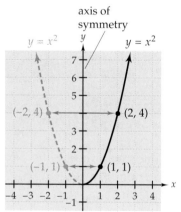

■ *NOTE   A vertical shift or a change in the width of a parabola does not change the axis of symmetry.*

When a parabola is shifted to the left or right, the axis of symmetry is also shifted to the left or right. Thus the axis of symmetry of $y = a(x - h)^2 + k$ is the vertical line containing the vertex $V(h, k)$, and its equation is $x = h$.

Figure 2.44 shows two examples of parabolas along with their vertices, intercepts, and axes of symmetry.

■ *NOTE In general, the axis of symmetry of a parabola can be vertical, horizontal, or diagonal. When the axis of symmetry is vertical, the parabola is a vertical parabola. Note again that only vertical parabolas represent functions.*

FIGURE 2.44

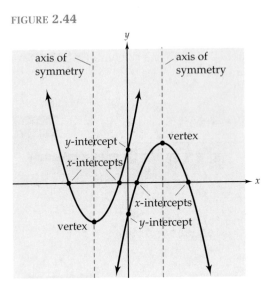

The Standard Form

When a function is written in the *standard form* $f(x) = a(x - h)^2 + k$, the values of h and k help us to determine information about the graph, such as its location relative to the graph of $y = x^2$, the vertex, and the axis of symmetry. For a quadratic function such as $g(x) = 2x^2 - 12x + 7$, which is not in standard form, such information may not be obvious. However, by completing the square, we can write the function g in the standard form.

EXAMPLE 3   **Writing a Quadratic Function in Standard Form**

By completing the square, write $g(x) = 2x^2 - 12x + 7$ in standard form. Describe the features of the graph.

SOLUTION

$$g(x) = 2x^2 - 12x + 7$$
$$g(x) = 2(x^2 - 6x) + 7 \qquad \text{Factor out 2 so that the coefficient of } x^2 \text{ is 1.}$$

To complete the square inside the parentheses, we add $\left(\dfrac{-6}{2}\right)^2 = 9$ to obtain $(x^2 - 6x + 9)$, which is a perfect square trinomial. However, we must also subtract 9 so that the value of the expression is not changed.

$$g(x) = 2(x^2 - 6x + 9 - 9) + 7 \qquad \text{Add and subtract 9 to complete the square.}$$
$$= 2(x^2 - 6x + 9) - 2(9) + 7 \qquad \text{Distributive Property}$$
$$= 2(x - 3)^2 - 18 + 7 \qquad \text{Factor the perfect square trinomial.}$$
$$= 2(x - 3)^2 - 11 \qquad \text{Now the function is in standard form.}$$

Note that $a = 2$, $h = 3$, and $k = -11$. Because a is positive, the graph of g opens upward, it is narrower than the graph of $y = x^2$, and it is shifted right 3 units and down 11 units. The vertex is $V(3, -11)$, and so the range is $[-11, \infty)$. The axis of symmetry is $x = 3$. (See Figure 2.45.)

FIGURE **2.45**

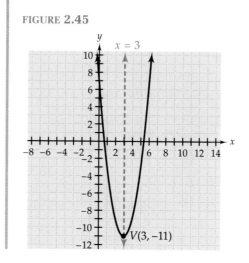

Each time we need to determine the vertex of a parabola, we could complete the square to write the function in standard form, as in Example 3. However, to avoid repeating the process, we can complete the square for the function in general form and use the resulting formula to determine the vertex.

DEVELOPING THE CONCEPT

A Formula for the Coordinates of a Vertex

We begin with a quadratic function $f(x) = ax^2 + bx + c$ and use the method of completing the square to write the function in standard form.

$f(x) = ax^2 + bx + c, a \neq 0$

$f(x) = a\left(x^2 + \dfrac{b}{a}x\right) + c$ Factor out a so that the coefficient of x^2 is 1.

$f(x) = a\left[x^2 + \dfrac{b}{a}x + \left(\dfrac{b}{2a}\right)^2 - \left(\dfrac{b}{2a}\right)^2\right] + c$ Add and subtract $\left(\dfrac{b}{2a}\right)^2$ to complete the square.

$f(x) = a\left[x^2 + \dfrac{b}{a}x + \left(\dfrac{b}{2a}\right)^2\right] - a\left(\dfrac{b}{2a}\right)^2 + c$ Distributive Property

$f(x) = a\left[x + \left(\dfrac{b}{2a}\right)\right]^2 - \dfrac{b^2}{4a} + c$ Factor.

$f(x) = a\left[x - \left(-\dfrac{b}{2a}\right)\right]^2 + \dfrac{4ac - b^2}{4a}$ Combine terms and write the function in standard form.

With the function now written in the standard form, we see that $h = -\dfrac{b}{2a}$.

This gives us a formula for the first coordinate of the vertex. Although the derivation also gives a formula for the second coordinate, we can determine that value simply by evaluating the function at $-\dfrac{b}{2a}$.

A Formula for the Vertex of a Parabola

The vertex of the graph of $f(x) = ax^2 + bx + c$, $a \neq 0$, is the point

$$V\left(-\frac{b}{2a}, f\left(-\frac{b}{2a}\right)\right).$$

VERTEX

Simply by entering a, b, and c, you can use this program to obtain the vertex of the graph of $f(x) = ax^2 + bx + c$.

EXAMPLE 4 **Using the Vertex to Write the Standard Form**

Determine the vertex of the graph of $f(x) = 2x^2 - 8x + 7$. Then write the function in standard form.

SOLUTION
The x-coordinate of the vertex is given by

$$-\frac{b}{2a} = -\frac{-8}{2(2)} = 2 \qquad a = 2, b = -8, c = 7$$

To determine the y-coordinate, evaluate $f(2)$.

$$f(x) = 2x^2 - 8x + 7$$
$$f(2) = 2(2)^2 - 8(2) + 7 = -1$$

The vertex is $(2, -1)$. Using this point and knowing that $a = 2$, we write the standard form.

$$f(x) = 2(x - 2)^2 - 1 \qquad f(x) = a(x - h)^2 + k$$

EXAMPLE 5 **Writing a Quadratic Function**

The vertex of a parabola is $V(-2, 3)$ and one x-intercept is $P(-1, 0)$. Write a quadratic function whose graph is the parabola.

SOLUTION

$f(x) = a(x - h)^2 + k$

$f(x) = a(x + 2)^2 + 3$ The vertex is $V(-2, 3)$, so replace h with -2 and k with 3.

$0 = a(-1 + 2)^2 + 3$ An x-intercept is $(-1, 0)$, so $f(-1) = 0$.

$a = -3$ Solve for a.

The function is $f(x) = -3(x + 2)^2 + 3$.

Real-Life Applications

Applications that are modeled with quadratic functions often involve determining the maximum or minimum value of the function.

A business receives *revenue* (or income) for sales of goods or services and incurs *cost* (or expenses) associated with operating the business. The *profit* is the difference between revenue and cost.

$$\text{Profit} = \text{Revenue} - \text{Cost}$$

The point at which the revenue and cost are equal is called the *break-even point*.

EXAMPLE 6   ## Shuttle Service

A small company provides airport shuttle bus service. With its five vans, the company's average daily revenue is $500 per van and the average daily cost of operating the business is $200 per van. For each additional van the company places in service, the average daily revenue per van will decrease by $5 and the average daily cost per van will increase by $15. How many vans should the company operate in order to maximize the profit?

SOLUTION

Let x = the number of *additional* vans that the company operates. Then $5 + x$ = the total number of vans.

$500 - 5x$ = the average revenue per van   The revenue is decreased by $5 per additional van.

$200 + 15x$ = the average cost per van   The cost is increased by $15 per additional van.

The total revenue $R(x)$ is the average revenue multiplied by the number of vans.

$$R(x) = (500 - 5x)(5 + x) = 2500 + 475x - 5x^2$$

Similarly, the total cost $C(x)$ is the average cost times the number of vans.

$$C(x) = (200 + 15x)(5 + x) = 1000 + 275x + 15x^2$$

The profit $P(x)$ is the difference between revenue $R(x)$ and cost $C(x)$.

$$P(x) = R(x) - C(x) = (2500 + 475x - 5x^2) - (1000 + 275x + 15x^2)$$
$$= 1500 + 200x - 20x^2$$

FIGURE **2.46**

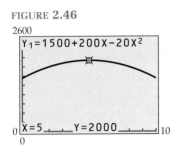

Because a is negative ($a = -20$), the parabola opens downward and the vertex is a maximum point of the graph. The estimated vertex is $V(5, 2000)$. (See Figure 2.46.)

The x-coordinate of the vertex is determined as follows:

$$x = -\frac{b}{2a} = -\frac{200}{2(-20)} = 5 \qquad a = -20 \text{ and } b = 200$$

Because x represents the number of *additional* vans, the maximum daily profit is achieved when 5 additional vans (or a total of 10 vans) are used. The profit is

$$P(5) = 1500 + 200(5) - 20(5)^2 = 2000$$

The maximum daily profit of $2000 is achieved when 10 vans are used.

2.5 Quick Reference

A New Model
- A **quadratic function** is a function of the form $f(x) = ax^2 + bx + c, a \neq 0$.
- The graph of a quadratic function is a **parabola.**
- The domain of a quadratic function is **R.**

The Intercepts of a Parabola
- The y-intercept of the graph of $f(x) = ax^2 + bx + c$ is $(0, c)$.
- A **zero** of a function f is a value of x for which $f(x) = 0$. For a quadratic function $f(x) = ax^2 + bx + c$, a zero of f is the same as a solution (or **root**) of the quadratic equation $ax^2 + bx + c = 0$. A zero of function f or a root of the corresponding equation is represented by an x-intercept of the graph of f.
- A vertical parabola has one y-intercept and zero, one, or two x-intercepts.

Graphs of Quadratic Functions
- The graph of $f(x) = ax^2 + bx + c$ opens upward if $a > 0$ and downward if $a < 0$.
- If $|a| > 1$, then the graph of f is narrower than the graph of $y = x^2$; if $0 < |a| < 1$, then the graph of f is wider than the graph of $y = x^2$.
- The graph of $v(x) = x^2 + k$ is the graph of $y = x^2$ shifted $|k|$ units vertically. The graph is shifted upward if $k > 0$ and downward for $k < 0$.
- The graph of $s(x) = (x - h)^2$ is the graph of $y = x^2$ shifted $|h|$ units horizontally. The graph is shifted right if $h > 0$ and left if $h < 0$.
- The graph of $f(x) = a(x - h)^2 + k$ is the graph of $y = ax^2$ with the vertex shifted to the point (h, k).

The Vertex of a Parabola
- The **vertex** is the highest ($a < 0$) or lowest ($a > 0$) point of a parabola.
- The graph of a quadratic function $f(x) = a(x - h)^2 + k$ has the vertex $V(h, k)$.
  - (a) For $a > 0$, f has a minimum value at $x = h$, and the minimum value is $f(h) = k$. The range is $[k, \infty)$.
  - (b) For $a < 0$, f has a maximum value at $x = h$, and the maximum value is $f(h) = k$. The range is $(-\infty, k]$.
- The **axis of symmetry** of a vertical parabola is the vertical line $x = h$ that contains the vertex $V(h, k)$.

The Standard Form ■ The *standard form* of a quadratic function is $f(x) = a(x - h)^2 + k$, where the vertex of the graph is $V(h, k)$.

■ For $f(x) = ax^2 + bx + c$, the function can be written in standard form by the method of completing the square.

(a) The first coordinate of the vertex can be found by using the formula

$$h = -\frac{b}{2a}.$$

(b) The second coordinate of the vertex can be found by evaluating

$$f\left(-\frac{b}{2a}\right).$$

2.5 Speaking the Language

1. A function of the form $f(x) = ax^2 + bx + c$, $a \neq 0$, is called a(n) ▨▨▨▨ function, and its graph is a(n) ▨▨▨▨ .

2. For a given function g, if $g(4) = 0$, then 4 is called a(n) ▨▨▨▨ of the function.

3. The lowest point of the graph of $h(x) = x^2 + x + 1$ is called the ▨▨▨▨ .

4. If the highest point of the graph of $f(x) = -x^2 + 3x - 7$ has the coordinates (h, k), then the line whose equation is $x = h$ is called the ▨▨▨▨ .

2.5 Exercises

Concepts and Skills

1. What feature of the graph of a quadratic function is associated with a zero of the function?

2. Suppose that $f(x) = a_1 x^2$ and $g(x) = a_2(x - 3)^2 + 5$. Describe the relationship between the graphs of f and g if $a_1 = a_2$.

In Exercises 3–8, produce the graphs of $y = x^2$ and the given function in the same coordinate system. Describe the graph of f relative to the graph of $y = x^2$.

3. $f(x) = 9 - x^2$

4. $f(x) = \frac{1}{2}(x - 4)^2$

5. $f(x) = (x + 2)^2 - 5$

6. $f(x) = (x - 3)^2 + 4$

7. $f(x) = -(x + 4)^2 + 2$

8. $f(x) = 7 - (x - 1)^2$

In Exercises 9–14, match the function with the given graph.

(A) $y = -x^2 - 1$

(B) $y = (x - 3)^2$

(C) $y = (x + 3)^2 + 1$

(D) $y = -(x + 1)^2$

(E) $y = -(x - 4)^2 - 3$

(F) $y = x^2 + 2$

9.

10.

11.

12.

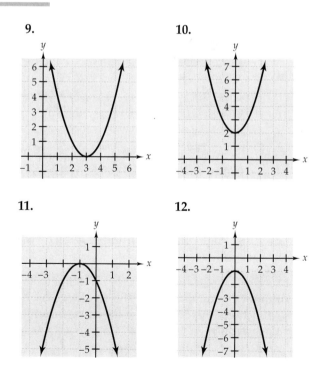

13.

14.

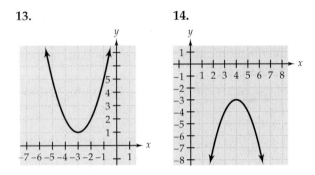

In Exercises 15–20, determine the vertex and the axis of symmetry of the graph of the given function.

15. $g(x) = \dfrac{1}{2}x^2 - 8$ **16.** $f(x) = 10 - x^2$

17. $y(x) = -(x - 4)^2 + 1$ **18.** $r(x) = (x + 3)^2 - 6$

19. $p(x) = 2(x + 1)^2 + 3$

20. $f(x) = -3(x - 2)^2 - 7$

In Exercises 21–24, use a graph to estimate the intercepts of the graph of the function. Then determine the intercepts algebraically.

21. $f(x) = 9x^2 - 12x + 4$

22. $g(x) = x^2 + 10x + 25$

23. $g(x) = x^2 + 8x + 7$ **24.** $h(x) = 12 + 4x - x^2$

In Exercises 25–28, use the method of completing the square to write the given function in the form $y = a(x - h)^2 + k$. Then determine the vertex and the axis of symmetry.

25. $h(x) = 3 + 4x - 2x^2$

26. $f(x) = -3x^2 + 6x - 4$

27. $f(x) = 2x^2 - x + 1$

28. $w(x) = -4x^2 + 5x + 7$

29. Explain why knowing the vertex of a parabola allows you to determine the range of the associated function.

30. For the function $f(x) = ax^2 + bx + c$, suppose that you know the vertex and the value of a. Describe the conditions under which the parabola has two x-intercepts.

In Exercises 31–36, use a graph to estimate the maximum or minimum value and the range of the given function. Then determine the same information algebraically.

31. $f(x) = 6 + 3x - x^2$ **32.** $g(x) = 3 + 2x + x^2$

33. $g(x) = x^2 - 5x - 6$

34. $h(x) = -x^2 + 8x - 4$

35. $f(x) = -\dfrac{2}{3}x^2 - 7x + 10$

36. $f(x) = \dfrac{1}{2}x^2 - x + 4$

In Exercises 37–42, write a quadratic function whose graph has the given characteristics.

37. Vertex $(-3, -2)$; y-intercept $(0, 7)$

38. Vertex $(1, 5)$; y-intercept $(0, 4)$

39. Vertex $(-2, -8)$; x-intercepts $(0, 0)$ and $(-4, 0)$

40. Vertex $(1, 16)$; x-intercepts $(-3, 0)$ and $(5, 0)$

41. Vertex $(-4, -5)$; point of the parabola $(-5, -7)$

42. Vertex $(6, 0)$; point of the parabola $(-2, 32)$

Concept Extension

In Exercises 43–48, for a function $f(x) = ax^2 + bx + c$, the value of a and the vertex are given. Determine the range and the number of zeros of the function.

43. $2, (3, -3)$ **44.** $-3, (4, 7)$

45. $-1, (-4, -1)$ **46.** $1, (-2, 5)$

47. $\dfrac{1}{2}, (5, 0)$ **48.** $-\dfrac{5}{3}, (-7, 0)$

In Exercises 49–52, information is given about the graph of a quadratic function. Determine the number of x-intercepts.

49. The vertex is $(2, -1)$ and the y-intercept is $(0, -8)$.

50. The vertex is $(3, 0)$.

51. The maximum value of the function occurs at $(7, 3)$.

52. The minimum value of the function occurs at $(-1, 2)$.

In Exercises 53–58, determine the value of k such that the statement is true.

53. 4 is a zero of $f(x) = kx^2 - 9x + 2k$.

54. The vertex of $f(x) = 1 + 2kx - x^2$ is $(3, 10)$.

55. The minimum value of $f(x) = kx^2 - 4kx + 4$ is -8.

56. The maximum value of $f(x) = kx^2 - 3kx - 1$ is 3.5.

57. $x = -\dfrac{3}{2}$ is the axis of symmetry of the function $f(x) = k^2x^2 + 6kx + 9$.

58. $\{y \mid y \le 9\}$ is the range of $f(x) = \dfrac{k}{2}x^2 - kx + 8$.

59. Suppose that the maximum value of the function $f(x) = k - 3kx - 3x^2$ is 5. Show that two functions satisfy the condition.

60. Suppose that the graph of $f(x) = ax^2 + bx + c$ contains the origin. What is the value of c?

61. How is the value of the discriminant related to the zeros of a quadratic function? to the number of x-intercepts of the graph of a quadratic function?

62. Is the graph of $y = (|x| - 2)^2 + 3$ a parabola? Show how the graph can be described in terms of the graphs of quadratic functions.

Applications

63. Holiday Shopping A gift store models its pre-Christmas sales with the quadratic equation $y = -0.02x^2 + 0.74x + 2.29$, where y is the sales (in thousands of dollars) and x is the number of days since November 24. In the following questions, round all coordinates to the nearest integer.

(a) Produce the graph of the equation and trace it to estimate the highest point of the graph. What is your interpretation of those coordinates?

(b) Trace the graph to estimate the store's sales on the day before Christmas.

(c) Trace the graph to estimate the date(s) on which the store's sales are predicted to be $5000.

64. Car Seat Production Costs The daily production cost for a manufacturer of infant car seats is modeled by $C(x) = 0.5x^2 - 70x + 5600$, where x is the number of seats produced. How many seats should be produced to minimize production cost?

65. Flight of a Ball When a soccer goalie kicks a ball, the height (in feet) of the ball after t seconds is modeled by $h(t) = -16t^2 + 64t$. What is the maximum height of the ball and how long is the ball in the air?

66. Mining Industry Employment In 1970, the mining industry had 623,000 employees. The number increased, and then declined to 605,000 in 1994. (*Source:* U.S. Bureau of Labor Statistics.) The number (in thousands) of employees can be modeled by $E(t) = -2.63t^2 + 61.8t + 623.8$, where t is the number of years since 1970.

(a) Produce the graph of E and trace it to estimate the year in which the number of employees was at its peak.

(b) Estimate the time period in which employment was greater than 900,000.

67. Foreign Tourism The number of foreign travelers to the United States increased from 39.5 million in 1990 and then decreased to 43 million in 1995. (*Source:* U.S. Travel and Tourism Association.) The quadratic function $F(t) = -0.86t^2 + 5t + 39.5$ models the number (in millions) of travelers, where t is the number of years since 1990. Use a graph to estimate the maximum number of visitors and the year in which the maximum occurred.

68. Tomato Production A commercial tomato grower knows that planting 300 tomato plants in his field will result in 4 pounds of tomatoes per plant. However, with less crowding, production will increase by 0.1 pound per plant. By what number should the 300 tomato plants be decreased to maximize production?

69. Charter Boat A charter boat company conducts sightseeing trips for groups of up to 50 passengers. The cost of operating the boat is $C(x) = 500 + 4.5x$ for x passengers. The price of a ticket is $20 plus 50 cents for each unsold seat.

(a) Write the revenue function R.

(b) Write a profit function P.

(c) Determine the maximum profit.

(d) Determine the number of passengers for maximum profit.

(e) How many passengers are needed to break even?

70. Sandwich Shop Chain The owner of a chain of sandwich shops has 20 locations in a city, and the business averages 300 customers per location per day. For each additional shop, the customer average per location would drop by 10. How many additional shops should the owner open to maximize the total number of customers?

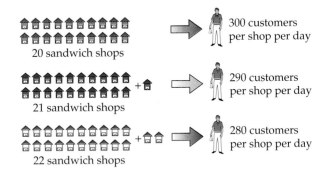
20 sandwich shops → 300 customers per shop per day

21 sandwich shops → 290 customers per shop per day

22 sandwich shops → 280 customers per shop per day

71. Exhibit Size At a technology conference, the setup crew uses 150 feet of dividers to form three exhibit areas in one corner of the building. (See figure.) Existing walls form the back and one end. What is the maximum area that can be enclosed?

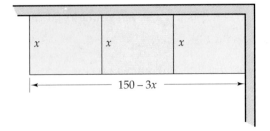
x x x
$150 - 3x$

72. Stadium Dimensions An athletic stadium is rectangular with a semicircle at each end. (See figure.) The perimeter of the stadium is 3000 feet. What dimensions will maximize the area of the rectangular portion of the stadium?

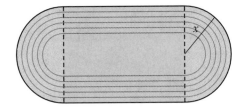
x

2.6 *Quadratic Regression Models*

Choosing a Model ■ *Quadratic Regression* ■
Real-Life Applications

When you attempt to model real-world data, your first decision must be the type of model to use. We have seen that a linear function is often the best model for linearly related data. If the plotted data points do not appear to approximate a straight line, then the data may not be linearly related. When data do not appear to be linear, other models should be considered.

The type of model that you choose should also be influenced by your knowledge and experience. For example, a business may have certain cyclical patterns or other historical tendencies that would make one model superior to another. You may also be aware of future planning that is likely to affect the data. The type of model that you choose may also be influenced by what you know about the behavior of the basic functions. For example, if the plotted data points appear to approximate a parabola, then we might choose a quadratic function as a model. In short, although models of data can be created, the validity of those models is always a matter of common sense and judgment.

Choosing a Model

Suppose that you are the owner of a medium-size business, and that the number of your employees has been increasing over the past four years. (See Figure 2.47.)

FIGURE **2.47**

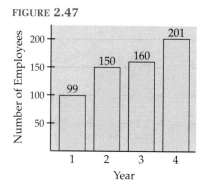

Figures 2.48, 2.49, and 2.50 show scatterplots of the data given in the bar graph. Because the data appear to be linearly related, we might choose a linear function to model the data. One such model, Figure 2.48, might be the line that contains the data points for the first and fourth years.

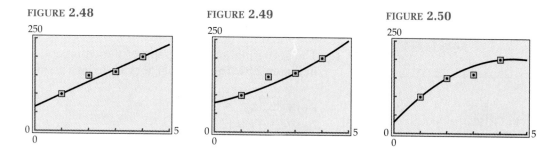

FIGURE **2.48**   FIGURE **2.49**   FIGURE **2.50**

Figure 2.49 shows another possible model that contains the data points for years 1, 3, and 4. Finally, the model in Figure 2.50 contains the data for years 1, 2, and 4. As you might suspect, these last two models are parabolas.

Which model you, as the business owner, would choose would depend on your vision about the growth of your company. The linear model would be a good choice if you anticipate a relatively steady increase in the number of employees. On the other hand, you may have reason to expect a rapid increase in your business, in which case the model in Figure 2.49 might be better. The model in Figure 2.50 might be more appropriate if you anticipate a flattening out or even a downturn in your business.

Quadratic Regression

Each of the preceding models ignores at least one of the data points, and the graphs are drawn so that they will contain exact points. If we want to take *all* of the data points into account, then we need a graph that fits the points as closely as possible.

If we believe that a parabola would fit the data best, then we need to create a **quadratic regression equation** of the parabola. This equation has the form $y = ax^2 + bx + c$.

Regression After you have entered data in the statistics list, you can select *quadratic regression* (sometimes listed as *QuadReg*) to create a quadratic regression equation. At least three data points must be entered.

EXAMPLE 1 **Producing a Quadratic Regression Equation**

(a) Referring again to the bar graph, which gives the number of employees by year, enter (1, 99), (2, 150), (3, 160), and (4, 201) in the statistics list.
(b) Produce the quadratic regression equation that models the data.
(c) Produce a scatterplot of the data along with the graph of the regression equation.
(d) Interpret the value of a in the regression equation.

SOLUTION

FIGURE 2.51

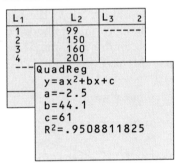

(a) Figure 2.51 shows the entry of the data in the statistics list.
(b) Figure 2.51 also shows the calculation of the values of a, b, and c for the quadratic regression. Substituting, we obtain the equation $y = -2.5x^2 + 44.1x + 61$.
(c) The scatterplot and the graph of the regression equation are shown in Figure 2.52. The parabola appears to be a good fit for the data.

FIGURE 2.52

$$y = -2.5x^2 + 44.1x + 61$$

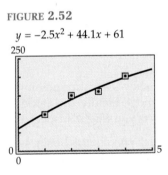

(d) Because $a < 0$, the parabola opens downward. This would indicate a slowing down in hiring and perhaps even some downsizing in the more distant future.

Your calculator may or may not display the correlation coefficient for quadratic regression equations. As shown in Figure 2.51, some calculators display the value of r^2 rather than r.

Real-Life Applications

Because the vertex of a vertical parabola is the highest or lowest point of the curve, we sometimes make use of a quadratic regression equation to estimate or predict a maximum or minimum value. The regression equation has the form $y = ax^2 + bx + c$. Knowing a and b allows us to determine the coordinates of the vertex.

EXAMPLE **2**   **Enrollment at Two-Year Colleges**

Figure 2.53 shows the fall enrollments (in millions) at two-year colleges for selected years in the period 1986–1996. (*Source*: National Center for Educational Statistics.)

FIGURE **2.53**

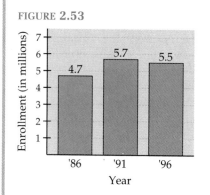

(a) Produce a quadratic regression equation to model the given data.
(b) According to your regression equation, in what year did enrollment peak?

SOLUTION

(a) We enter ordered pairs of the form (year, enrollment) in the statistics list. For convenience, we let the first coordinate be the number of years since 1986:

$$(0, 4.7), (5, 5.7), (10, 5.5)$$

Then we select the quadratic regression option and obtain the equation $y = -0.024x^2 + 0.32x + 4.7$.

(b) Note that $a = -0.024$ and $b = 0.32$. Because $a < 0$, the parabola opens downward and the function has a maximum (peak) value at the vertex. The first coordinate of the vertex is

$$-\frac{b}{2a} = -\frac{0.32}{2(-0.024)} = 6.\overline{6}$$

Thus we estimate the peak year to be 7 years after 1986, or 1993.

EXAMPLE 3 **Bankruptcy Filings**

Huge credit card debts, attorney advertising, and legal loopholes are just a few of the many causes cited for the explosion in bankruptcy filings in the past 50 years. The accompanying table shows the number of filings from 1950 to 1997. (*Source*: Administrative Office of the United States Courts.)

Year	Bankruptcy Filings
1950	20,000
1960	100,000
1970	200,000
1980	300,000
1990	700,000
1997	1,300,000

(a) Let x represent the number of years since 1940, and let y represent the number of bankruptcy filings. Produce a quadratic regression equation for the data in the bar graph. Round the values of a, b, and c to the nearest integer.
(b) Determine the correlation coefficient. How well does the regression equation model the data?
(c) Use the regression equation to estimate the number of bankruptcy filings in the year 2000.
(d) Use the regression equation to estimate the year when there were 450,000 bankruptcy filings.

SOLUTION

(a) The ordered pairs are (10, 20,000), (20, 100,000), and so on. The quadratic regression equation is

$$f(x) = 816x^2 - 30{,}792x + 309{,}234$$

(b) The correlation coefficient $r \approx 0.98$, which indicates that the regression equation is a good model of the data.

(c) For the year 2000, $x = 60$, and $f(60) \approx 1{,}400{,}000$. Thus the estimated number of bankruptcy filings in the year 2000 is 1,400,000.

(d) We can trace the graph of f to the point whose second coordinate is 450,000. (See Figure 2.54.) The first coordinate is approximately 42, which indicates that the estimated year is $1940 + 42 = 1982$.

Alternatively, we can use the Quadratic Formula to solve the following equation:

$$816x^2 - 30{,}792x + 309{,}234 = 450{,}000$$

The positive solution is approximately 42, which again indicates that the estimated year is 1982.

FIGURE 2.54

$y_1 = 816x^2 - 30{,}792x + 309{,}234$

$y_2 = 450{,}000$

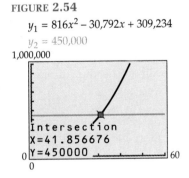

Linear and quadratic regression equations are not the only models for data. As we discuss other types of functions, we will see that other kinds of regression models may be more appropriate for certain data patterns.

However you choose to model data, keep in mind that a model is only as good as the known data. There are limitations on the extent to which estimations and predictions can be made with any model. Nevertheless, modeling can be a valuable tool in the analysis and interpretation of information, which is the essential basis for decision making.

2.6 Quick Reference

Choosing a Model
- Scatterplots of data often reveal a pattern that helps us decide whether a linear, quadratic, or other kind of model is most appropriate.
- The choice of a model should also be based on experience and on knowledge concerning factors that are likely to affect the trends in the data.

Quadratic Regression
- When we believe that a parabola would be a good model of data, we can use a calculator to produce a **quadratic regression equation.**
- Superimposing the graph of a quadratic regression equation on a scatterplot helps us to judge how well the curve fits the data.
- Your calculator may report the correlation coefficient r, or it may report the value of r^2. Although r is an indicator of how well a regression equation models data, it does not suggest a cause-and-effect relationship.

2.6 Exercises

In Exercises 1–4, produce a scatterplot of the three given data points and a regression equation of the parabola that contains the points. Graph the equation on the same coordinate system as the scatterplot and verify that the points belong to the parabola.

1. $(2, 5), (6, 9), (8, 35)$

2. $(-3, -24), (1, 16), (8, 9)$

3. $(-7, 23), (-3, 2.2), (9, 16.6)$

4. $(-6, 9.2), (3, -0.7), (6, -14.8)$

In Exercises 5–8, you are given a scatterplot of a certain set of data. By observing the pattern of the data, judge whether a linear regression equation or a quadratic regression equation might be an appropriate model.

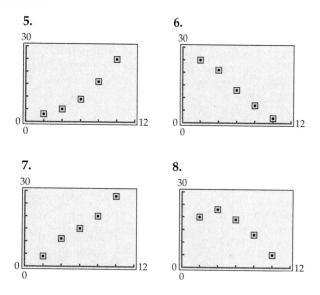

5.

6.

7.

8.

In Exercises 9–14, produce a first-quadrant scatterplot of the data in the table and decide whether a quadratic model of the data is reasonable. If you think so, produce and graph a quadratic regression equation on the same screen as the scatterplot. Judging from the correlation coefficient, would you say that a quadratic model still seems reasonable?

9.
x	0	2	4	6	8	10
y	25	16	9	12	20	27

10.
x	5.0	10.0	15.0	20.0	25.0	30.0
y	17.1	36.5	48.3	40.5	26.8	20.3

11.
x	2	5	8	11	14	17
y	11	14	18	21	24	26

12.
x	10.0	16.0	22.0	28.0	34.0	40.0
y	87.1	79.3	73.7	67.1	62.6	55.4

13.
x	3	4	5	6	7	8
y	56	151	213	187	144	78

14.
x	6.0	12.0	15.0	20.0	23.0	30.0
y	89.3	52.1	45.6	60.2	65.7	97.9

15. Cellular Phone Rates In the early 1990s, only two cellular telephone carriers were allowed in each U.S. city. As shown in the bar graph, deregulation has increased competition and lowered rates. (*Source*: Herschel Shosteck Associates.)

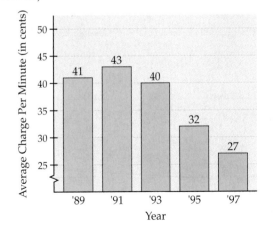

(a) If the data can be modeled by some quadratic function $f(x) = ax^2 + bx + c$, what is your prediction about the sign of a?

(b) Just by examining the bar graph, estimate the vertex of the graph of f.

(c) Letting x represent the number of years since 1985, produce a quadratic regression equation of the data. Produce the graph of the equation and trace it to estimate the vertex. What do the coordinates of that point represent?

16. Carbon Dioxide Emissions In 1983, the atmosphere in the United States was polluted by 1.2 billion tons of carbon dioxide emissions. By 1995, the emissions had reached 1.45 billion tons. (*Source*: U.S. Department of Energy.) The goal set by environmentalists for the year 2010 is to reduce emissions to 1.38 billion tons.

(a) Letting x represent the number of years since 1980, write a quadratic regression equation to model the emissions.

(b) What emissions level does your model project for the year 2005?

17. Internet Service Providers The bar graph shows the number (and projected number) of Internet service providers through the year 2002. (*Source*: Gartner Group.)

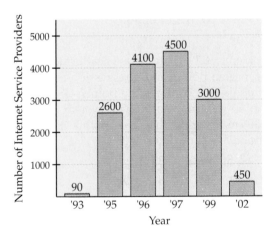

(a) Use a scatterplot of the data to help you decide what type of model is reasonable.

(b) Let x represent the number of years since 1990 and determine a quadratic regression equation to model the data. (Round coefficients to the nearest integer.)

(c) Produce the scatterplot and regression equation graph on the same coordinate system. For which two years does the model most underestimate the data?

18. **Stock and Bond Funds** The table gives the total number of stock funds and bond funds in selected years. (*Source*: Investment Company Institute.)

Year	Number of funds
1991	244
1993	653
1994	756
1996	541

(a) By examining a scatterplot of the data, decide what type of model is appropriate.
(b) Let x represent the number of years since 1990 and determine a quadratic regression equation to model the data. (Round coefficients to the nearest integer.)
(c) Use the model to estimate the year(s) in which the number of funds is approximately 400.
(d) Estimate and interpret the vertex of the graph of the model function.

19. **Grants to Museums** The table shows the amount (in millions of dollars) awarded in grants to museums by the National Endowment for the Arts for selected years. (*Source*: National Endowment for the Arts.)

Year	Amount
1980	11.2
1985	11.9
1989	12.7
1991	11.3
1994	9.4

(a) Letting x represent the number of years since 1980, determine an appropriate regression model for the data.
(b) Does the value of the correlation coefficient indicate that the model is reasonable?
(c) Interpret the x-intercept that has a positive x-coordinate.

20. **Business Majors** The table shows the percentage of first-year college students who say that business is their probable major. (*Source*: Higher Education Research Institute.)

Year	Percentage
1970	16
1985	27
1990	21
1995	16

(a) Letting x represent the number of years since 1970, determine an appropriate regression model for the data.
(b) Does the value of the correlation coefficient indicate that the model is reasonable?
(c) Determine and interpret the vertex of the parabola.

21. **Sporting Equipment Sales** The table shows sales (in millions of dollars) of baseball and softball equipment in the United States for selected years. (*Source*: National Sporting Goods Association.)

Year	Sales
1991	214
1992	245
1993	323
1994	305
1995	317

(a) Let x represent the number of years since 1990 and produce both a linear regression equation and a quadratic regression equation.
(b) Which model do you think best fits the data?

22. **Imported Cars** The table shows the number (in thousands) of cars imported from Germany to the United States for selected years. (*Source*: American Automobile Manufacturing Association.)

Year	Imports
1980	339
1985	473
1990	245
1992	206
1994	188

(a) Let x represent the number of years since 1980 and produce both a linear regression equation and a quadratic regression equation.
(b) Which model do you think best fits the data?

2.7 *Graphs of Basic Functions*

Intercepts ■ *Increasing and Decreasing Functions* ■
Relative Maxima and Minima ■ *Symmetry* ■
Common Functions ■ *Discontinuous Functions*

We can often observe or estimate important properties of functions by examining the graph of the function. In this section we introduce some characteristics of graphs and illustrate them with basic functions that will be used throughout the subsequent chapters.

Intercepts

In Figure 2.55, the graph crosses the x-axis at points A, B, and C, and it crosses the y-axis at point P.

FIGURE **2.55**

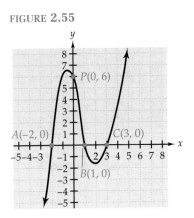

Recall that a point at which a graph intersects an axis is called an intercept. Note again that the y-coordinate of an x-intercept is 0, and the x-coordinate of a y-intercept is 0.

The graph in Figure 2.55 represents a function because it passes the Vertical Line Test. Note that the graph of a function may have any number of x-intercepts but at most one y-intercept.

Referring to the x-intercepts in Figure 2.55, we see that $f(-2) = 0, f(1) = 0,$ and $f(3) = 0$. Recall that the numbers -2, 1, and 3 are called zeros of the function because they are all values of x for which $f(x) = 0$.

EXAMPLE **1**   **Intercepts and Zeros**

Produce the graph of $f(x) = \frac{1}{4}x^3 - x^2 - 5x + 12$.

(a) Use the graph to estimate the zeros of function f.
(b) Verify your estimates in part (a).

SOLUTION

(a) VISUALIZING THE ZEROS

Figure 2.56 shows the graph of function f with the tracing cursor on one x-intercept, $(6, 0)$. This means that $f(6) = 0$, and so 6 is a zero of function f.

FIGURE **2.56**

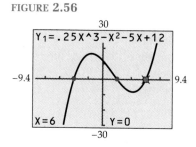

We can also trace to the other two x-intercepts, $(-4, 0)$ and $(2, 0)$. Thus the apparent zeros of function f are $-4, 2,$ and 6.

(b) DETERMINING THE ZEROS

Figure 2.57 shows function f evaluated at $x = -4, 2,$ and 6.

Because $f(x) = 0$ for these three values of x, the zeros $-4, 2,$ and 6 are verified.

FIGURE **2.57**

Y₁(-4)	
	0
Y₁(2)	
	0
Y₁(6)	
	0

Increasing and Decreasing Functions

When we analyze a function, we are often interested in knowing the intervals in which the graph rises or falls from left to right.

DEVELOPING THE CONCEPT

Graphs That Rise and Fall

Consider the graph in Figure 2.58.

FIGURE **2.58**

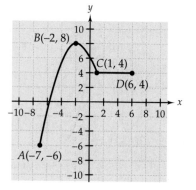

If we think of moving from left to right along the graph, we observe that the graph rises from point *A* to point *B*, then falls from point *B* to point *C*, and finally becomes horizontal from point *C* to point *D*. Correspondingly, we say that the function is **increasing,** then **decreasing,** and then **constant.**

We specify where the function exhibits these properties by describing the associated intervals of *x*-values (that is, intervals of the domain). In Figure 2.58 we note the *x*-coordinates of points *A*, *B*, *C*, and *D* to produce the following summary.

The function is	*In the interval*
Increasing	$(-7, -2)$
Decreasing	$(-2, 1)$
Constant	$(1, 6)$

The following is a general summary of these properties of a function.

> **Increasing, Decreasing, and Constant Functions**
>
> 1. If the graph of a function *f* rises from left to right throughout the interval (a, b), then we say that *f* is **increasing** in that interval.
> 2. If the graph of a function *f* falls from left to right throughout the interval (a, b), then we say that *f* is **decreasing** in that interval.
> 3. If the graph of a function *f* is horizontal throughout the interval (a, b), then we say that *f* is **constant** in that interval.

EXAMPLE 2   **Increasing and Decreasing Functions**

Figure 2.59 shows the graph of a function *g*. Identify the intervals in which *g* is (a) increasing and (b) decreasing.

FIGURE 2.59

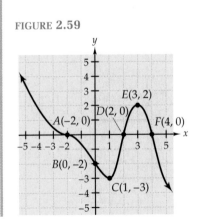

SOLUTION

(a) The function is increasing on the interval $(1, 3)$.
(b) The function is decreasing on the intervals $(-\infty, 1)$ and $(3, \infty)$.

Relative Maxima and Minima

As we saw when we studied quadratic functions, applications often involve determining the maximum or minimum value of the function. For quadratic functions, those values are associated with the vertex of a parabola. Many other types of functions have graphs that exhibit peaks and dips that are associated with maximum or minimum values.

DEVELOPING THE CONCEPT

Graphs with Peaks and Dips

Consider the function whose graph is shown in Figure 2.60.

FIGURE **2.60**

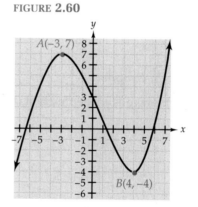

Looking from left to right, we see that the graph increases to a peak and then begins to decrease. The point $(-3, 7)$, where the graph changes from increasing to decreasing, corresponds to a **relative maximum** (plural: maxima) of the function.

Similarly, the point $(4, -4)$, where the graph changes from decreasing to increasing, corresponds to a **relative minimum** (plural: minima) of the function.

■ *NOTE   If the graph of a function f peaks (or dips) at some point P(x, y), then the relative maximum (or minimum) of f is the y-coordinate of point P.*

Although there are special cases that are exceptions, we can generally think of a peak of a graph as indicating a relative maximum and a dip as indicating a relative minimum.

EXAMPLE 3   **Properties of a Graph**

Referring to Figure 2.61, estimate the following.

(a) Coordinates of all intercepts
(b) Relative maxima and minima
(c) Intervals where the function is increasing and decreasing

FIGURE 2.61

SOLUTION

(a) x-intercepts: $(-5, 0), (-1, 0), (3, 0)$
    y-intercept: $(0, 1.5)$

(b) Relative maximum: $(1, 2)$
    Relative minima: $(-3, -2), (3, 0)$

(c) Increasing: $(-3, 1), (3, \infty)$
    Decreasing: $(-\infty, -3), (1, 3)$

Symmetry

The shape of a graph may be such that one part is a mirror image of the other part. We refer to this property as *symmetry*.

DEVELOPING THE CONCEPT

Symmetry in Graphs

Consider the graphs of the functions in Figure 2.62.

FIGURE 2.62

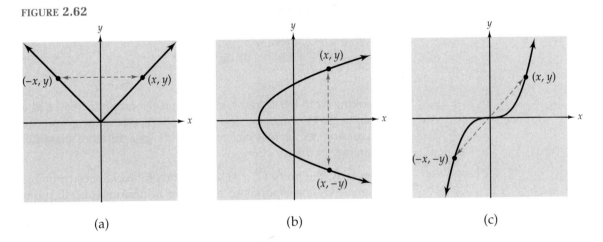

(a)   (b)   (c)

In Figure 2.62(a), the left and right sides of the graph are mirror images. More specifically, for each point (x, y) of the graph, $(-x, y)$ is also a point of the graph. We refer to such graphs as **symmetric with respect to the y-axis.**

The upper branch of the graph in Figure 2.62(b) is a mirror image of the lower branch. For each point (x, y) of the graph, $(x, -y)$ is also a point of the graph. Such graphs are said to be **symmetric with respect to the x-axis.**

In Figure 2.62(c), the portion of the graph in Quadrant I is a reversed reflection of the portion in Quadrant III. Observe that for each point (x, y) of the graph, $(-x, -y)$ is also a point of the graph. We refer to such graphs as **symmetric with respect to the origin.**

The graph in Figure 2.62(a) is an example of an **even** function, and the graph in Figure 2.62(c) is an example of an **odd** function. (Note that the graph in Figure 2.62(b) is not the graph of a function.)

Definitions of Even and Odd Functions

1. A function is **even** if for every x in the domain, $-x$ is also in the domain and $f(-x) = f(x)$. An even function is symmetric with respect to the y-axis.
2. A function is **odd** if for every x in the domain, $-x$ is also in the domain and $f(-x) = -f(x)$. An odd function is symmetric with respect to the origin.

EXAMPLE **4**   **Identifying Symmetry in a Graph**

Identify the type of symmetry exhibited by each of the graphs in Figure 2.63. If the graph represents a function, determine whether the function is even or odd.

FIGURE **2.63**

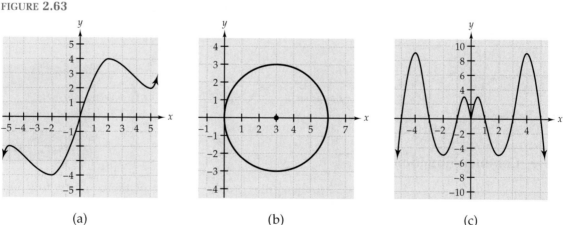

(a)   (b)   (c)

SOLUTION

(a) The graph is symmetric with respect to the origin, and the function is odd.
(b) The graph is symmetric with respect to the x-axis, but the graph does not represent a function.
(c) The graph is symmetric with respect to the y-axis, and the function is even.

EXAMPLE 5 **Identifying Even and Odd Functions**

Determine whether the given function is even, odd, or neither.

(a) $f(x) = \dfrac{1}{x^2}, x \neq 0$

(b) $g(x) = x + 4$

(c) $h(x) = x$

SOLUTION

(a)
$$f(x) = \frac{1}{x^2}$$

$$f(-x) = \frac{1}{(-x)^2} = \frac{1}{x^2}$$

Because $f(-x) = f(x)$, f is an even function.

(b)
$$g(x) = x + 4$$
$$g(-x) = -x + 4$$

Because $g(-x) \neq g(x)$, g is not an even function.

$$g(x) = x + 4$$
$$g(-x) = -x + 4$$
$$-g(x) = -(x + 4) = -x - 4$$

Because $g(-x) \neq -g(x)$, g is not an odd function.

(c)
$$h(x) = x$$
$$h(-x) = -x$$
$$-h(x) = -x$$

Because $h(-x) = -h(x)$, h is an odd function.

Common Functions

Figure 2.64 shows the graphs of some of the common functions that we encounter in algebra. In each case, the name of the function is given along with its domain and range.

FIGURE **2.64**

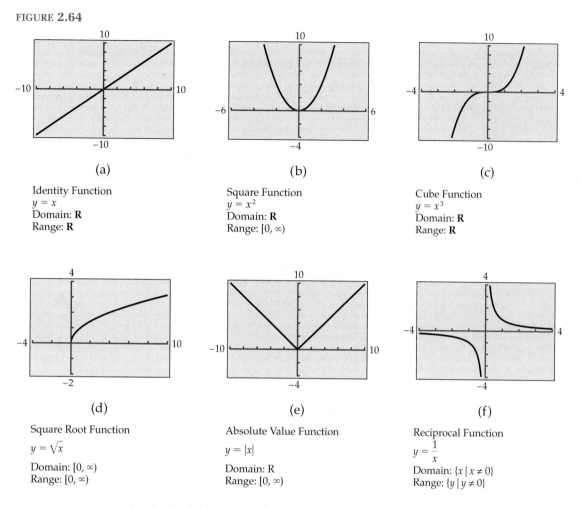

(a)

Identity Function
$y = x$
Domain: **R**
Range: **R**

(b)

Square Function
$y = x^2$
Domain: **R**
Range: $[0, \infty)$

(c)

Cube Function
$y = x^3$
Domain: **R**
Range: **R**

(d)

Square Root Function
$y = \sqrt{x}$
Domain: $[0, \infty)$
Range: $[0, \infty)$

(e)

Absolute Value Function
$y = |x|$
Domain: R
Range: $[0, \infty)$

(f)

Reciprocal Function
$y = \dfrac{1}{x}$
Domain: $\{x \mid x \neq 0\}$
Range: $\{y \mid y \neq 0\}$

In Figure 2.64, no graph is symmetric with respect to the x-axis. If it were, the graph would not represent a function. However, some of the graphs exhibit symmetry with respect to the y-axis or the origin.

EXAMPLE **6** **Symmetry of Graphs**

Which of the six graphs shown in Figure 2.64 exhibit symmetry with respect to the

(a) y-axis?   (b) origin?

SOLUTION

(a) The square function and the absolute value function are symmetric with respect to the y-axis.
(b) The identity function, the cube function, and the reciprocal function are symmetric with respect to the origin.

Note that the square root function has no symmetry.

EXAMPLE 7 **Increasing and Decreasing Functions**

For each of the functions whose graphs are shown in Figure 2.64, identify the intervals in which the function is

(a) increasing.
(b) decreasing.

SOLUTION

(a) The identity function, the cube function, and the square root function are all increasing over their entire domains.

The square function and the absolute value function are increasing in the interval $(0, \infty)$.

(b) The reciprocal function is decreasing over its entire domain.

The square function and the absolute value function are decreasing in the interval $(-\infty, 0)$.

Discontinuous Functions

The graphs in parts (a) through (e) of Figure 2.64 are graphs that we can draw by hand with one continuous motion; that is, we do not have to lift the pencil from the paper. These functions are examples of **continuous** functions.

By contrast, drawing the graph of the reciprocal function [see Figure 2.64(f)] requires us to lift the pencil as we cross the *y*-axis. Therefore, the reciprocal function is not continuous.

Although a precise definition of a continuous function is beyond the scope of this book, we can use the informal definition that a function is continuous over an interval of its domain if we can draw the graph in that interval without lifting the pencil. Any point at which the pencil must be lifted is a *point of discontinuity*.

The basic functions may be used to define a piecewise function, which may or may not be continuous. Points of discontinuity, if any, often occur where one function ends and another function begins. The graph of a piecewise function will usually reveal such points.

EXAMPLE 8 **A Piecewise Function**

Consider the following piecewise function:

$$f(x) = \begin{cases} x^2 & -5 \le x < 4 \\ x & x \ge 4 \end{cases}$$

(a) Sketch the graph of f.
(b) What are the domain and range of f?
(c) Is the function continuous? If not, identify the point(s) of discontinuity.

SOLUTION

(a) Figure 2.65 shows the graph of function f.

FIGURE **2.65**

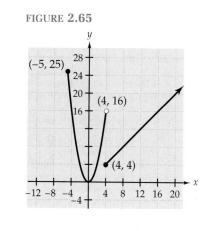

(b) Domain: $[-5, \infty)$; Range: $[0, \infty)$

(c) The function is discontinuous at $x = 4$.

Mode   When, for example, you produce the graph of a piecewise function, extraneous lines may appear in the display. These can be prevented by selecting the *Dot* mode.

Figure 2.66 shows a *Dot* mode display of function f in Example 8.

FIGURE **2.66**

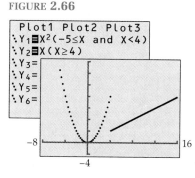

2.7   Quick Reference

Intercepts   ■   The previously stated definitions of intercepts and zeros can be extended to all functions and their graphs.

1. An *x-intercept* of a graph is a point at which the graph intersects the *x*-axis. A *y-intercept* of a graph is a point at which the graph intersects the *y*-axis.

2. The *zeros* of a function f are all values of x for which $f(x) = 0$.

Increasing and Decreasing Functions

■ If the graph of function f rises from left to right throughout the interval (a, b), then we say that f is **increasing** in that interval.

■ If the graph of a function f falls from left to right throughout the interval (a, b), then we say that f is **decreasing** in that interval.

■ If the graph of a function f is horizontal throughout the interval (a, b), then we say that f is **constant** in that interval.

Relative Maxima and Minima

■ Roughly, a peak of a graph indicates a relative maximum of a function, and a dip indicates a relative minimum.

Symmetry

■ A graph is **symmetric with respect to the y-axis** if for each point (x, y) of the graph, $(-x, y)$ is also a point of the graph.

■ A graph is **symmetric with respect to the x-axis** if for each point (x, y) of the graph, $(x, -y)$ is also a point of the graph.

■ A graph is **symmetric with respect to the origin** if for each point (x, y) of the graph, $(-x, -y)$ is also a point of the graph.

■ A function is **even** if for every x in the domain, $-x$ is also in the domain and $f(-x) = f(x)$. An even function is symmetric with respect to the y-axis.

■ A function is **odd** if for every x in the domain, $-x$ is also in the domain and $f(-x) = -f(x)$. An odd function is symmetric with respect to the origin.

Common Functions

■ The graphs of some common functions are shown in Figure 2.64.

Discontinuous Functions

■ Roughly, a function is continuous over an interval of its domain if we can draw the graph in that interval without lifting the pencil. Any point at which the pencil must be lifted is a *point of discontinuity*.

■ The basic functions may be used to define a piecewise function, which may or may not be continuous.

2.7 Speaking the Language

1. If a function rises from left to right throughout the interval (a, b), then we say that f is ▨▨▨▨ in that interval.

2. A peak of the graph of a function f represents a(n) ▨▨▨▨ of the function, whereas a dip represents a(n) ▨▨▨▨.

3. If for each point (x, y) of a graph, $(x, -y)$ is also a point of the graph, then we say that the graph is ▨▨▨▨ with respect to the ▨▨▨▨.

4. If for every x in the domain of a function f, $-x$ is also in the domain and $f(x) = f(-x)$, then we call f a(n) ▨▨▨▨ function.

2.7 Exercises

Concepts and Skills

1. Describe the coordinates of the intercepts of a graph.

2. Suppose that function f is increasing in the interval $(1, 7)$. Compare the values of $f(2)$ and $f(5)$.

In Exercises 3–10, produce a graph of the given function. Use the graph to estimate the intercepts.

3. $f(x) = 2x - 7$

4. $h(x) = -3x$

5. $g(x) = -x^2 + x - 2$

6. $q(x) = 2x^2 + 7x - 4$

7. $p(x) = x^3 - 7x - 6$

8. $f(x) = x^4 - x^3 - 9x^2 + 9x$

9. $c(x) = 4 - \sqrt{25 - x^2}$

10. $h(x) = \sqrt{x^2 - 144} - 5$

11. Suppose that function g is decreasing in the interval $(-5, -1)$ and is increasing in the interval $(-1, 6)$. What can you conclude about $g(-1)$?

12. If the graph of a function f has one peak at $P(-3, 6)$, explain how to determine the relative maximum of f.

In Exercises 13 and 14, the graph of a function is given. Estimate the intervals in which the function is increasing, decreasing, and constant.

13.   **14.**

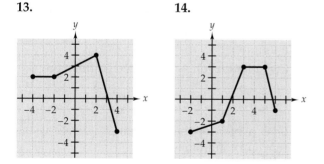

In Exercises 15 and 16, the graph of a function is given. Estimate the intercepts, relative maxima, and relative minima. Then estimate the intervals in which the function is increasing and decreasing.

15.   **16.**

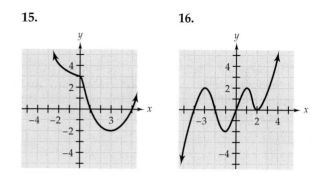

In Exercises 17–24, begin with the standard setting to produce a graph of the given function. Then, if necessary, adjust the settings for the y-axis so that peaks and dips are visible. Estimate the relative maxima and relative minima. Then estimate the intervals in which the function is increasing, decreasing, and constant.

17. $f(x) = x^2 + 2x - 8$

18. $g(x) = -x^2 + 8x - 16$

19. $h(w) = w^3 - 6w^2 - 15w + 12$

20. $P(x) = -x^3 + 9x^2 - 20x - 5$

21. $f(t) = 4t^2 - t^4 - 2$

22. $h(x) = \dfrac{1}{4}x^4 + x^3 - 2x^2 + 6$

23. $g(t) = 3 - |t - 1|$

24. $h(x) = |1 - x| - 4$

In Exercises 25–28, use the standard setting to produce a graph of the given function. Estimate the relative maxima and relative minima. Then estimate the intervals in which the function is increasing, decreasing, and constant.

25. $f(x) = \dfrac{3 - x}{x^2 + 1}$

26. $T(s) = \dfrac{s^2}{s^2 + 4}$

27. $r(t) = t\sqrt{16 - t^2}$

28. $g(x) = \dfrac{x}{\sqrt{x^2 - 4}}$

29. Describe the symmetries, if any, of even and odd functions.

30. Why is a graph that is symmetric with respect to the *x*-axis not a graph of a function?

In Exercises 31–34, sketch the geometric figure with the given vertices. Then sketch the figure that is symmetric with respect to (a) the *x*-axis and (b) the *y*-axis. Give the vertices of the resulting figure.

31. $(2, 3), (-3, -1), (2, -1)$

32. $(2, 5), (4, -1), (-2, 1)$

33. $(1, -4), (3, -2), (4, 6), (-6, -4)$

34. $(0, 1), (0, 9), (10, -1), (4, -3)$

In Exercises 35–38, a graph is shown on the interval $[0, 3]$. Complete the graph if the graph is symmetric with respect to (a) the *y*-axis, (b) the *x*-axis, and (c) the origin. (*Hint*: For parts (a) and (c), sketch the graph in the interval $[-3, 3]$, and for part (b), sketch the graph in the interval $[0, 3]$.)

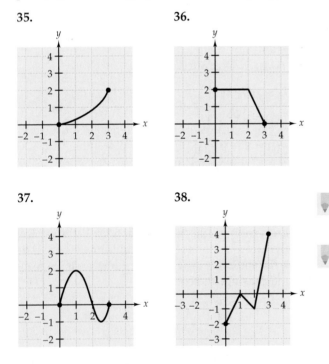

35.

36.

37.

38.

In Exercises 39–48, determine whether the function is even, odd, or neither. Begin by producing the graph of the function to help you decide. Then verify your conclusion algebraically.

39. $f(x) = -\dfrac{3}{x}$

40. $h(x) = -2x^3$

41. $h(x) = x^2 - 5$

42. $g(x) = 5 - 2x$

43. $f(x) = |x - 2|$

44. $g(x) = 3 - |x|$

45. $f(x) = -2x^4 + 3x^2 + 1$

46. $h(x) = 2x^3 - x$

47. $g(x) = \dfrac{-x}{x^2 + 1}$

48. $g(x) = \dfrac{x^2}{x^2 + 1}$

In Exercises 49 and 50, determine whether the graph of the equation is symmetric with respect to the *x*-axis. (*Hint*: The graph of an equation is symmetric with respect to the *x*-axis if replacing *y* with $-y$ results in the same equation.)

49. $x = y^2 - 5$

50. $x + 2 = |y|$

In Exercises 51–56, describe all symmetries.

51. $x = y^2$

52. $x^2 = y^2$

53. $xy = 1$

54. $xy^2 = 1$

55. $|x| + |y| = 5$

56. $y = x|x|$

57. For the functions $f(x) = x$, $g(x) = x^2$, and $h(x) = x^3$, explain why only one has a relative minimum.

58. Consider the following piecewise function:

$$f(x) = \begin{cases} x & x < 3 \\ x^2 & x \geq 3 \end{cases}$$

Over which of the following intervals is the function continuous? Why?

(i) **R**   (ii) $[0, \infty)$ (iii) $(-\infty, 0]$

In Exercises 59–62, use the basic functions described in this section to write a piecewise function whose graph is given.

59.

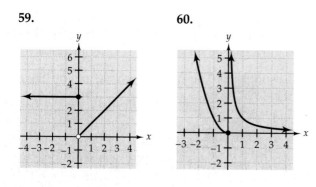

60.

61.

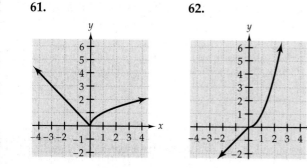

62.

69. $g(x) = \begin{cases} x^2 & x \le 0 \\ x & 0 < x < 3 \\ -2 & x \ge 3 \end{cases}$

70. $p(x) = \begin{cases} |x| & x < -4 \\ 4 & -4 \le x \le 2 \\ x^2 & x > 2 \end{cases}$

Applications

71. Vehicle Thefts The bar graph shows the number of vehicle thefts (in millions) in the United States for three selected years. (*Source*: Federal Bureau of Investigation.)

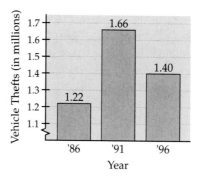

The number of thefts T (in millions) can be modeled by $T(x) = -0.014x^2 + 0.158x + 1.22$, where x is the number of years since 1986.

(a) Produce the graph of T and trace it to estimate the relative maximum of the function. Compare your result to the data shown in the bar graph.
(b) How would you describe T in the interval $[0, 4]$? in the interval $[6, 10]$?

In Exercises 63–70, sketch the graph of the given piecewise function. Identify all points of discontinuity.

63. $f(x) = \begin{cases} \dfrac{1}{2}x + 1 & x \le 2 \\ x & x > 2 \end{cases}$

64. $g(x) = \begin{cases} 4 & x < 0 \\ -x + 1 & x \ge 0 \end{cases}$

65. $h(x) = \begin{cases} x + 2 & x \le -3 \\ -1 & -3 < x < 2 \\ 3 - x & x \ge 2 \end{cases}$

66. $f(x) = \begin{cases} x^2 & x \le 0 \\ x + 2 & 0 < x \le 3 \\ x - 1 & x > 3 \end{cases}$

67. $f(x) = \begin{cases} \dfrac{x^2 - 4}{x + 2} & x \ne -2 \\ 5 & x = -2 \end{cases}$

68. $f(x) = \begin{cases} \dfrac{x^2 - 1}{x - 1} & x \ne 1 \\ 2 & x = 1 \end{cases}$

72. Education and Success In 1978, 49% of the population believed that a college education was needed to succeed. By 1997, the percentage had risen to 75%. (*Source*: CBS News Poll.) Suppose that you plotted these two data points and drew a straight line through the points.

(a) Would the resulting graph represent an increasing or decreasing function?
(b) Because the line would not be horizontal, it would have an x-intercept. What would be the significance of the coordinates of that point? How meaningful do you think that point would be?

73. Sorority Membership A new college sorority was founded in 1995. The following table shows the membership for the years 1994 to 1998.

Year	Membership
1994	0
1995	12
1996	24
1997	35
1998	47

If you were to sketch a continuous graph of the membership data, which of the graphs of common functions in Figure 2.64 would your graph most resemble?

74. Book Sales When a well-known novelist published a new book, sales increased rapidly for the first four months. During the rest of the year, sales continued to increase, but at a slower rate. If you were to sketch a continuous graph of the sales, which of the graphs of common functions in Figure 2.64 would your graph most resemble?

75. Speed and Time Suppose that you drive your car 100 miles. As the following table shows, increasing your speed reduces the time of the trip.

Speed	Time
20 mph	5 hours
25 mph	4 hours
40 mph	2.5 hours
50 mph	2 hours
60 mph	$1\frac{2}{3}$ hours

If you were to sketch a continuous graph of the time data, which of the graphs of common functions in Figure 2.64 would your graph most resemble?

76. Software Sales A software company's sales of a spreadsheet product were high in 1994, but sales declined rapidly until there were no sales at all in 1997. The company then introduced a new version of the product, and sales increased rapidly over the next three years. If you were to sketch a continuous graph of the sales data, which of the graphs of common functions in Figure 2.64 would your graph most resemble?

77. Cost of Mail The accompanying graph shows the 1997 charges for first-class mail up to 6 ounces.

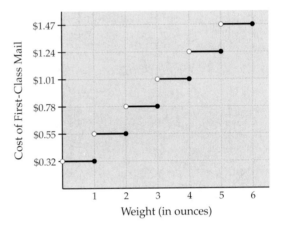

(a) Write a piecewise function C for the cost (in cents) of a letter weighing x ounces.
(b) Evaluate function C for the following weights and compare your results with the data shown in the graph.
 (i) 0.6 oz (ii) 2.9 oz (iii) 5.1 oz

78. Floor Area One of the stores in a shopping mall is to be rectangular with a total floor area of 6000 square feet.

(a) Letting W represent the width of the store, write a function L for the length.
(b) What is the domain of L?
(c) Is L increasing or decreasing over its domain?
(d) Explain why L has no relative maximum or minimum.

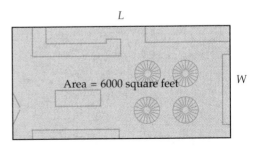

2.8 *Transformations*

Transforming a Graph ■ *Translations* ■ *Distortions* ■
Reflections ■ *Combined Transformations*

In our discussion of quadratic functions and their graphs, we saw how a parabola can open upward or downward, how the breadth of a parabola can vary, and how a parabola can be shifted horizontally and/or vertically.

We can extend those concepts to the graphs of the basic functions that we introduced in Section 2.7. We can change the location, the shape, or the orientation of those graphs. Any such change to the graph of a basic function is called a **transformation** of the graph.

Understanding how the graph of a basic function is transformed increases our ability to visualize or sketch the graphs of many types of functions.

Transforming a Graph

In this section we consider three kinds of transformations of the graphs of basic functions.

1. The result of shifting a graph vertically or horizontally without changing the shape of the graph is called a **translation** of the graph. A translation is also called a *rigid transformation* because the shape of the graph remains unchanged.
2. A **distortion** results from making a graph wider or narrower or from increasing or decreasing the rate at which the graph rises or falls. A distortion is also called a *nonrigid transformation* because the shape of the graph is changed.
3. A mirror image of a graph is called a **reflection.** Usually the "mirror" is the x-axis or the y-axis. A graph and its reflection exhibit symmetry with respect to the x- or y-axis.

Although we will examine changes to many of the basic functions in more detail in subsequent chapters, this section is an introduction to common transformations that are applicable to all functions.

Translations

We translate a graph by copying it at a new location.

EXPLORATION

Translating the Graph of $y = |x|$

(a) Consider several absolute value functions of the form $y = |x| + c$. For each function that you select, produce its graph in the same coordinate system as the basic absolute value function $y = |x|$. In general, what appears to be the effect of adding c to a function?

(b) Consider several absolute value functions of the form $y = |x + c|$ and repeat the experiment in part (a). In general, what appears to be the effect of adding c to the variable?

(c) Are your observations in parts (a) and (b) consistent with the results that we obtained for parabolas of the form $y = x^2 + k$ and $y = (x - h)^2$?

DEVELOPING THE CONCEPT

Translating the Graph of $y = |x|$

In Figure 2.67(a) we see the graph of $y = |x|$ along with the results of shifting (translating) the graph up 4 units and down 6 units. The same graph has been translated 3 units to the right and 5 units to the left in Figure 2.67(b).

FIGURE 2.67

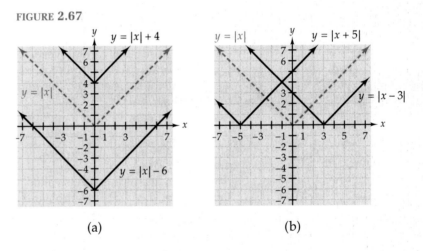

(a)   (b)

Note that in Figure 2.67(a), a constant is added to or subtracted from $|x|$ in the original function $y = |x|$. The new function has the form $y = |x| \pm c$, and the resulting graph is translated vertically.

In Figure 2.67(b), a constant is added to or subtracted from x in the original function $y = |x|$. The new function has the form $y = |x \pm c|$, and the resulting graph is translated horizontally.

The following summary lists the ways in which the graph of a function f can be translated horizontally or vertically.

> ***Translations of Graphs***
>
> Suppose that c is a positive number.
>
> *Vertical translations*
>
> 1. The graph of $f(x) + c$ is obtained by shifting the graph of f *upward* c units.
> 2. The graph of $f(x) - c$ is obtained by shifting the graph of f *downward* c units.
>
> *Horizontal translations*
>
> 1. The graph of $f(x + c)$ is obtained by shifting the graph of f to the *left* c units.
> 2. The graph of $f(x - c)$ is obtained by shifting the graph of f to the *right* c units.

EXAMPLE 1   **Translations of the Cube Function**

(a) Use the graph of the cube function $f(x) = x^3$ to sketch the graphs of $g(x) = x^3 + 6$ and $h(x) = x^3 - 4$.

(b) Write the function k whose graph is the result of shifting the graph of the cube function upward 2 units.

SOLUTION

(a) Because $g(x) = x^3 + 6 = f(x) + 6$, the graph of g is the graph of f translated 6 units upward. Similarly, $h(x) = x^3 - 4 = f(x) - 4$, so the graph of h is the graph of f translated 4 units downward. (See Figure 2.68.)

FIGURE **2.68**

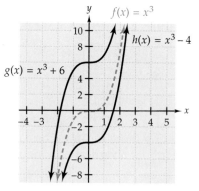

(b) Shifting $f(x) = x^3$ upward 2 units results in the new function $k(x) = f(x) + 2 = x^3 + 2$.

EXAMPLE 2 **Translations of the Square Root Function**

(a) Use the graph of the square root function $f(x) = \sqrt{x}$ to sketch the graphs of $g(x) = \sqrt{x+8}$ and $h(x) = \sqrt{x-6}$.

(b) Write the functions whose graphs are obtained by shifting the graph of $f(x) = \sqrt{x}$

  (i) to the left 7 units.   (ii) to the right 2 units.

SOLUTION

(a) Because $g(x) = \sqrt{x+8} = f(x+8)$, the graph of g is the graph of f shifted 8 units to the left. Similarly, $h(x) = \sqrt{x-6} = f(x-6)$, so the graph of h is the graph of f shifted 6 units to the right. (See Figure 2.69.)

FIGURE **2.69**

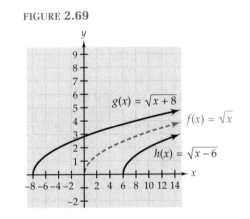

(b) Shifting the graph of $f(x)$ to the left 7 units results in the new function $g(x) = f(x+7) = \sqrt{x+7}$.

Shifting the graph of $f(x)$ to the right 2 units results in the new function $h(x) = f(x-2) = \sqrt{x-2}$.

Distortions

A distortion (nonrigid transformation) results in a change in the shape of a graph without a vertical or horizontal change in its location.

In our earlier work with parabolas of the form $y = ax^2$, we found that increasing $|a|$ narrows the parabola, whereas decreasing $|a|$ widens the parabola. The same general principle applies to the graphs of other basic functions. In particular, we will examine the graph of $y = cf(x)$, where c is a positive number.

EXPLORATION

Distortions of the Graph of f(x) = $\sqrt{x}$

Consider functions of the form $y = c\sqrt{x}$.

(a) For several values of $c > 1$, produce the graph of the function in the same coordinate system with $f(x) = \sqrt{x}$. What appears to be the effect of c on the graph?

(b) Use several values of c, $0 < c < 1$, and repeat the experiment in part (a). What appears to be the effect of c on the graph?

(c) Are your observations in parts (a) and (b) consistent with the results for parabolas of the form $y = ax^2$?

DEVELOPING THE CONCEPT

Distortions of the Graph of f(x) = $\sqrt{x}$

Figure 2.70 shows some distortions (nonrigid transformations) of the graph of the square root function, $f(x) = \sqrt{x}$.

FIGURE **2.70**

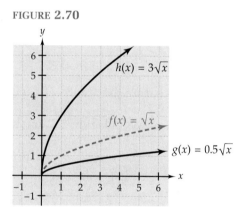

Note that $g(x) = 0.5\sqrt{x} = 0.5 \cdot f(x)$ and that the graph of g rises more gradually than the graph of f.

We also observe that $h(x) = 3\sqrt{x} = 3 \cdot f(x)$ and that the graph of h rises more rapidly than the graph of f.

These observations support the following generalization.

> **Distortions of Graphs**
>
> 1. If $c > 1$, then the graph of $y = c \cdot f(x)$ rises more rapidly than the graph of f.
> 2. If $0 < c < 1$, then the graph of $y = c \cdot f(x)$ rises more gradually than the graph of f.

EXAMPLE **3**   **A Distortion of the Cube Function**

Figure 2.71 shows the graphs of $f(x) = x^3$ and $g(x) = cx^3$.

FIGURE **2.71**

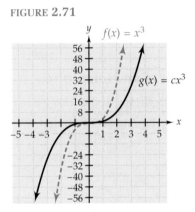

(a) By comparing the graphs of f and g, what can you predict about the value of c?

(b) Use the graph in Figure 2.71 to complete the following table.

x	$f(x)$	$g(x)$
2		
4		

(c) Use the results in part (b) to write the specific function g and to determine the value of c.

SOLUTION

(a) Because the graph of g rises more gradually than the graph of f, $0 < c < 1$.

(b)

x	$f(x)$	$g(x)$
2	8	1
4	64	8

(c) From the table, we see that $f(x) = 8 \cdot g(x)$. Solving for $g(x)$, we obtain $g(x) = \frac{1}{8} \cdot f(x)$. Thus $c = \frac{1}{8}$.

Reflections

The last transformations that we consider are **reflections** of a graph across the x- and y-axes.

DEVELOPING THE CONCEPT

Reflecting Graphs Across the Axes

Figure 2.72 shows the graphs of $f(x) = |x|$ and $g(x) = -|x|$. Note that $g(x) = -f(x)$ and that, considered together, the graphs appear to be symmetric with respect to the x-axis. We say that the graph of g is a reflection of the graph of f across the x-axis.

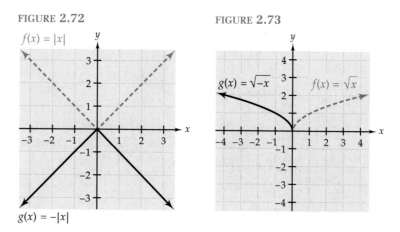

FIGURE **2.72**

$f(x) = |x|$

$g(x) = -|x|$

FIGURE **2.73**

$g(x) = \sqrt{-x}$ $f(x) = \sqrt{x}$

In Figure 2.73, we see the graphs of $f(x) = \sqrt{x}$ and $g(x) = \sqrt{-x}$. Note that $g(x) = f(-x)$ and that, considered together, the graphs appear to be symmetric with respect to the y-axis. We say that the graph of g is a reflection of the graph of f across the y-axis.

These results suggest the following methods for reflecting graphs across the axes.

Reflections of Graphs

1. To obtain the graph of $-f(x)$, reflect the graph of f across the x-axis.
2. To obtain the graph of $f(-x)$, reflect the graph of f across the y-axis.

EXAMPLE 4   **Reflections of the Cube Function**

Use the graph of the cube function $f(x) = x^3$ to obtain the graphs of $g(x) = -x^3$ and $h(x) = (-x)^3$. Why do we obtain the same results?

SOLUTION

Because $g(x) = -x^3 = -f(x)$, the graph of g is the reflection of the graph of f across the x-axis. [See Figure 2.74(a).]

FIGURE 2.74

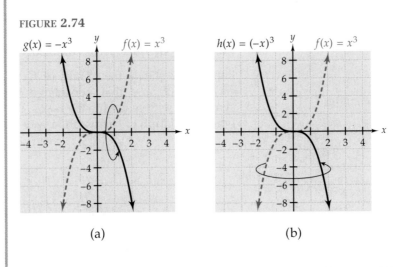

(a)   (b)

Also, $h(x) = (-x)^3 = f(-x)$, so the graph of h is the reflection of the graph of f across the y-axis. [See Figure 2.74(b).] Because $(-x)^3 = -x^3$, the graphs of g and h are the same.

Combined Transformations

The graph of a basic function may be transformed in two or more ways at once.

EXAMPLE 5   **Combined Transformations**

Explain how to obtain the graphs of the following functions. Then sketch the graphs.

(a) $g(x) = |x + 3| - 6$ (b) $h(x) = -\sqrt{x} + 3$

SOLUTION

(a) To obtain the graph of $g(x) = |x + 3| - 6$, shift the graph of the basic absolute value function $f(x) = |x|$ left 3 units and down 6 units. [See Figure 2.75(a).]

FIGURE **2.75**

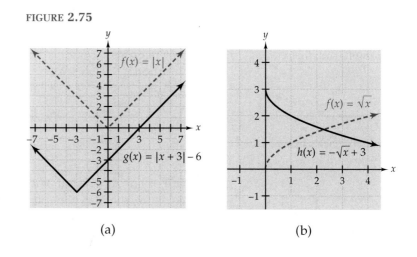

(a)   (b)

(b) To obtain the graph of $h(x) = -\sqrt{x} + 3$, reflect the graph of the basic square root function $f(x) = \sqrt{x}$ across the x-axis and shift the result up 3 units. [See Figure 2.75(b).]

2.8 Quick Reference

Transforming a Graph
- A **transformation** of a graph is a change in the location, shape, or orientation of the graph.
- A **translation** (rigid transformation) is a vertical or horizontal shift of a graph.
- A **distortion** (nonrigid transformation) is a change in the shape of a graph.
- A **reflection** is a mirror image of a graph, usually across the x-axis or the y-axis.

Translations
- If c is a positive number, then the graph of $f(x) + c$ is obtained by shifting the graph of f *upward* c units; the graph of $f(x) - c$ is obtained by shifting the graph of f *downward* c units.
- If c is a positive number, then the graph of $f(x + c)$ is obtained by shifting the graph of f to the *left* c units; the graph of $f(x - c)$ is obtained by shifting the graph of f to the *right* c units.

Distortions
- A distortion (nonrigid transformation) results in a change in the shape of a graph without a vertical or horizontal change in its location.
  1. If $c > 1$, then the graph of $y = c \cdot f(x)$ rises more rapidly than the graph of f.
  2. If $0 < c < 1$, then the graph of $y = c \cdot f(x)$ rises more gradually than the graph of f.

Reflections	■ To obtain the graph of $-f(x)$, reflect the graph of f across the x-axis; to obtain the graph of $f(-x)$, reflect the graph of f across the y-axis.
Combined Transformations	■ A graph may be a combination of transformations of the graph of a basic function.

2.8 Speaking the Language

1. A vertical or horizontal shift of a graph is called a(n) ▨▨▨▨▨, whereas a change in the shape of a graph is called a(n) ▨▨▨▨▨.

2. If $f(x) = x^2$, then the graph of $g(x) = -x^2$ is the ▨▨▨▨▨ of the graph of f across the x-axis.

3. If $c > 0$, then the graph of $g(x) = f(x) + c$ is the graph of function f shifted ▨▨▨▨▨ c units.

4. If $c > 0$, then the graph of $g(x) = f(x + c)$ is the graph of function f shifted ▨▨▨▨▨ c units.

2.8 Exercises

Concepts and Skills

1. What is the difference between a rigid transformation and a nonrigid transformation?

2. What is the difference between the graphs of $f(x)$ and $g(x) = f(x) + k$, where k is some real number?

In Exercises 3–8, a function f is given. In parts (a) and (b), produce the graphs of the associated functions in the same coordinate system and describe the graphs in comparison to the graph of f. In part (c), write a function whose graph is described.

3. $f(x) = x$
   (a) $g(x) = cx$ for $c = 2, 3, 4$
   (b) $h(x) = x + c$ for $c = -7, -3, 5$
   (c) The graph of $y = 3x$ shifted down 3 units

4. $f(x) = x$
   (a) $g(x) = x + c$ for $c = 3, 6, 10$
   (b) $h(x) = cx$ for $c = -1, -2, -4$
   (c) The graph of $y = -5x$ shifted up 4 units

5. $f(x) = x^2$
   (a) $g(x) = x^2 + c$ for $c = -3, -1, 5$
   (b) $h(x) = (x + c)^2$ for $c = -2, 4, 6$
   (c) The graph of $y = (x + 2)^2$ shifted up 3 units

6. $f(x) = x^3$
   (a) $g(x) = (x + c)^3$ for $c = -4, 1, 2$
   (b) $h(x) = -x^3 + c$ for $c = -8, -2, 3$
   (c) The graph of $y = -x^3$ shifted left 4 units

7. $f(x) = |x|$
   (a) $g(x) = |x + c|$ for $c = -5, -3, 7$
   (b) $h(x) = -|x| + c$ for $c = -6, 4, 9$
   (c) The graph of $y = -|x| - 3$ shifted right 2 units

8. $f(x) = \sqrt{x}$
   (a) $g(x) = \sqrt{x + c}$ for $c = -9, -4, 4$
   (b) $h(x) = -\sqrt{x} + c$ for $c = -4, 5, 8$
   (c) The graph of $y = \sqrt{6 - x}$ shifted up 2 units

In Exercises 9–20, identify a basic function f that is associated with the given functions. Produce the graphs of f, g, and h in the same coordinate system. Describe how the graph of function g is related to the graph of function f. Then describe how the graph of function h is related to the graph of function g.

9. $g(x) = -x$ and $h(x) = -x + 2$

10. $g(x) = -x^2$ and $h(x) = -(x + 4)^2$

11. $g(x) = |x| + 2$ and $h(x) = |x + 3| + 2$

12. $g(x) = \dfrac{1}{2}x^3$ and $h(x) = \dfrac{1}{2}(x - 4)^3$

13. $g(x) = \sqrt{x + 9}$ and $h(x) = \sqrt{x + 9} - 4$

14. $g(x) = \dfrac{1}{x + 5}$ and $h(x) = 3 + \dfrac{1}{x + 5}$

15. $g(x) = x^3 - 2$ and $h(x) = 2 - x^3$

16. $g(x) = 2x$ and $h(x) = 2x - 1$

17. $g(x) = -\dfrac{1}{x}$ and $h(x) = \dfrac{-1}{x - 6}$

18. $g(x) = (x - 1)^2$ and $h(x) = (x - 1)^2 + 3$

19. $g(x) = \sqrt{-x}$ and $h(x) = \sqrt{3 - x}$

20. $g(x) = -|x|$ and $h(x) = 7 - |x|$

21. Suppose that the zeros of a function f are -2 and 5. What are the zeros of $f(x + 3)$?

22. Suppose that the y-intercept of the graph of a function f is $(0, -3)$. What is the y-intercept of the graph of $-f(x)$?

In Exercises 23–26, the graph of a function f is given. Sketch the graph of function g.

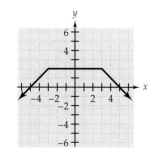

23. $g(x) = -f(x)$ **24.** $g(x) = f(x) - 5$

25. $g(x) = f(x + 3)$

26. $g(x) = -2 + f(x - 4)$

In Exercises 27–32, the graph of a function f is given. Sketch the graph of the function g.

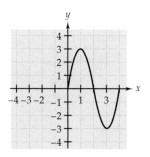

27. $g(x) = 2f(-x)$ **28.** $g(x) = f(x) + 1$

29. $g(x) = 1 - f(x)$ **30.** $g(x) = f(x + 2)$

31. $g(x) = f(x - 4) - 1$ **32.** $g(x) = -\dfrac{1}{2}f(x)$

33. Consider the function $f(x) = kx$, where k is a positive number. What is the effect on the graph of f as k becomes larger?

34. If f is a function, compare the graphs of $f(-x)$ and $-f(x)$ to the graph of f.

In Exercises 35–40, match the function to its graph.

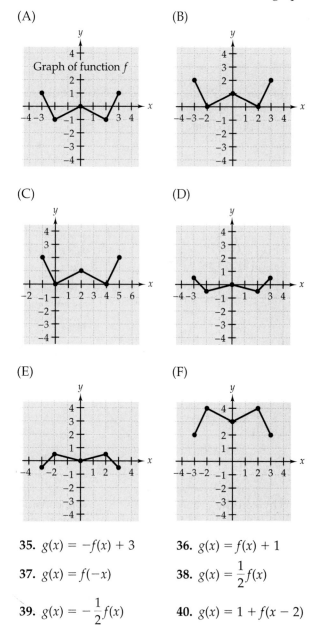

35. $g(x) = -f(x) + 3$ **36.** $g(x) = f(x) + 1$

37. $g(x) = f(-x)$ **38.** $g(x) = \dfrac{1}{2}f(x)$

39. $g(x) = -\dfrac{1}{2}f(x)$ **40.** $g(x) = 1 + f(x - 2)$

In Exercises 41–46, write a function whose graph is described.

41. $f(x) = x^2$ shifted to the left 6 units and upward 5 units

42. $f(x) = |x|$ shifted to the right 2 units and downward 4 units

43. $f(x) = \dfrac{1}{x}$ reflected across the y-axis

44. $f(x) = x^3$ reflected across the x-axis

45. $f(x) = \sqrt{x}$ reflected across the y-axis and shifted to the right 2 units

46. $f(x) = \sqrt[3]{x}$ shifted upward 3 units and reflected across the x-axis

In Exercises 47–50, identify the basic function associated with the graph. Then write the function whose graph is given.

47.

48.

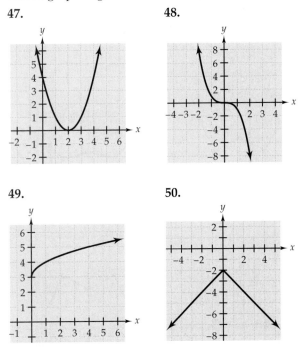

49.

50.

51. Why can the graph of $g(x) = x - 3$ be considered the graph of $f(x) = x$ shifted either to the right 3 units or downward 3 units?

52. Suppose that a graph is shifted horizontally and then reflected across the x-axis. Does reversing the order of those transformations change the result?

In Exercises 53 and 54, produce the graph of the given function. Then sketch the graph of its reflection across the y-axis. Write a function for the reflected graph and produce the graphs of the two functions on the same coordinate system.

53. $f(x) = \sqrt{5 - x}$ **54.** $g(x) = -x^2 - 3x$

In Exercises 55 and 56, produce the graph of the given function. Then sketch the graph of its reflection across the x-axis. Write a function for the reflected graph and produce the graphs of the two functions on the same coordinate system.

55. $c(x) = \sqrt{16 - x^2}$ **56.** $h(x) = |x - 2| - 3$

In Exercises 57 and 58, produce the graph of the given function. Then sketch the graph obtained by first reflecting the graph across the x-axis and then reflecting the result across the y-axis. Write a function for the reflected graph and produce the graphs of the two functions on the same coordinate system.

57. $g(x) = \dfrac{1}{2}x + 3$ **58.** $f(x) = |x + 4|$

Applications

59. Coed Soccer Suppose that a coed soccer team has 16 players.

  (a) Write a function f that describes the number of girls on the team if x of the players are boys.

  (b) State two transformations that would need to be made to the graph of $y = x$ that would result in the graph of f.

60. Flight of a Baseball Suppose that the height above the ground of a hit baseball is modeled with the function $h(x) = -0.005x^2 + x$, where x is the horizontal distance (in feet) that the ball travels.

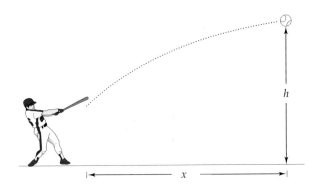

(a) Determine $h(0)$. How does the result indicate that the model is flawed?

(b) Verify that $h(200) = 0$. Interpret the result.

(c) Suppose that the batter hit the ball when it was 3 feet above the ground. What adjustment would you make to the model function? What effect would this adjustment have on the graph of h?

(d) How would the adjustment in part (c) affect the distance that the ball traveled?

61. Automobile Speeds In the following two situations, functions can be written to model the information; however, you do not need to write the specific functions to answer the question.

(i) From a standing start, a car increases its speed by 5 mph every second until it reaches a speed of 60 mph.

(ii) Starting at 60 mph, a car decreases its speed by 5 mph every second until it comes to a stop.

If situation (i) is modeled by a function f, and situation (ii) is modeled by a function g, then the graph of g can be obtained by reflecting the graph of f across the x-axis and then shifting the resulting graph upward by 60. Write function g in terms of f.

62. Copier Temperatures The figure shows the graph of a piecewise function f that models the internal temperature of a copier from the time it was turned on until the time it was turned off and cooled.

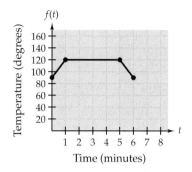

If the room is air-conditioned, the internal temperature of the copier is reduced by 10°, even when it is in use.

(a) Sketch the temperature graph for an air-conditioned room.

(b) If the graph in part (a) is the graph of a piecewise function g, is the graph of g a reflection, a distortion, or a translation of the graph of f?

(c) Write function g in terms of f.

Chapter Review Exercises

Section 2.1

1. The graph of a relation is given. Is the relation a function?

(a)   (b)

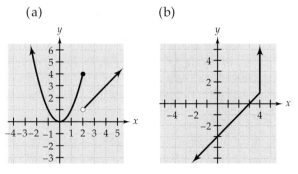

In Exercises 2 and 3, determine whether the given equation defines a function.

2. $x^2 = \sqrt{y^2 - 25}$ **3.** $|x^2 - 3| = y - 1$

In Exercises 4–6, evaluate the given function and simplify.

4. $f(x) = 2x - x^2$

   (a) $f\left(\dfrac{1}{2}\right)$ (b) $f(-3)$ (c) $f(t - 1)$

5. $g(t) = \dfrac{t + 8}{t^2 - 16}$

   (a) $g(0)$ (b) $g(4)$ (c) $g(2a)$

6. $f(x) = \begin{cases} 2 - |x| & x \le 4 \\ 3x - 10 & x > 4 \end{cases}$

   (a) $f(-5)$ (b) $f(4)$ (c) $f(6)$

In Exercises 7 and 8, produce a graph of the given function. Use the graph to estimate the domain and range of the function.

7. $h(x) = |x - 8| - |x|$ 8. $g(x) = \sqrt[3]{x^2 - 8}$

In Exercises 9 and 10, determine the domain of the given function.

9. $r(x) = \dfrac{x + 2}{5x - x^2}$ 10. $s(t) = \sqrt{t^2 + 1}$

11. Which point(s) of the graph of f corresponds to the value(s) of x for which $f(x) = 0$?

12. Use a graph of $f(x) = \dfrac{x}{x - 2}$ to estimate the value of x so that (a) $f(x) = -4$ and (b) $f(x) = 2.25$.

Section 2.2

13. Determine the slope of the line containing $(7, -2)$ and $(5, 1)$.

14. Write the equation $2y + x = 14$ in slope-intercept form. What are the slope and y-intercept of the graph of the equation?

In Exercises 15 and 16, write an equation of the line that satisfies the given conditions.

15. Contains $A(1, -3)$, $m = -4$

16. Contains $P(-4, 2)$ and $Q(-4, 0)$

In Exercises 17 and 18, write a linear function satisfying the given conditions.

17. $f(0) = 0$, $m = -2$

18. $f(-3) = 6$, $f(-1) = 6$

19. Determine the intercepts of the line $3y - x = 15$.

20. Write a linear function that intersects the graph of $f(x) = \sqrt{x - 3}$ where $x = 4$ and $x = 12$.

21. Use graphs of linear functions to estimate the solutions of each equation or inequality. Then solve algebraically.

    (a) $1 - 3x = x + 13$ (b) $1 - 3x < x + 3$

22. **Military Aircraft Losses** The number of military aircraft lost in crashes decreased from 132 in 1989 to 54 in 1997. (*Source*: Department of Defense.) Let t represent the number of years since 1985 and write a linear function $A(t)$ to model the data. According to the model, what was the number of aircraft losses in 1987?

Section 2.3

23. **Sport Utility Vehicles** The table shows the percentage of vehicles sold for over $28,000 that were sport utility vehicles. (*Source*: Mercedes-Benz.)

Year	Percentage
1993	33%
1995	38%
1997	46%

Use a calculator to determine a linear regression equation to model the data. (Let t represent the number of years since 1990.) According to the model, in what year will the percentage exceed 60%?

24. Suppose that a table of values for a function is given. Explain how to decide if the function could be linear.

25. **Women's Earnings** The accompanying table shows women's annual earnings as a percentage of men's annual earnings for the period 1980–1992. (*Source*: Department of Labor, Women's Bureau.)

Year	Percentage
1980	60.2
1982	61.7
1984	63.7
1986	64.3
1988	66.0
1990	71.6
1992	70.6

(a) Letting x represent the number of years since 1980, produce a linear regression equation to model the data.

(b) Produce a scatterplot of the data and the graph of the equation in part (a). How well does the graph appear to fit the data?

(c) Use the model to estimate the year in which women's and men's annual earnings are projected to be equal.

26. In Exercise 25, what is the correlation coefficient? What does this value indicate about the quality of the linear regression equation as a model of the data?

Section 2.4

In Exercises 27–29, lines L_1 and L_2 are described. Determine whether the lines are parallel, perpendicular, or neither.

27. L_1 contains $(4, -4)$ and $(6, -3)$
L_2 contains $(-4, -5)$ and $(0, -3)$

28. L_1: $4y + 3x - 3 = 0$
L_2: $3x = 4y + 8$

29. L_1: $4 - x = 7$
L_2 contains $(-3, 8)$ and $(2, 8)$

In Exercises 30 and 31, write a linear function f whose graph is described.

30. Parallel to the line $2x - y = 4$ and $f(3) = -2$

31. Perpendicular to the line containing $(-3, 5)$ and $(0, 7)$ and having the same y-intercept as the line $x - 2y - 8 = 0$

32. For points $P(0, -4)$ and $Q(9, 8)$, determine (a) PQ and (b) the midpoint of $\overline{PQ}$.

33. Describe two ways to show that three points are the vertices of a right triangle.

34. Show that $A(6, 5)$, $B(14, 1)$, and $C(1, -5)$ are the vertices of a right triangle.

35. Write an equation of the perpendicular bisector of $\overline{AB}$ where A is $(-4, 3)$ and B is $(2, -7)$.

36. The points $A(4, 1)$, $B(-2, 7)$, $C(-6, -9)$, and $D(-7, 2)$ are the vertices of a trapezoid. Determine (a) the area and (b) the perimeter of the trapezoid.

37. The Gap Sales From 1996 to 1998, sales at The Gap rose from $5.3 billion to $7.1 billion. (*Source*: The Gap.) What was the rate of change in sales? Based on the rate of change, what are the anticipated sales in 2001?

38. Online Service Users The number of Internet and online service users in the United States increased from 4.6 million in 1993 to 46.0 million in 1997. (*Source*: Yankee Group.) Determine and interpret the midpoint of the line segment that connects the two data points.

Section 2.5

39. Produce the graphs of the functions $y = x^2$ and $f(x) = (x - 5)^2 + 1$ in the same coordinate system. Describe the graph of f relative to the graph of $y = x^2$.

40. What are the vertex and the axis of symmetry of the graph of $f(x) = -(x + 3)^2 + 6$?

In Exercises 41 and 42, determine the intercepts of the function. Then write the function in the form $y = a(x - h)^2 + k$ and give the vertex.

41. $g(x) = x^2 - 6x - 10$ **42.** $h(x) = 3 - x - x^2$

In Exercises 43 and 44, determine the range and the maximum or minimum value of the function.

43. $f(x) = -x^2 + 4x - 5$ **44.** $g(x) = x^2 - 8x + 5$

In Exercises 45 and 46, determine the number of zeros of the quadratic function.

45. The minimum value of the function occurs at $(-4, 5)$.

46. The vertex is $(1, 7)$ and the y-intercept is $(0, 2)$.

47. Suppose that the graph of a quadratic function has only one x-intercept. Where is the vertex?

48. The vertex of the graph of a quadratic function is $V(2, -1)$ and the y-intercept is $(0, -5)$. Write the quadratic function.

Section 2.6

49. Use a calculator to write an equation of a parabola that contains $A(-3, 26)$, $B(6, 8)$, and $C(12, 56)$.

50. Per-Share Earnings The per-share earnings for Delta Airlines for 1990, 1993, 1996, and 1998 were $5.28, -$10.54, $1.42, and $12.40, respectively. (*Source*: Delta Airlines.) Let x represent the number of years since 1990, and use a calculator to find a quadratic regression equation to model the data. (Round coefficients to the nearest hundredth.) What does the vertex of the function represent?

51. Age Discrimination The accompanying table shows the number (in thousands) of age discrimination complaints that were filed in the period 1990–1996. (*Source*: U.S. Equal Employment Opportunity Commission.)

Year	Complaints
1990	14.7
1991	17.1
1992	19.3
1993	19.9
1994	19.6
1995	17.4
1996	15.7

(a) Given the trend shown in the table, what kind of regression equation would seem most appropriate to model the data?

(b) Let x represent the number of years since 1990 and produce an appropriate regression equation.

(c) If you chose a quadratic regression equation, determine and interpret the vertex of the parabola.

52. In Exercise 51, determine and interpret the value of the correlation coefficient.

Section 2.7

In Exercises 53 and 54, produce a graph of the function. Use the graph to estimate the intercepts.

53. $g(t) = \sqrt[3]{t^2 - 7t - 8}$

54. $T(w) = w^3 + 10w^2 + 21w$

In Exercises 55 and 56, produce a graph of the function. Estimate the intervals in which the function is increasing, decreasing, or constant. Estimate any relative maxima and relative minima.

55. $p(x) = \dfrac{1}{8}x^3 + \dfrac{1}{4}x^2 - 3x$

56. $s(t) = t\sqrt{4 - t}$

In Exercises 57–60, determine the symmetries of the graph.

57. $f(x) = \sqrt[3]{x} - \sqrt[5]{x}$

58. $g(x) = \dfrac{|x|}{x^2 + 3}$

59. $y + 3x = x^2$

60. $xy^2 = 10$

61. Can the graph of a function be symmetric with respect to the x-axis? Why?

62. Graph the function and identify points of discontinuity.

$$f(x) = \begin{cases} x^3 & x \le 0 \\ -|x| & 0 < x \le 4 \\ 3 & x > 4 \end{cases}$$

Section 2.8

In Exercises 63–68, identify a basic function associated with the given function. Produce the graph of both functions in the same coordinate system. Then describe the graph of the given function relative to the basic function.

63. $h(x) = |3 - x|$

64. $q(x) = \dfrac{1}{2}x^2 + 4$

65. $c(x) = -1 + (x + 4)^3$

66. $s(x) = -\sqrt{9 - x}$

67. $r(x) = 5 + \dfrac{1}{x - 3}$

68. $f(x) = \sqrt[3]{x + 8}$

In Exercises 69 and 70, write a function whose graph is the graph of the given basic function with the indicated transformations.

69. $y = \dfrac{1}{x}$; reflected in the x-axis and then shifted to the right 7 units

70. $y = |x|$; shifted to the left 4 units, and upward 6 units

In Exercises 71 and 72, the graph of function g is given. Sketch the graph of function G.

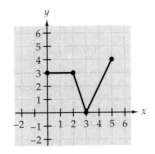

71. $G(x) = g(x + 2) - 3$

72. $G(x) = g(-x)$

In Exercises 73 and 74, identify the basic function associated with the graph. Then write the function whose graph is given.

73.

74.

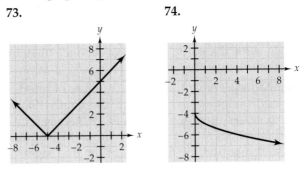

Chapter 3

Polynomial and Rational Functions

Some natural and scientific phenomena can be modeled with polynomial functions whose graphs are often characterized by peaks and dips. For example, a map of the jet stream track exhibits many of the properties of the graph of a polynomial function. The analysis of such a function can assist meteorologists in predicting the progress of weather fronts and in preparing long-range forecasts.

This chapter presents a detailed discussion of polynomial functions, including graphical methods for estimating relative maxima and minima and methods for analyzing the end behavior of such functions. Related to polynomial functions are two important theorems, the Remainder Theorem and the Factor Theorem. We then turn to rational functions and the special properties of their graphs. Finally, we introduce the complex number system and conclude with the Fundamental Theorem of Algebra.

Linear and quadratic functions are special cases of the family of polynomial functions. Thus, this chapter represents an extension of many of the concepts you have learned before.

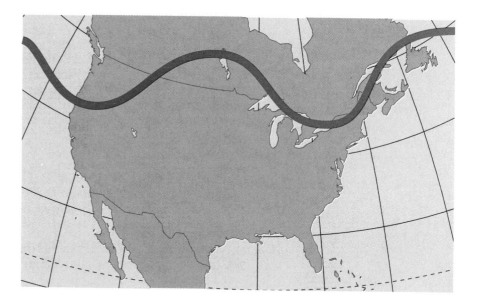

3.1 Polynomial Functions

Graphs of Polynomial Functions ■ *Intercepts and*
Relative Extrema ■ *The Range of a Polynomial Function* ■
Modeling with Real Data

Graphs of Polynomial Functions

The linear and quadratic functions that we studied in the chapter on basic functions and graphs are members of a broader family of functions called **polynomial functions.** The following are all examples of polynomial functions.

$$L(x) = 2x + 8 \qquad\qquad Q(x) = x^2 - 5x + 3$$
$$f(x) = 2x^3 + 7x^2 - 3x + 1 \qquad g(x) = x^4 - 4x^2 + x - 9$$

■ *NOTE   Except for the*
coefficient a_n, any or all of the
coefficients may be 0.

> ### Definition of a Polynomial Function
> A **polynomial function** of degree n in the variable x is a function defined by
> $$P(x) = a_n x^n + a_{n-1} x^{n-1} + \cdots + a_1 x + a_0$$
> where $a_n \neq 0$ and n is a whole number.

The coefficient a_n is called the **leading coefficient,** and a_0 is called the **constant term.**

Because x can be replaced with any real number, the domain of a polynomial function is **R.**

We have already studied polynomial functions for degrees $n = 0, 1,$ and 2.

Degree	Example	Type	Polynomial form
0	$P(x) = 5$	Constant	$P(x) = a_0$
1	$P(x) = 2x + 1$	Linear	$P(x) = a_1 x + a_0$
2	$P(x) = 3x^2 - 4x + 2$	Quadratic	$P(x) = a_2 x^2 + a_1 x + a_0$

In this section we focus on the graphs of polynomial functions of degree greater than 2. The graphs of such polynomial functions are smooth, continuous curves that are often characterized by peaks and dips. (See Figure 3.1.)

In Figure 3.1(a), the right side of the graph rises up off the viewing screen and continues to rise as x increases in value. We say that the graph rises *without bound.* Similarly, the left side of the graph falls without bound as x decreases in value.

As shown in Figure 3.1(b), both sides of the graph may rise without bound. In other cases, both sides may fall without bound. We refer to such properties as the *end behavior* of the graph.

FIGURE **3.1**

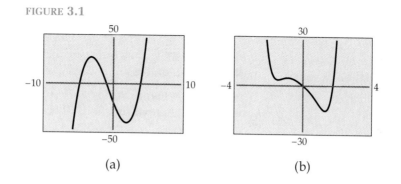

(a)   (b)

When we display the graph of a polynomial function, we want to show the most important features of the graph. To do that, we may need to adjust the viewing window so that we can see the end behavior of the graph as well as peaks, dips, and intercepts that may occur in between. An understanding of end behavior is a valuable aid in selecting window settings to produce a graph of a polynomial function.

The simplest examples of polynomial functions of degree greater than 2 are functions of the form $y = a_n x^n$. We begin by examining the end behavior of the graphs of such functions, with $a_n = 1$ and $a_n = -1$.

EXPLORATION

The End Behavior of the Graphs of $y = x^n$

Use the decimal setting to produce the graphs of functions of the form $y = x^n$ for several positive integers n. What do you observe about the end behavior of the graphs of the functions when n is even? when n is odd? Repeat the experiment for functions of the form $y = -x^n$.

Developing the Concept

The End Behavior of the Graphs of $y = x^n$

Figure 3.2(a) and (b) shows the graphs of two groups of polynomial functions of the form $y = x^n$.

FIGURE **3.2**

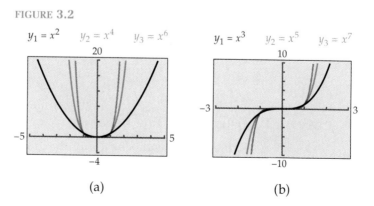

$y_1 = x^2 \qquad y_2 = x^4 \qquad y_3 = x^6$ $y_1 = x^3 \qquad y_2 = x^5 \qquad y_3 = x^7$

(a)   (b)

If n is an even integer, as in Figure 3.2(a), then the graphs of $y = x^n$ all rise without bound on both the left and the right.

If n is an odd integer, as in Figure 3.2(b), then the graphs of $y = x^n$ all fall without bound on the left and rise without bound on the right.

Now consider Figure 3.3(a) and (b), which shows the graphs of two groups of polynomial functions of the form $y = -x^n$.

FIGURE 3.3

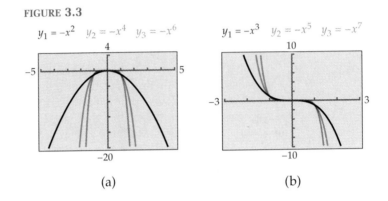

$y_1 = -x^2 \quad y_2 = -x^4 \quad y_3 = -x^6$

$y_1 = -x^3 \quad y_2 = -x^5 \quad y_3 = -x^7$

(a)   (b)

If n is an even integer, as in Figure 3.3(a), then the graphs of $y = -x^n$ all fall without bound on both the left and the right.

If n is an odd integer, as in Figure 3.3(b), then the graphs of $y = -x^n$ all rise without bound on the left and fall without bound on the right.

For a polynomial function, the leading term $a_n x^n$ has the dominating influence on the value of the function. For this reason, the end behavior of the graphs of *all* polynomial functions follows exactly the same patterns as those exhibited by the graphs of $y = x^n$ and $y = -x^n$. Figure 3.4 summarizes the end behavior patterns.

FIGURE 3.4

The End Behavior of the Graph of a Polynomial Function with Leading Coefficient a_n

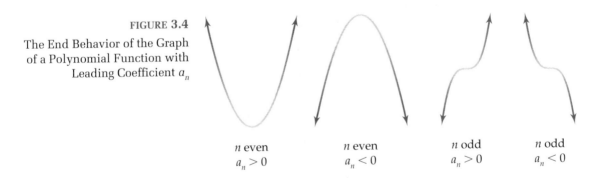

n even n even n odd n odd
$a_n > 0$ $a_n < 0$ $a_n > 0$ $a_n < 0$

EXAMPLE 1   **Determining End Behavior**

Predict the end behavior of the graph of each given function. Then produce a graph of the function to verify your prediction.

(a) $f(x) = x - x^3$
(b) $g(x) = x^3 - 5x^2 + x + 5$

(c) $h(x) = x^4 - 5x^2 + 4$
(d) $p(x) = 1 + 5x^3 - 2x^4$

SOLUTION

(a) The term $-x^3$ determines the end behavior. The coefficient is negative, and the degree is odd. The graph should rise on the left and fall on the right. [See Figure 3.5(a).]
(b) The term x^3 determines the end behavior. The coefficient is positive, and the degree is odd. The graph should fall on the left and rise on the right. [See Figure 3.5(b).]
(c) The term x^4 determines the end behavior. The coefficient is positive, and the degree is even. The graph should rise on both the left and the right. [See Figure 3.5(c).]
(d) The term $-2x^4$ determines the end behavior. The coefficient is negative, and the degree is even. The graph should fall on both the left and the right. [See Figure 3.5(d).]

FIGURE **3.5**

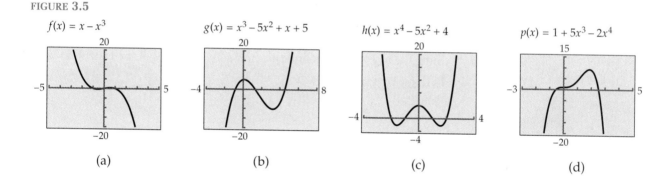

$f(x) = x - x^3$

(a)

$g(x) = x^3 - 5x^2 + x + 5$

(b)

$h(x) = x^4 - 5x^2 + 4$

(c)

$p(x) = 1 + 5x^3 - 2x^4$

(d)

Intercepts and Relative Extrema

For any polynomial function, $P(0) = a_0$. Therefore, the y-intercept of the graph of a polynomial function is $(0, a_0)$.

Recall that determining the x-intercepts of a graph is equivalent to determining the zeros of the function. For polynomial functions, these are both equivalent to finding a solution (root) of the equation $P(x) = 0$. We know that a linear function has one zero and a quadratic function has zero, one, or two zeros. In general, an nth-degree polynomial function has at most n distinct real zeros.

The graphs of polynomial functions often include peaks and dips. Roughly speaking, the top of a peak is the highest point in that neighborhood of the graph. Similarly, the bottom of a dip is the lowest point in that neighborhood of the graph. Recall that we say that the function has a relative maximum value at a peak and a relative minimum value at a dip. We refer to relative maxima and relative minima as **relative extrema.**

Referring again to Figure 3.5, note that the degree of the function in Figure 3.5(b) is 3 and there are two relative extrema. The degree of the function in Figure 3.5(c) is 4 and there are three relative extrema. In general, it can be shown that the number of relative extrema in the graph of a polynomial function is *at most* one less than the degree of the function.

> **Number of Relative Extrema of a Polynomial Function**
>
> For a polynomial function of degree n, the number of relative extrema of the function is at most $n - 1$.

We will study several techniques for finding the zeros of a polynomial. The following example includes a review of the factoring method.

EXAMPLE 2 **Identifying Features of the Graph of a Polynomial Function**

Produce the graph of each given function and use your calculator's *Minimum* and *Maximum* features to estimate any relative extrema. Determine the intercepts algebraically and state the zeros of the function.

(a) $f(x) = 12 + 4x - 3x^2 - x^3$
(b) $g(x) = -x^5 + 6x^4 - 9x^3$

SOLUTION

(a) VISUALIZING THE RELATIVE EXTREMA

FIGURE 3.6

$f(x) = 12 + 4x - 3x^2 - x^3$

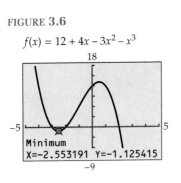

The graph in Figure 3.6 shows the use of the calculator's *Minimum* feature for approximating the relative minimum, $(-2.55, -1.13)$. We can also use the *Maximum* feature at the peak to approximate the relative maximum, $(0.53, 13.13)$.

DETERMINING THE INTERCEPTS AND ZEROS

Because $f(0) = 12$, the y-intercept is $(0, 12)$. To determine the x-intercepts, we solve $f(x) = 0$.

$$
\begin{array}{ll}
12 + 4x - 3x^2 - x^3 = 0 & f(x) = 0 \\
4(3 + x) - x^2(3 + x) = 0 & \text{Factor by grouping.} \\
(3 + x)(4 - x^2) = 0 & \text{The common factor is } 3 + x. \\
(3 + x)(2 + x)(2 - x) = 0 & \text{Factor completely.}
\end{array}
$$

$$
\begin{array}{llll}
3 + x = 0 & 2 + x = 0 & 2 - x = 0 & \text{Zero Factor Property} \\
x = -3 & x = -2 & x = 2 & \text{Solve for } x.
\end{array}
$$

The x-intercepts are $(-3, 0)$, $(-2, 0)$, and $(2, 0)$. The third-degree function has three zeros, $-3, -2$, and 2.

(b) VISUALIZING THE RELATIVE EXTREMA

FIGURE 3.7

$g(x) = -x^5 + 6x^4 - 9x^3$

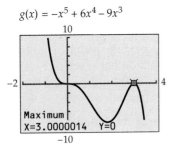

Using the calculator's *Maximum* feature, we estimate the relative maximum to be $(3, 0)$. (See Figure 3.7.) We can also use the *Minimum* feature near the dip to approximate the relative minimum, $(1.80, -8.40)$.

DETERMINING THE INTERCEPTS AND ZEROS

Because $g(0) = 0$, the y-intercept is $(0, 0)$. Solve $g(x) = 0$ to determine the x-intercepts.

$$-x^5 + 6x^4 - 9x^3 = 0 \qquad g(x) = 0$$
$$-x^3(x^2 - 6x + 9) = 0 \qquad \text{Factor the polynomial.}$$
$$-x^3(x - 3)^2 = 0 \qquad \text{The trinomial is a perfect square.}$$
$$x = 0 \qquad x - 3 = 0 \qquad \text{Set each factor equal to 0.}$$
$$x = 0 \qquad\qquad x = 3 \qquad \text{Solve for } x.$$

The x-intercepts are $(0, 0)$ and $(3, 0)$. The zeros of the function are 0 and 3.

In Example 2(b), the factored expression could be written as the product $-1 \cdot x \cdot x \cdot x \cdot (x - 3) \cdot (x - 3)$. The zero 3 results from the factor $x - 3$ that appears twice. We say that 3 is a zero of *multiplicity* 2. Note that the multiplicity is even and that the graph just touches (is *tangent* to) the x-axis at $x = 3$. Similarly, 0 has multiplicity 3 because the factor x appears three times. Note that the multiplicity is odd and that the graph crosses the x-axis at $x = 0$.

These observations can be generalized as follows:

> If a polynomial function has n factors of the form $x - a$, then the number a is called a zero of *multiplicity n*.
> 1. At a zero with even multiplicity, the graph is tangent to the x-axis.
> 2. At a zero with odd multiplicity, the graph crosses the x-axis.

EXAMPLE 3 **Multiplicity of Zeros**

Predict the appearance of the graph of $f(x) = (x + 2)^3(x - 1)^4$. Then produce the graph of f to verify your predictions.

SOLUTION

The degree of the polynomial is 7, and the leading coefficient is positive. Thus the end behavior of the graph of f is like that of the graph of $y = x^7$. The graph should fall on the left and rise on the right.

Because the factor $(x + 2)$ appears three times, -2 is a zero of multiplicity 3. Thus the graph of f crosses the x-axis at $(-2, 0)$. Because 1 is a zero of multiplicity 4, the graph is tangent to the x-axis at $(1, 0)$. (See Figure 3.8.)

FIGURE 3.8

$f(x) = (x + 2)^3(x - 1)^4$

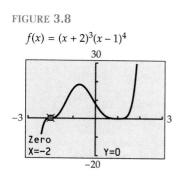

▮ Zero   If the graph of a function intersects the x-axis, then you can use your calculator to compute the corresponding zero of the function. (See Figure 3.8.)

EXAMPLE **4**   **Hidden Intercepts**

Estimate the x-intercepts of the function $f(x) = x^6 - 5x^5 + 5x^4 + 3x^3$.

SOLUTION

FIGURE **3.9**

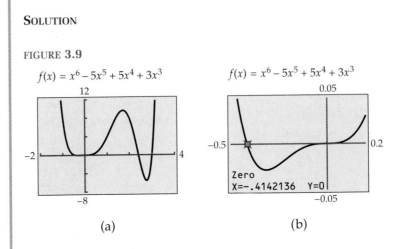

From Figure 3.9(a), the graph appears to have three intercepts. By tracing, we estimate the two intercepts on the right to be $(2.41, 0)$ and $(3, 0)$. However, if we zoom in on the apparent intercept near the origin, we see that there are actually two intercepts in that vicinity. [See Figure 3.9(b)]. We estimate those intercepts to be $(-0.41, 0)$ and $(0, 0)$.

The Range of a Polynomial Function

For any polynomial function with an odd degree, the graph will rise at one end and fall at the other end. Therefore, the range of such functions is **R**.

The graphs of polynomial functions with an even degree rise at both ends or fall at both ends. In this case, the range is dependent on relative extrema.

EXAMPLE **5**   **The Range of Polynomial Functions**

If necessary, for each given polynomial function, use a graph of the function to estimate the range of the function.

(a) $f(x) = x^4 - 3x^3 + 4$
(b) $g(x) = -x^5 + 2x^3 - x^2 + 1$
(c) $h(x) = -2x^6 + x^3 + 3x^2$

FIGURE **3.10**

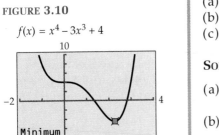

SOLUTION

(a) From Figure 3.10, we estimate the relative minimum of f to be -4.5. Thus the range is approximately $[-4.5, \infty)$.
(b) Because the degree of function g is odd and the leading coefficient is negative, the graph rises without bound on the left and falls without bound on the right. Thus the range of function g is **R**.

(c) From Figure 3.11, we estimate the relative maximum of h to be 2.1. Thus the range is approximately $(-\infty, 2.1]$.

FIGURE **3.11**

$h(x) = -2x^6 + x^3 + 3x^2$

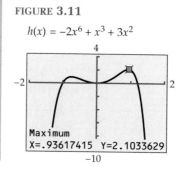

Modeling with Real Data

A polynomial function can sometimes be used to model data. However, because real data usually do not increase or decrease without bound, the end behavior of polynomial functions limits the extent to which the model might be valid. We refer to the interval of x-values for which a model might be considered to be reasonable as the *domain of validity*. Because we have no guarantee that the trends indicated by the data will continue indefinitely, using the model for x-values outside the domain of validity may lead to unrealistic and invalid results.

■ *NOTE Recall that producing a scatterplot of the data for an application and observing the general shape of the data are often helpful when choosing an appropriate model.*

Regression   The typical menu selections for third-degree and fourth-degree regression equations are *CubicReg* and *QuartReg*, respectively.

EXAMPLE **6**   **The Cost of Nursing Home Care**

The climbing costs of nursing home care (in billions of dollars) are shown in the following table. (*Source:* Health Care Financing Administration.)

Year	Nursing home costs
1975	10
1980	19
1985	30
1990	50
1995	75

(a) Produce a scatterplot of the data, with x representing the number of years since 1970. Then produce and graph a third-degree regression equation. How well does the graph appear to model the data?

(b) Produce a fourth-degree regression equation and observe the leading coefficient. What conclusion do you draw?

(c) Based on the model, what are the projected nursing home costs for the year 2050? Do you think 2050 is in the domain of validity?

SOLUTION

(a) Using the cubic regression option [see Figure 3.12(a)], we obtain a third-degree regression equation $y = 0.002x^3 + 0.027x^2 + 0.886x + 4.8$. Figure 3.12(b) shows a scatterplot of the data and the graph of the regression equation, which appears to model the data well.

FIGURE 3.12

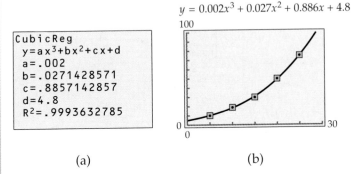

(a)   (b)

(b) Figure 3.13 shows the coefficients of a fourth-degree regression equation. The leading coefficient of the equation is nearly 0, so the value of the fourth-degree term is negligible. This suggests that a third-degree equation is a better model.

FIGURE 3.13

```
QuarticReg
y=ax⁴+bx³+...+e
a=-7.333333E-4
b=.046
c=-.8816666667
d=8.35
e=-15
```

(c) The model's projected costs for the year 2050 are about 1.2 trillion dollars, an amount that our nation is unlikely to be able to afford. However, the year 2050 is far beyond the years for which we have data. Thus the year is well outside the domain of validity, and the prediction should be considered unreliable.

3.1 Quick Reference

Graphs of Polynomial Functions

■ A **polynomial function** of degree n in the variable x is a function defined by $P(x) = a_n x^n + a_{n-1} x^{n-1} + \cdots + a_1 x + a_0$, where $a_n \neq 0$ and n is a whole number.

■ The coefficient a_n is called the **leading coefficient**, and a_0 is the **constant term.**

■ The domain of a polynomial function is **R.**

■ The *end behavior* of the graphs of polynomial functions is summarized in Figure 3.4.

Intercepts and Relative Extrema

■ The y-intercept of the graph of a polynomial function is $(0, a_0)$.

■ The x-intercepts of the graph of a polynomial function represent the zeros of the function.

■ An nth-degree polynomial function has at most n distinct real zeros, which are the solutions of the polynomial equation $P(x) = 0$.

■ The top of a peak (bottom of a dip) represents a relative maximum (minimum) of a polynomial function. Relative minima and relative maxima are called **relative extrema.**

■ The number of relative extrema of an nth-degree polynomial function is at most $n - 1$.

■ If a polynomial expression has *n* factors of the form $x - a$, then *a* is called a zero of *multiplicity n*.

■ The graph of a polynomial function is tangent to the *x*-axis at a zero with even multiplicity and crosses the *x*-axis at a zero with odd multiplicity.

The Range of a Polynomial Function

■ The range of an odd-degree polynomial function is **R**.

■ The range of an even-degree polynomial function is dependent on the relative extrema.

Modeling with Real Data

■ The end behavior of a polynomial function may invalidate the function as a model of data that are outside the *domain of validity*.

3.1 Speaking the Language

1. The function $f(x) = 2x^3 - 5x^2 + x - 7$ is an example of a(n) _____ function.

2. For the function $g(x) = (x - 5)^3$, we say that 5 is a zero of _____ 3.

3. The graph of $y = x^3$ falls without bound on the left and rises without bound on the right. We refer to this as the _____ of the graph.

4. The peaks and dips of the graph of a function are associated with the relative _____ of the function.

3.1 Exercises

Concepts and Skills

1. Describe the symmetry of the graphs of $y = x^n$.

2. Describe the conditions under which $y = x^n$ is an even function and an odd function.

In Exercises 3–6, match the given function to the calculator display of its graph.

3. $f(x) = -\dfrac{2}{3}x^3 + 3x^2 - 5$

4. $p(x) = \dfrac{1}{2}x^4 - 4x^2 + x$

5. $f(x) = \dfrac{1}{4}x^5 - 2x^3 - x^2 + 3x$

6. $f(x) = -\dfrac{1}{4}x^6 + x^3 + 4x^2$

(A)

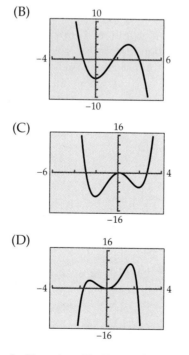

(B)

(C)

(D)

In Exercises 7–12, produce the graph of each function in the same coordinate system. Describe the graph of the second function relative to the first function.

7. $y = x^3$; $f(x) = (x - 4)^3$

8. $y = x^5$; $f(x) = x^5 + 2$

9. $y = x^4$; $f(x) = x^4 - 7$

10. $y = x^6$; $f(x) = (x + 1)^6$

11. $y = x^4$; $f(x) = (x + 1)^4 - 7$

12. $y = x^5$; $f(x) = (x - 2)^5 + 4$

In Exercises 13–18, identify the term that determines the end behavior. Predict the end behavior. Use an appropriate window to produce the graph of the function to verify your prediction.

13. $f(x) = -2x^3 + 4x^2 + 5x - 12$

14. $g(x) = 9 + 8x - x^3 + \dfrac{1}{2}x^4$

15. $H(x) = 3x^5 + x^6$

16. $F(x) = -\dfrac{2}{3}x^5 + 7x^4 - 6x^2$

17. $p(x) = -x^4 + x^3 - 12$

18. $q(x) = 3x^2 + 0.2x^3$

19. Explain why a polynomial function with odd degree must have at least one zero.

20. Suppose $F(x) = -f(x)$. If you know the zeros of f, explain how you know the zeros of F.

In Exercises 21–32, use a graph to estimate the zeros of the function. Then determine the zeros algebraically and state the multiplicity of each one.

21. $f(x) = x^3 + x^2 - 6x$

22. $g(x) = x^5 - 5x^3 + 4x$

23. $q(x) = -3x(x + 2)^4(x - 1)^3$

24. $h(x) = 5(x + 1)^5(x - 6)^2(x - 4)$

25. $p(x) = (x^2 - 6x + 8)^2$

26. $f(x) = x^7 - 8x^5 + 16x^3$

27. $h(x) = x^3 + 4x^2 - 2x - 8$

28. $g(x) = 2x^3 - x^2 - 16x + 8$

29. $p(x) = x^4 - 2x^3 + x - 2$

30. $f(x) = x^6 - 14x^4 + 45x^2$

31. $h(x) = x^5 + 3x^4 - 6x^3$

32. $f(x) = x^2 + 5x^3 - 2x^4$

33. Suppose that a fourth-degree polynomial has three distinct zeros. Explain why none of the zeros can have a multiplicity greater than 2.

34. Use the end behavior of the graph of a fifth-degree polynomial function to explain why the graph cannot have five peaks and dips.

In Exercises 35–40, determine the x-intercepts of the graph of the function and predict whether the graph crosses the x-axis or is tangent to the x-axis at each intercept. Produce the graph to check your prediction.

35. $f(x) = x^3(x + 2)(x - 3)^2$

36. $g(x) = -2x(3x - 1)^2(x + 4)^3$

37. $F(x) = (4 - 3x)^2(x + 5)^3$

38. $P(x) = (1 - x)(3 + 4x)^5(x + 7)^2$

39. $h(x) = x(x + 5)(x + 2)^2$

40. $f(x) = (4x + 1)^3(x - 6)(2x + 3)$

In Exercises 41–44, use a graph to estimate the x-intercepts. (Be careful—some of the x-intercepts are close together.)

41. $g(x) = 18x^3 - 81x^2 + 121x - 60$

42. $f(x) = 48x^3 + 100x^2 + 68x + 15$

43. $f(x) = 6x^4 - x^3 - 24x^2 + 2x + 24$

44. $h(x) = 540x^4 + 18x^3 - 48x^2 - x + 1$

45. Suppose that you are using a calculator graph of a polynomial function to estimate the zeros of the function. If the graph appears to have a relative minimum at $(4, 0)$, before you conclude that 4 is a zero, why is zooming in at that point a good idea?

46. Describe where a function is increasing and decreasing in the neighborhoods of relative extrema.

In Exercises 47–52, determine the intercepts of the graph of the function. Produce the graph of the function and estimate the extrema.

47. $f(x) = 3x^2 - x^3$ 48. $g(x) = x^3 - 8$

49. $g(x) = x^4 - 5x^2 + 4$ 50. $f(x) = x^2(5 - x^2)$

51. $g(x) = x(4 - x)(x + 2)$

52. $f(x) = x(x + 3)(x - 1)(x + 5)$

In Exercises 53–56, estimate the relative extrema of the given function.

53. $F(x) = 20 + 36x + 6x^2 - x^3$

54. $H(x) = x^3 + 3x^2 - 9x + 9$

55. $P(x) = \dfrac{6}{5}x^5 + 3x^4 - 2x^3 - 6x^2 + 6$

56. $Q(x) = \dfrac{3}{2}x^4 + 2x^3 - 3x^2 - 6x - 15$

In Exercises 57–60, produce the graph and estimate the x-intercepts and the range of the given function. (Round to the nearest hundredth.)

57. $f(x) = 2 - 3x + 6x^2 - 2x^3$

58. $G(x) = 3x^3 + 4x - 8$

59. $F(x) = x^6 - 5x^4 - 4x^2 + 3$

60. $f(x) = 2 - 9x^2 - 6x^3 - 6x^5 - x^8$

In Exercises 61 and 62, produce a scatterplot of the given data and a third-degree (cubic) regression equation to model the data. Graph the equation with the scatterplot to test the accuracy of the model.

61. $(-4, 35), (1, 10), (3, 28), (5, -10)$

62. $(-3, -32), (1, 0), (2, 8), (4, 66)$

In Exercises 63 and 64, produce a scatterplot of the given data and a fourth-degree (quartic) regression equation to model the data. Graph the equation with the scatterplot to test the accuracy of the model.

63. $(-2, 35), (-1, 17), (1, -7), (2, -13), (3, 25)$

64. $(-2, 11), (0, 1), (2, 23), (3, 38.5), (4, 29)$

Concept Extension

65. Suppose that $F(x) = f(x + 2)$. If the zeros of f are $-2, 0$, and 5, what are the zeros of F?

66. Suppose that c is a zero of the polynomial function $P(x) = 2x^4 - 4x^3 + 3x^2 - 4x + 2$. Show that $\dfrac{1}{c}$ is also a zero. $\left[\textit{Hint: }\right.$ Write the expression for $P\left(\dfrac{1}{c}\right)$. Then show that $c^4 \cdot P\left(\dfrac{1}{c}\right) = P(c).\left.\right]$

Applications

67. Kite Flying   The height h (in feet) of a kite after t minutes of flying time is modeled by the following function:

$$h(t) = -0.005t^4 + 0.3t^3 - 6t^2 + 43t$$

Produce a first-quadrant graph of function h and use it to answer each part. Round your answers to the nearest whole number.

(a) After its first ascent, to what height does the kite dip?

(b) After how many minutes does the kite achieve its maximum height? What is the maximum height?

(c) What is the kite's total flying time?

ART FOR EXERCISE **67**

68. Allergy Medication   Suppose that the amount (in milligrams) of a time-release allergy medication in a patient's bloodstream t hours after administering the drug is modeled by the following function:

$$M(t) = -0.07t^4 + 1.2t^3 - 6.7t^2 + 12.6t$$

Following each release of the medication, the number of milligrams in the bloodstream initially increases and then decreases until the next release of medication.

Produce a first-quadrant graph of function M and use it to answer each part. Round your answers to the nearest tenth.

(a) After how many hours will the amount of medication in the bloodstream be at its maximum?

(b) Estimate and interpret the x-intercepts of the graph of M.

(c) To what level does the amount of medication decrease after the first release of medication?

(d) To what level does the amount of medication increase after the second release of medication?

69. Plywood Bin   A warehouse operator has 100 square feet of plywood with which to construct an open storage bin with a square base. (See figure.)

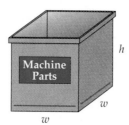

(a) Write an equation to model the condition that the surface area of the box is 100 square feet.

(b) Solve the equation for h to obtain an expression for the height of the box.

(c) Use the expression in part (b) to write the function $V(w)$, which is the volume of the box as a function of the width w of the base.

(d) Produce a first-quadrant graph of function V and use it to estimate the width of the base that would maximize the volume of the bin.

70. Metal Drums A company manufactures metal drums that have a total surface area of 24π square feet. (See figure.) The total surface area of a drum is given by $S = 2\pi r(r + h)$, and the volume of a drum is given by $V = \pi r^2 h$.

(a) Write an equation that models the condition that the total surface area is 24π square feet.

(b) Solve the equation in part (a) for h to obtain an expression for the height of a drum.

(c) Use the expression for h to write the function $V(r)$, which is the volume of a drum as a function of the radius r.

(d) Produce a first-quadrant graph of V and use it to estimate the radius of a drum that would maximize the volume.

71. Gasoline Prices The table shows the retail price per gallon (in cents) for unleaded gasoline for selected years. (*Source*: Energy Information Administration.)

Year	Cost
1976	61.4
1980	124.5
1984	121.2
1988	94.6
1992	112.7
1994	111.2

(a) Letting x represent the number of years since 1970, produce a scatterplot of the data.

(b) Produce a third-degree (cubic) regression equation to model the data and graph the equation on the same coordinate system as the scatterplot.

(c) Produce a fourth-degree (quartic) regression equation to model the data and graph the equation on the same coordinate system as the scatterplot.

(d) Which model appears to better fit the data?

(e) What trend does each model indicate?

72. (a) Use the model equation in part (c) of Exercise 71 to estimate the retail price of gasoline in the year 2000.

(b) Given that international circumstances can quickly and dramatically affect the price of gasoline, what domain of validity would you choose for the model regression equation?

73. Crude Oil Production Crude oil produced from federally leased offshore sites was 259 million barrels in 1980 and 307 million barrels in 1993. However, the amount fluctuated dramatically in the intervening years. (*Source*: U.S. Department of Interior.) The production (in millions of barrels) can be modeled by the following function:

$$B(t) = 0.61t^3 - 12.7t^2 + 66.4t + 259.5$$

For this model, t is the number of years since 1980.

(a) According to the model, what was the production in 1980? in 1993?

(b) Estimate and interpret the relative maximum.

(c) Estimate and interpret the relative minimum.

3.2 *The Remainder and Factor Theorems*

Division of Polynomials   ■   *The Remainder Theorem*   ■
The Factor Theorem   ■   *The Rational Zero Theorem*   ■
Equations and Inequalities

Division of Polynomials

In this section we introduce additional methods for determining the zeros of a polynomial function. The methods rely on division of polynomials. We begin with a brief review of the procedure and vocabulary associated with long division of numbers and polynomials.

$$
\begin{array}{r}
5 \quad\quad \text{Quotient} \\
\text{Divisor} \quad 53\overline{)287} \quad\quad \text{Dividend} \\
\underline{265} \quad\quad\quad\quad \\
22 \quad\quad \text{Remainder}
\end{array}
$$

We can write the result as $\frac{287}{53} = 5 + \frac{22}{53}$, and we can check the result with the relation

$$
\begin{array}{ccccccc}
\text{Dividend} & = & \text{(Quotient)} & \text{(Divisor)} & + & \text{Remainder} \\
\downarrow & & \downarrow & \downarrow & & \downarrow \\
287 & = & 5 & \cdot \quad 53 & + & 22
\end{array}
$$

If we divide a polynomial $P(x)$ by a divisor $D(x)$, we obtain a quotient $Q(x)$ and a remainder $R(x)$. The result can be written in either of the following ways:

$$P(x) = Q(x) \cdot D(x) + R(x)$$

or

$$\frac{P(x)}{D(x)} = Q(x) + \frac{R(x)}{D(x)}$$

If the remainder is 0, we say that $P(x)$ is *divisible* by $D(x)$.

Example 1 reviews the long division process.

EXAMPLE 1   Long Division of Polynomials

Divide $2x^3 - 5x^2 + 2x - 7$ by $x - 3$.

SOLUTION

■ *NOTE   When we use long division, we stop dividing when the remainder is 0 or when the degree of the remainder is less than the degree of the divisor.*

$$
\begin{array}{r}
2x^2 + x + 5 \\
x - 3\overline{)2x^3 - 5x^2 + 2x - 7} \\
\underline{2x^3 - 6x^2} \\
x^2 + 2x \\
\underline{x^2 - 3x} \\
5x - 7 \\
\underline{5x - 15} \\
8
\end{array}
$$

Divide $2x^3$ by x and write the result on the top row.
Multiply the divisor $x - 3$ by $2x^2$.
Subtract and bring down $2x$. Divide x^2 by x.
Multiply $x - 3$ by x.
Subtract and bring down -7. Divide $5x$ by x.
Multiply $x - 3$ by 5.
Subtract.

The quotient is $2x^2 + x + 5$ and the remainder is 8. We can write the results of the division in the following way:

$$
\begin{array}{ccccccc}
P(x) & = & D(x) & \cdot & Q(x) & + & R(x) \\
\downarrow & & \downarrow & & \downarrow & & \downarrow \\
2x^3 - 5x^2 + 2x - 7 & = & (x - 3) & & (2x^2 + x + 5) & + & 8
\end{array}
$$

Synthetic division is a numerical method for dividing a polynomial by a binomial in the form $x - c$. This method omits the variable in the long division process and records only the essential coefficients.

DEVELOPING THE CONCEPT

Synthetic Division

Compare the long division on the left with the corresponding synthetic division format on the right. Note that we inserted $0x^2$ for the missing x^2 term.

$$
\begin{array}{r}
3x^2 + 6x + 5 \\
x - 2 \overline{) 3x^3 + 0x^2 - 7x - 8} \\
\underline{3x^3 - 6x^2 } \\
6x^2 - 7x \\
\underline{6x^2 - 12x} \\
5x - 8 \\
\underline{5x - 10} \\
2
\end{array}
$$

$$
\begin{array}{r|rrrr}
2 & 3 & 0 & -7 & -8 \\
  &   & 6 & 12 & 10 \\
\hline
  & 3 & 6 & 5 & 2
\end{array}
$$

Referring to the synthetic division format, we can make the following observations:

1. The 2 is the same as the 2 in the divisor $x - 2$. We refer to this number as the divider.
2. The numbers 3, 0, -7, and -8 along the first row are the coefficients of the dividend.
3. Along the bottom row, the numbers 3, 6, and 5 are the coefficients of the quotient, and 2 is the remainder.
4. Each number in the bottom row is the sum of the two numbers above it.
5. Each number in the middle row is the product of the divider 2 and the number in the bottom row of the preceding column.

Note that certain requirements must be met to use synthetic division.

1. The divisor must be of the form $x - c$. For a divisor such as $x + 7$, we write $x - (-7)$, and so the divider is -7.
2. The dividend must be in descending order with 0 as the coefficient of any missing term.

The following example illustrates the use of synthetic division.

EXAMPLE 2   **Synthetic Division of Polynomials**

Use synthetic division to divide $2x^3 + 3x^2 - x + 4$ by $x + 2$.

SOLUTION

$$
\begin{array}{r|rrrr}
-2 & 2 & 3 & -1 & 4 \\
  & & \downarrow & & \\ \hline
  & 2 & & &
\end{array}
$$

Write the divider and the coefficients of the dividend. Because $x + 2 = x - (-2)$, the divider is -2.

Bring the first number in the top row down to the bottom row.

$$
\begin{array}{r|rrrr}
-2 & 2 & 3 & -1 & 4 \\
  & & -4 & & \\ \hline
  & 2 & -1 & &
\end{array}
$$

Multiply the first number, 2, in the bottom row by the divider, -2, to obtain the first number, -4, in the middle row.

Add the 3 and -4 to obtain the second number, -1, in the bottom row.

$$
\begin{array}{r|rrrr}
-2 & 2 & 3 & -1 & 4 \\
  & & -4 & 2 & \\ \hline
  & 2 & -1 & 1 &
\end{array}
$$

Similarly, multiply the -1 in the bottom row by the divider, -2, to obtain 2, the second number in the middle row.

Then add the column to obtain 1.

$$
\begin{array}{r|rrrr}
-2 & 2 & 3 & -1 & 4 \\
  & & -4 & 2 & -2 \\ \hline
  & 2 & -1 & 1 & 2
\end{array}
$$

Finally, multiply this 1 by the divider, -2, to obtain -2, the last number in the middle row.

Then add the column to obtain 2.

The entries 2, -1, and 1 in the last row are the coefficients of the quotient, and the last entry, 2, is the remainder. Because the degree of the quotient is 1 less than the degree of the dividend, the quotient is $2x^2 - x + 1$ and the remainder is 2. We can write the result as follows:

$$\frac{2x^3 + 3x^2 - x + 4}{x + 2} = 2x^2 - x + 1 + \frac{2}{x + 2}$$

This relationship can also be written

$$2x^3 + 3x^2 - x + 4 = (2x^2 - x + 1)(x + 2) + 2$$

The Remainder Theorem

When we divide a polynomial $P(x)$ by $x - c$, we can observe a relationship between the remainder and the value of $P(c)$.

EXPLORATION

Dividing P(x) by x − c and Comparing Remainders with P(c)

Consider the polynomial $P(x) = x^4 + 2x^3 - 3x^2 + x - 1$ and the binomial $x - c$. Choose several integer values for c and use your calculator to evaluate $P(c)$. Then for each value of c that you selected, use synthetic division to obtain the remainder R when the polynomial is divided by $x - c$. What relationship between $P(c)$ and R do you observe?

DEVELOPING THE CONCEPT

Dividing P(x) by x − c and Comparing Remainders with P(c)

The following summarizes the results of determining the remainder when $P(x)$ is divided by $x - c$. The remainders can be verified by synthetic division. Verify the entries in the last column with your calculator.

$P(x)$	$x - c$	c	Remainder	$P(c)$
$x^3 - 2x^2 + 4x - 3$	$x - 1$	1	0	0
$2x^4 + 5x^3 + 3x - 10$	$x + 3$	−3	8	8
$x^3 - 5x^2 + 9$	$x - 4$	4	−7	−7
$x^3 + 8$	$x + 2$	−2	0	0

In each part, dividing $P(x)$ by $x - c$ results in a remainder that is equal to $P(c)$.

The results suggest the following theorem:

> **The Remainder Theorem**
>
> If a polynomial $P(x)$ is divided by $x - c$, where c is a constant, then the remainder is $P(c)$.

The Remainder Theorem can be justified algebraically. Suppose that when a polynomial $P(x)$ is divided by $x - c$, the quotient is $Q(x)$, and the remainder is R. The relationship can be written in the following way:

$$P(x) = Q(x) \cdot (x - c) + R$$

Now we evaluate $P(c)$.

$$P(c) = Q(c) \cdot (c - c) + R = Q(c) \cdot 0 + R = 0 + R = R$$

Observe that $P(c)$ is the same as the remainder when $P(x)$ is divided by $x - c$.

EXAMPLE 3   **Using the Remainder Theorem**

Given $P(x) = 2x^3 - 9x^2 + 8x - 7$, use the Remainder Theorem to evaluate $P(3)$.

SOLUTION

The value of $P(3)$ is the remainder when $P(x)$ is divided by $x - 3$. To determine the remainder, we use synthetic division.

```
3 |  2   -9    8    -7
  |       6   -9    -3
  ----------------------
     2   -3   -1   -10
```

Because the remainder is −10, $P(3) = -10$. The result can be verified by evaluating the function directly.

The Factor Theorem

When a polynomial is divided by $x - c$ and the remainder is 0, we can write $P(x) = Q(x) \cdot (x - c)$. Thus $x - c$ is a factor of $P(x)$. This result and the Remainder Theorem lead to the Factor Theorem.

The Factor Theorem

For a polynomial $P(x)$ and a constant c, $P(c) = 0$ if and only if $x - c$ is a factor of $P(x)$.

EXAMPLE **4**   **Using the Factor Theorem**

Show that $x + 1$ is a factor of $P(x) = x^{20} - 2x^{14} + x^7 + 2$.

SOLUTION

To determine the remainder when $P(x)$ is divided by $x + 1$, we evaluate $P(-1)$.

$$P(-1) = (-1)^{20} - 2(-1)^{14} + (-1)^7 + 2 = 1 - 2 - 1 + 2 = 0$$

Because $P(-1) = 0$, the remainder is 0. Thus $x + 1$ is a factor of $P(x)$.

The Factor Theorem implies all of the following equivalent statements; that is, if any one of the statements is true, then all of the other statements are also true.

Equivalent Statements Implied by the Factor Theorem

1. The number c is a solution (or root) of $P(x) = 0$.
2. The number c is a zero of the function $P(x)$.
3. An x-intercept of the graph of P is $(c, 0)$.
4. A factor of $P(x)$ is $x - c$.
5. When $P(x)$ is divided by $x - c$, the remainder is 0.

These equivalent statements give us great flexibility in determining certain properties of a polynomial function. For example, if we know that an x-intercept of the graph of function P is $(1, 0)$, then we immediately know that $x - 1$ is a factor of P and that 1 is a zero of P.

EXAMPLE **5**   **Intercepts, Factors, and Zeros**

Produce the graph of $P(x) = x^3 + 3x^2 - 28x - 60$ and estimate the x-intercepts. Then use synthetic division to factor the polynomial and to determine the zeros of P.

DETERMINING THE SOLUTIONS

We use synthetic division to test the apparent zeros of the function.

$$
\begin{array}{r|rrrrrr}
-1 & 1 & -3 & -4 & 0 & 27 & 27 \\
   &   & -1 & 4 & 0 & 0 & -27 \\
\hline
\end{array}
$$

$$
\begin{array}{r|rrrrrr}
3 & 1 & -4 & 0 & 0 & 27 & 0 \\
  &   & 3 & -3 & -9 & -27 & \\
\hline
\end{array}
$$

The remainder is 0, so -1 is a zero.

$$
\begin{array}{r|rrrrr}
3 & 1 & -1 & -3 & -9 & 0 \\
  &   & 3 & 6 & 9 & \\
\hline
  & 1 & 2 & 3 & 0 & \\
\end{array}
$$

The remainder is 0, so 3 is a zero.

The number 3 is a zero of multiplicity 2.

Now we write the equation in factored form and solve with the Zero Factor Property.

$$(x + 1)(x - 3)^2(x^2 + 2x + 3) = 0$$

The first factors lead to the solutions -1 and 3. Using the Quadratic Formula, we find that $x^2 + 2x + 3 = 0$ has no real number solutions. Thus, as the graph suggests, the only real number solutions of the original equation are -1 and 3.

(b) VISUALIZING THE SOLUTIONS

The inequality $x^5 - 3x^4 - 4x^3 + 27x + 27 \leq 0$ is true for all values of x for which the graph of Q is on or below the x-axis. Referring to Figure 3.18, we see that the inequality is satisfied by all values of x in the interval $(-\infty, -1]$ and by the single value 3. The solution set is $(-\infty, -1] \cup \{3\}$.

3.2 Quick Reference

Division of Polynomials

■ To divide a polynomial by a polynomial, we use a method similar to long division in arithmetic.

■ Synthetic division is a numerical method for dividing a polynomial by $x - c$, where c is a constant called the *divider*. The polynomial must be written in descending order with 0 as the coefficient of any missing term.

The Remainder Theorem

■ The Remainder Theorem states that if a polynomial $P(x)$ is divided by $x - c$, where c is a constant, then the remainder is $P(c)$.

The Factor Theorem

■ The Factor Theorem states that for a polynomial $P(x)$ and a constant c, $P(c) = 0$ if and only if $x - c$ is a factor.

■ The following are all equivalent statements:
1. The number c is a solution (or root) of $P(x) = 0$.
2. The number c is a zero of the function $P(x)$.
3. An x-intercept of the graph of P is $(c, 0)$.
4. A factor of $P(x)$ is $x - c$.
5. When $P(x)$ is divided by $x - c$, the remainder is 0.

■ The following procedure can be used to factor a polynomial $P(x)$:

1. Produce the graph of P and estimate an x-intercept $(c, 0)$.
2. Use synthetic division to verify that $x - c$ is a factor and to determine the quotient $Q(x)$.
3. Write $P(x) = Q(x) \cdot (x - c)$, and then factor $Q(x)$.

The Rational Zero Theorem   ■ The Rational Zero Theorem states that for a polynomial function $P(x) = a_nx^n + a_{n-1}x^{n-1} + \cdots + a_1x + a_0$ with integral coefficients $(a_n \neq 0$ and $a_0 \neq 0)$, if $\dfrac{p}{q}$ is a rational zero of $P(x)$, then p is a factor of a_0 and q is a factor of a_n.

■ The following procedure can be used to locate a rational zero of a polynomial function:

1. List all possible rational zeros by writing the quotients of all factors of the constant term and all factors of the leading coefficient.
2. Use a graph or synthetic division to determine which numbers in the list are zeros of the function.

Equations and Inequalities   ■ A combination of graphing and factoring methods can be used to solve polynomial equations and inequalities.

3.2 Speaking the Language

1. A numerical method for dividing a polynomial by a binomial $x - c$ is called ▨▨▨▨ .

2. According to the ▨▨▨▨ Theorem, if $x - c$ is a factor of $P(x)$, then $P(c) = 0$.

3. According to the ▨▨▨▨ Theorem, if $\dfrac{P(x)}{x - c} = Q(x) + \dfrac{R}{x - c}$, then $P(c) = R$.

4. If $P(x) = x^3 - 2x^2 + 5x - 1$, then the ▨▨▨▨ Theorem states that the only rational number zeros of P are ± 1.

3.2 Exercises

Concepts and Skills

1. Suppose that you only want to know the remainder R when $P(x)$ is divided by $x - 7$. How can you determine R without performing the division?

2. Describe two ways to evaluate a polynomial P for $x = 3$.

In Exercises 3–10, divide $P(x)$ by $D(x)$ and write $P(x)$ in the form $Q(x) \cdot D(x) + R$.

3. $P(x) = x^3 - x^2 + 2x + 4$
   $D(x) = x - 2$

4. $P(x) = x^3 - 2x^2 + 3x + 1$
   $D(x) = x + 1$

5. $P(x) = 2x^3 - 11x^2 + 4x + 5$
   $D(x) = 2x + 1$

6. $P(x) = 27x^3 - 8$
   $D(x) = 3x - 2$

7. $P(x) = x^5 + 5x^3 - 10$
   $D(x) = x + 1$

8. $P(x) = x^4 + 2x^2 - 4$
   $D(x) = x - 2$

9. $P(x) = x^4 + x^3 - x - 8$
   $D(x) = x^2 - 5$

10. $P(x) = x^4 - x^2 - 12$
   $D(x) = x^2 + 3$

In Exercises 11–16, use synthetic division to determine the quotient $Q(x)$ and the remainder $R(x)$.

11. $(4x^3 - 3x^2 - 8x + 4) \div (x + 2)$

12. $(3x^4 - 6x^3 - 5x + 10) \div (x - 2)$

13. $(3x^3 - 10x^2 + x + 6) \div \left(x + \dfrac{2}{3}\right)$

14. $(4x^3 - 4x - 1) \div \left(x - \dfrac{1}{2}\right)$

15. $(2x^4 - 5x^3 - x - 15) \div (x - 3)$

16. $(3x^3 + 9x^2 - 10x + 9) \div (x + 4)$

In Exercises 17 and 18, use synthetic division to evaluate the function as indicated.

17. $P(x) = 2x^4 - x^3 + 5x^2 - 3x + 1;\quad P(-2)$

18. $Q(x) = x^5 + 2x^3 - 3x^2 - 4;\quad Q(3)$

In Exercises 19 and 20, determine the remainder if $P(x)$ is divided by $Q(x)$.

19. $P(x) = x^{12} - 3x^7 + 5$
   $Q(x) = x - 1$

20. $P(x) = x^{15} + x^8 - x^4 + x$
   $Q(x) = x + 1$

In Exercises 21 and 22, determine whether the number given in each part is a zero of the polynomial.

21. $P(x) = 2x^4 + 3x^3 - 12x^2 - 7x + 6$

   (a) 1   (b) $\dfrac{1}{2}$ (c) 2

22. $P(x) = 3x^3 + 16x^2 + 23x + 6$

   (a) -3 (b) 3   (c) $-\dfrac{1}{3}$

In Exercises 23 and 24, determine whether the binomial given in each part is a factor of the polynomial.

23. $P(x) = 3x^3 - 7x^2 - 2x + 8$
   (a) $x - 1$ (b) $3x - 4$ (c) $x - 2$

24. $P(x) = 2x^4 - x^3 - 8x^2 + x + 6$
   (a) $x - 1$ (b) $x + 2$ (c) $2x + 3$

In Exercises 25–30, determine all rational number zeros of the given function.

25. $P(x) = x^3 - x^2 - 10x - 8$

26. $F(x) = x^3 + x^2 - 17x + 15$

27. $R(x) = 2x^3 - x^2 - 11x + 10$

28. $G(x) = 6x^3 - x^2 - 27x - 20$

29. $p(x) = 16x^3 - 20x^2 - 56x + 15$

30. $r(x) = 24x^3 - 50x^2 - 11x + 30$

31. If P is a polynomial function, describe the relationship among a solution of an equation $P(x) = 0$, a zero of P, and an x-intercept of the graph of P.

32. Explain how a polynomial function P can have a real number zero even if the graph of P never crosses the x-axis.

In Exercises 33–36, determine all real number zeros of the given function and state the multiplicity of each one.

33. $f(x) = x^3 + 8x^2 + 13x + 6$

34. $g(x) = x^3 - x^2 - 8x + 12$

35. $P(x) = x^5 + 4x^4 - 2x^3 - 8x^2 + x + 4$

36. $F(x) = x^4 + 4x^3 - 2x^2 - 12x + 9$

In Exercises 37–42, determine all real number zeros of the given function.

37. $h(x) = x^3 - 4x^2 - 11x + 44$

38. $r(x) = x^3 - 5x^2 - 7x + 6$

39. $Q(x) = 8x^4 - 10x^3 - 59x^2 + 70x + 21$

40. $G(x) = 120x^4 - 42x^3 - 56x^2 + 7x + 6$

41. $C(x) = x^3 - 6x^2 - 6x - 7$

42. $W(x) = 6x^4 + 7x^3 + x^2 + 7x - 5$

In Exercises 43–50, factor the polynomial function completely.

43. $G(x) = x^4 - 6x^3 + 7x^2 + 6x - 8$

44. $F(x) = x^4 - 2x^3 - 20x^2 - 24x$

45. $f(x) = 6x^3 + 17x^2 + 4x - 12$

46. $p(x) = 3x^3 - 4x^2 - 17x + 6$

47. $R(x) = x^3 - x + 6$

48. $q(x) = x^3 + 3x^2 + 7x + 21$

49. $P(x) = 5x^3 - 7x^2 + 22x - 8$

50. $Q(x) = 40x^4 - 12x^3 - 46x^2 + 9x + 12$

51. Use the Rational Zero Theorem to explain why all possible rational number zeros of the function $P(x) = x^3 - 5x^2 + x - 7$ are integers.

52. If $P(x)$ is a polynomial function, what is the first step in solving $P(x) > 0$?

In Exercises 53–62, determine all real number solutions of the given equation. (*Hint*: In some cases you will need to simplify both sides and write the equation with 0 on one side.)

53. $x^4 + 2x^3 - 13x^2 - 14x + 24 = 0$

54. $x^5 - 7x^4 + 8x^3 + 16x^2 = 0$

55. $10x^3 + 49x + 12 = 53x^2$

56. $18x^2 - 8 = 3x^3 + 22x$

57. $7x^2 + 6x = 20 - x^3$

58. $4 - 5x^2 = 9x - 3x^3$

59. $15 - 13x^2 = -4(x^3 + 2x)$

60. $2x(x^2 - 16) + 14 = -x^2$

61. $3(2x^4 - 23x - 20) = -(23x^3 + 2x^2)$

62. $2x^2(3x^2 + 7) - 12 = 5x(3 + x^2)$

In Exercises 63–72, solve the given inequality.

63. $x^3 - 21x - 20 > 0$

64. $x^3 - 2x^2 - 5x + 6 < 0$

65. $2x^3 - x^2 - 21x \leq 0$

66. $3x^3 - 2x^2 - 3x + 2 \geq 0$

67. $x^4 - 16x^2 \leq 0$

68. $x^4 - x^3 - 12x^2 - 4x + 16 \geq 0$

69. $4x^4 + 15x^3 - 7x^2 - 12x > 0$

70. $x^4 + 4x^3 - 15x^2 - 19x + 30 < 0$

71. $x^5 + 3x^4 + 3x^3 + x^2 \leq 0$

72. $x^5 + 8x^4 + 16x^3 \geq 0$

Concept Extension

73. If $x^3 + kx - 4$ is divided by $x + 2$, the remainder is -8. Determine k.

74. Determine k so that $x - 3$ is a factor of the polynomial $2x^3 - kx^2 - 18$.

In Exercises 75–78, write a polynomial function P with the given characteristics.

75. Degree 3; zeros $-2, 1, 4$; the graph of P contains the point $(3, 30)$

76. Degree 4; zeros 3 (multiplicity 2), -1, 1; the graph of P contains the point $(4, 5)$

77. Degree 4; zeros -4 (multiplicity 3), 0; $P(-5) = 25$

78. Degree 5; zeros 4 (multiplicity 2), 0 (multiplicity 2), -5; $P(3) = 36$

79. Show that $x - c$ is a factor of $x^n - c^n$.

80. Consider the function $f(x) = x^n - c^n$. For what whole number values of n is (a) c a zero of f? (b) $-c$ a zero of f?

81. Suppose a polynomial function $P(x)$ has three zeros: -2, 4, and 7. Determine the zeros of (a) $P(x + 3)$; (b) $P(2x)$.

82. Consider the equation $x^n = x^m$ for positive integers n and m, $n \neq m$. What are the solutions (a) if both m and n are odd? (b) if either m is even or n is even but not both?

Applications

83. Cost of Weddings   With 1993 as the base year, the following table shows the approximate percent increases in the cost of weddings. (*Source*: Moet Matrimonial Matrix.)

Year	Percent
1994	0
1995	5
1996	3
1997	8

(a) Letting x represent the number of years since the base year, produce a cubic regression equation to model the data. Show that the equation and the following function have the same graphs.

$$P(x) = \frac{1}{6}(14x^3 - 105x^2 + 247x - 156)$$

(b) Produce the graph of P. Tracing the graph suggests that $(1, 0)$ is an x-intercept. To what data does this point correspond?

(c) Use synthetic division to verify that 1 is a zero of function P.

84. Referring to Exercise 83, use the result in part (c) to write P in the form $P(x) = Q(x) \cdot (x - 1)$. Now produce the graph of function Q. How does this graph provide further evidence that function P has no other zeros?

3.3 *Rational Functions*

Domains and Vertical Asymptotes ■ *Horizontal Asymptotes* ■ *Graphs of Rational Functions* ■ *Rational Inequalities* ■ *Real-Life Applications*

Domains and Vertical Asymptotes

A rational function is the quotient of two polynomial functions.

> **Definition of a Rational Function**
>
> A **rational function** is a function that can be written in the form $R(x) = \dfrac{P(x)}{Q(x)}$, where $P(x)$ and $Q(x)$ are polynomials with no common factors other than 1 and -1, and $Q(x) \neq 0$.

The following are some examples of rational functions.

$$r(x) = \frac{1}{x} \qquad f(x) = \frac{x + 1}{3 - x} \qquad g(x) = \frac{x^2 - 3}{x^3 - 2x + 2} \qquad h(x) = \frac{x^4 + x^2 + 2}{6 + x}$$

Consider the function $f(x) = \dfrac{2x + 1}{3 - x}$. From the definition of a rational function, $3 - x$ cannot be 0; that is, $x \neq 3$. Recall that we refer to 3 as a *restricted value*. In general, the domain of a rational function is the set of all real numbers, excluding restricted values.

The behavior of a rational function near a restricted value is of particular interest as we analyze the function.

EXPLORATION

Graphs and Restricted Values

Consider the function $f(x) = \dfrac{2x + 1}{x - 4}$.

(a) Complete the following table.

x	5.0	4.3	4.1	4.01	4.001	4.0001
$f(x)$						

What do you observe about the value of the function as x becomes closer and closer to 4?

(b) Complete the following table.

x	10	100	1000	10,000	100,000
$f(x)$					

What do you observe about the value of the function as x becomes larger and larger?

DEVELOPING THE CONCEPT

Graphs and Restricted Values

Returning to the rational function $f(x) = \dfrac{2x + 1}{3 - x}$, we will investigate the behavior of f near the restricted value 3. To do this, we evaluate f for x-values that are closer and closer to 3.

In the following tables, we use the notation $x \to 3$, which is read "x approaches 3." We begin with values of x that are located just to the right of 3 on a number line. Note that the x-values become closer and closer to 3. We say that *x approaches 3 from the right.*

x	3.1	3.01	3.001	3.0001	3.00001	As $x \to 3$ from the right, $f(x)$ decreases in value.
$f(x)$	-72	-702	-7002	$-70,002$	$-700,002$	

We see that as $x \to 3$ from the right, the values of f decrease without bound, which we indicate by $f(x) \to -\infty$. Similarly, in the following table, we see that as $x \to 3$ from the left, the values of f increase without bound, which we indicate by $f(x) \to \infty$.

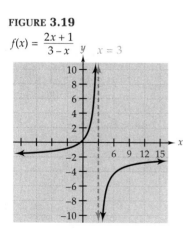

FIGURE 3.19

$f(x) = \dfrac{2x + 1}{3 - x}$

x	2.9	2.99	2.999	2.9999	2.99999	As $x \to 3$ from the left, $f(x) \to \infty$.
$f(x)$	68	698	6998	69,998	699,998	

This same behavior can also be observed in the graph of f. (See Figure 3.19.)

Note that the graph falls as x approaches 3 from the right and rises as x approaches 3 from the left.

In Figure 3.19, the vertical line $x = 3$ is drawn at the restricted value. This line partitions the graph into two *branches*. We call this line a **vertical asymptote.**

Definition of a Vertical Asymptote

The line $x = a$ is a vertical asymptote of the graph of f if $f(x) \to \infty$ or $f(x) \to -\infty$ as $x \to a$ from either the left or the right.

The vertical asymptotes of a rational function are associated with the restricted values of the function.

■ *NOTE   When you produce the graph of a rational function, your calculator may attempt to connect the branches. The result is a line that looks like a vertical asymptote. You can eliminate this line by producing the graph in the Dot mode.*

Vertical Asymptote of a Rational Function

If c is a restricted value of a rational function R, then $x = c$ is a vertical asymptote of the graph of R.

EXAMPLE 1   **Restricted Values and Vertical Asymptotes**

Determine the restricted values and the vertical asymptotes of each given function.

(a) $f(x) = \dfrac{5}{x - 2}$ (b) $g(x) = \dfrac{x}{x^2 - 9}$ (c) $h(x) = \dfrac{2}{x^2 + 1}$

SOLUTION

(a) The function f is not defined if x is 2. Thus 2 is a restricted value and the line $x = 2$ is a vertical asymptote. [See Figure 3.20(a).]

FIGURE **3.20**

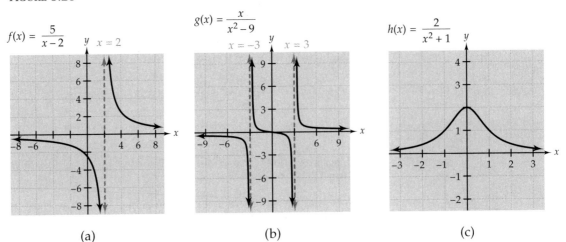

(a)   (b)   (c)

(b) The denominator $x^2 - 9 = (x + 3)(x - 3)$ cannot be 0. Thus, the restricted values are -3 and 3. The graph of g has two vertical asymptotes: $x = -3$ and $x = 3$. [See Figure 3.20(b).]

(c) Because $x^2 + 1$ is never 0, h has no restricted values and its graph has no vertical asymptotes. [See Figure 3.20(c).]

Horizontal Asymptotes

A thorough analysis of a rational function should include the end behavior of the function.

DEVELOPING THE CONCEPT

The End Behavior of $f(x) = \dfrac{1}{x}$

To investigate the end behavior of $f(x) = \dfrac{1}{x}$, we evaluate f for large and small values of x.

FIGURE **3.21**

$f(x) = \frac{1}{x}$

x	10	100	1000	10,000	100,000	
$f(x)$	0.1	0.01	0.001	0.0001	0.00001	As $x \to \infty$, $f(x) \to 0$.

x	-10	-100	-1000	$-10,000$	$-100,000$	
$f(x)$	-0.1	-0.01	-0.001	-0.0001	-0.00001	As $x \to -\infty$, $f(x) \to 0$.

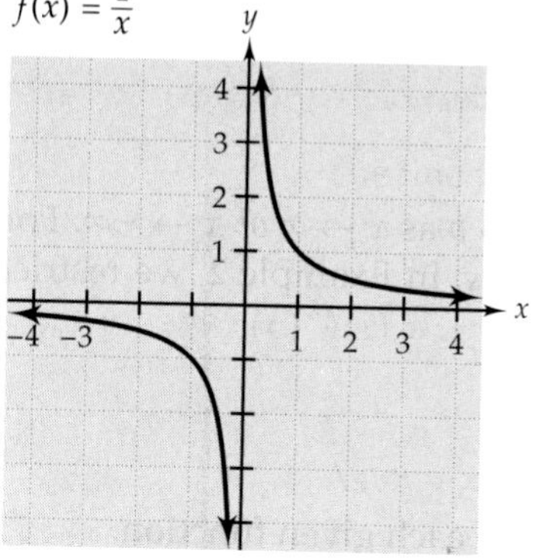

These tables and the graph of $f(x) = \dfrac{1}{x}$ (see Figure 3.21) indicate that $\dfrac{1}{x} \to 0$ as x becomes larger and larger and as x becomes smaller and smaller.

Note that the graph becomes closer and closer to the x-axis as $x \to \infty$ and as $x \to -\infty$.

The same observations can be made if the numerator is any nonzero constant and/or if the denominator is any natural number power of x.

The End Behavior of $f(x) = \dfrac{a}{x^n}$

If a is a nonzero constant and n is a natural number, then $\dfrac{a}{x^n} \to 0$ as $x \to \infty$ and as $x \to -\infty$.

(c) The degree of the numerator is greater than the degree of the denominator, and there is no horizontal asymptote.

These observations can be generalized.

■ *NOTE   When we sketch a graph, we usually draw the asymptotes as dashed lines to emphasize that asymptotes are lines that the graph approaches and are not part of the graph itself.*

> ### Tests for Horizontal Asymptotes
>
> Let $R(x)$ be a rational function, where the numerator has the leading coefficient a_n and degree n, and the denominator has the leading coefficient b_m and degree m.
> 1. If $n < m$, then the horizontal asymptote is the line $y = 0$.
> 2. If $n = m$, then the horizontal asymptote is the line $y = \dfrac{a_n}{b_m}$.
> 3. If $n > m$, then the graph has no horizontal asymptote.

Graphs of Rational Functions

The important aspects of the graph of a rational function are the intercepts, the asymptotes, the behavior of the function near vertical asymptotes, and the end behavior. The following is a summary of the procedure for sketching the graph of a rational function.

■ *NOTE   Displaying the graph of a rational function may require significant adjustments in the window settings.*

> ### Sketching the Graph of a Rational Function
>
> 1. Create a "template" for the graph by plotting the intercepts and drawing the asymptotes.
>
>   (a) The zeros of the denominator correspond to the vertical asymptotes.
>   (b) The zeros of the numerator correspond to the x-intercepts of the graph.
>
> 2. Produce a calculator graph of the function to observe the behavior of the function in the vicinity of the asymptotes.
> 3. Use the template and the visual guidance of the calculator graph to complete the sketch of the graph.

EXAMPLE 3   **The Graph of a Rational Function**

Sketch the graph of $f(x) = \dfrac{2x + 3}{x - 3}$.

SOLUTION

Because $f(0) = -1$, the y-intercept is $(0, -1)$.

The numerator is 0 when $x = -\frac{3}{2}$, so the x-intercept is $\left(-\frac{3}{2}, 0\right)$.

Because 3 is a restricted value, the line $x = 3$ is a vertical asymptote.

The degrees of the numerator and denominator are equal, so the horizontal asymptote is determined by the ratio of the leading coefficients: $y = \frac{2}{1} = 2$.

Figure 3.23(a) shows these intercepts and asymptotes.

The calculator graph in Figure 3.23(b) assists us in completing the sketch in Figure 3.23(c).

FIGURE 3.23

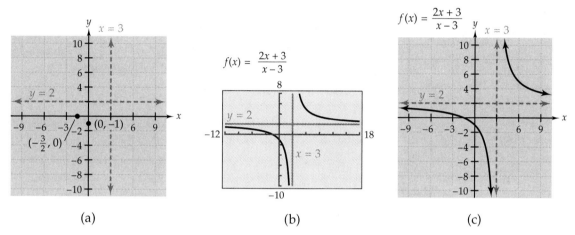

(a)   (b)   (c)

EXAMPLE 4   **The Graph of a Rational Function**

Sketch the graph of $g(x) = \dfrac{x^2 - 36}{x^2 + x - 12}$.

SOLUTION

Because $g(0) = 3$, the y-intercept is $(0, 3)$.

To facilitate determining the x-intercepts and vertical asymptotes, factor the numerator and denominator of the function.

$$g(x) = \frac{x^2 - 36}{x^2 + x - 12}$$

$$= \frac{(x + 6)(x - 6)}{(x + 4)(x - 3)}$$

The zeros of the numerator are ± 6. Thus the x-intercepts are $(-6, 0)$ and $(6, 0)$.

Because the zeros of the denominator are -4 and 3, the vertical asymptotes are $x = -4$ and $x = 3$.

To determine the horizontal asymptotes, we note that the degrees of the numerator and the denominator are equal. Therefore, the horizontal asymptote is $y = \frac{1}{1} = 1$.

We use this information to create the template shown in Figure 3.24(a).

FIGURE **3.24**

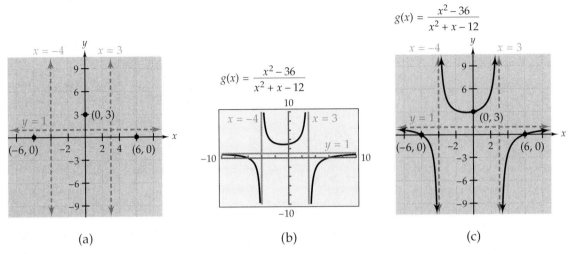

(a) (b) (c)

Finally, the calculator graph in Figure 3.24(b) guides us in completing the sketch of the graph in Figure 3.24(c).

EXAMPLE 5 Analyzing a Rational Function

Determine the intercepts and asymptotes of $h(x) = \dfrac{x}{x^2 + 9}$. Then produce the graph of h and check your results.

SOLUTION

The only intercept is $(0, 0)$.

Because the denominator is never 0, the graph has no vertical asymptote.

The degree of the numerator is less than the degree of the denominator, and so the horizontal asymptote is $y = 0$.

These results are confirmed in Figure 3.25.

FIGURE **3.25**

$$h(x) = \frac{x}{x^2 + 9}$$

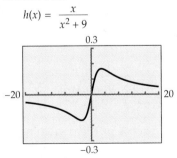

■ *NOTE As shown in Figure 3.25, a graph may intersect a horizontal asymptote. However, a graph can never intersect a vertical asymptote.*

We have seen that when the degree of the numerator is larger than the degree of the denominator, the graph has no horizontal asymptote. For such functions, a line that is neither vertical nor horizontal may be an asymptote.

Such a line is called a **slant asymptote.** In the next example we illustrate a technique for analyzing this kind of function.

EXAMPLE **6**   **A Graph with a Slant Asymptote**

Discuss the features of the graph of $f(x) = \dfrac{x^2 - 2x + 2}{2x - 4}$.

SOLUTION

Because $f(0) = -\frac{1}{2}$, the y-intercept is $\left(0, -\frac{1}{2}\right)$.

The equation $x^2 - 2x + 2 = 0$ has no real number solution, so the graph has no x-intercepts.

Because 2 is a restricted value, the graph has a vertical asymptote at $x = 2$. However, the graph has no horizontal asymptote because the degree of the numerator is greater than the degree of the denominator.

To investigate the end behavior, we can use long division to write function f in the following form:

$$f(x) = \frac{x^2 - 2x + 2}{2x - 4} = \frac{1}{2}x + \frac{1}{x - 2}$$

As $x \to \infty$, $\dfrac{1}{x - 2} \to 0$. Thus, as $x \to \infty$, the function behaves like $y = \dfrac{1}{2}x$.

The line $y = \dfrac{1}{2}x$ is called a *slant asymptote*. Figure 3.26 shows the graph of f.

FIGURE **3.26**

$f(x) = \dfrac{x^2 - 2x + 2}{2x - 4}$

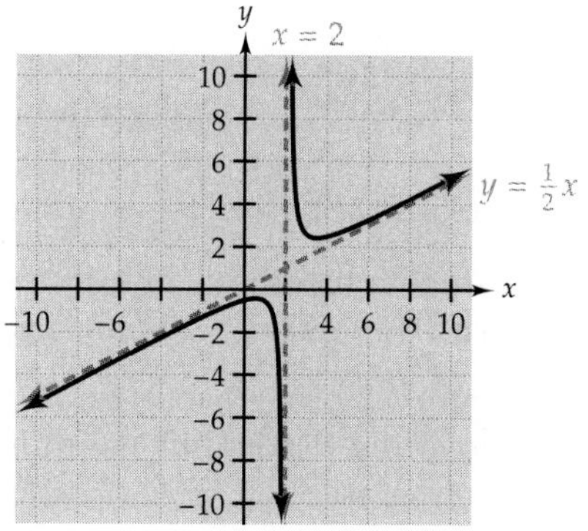

Rational Inequalities

For an inequality with rational expressions, restricted values and zeros of the associated function divide the number line into intervals. A graph shows which intervals contain solutions.

EXAMPLE 7 **Solving Rational Inequalities**

Solve the given inequalities.

(a) $\dfrac{x - 2}{x(x + 4)} < 0$ (b) $\dfrac{4x + 3}{x + 2} \geq 3$

SOLUTION

(a) The function $y = \dfrac{x - 2}{x(x + 4)}$ has a zero at 2 and restricted values 0 and −4. [See Figure 3.27(a).] The graph of the function is below the x-axis to the left of the line $x = -4$ and between the line $x = 0$ and the zero 2. Thus the solution set is $(-\infty, -4) \cup (0, 2)$.

FIGURE 3.27

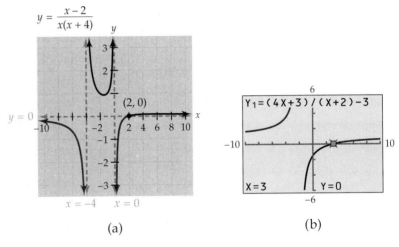

$$y = \frac{x - 2}{x(x + 4)}$$

(a)

$Y_1 = (4X+3)/(X+2)-3$

(b)

■ *NOTE As shown in Example 7(b), writing an inequality with 0 on one side is convenient for analyzing the graph with respect to the x-axis.*

(b) $\dfrac{4x + 3}{x + 2} \geq 3$

$\dfrac{4x + 3}{x + 2} - 3 \geq 0$ **Write the inequality with 0 on one side.**

Figure 3.27(b) shows the graph of $y = \dfrac{4x + 3}{x + 2} - 3$. From the graph, we estimate the one x-intercept to be $(3, 0)$. (We can use substitution to verify this.) Also, −2 is a restricted value.

In Figure 3.27(b), we look for those portions of the graph that are on or above the x-axis, but we must exclude the restricted value. The solution set is $(-\infty, -2) \cup [3, \infty)$.

Real-Life Applications

Some applications involve ratios, fractions, or percentages, and the equation that models the conditions of the problem could involve rational expressions. Your ability to solve such applications often depends on your skill in analyzing the features of a rational function.

EXAMPLE **8** **Campaign Spending**

An advertising consultant for a candidate in a local election predicts that the fraction of the total votes cast that the candidate will receive can be modeled by the function

$$V(x) = \frac{800 + \frac{2}{3}x}{2000 + x}$$

where x is the number of dollars spent on advertising.

(a) Assuming that the consultant is correct, what spending level will ensure the candidate's election?
(b) Determine the horizontal asymptote and interpret its meaning in terms of the application.

SOLUTION

(a) To win, the candidate must receive more than $\frac{1}{2}$ the votes cast.

$$\frac{800 + \frac{2}{3}x}{2000 + x} > \frac{1}{2} \qquad \text{The fraction of votes must exceed } \tfrac{1}{2}.$$

$$\frac{800 + \frac{2}{3}x}{2000 + x} - \frac{1}{2} > 0 \qquad \text{Write the inequality with 0 on one side.}$$

Figure 3.28 shows the graph of $y = \dfrac{800 + \frac{2}{3}x}{2000 + x} - \dfrac{1}{2}$. We estimate the x-intercept to be $(1200, 0)$. This can be verified by storing 1200 for x and evaluating the function.

All points of the graph from the x-intercept to the right lie *above* the x-axis, so the solution set is $(1200, \infty)$. Thus, spending at least \$1200 will ensure that the candidate will win the election.

(b) Note that the degrees of the numerator and the denominator are equal. Thus the horizontal asymptote is $y = \dfrac{\frac{2}{3}}{1} = \frac{2}{3}$. This means that as $x \to \infty$, $V(x) \to \frac{2}{3}$; that is, the fraction of votes that the candidate will receive approaches $\frac{2}{3}$. According to the model, even with a very large advertising expenditure, the candidate would receive no more than $\frac{2}{3}$ of the votes. We refer to this as the *limiting* fraction of votes.

FIGURE **3.28**

$$y = \frac{800 + \frac{2}{3}x}{2000 + x} - \frac{1}{2}$$

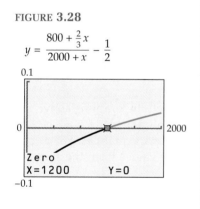

3.3 Quick Reference

Domains and Vertical Asymptotes

■ A rational function is a function that can be written in the form $R(x) = \dfrac{P(x)}{Q(x)}$, where $P(x)$ and $Q(x)$ are polynomials with no common factors other than 1 and -1 and $Q(x) \neq 0$.

■ The line $x = a$ is a vertical asymptote of the graph of f if $f(x) \to \infty$ or $f(x) \to -\infty$ as $x \to a$ from either the left or the right.

■ If c is a restricted value for a rational function R, then the line $x = c$ is a vertical asymptote of the graph of R.

Horizontal Asymptotes ■ If a is a nonzero constant and n is a natural number, then
$$\frac{a}{x^n} \to 0 \text{ as } x \to \infty \text{ and as } x \to -\infty.$$

■ The line $y = b$ is a horizontal asymptote of f if $f(x) \to b$ as $x \to \infty$ or $x \to -\infty$.

■ For a rational function R whose numerator has the leading coefficient a_n and degree n and whose denominator has the leading coefficient b_m and degree m,

1. If $n < m$, then the horizontal asymptote is the line $y = 0$.

2. If $n = m$, then the horizontal asymptote is the line $y = \dfrac{a_n}{b_m}$.

3. If $n > m$, then the graph has no horizontal asymptote.

Graphs of Rational Functions ■ To sketch the graph of a rational function, follow these steps.

1. Create a guide for the graph by plotting the intercepts and drawing the asymptotes. The zeros of the denominator correspond to the vertical asymptotes, and the zeros of the numerator correspond to the x-intercepts.
2. Produce the graph of the function and observe the behavior of the function in the vicinity of the asymptotes.
3. Use the guide along with the calculator graph to complete the sketch of the graph.

■ When the degree of the numerator of a rational function is larger than the degree of the denominator, the graph may have a slant asymptote. Divide the numerator by the denominator to identify a slant asymptote.

Rational Inequalities ■ The following is a summary of a method for solving a rational inequality $R(x) > 0$.

1. Produce the graph of R and determine the x-intercepts and restricted values.
2. Determine the intervals of x-values for which the graph is above the x-axis, and use those intervals to write the solution set.
3. Similar methods can be used for $<$, $\leq$, and $\geq$ inequalities. For $\leq$ and $\geq$ inequalities, be sure to exclude restricted values from the solution set.

3.3 Speaking the Language

1. If the graph of a function f approaches ∞ or $-\infty$ as x approaches a from the left or the right, then the line $x = a$ is a(n) _____ of the graph of f.

2. A function of the form $R(x) = \dfrac{P(x)}{Q(x)}$, $Q(x) \neq 0$, is called a rational function if $P(x)$ and $Q(x)$ are _____ .

3. The line $y = 2$ is a(n) _____ of the graph of $f(x) = \dfrac{2x^2 + 3x - 5}{x^2 - 1}$.

4. The domain of a rational function includes all real numbers except _____ .

3.3 Exercises

Concepts and Skills

1. Explain why the graph of a rational function cannot intersect a vertical asymptote.

2. For the graph of a rational function, compare the possible number of horizontal asymptotes to the possible number of vertical asymptotes.

In Exercises 3–8, determine the intercepts and the asymptotes of the given function.

3. $r(x) = \dfrac{7}{2x - 9}$

4. $f(x) = \dfrac{x + 1}{x^2 - 16}$

5. $g(x) = \dfrac{2x + 1}{x - 5}$

6. $w(x) = \dfrac{4 - x}{x + 3}$

7. $f(x) = \dfrac{3x^2 + 5x}{2x^2 - x - 1}$

8. $h(x) = \dfrac{3 + 2x - x^2}{2x^2 + x}$

In Exercises 9–12, match the graph to the function.

(A) $f(x) = \dfrac{2}{2 - x}$

(B) $h(x) = \dfrac{x}{x^2 + 4}$

(C) $r(x) = \dfrac{x}{x^2 - 4}$

(D) $g(x) = \dfrac{1}{2x - x^2}$

9.

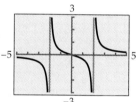

10.

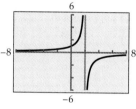

11.

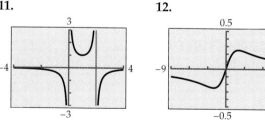

12.

13. Consider the rational function $R(x) = \dfrac{P(x)}{Q(x)}$. To what feature of the graph of R does a zero of function Q pertain?

14. Given that $\dfrac{1}{x} \to 0$ as $x \to \infty$, show that $\dfrac{a}{x^n} \to 0$ as $x \to \infty$.

In Exercises 15–36, sketch the graph of the given rational function.

15. $f(x) = \dfrac{3}{x - 6}$

16. $g(x) = \dfrac{1}{x + 3}$

17. $g(x) = \dfrac{-1}{x + 4}$

18. $h(x) = \dfrac{2}{1 - x}$

19. $f(x) = \dfrac{x + 3}{x}$

20. $r(x) = \dfrac{x - 4}{x - 2}$

21. $h(x) = \dfrac{1 - x}{x + 2}$

22. $f(x) = \dfrac{x - 4}{3 - x}$

23. $r(x) = \dfrac{1}{x^2 + 4}$

24. $w(x) = \dfrac{x}{x^2 + 5}$

25. $h(x) = \dfrac{x}{(x + 1)^2}$

26. $f(x) = \dfrac{1}{(x - 3)^2}$

27. $g(x) = \dfrac{2x - 5}{x + 1}$

28. $r(x) = \dfrac{6x + 1}{1 - 2x}$

29. $w(x) = \dfrac{x^2}{x^2 + 4}$

30. $f(x) = \dfrac{x^2}{x^2 - 1}$

31. $h(x) = \dfrac{x}{x^2 - 4}$

32. $g(x) = \dfrac{x - 1}{2x - x^2}$

33. $r(x) = \dfrac{2 - x}{x^2 + 4x + 3}$

34. $g(x) = \dfrac{x + 3}{x^2 - 3x - 4}$

35. $f(x) = \dfrac{x^2 - x - 1}{x^2 + 3}$

36. $h(x) = \dfrac{x^2 + 5x + 4}{(x - 2)^2}$

37. Suppose that the graph of a rational function has vertical asymptotes $x = 2$ and $x = -3$ and a horizontal asymptote $y = 1$. From the following list, identify the function that satisfies those conditions and explain why the other functions do not.

   (i) $f(x) = \dfrac{x^2}{x^2 - x - 6}$

   (ii) $g(x) = \dfrac{x}{x^2 + x - 6}$

   (iii) $h(x) = \dfrac{x^2}{x^2 + x - 6}$

38. Suppose that $R(x) = \dfrac{x - 5}{x - 3}$. Compare the behavior of R as $x \to 3$ to the behavior of R as $x \to 5$.

In Exercises 39–46, sketch the graph of the given function.

39. $h(x) = \dfrac{x^2 + x - 2}{x}$

40. $f(x) = \dfrac{x^2 + 2x - 3}{2x}$

41. $f(x) = \dfrac{x^3 + 8}{x + 4}$

42. $w(x) = \dfrac{x^3 - 9x}{2 - x}$

43. $g(x) = \dfrac{x^3 + 8x^2}{x^2 + 2x - 8}$

44. $r(x) = \dfrac{x^2 + x - 6}{x + 6}$

45. $f(x) = \dfrac{-x^3 + 2}{x^2}$

46. $g(x) = \dfrac{1 - x^2}{x + 2}$

47. Explain why the inequality $\dfrac{x^2}{x^2 - 1} > 1$ is satisfied by all x-values for which the graph of $R(x) = \dfrac{x^2}{x^2 - 1}$ is above the horizontal asymptote.

48. Explain how you would use the graphing method to estimate the solutions of the double inequality $\dfrac{1}{x} \le \dfrac{3x}{x + 2} \le 2$.

In Exercises 49–64, solve the inequality.

49. $\dfrac{2}{x - 5} > 0$

50. $\dfrac{3}{2 - x} < 0$

51. $\dfrac{x + 1}{4 - x} \le 3$

52. $\dfrac{x + 2}{x} \ge 5$

53. $\dfrac{x + 1}{x + 5} \ge \dfrac{x}{x - 1}$

54. $\dfrac{1 - 2x}{x + 1} \le \dfrac{x + 2}{x + 4}$

55. $3 + \dfrac{2 - x}{x} > \dfrac{x + 1}{x - 2}$

56. $\dfrac{x + 5}{x + 1} > \dfrac{x}{x - 3} + 5$

57. $x + 1 < \dfrac{1}{x}$

58. $5 \ge \dfrac{2}{x} + x$

59. $\dfrac{1}{x^2 - 4} \ge \dfrac{2}{x^2 - 3x - 10}$

60. $\dfrac{x - 1}{x^2 + 4x + 3} > \dfrac{x}{x^2 - x - 12}$

61. $\dfrac{4}{x + 1} < \dfrac{x}{x^2 + 5x + 4}$

62. $\dfrac{x}{x + 2} < \dfrac{2}{x^2 - x - 6}$

63. $\dfrac{x^2 + x - 6}{x - 4} \le 0$

64. $x \ge \dfrac{x + 10}{x - 2}$

Concept Extension

65. At first glance, we might conclude that the graph of $R(x) = \dfrac{x^2 - 10x}{x - 10}$ would have a vertical asymptote $x = 10$.

(a) Use the integer setting to produce the graph of R. Why does the graph appear to be a line? (*Hint*: Simplify the rational expression.)

(b) Trace the graph to $x = 10$. Explain the result.

(c) Compare the domains of $R(x) = \dfrac{x^2 - 10x}{x - 10}$ and $f(x) = x$.

66. Compare the graphs of $f(x) = \dfrac{x^2 + 4x + 3}{x^2 - 9}$ and $g(x) = \dfrac{x + 1}{x - 3}$. (*Hint*: Simplify the expression for function f.)

In Exercises 67–70, sketch the graph of the given function.

67. $H(x) = \dfrac{x^3 - 3x^2}{3 - x}$

68. $g(x) = \dfrac{2x^3 + 2x^2}{x^3 - 2x^2 - 3x}$

69. $f(x) = \dfrac{x^3 + 4x^2 + x - 6}{x + 2}$

70. $r(x) = \dfrac{x^4 - 1}{x^2 - 1}$

Applications

71. Growth Habit After its second year, the growth habit of Amur River privet can be modeled by $h(x) = \dfrac{11x - 17}{x}$, where $h(x)$ is the height (in feet) of the shrub after x years.

(a) What is the expected height of a five-year-old shrub?

(b) The behavior of the model function near the vertical asymptote invalidates the model for the first two years of the shrub's life. Why?

(c) According to the model function, what is the limiting height of the shrub?

ART FOR EXERCISE 71

72. **Dog's Weight** After a puppy is two months old, its weight w (in pounds) increases according to the model

$$w(x) = \frac{80x - 127}{2x}$$

where x is the dog's age in months.

(a) Why is the x-intercept of the graph of w an invalid data point?
(b) What value does $w(x)$ approach as $x \to \infty$?
(c) According to the model function, what is the limiting weight of the dog?

73. **Illiteracy** The probability p (expressed as a percentage) that a person will be functionally illiterate is

$$p(x) = \frac{200}{x}$$

where x is the number of years of formal schooling that the person has completed.

(a) What are the asymptotes of the graph of p?
(b) Why is the model invalid for a person who never attended school at all?
(c) What is the probability that a recent high school graduate is functionally illiterate?
(d) What does the end behavior of function p suggest about the validity of the model beyond high school?

74. **County Revenue** Over a ten-year period, the population growth of a certain county has increased the tax base for county revenues. The following functions could be used to model the tax base (in millions of dollars), where x is the number of years since the population growth began.

$$q(x) = 0.1x^2$$

$$r(x) = \frac{x}{10 - x}$$

Produce the graphs of functions q and r in the same coordinate system.

(a) Which function is probably the better model if the population growth was relatively steady?
(b) Which function is probably the better model if the population growth was slow during the first half of the period and very rapid during the second half?
(c) As $x \to 10$ from the left, what values do the two functions approach? Which of the two functions is invalid near the end of the ten-year period?

75. **Parallel Circuit Resistance** If the total resistance in a parallel circuit is 100 ohms, then the resistance r_1 in one branch is given by the function

$$r_1 = \frac{100r_2}{r_2 - 100}$$

where r_2 is the resistance in the other branch of the circuit.

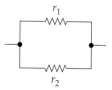

(a) What are the asymptotes of the graph of r_1?
(b) If resistance cannot be negative, what is the domain of the function?
(c) What happens to the values of r_1 as $r_2 \to 100$ from the right? as $r_2 \to \infty$?

76. **Focal Length of a Lens** If the focal length of the lens in your camera is 3, then the distance d from the lens to the film is

$$d = \frac{3x}{x - 3}$$

where x is the distance from the lens to the object that you are photographing.

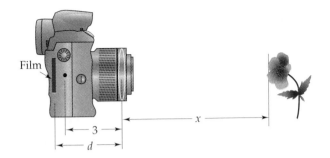

(a) What is the vertical asymptote of the graph of d?

(b) On which side of the vertical asymptote is the graph of d meaningful for this application? Why?

(c) As you move closer to the object that you are photographing, what happens to the distance d?

(d) Determine the horizontal asymptote and compare your result to the focal length of the lens.

77. Campaign Poster A campaign poster is designed to have 600 square inches of print area. The top and bottom margins are 4 inches, and the side margins are 3 inches.

(a) Referring to the figure, write expressions for the width and height of the poster.

(b) Write an equation to model the condition that the area of the printed portion is 600 square inches.

(c) Solve the equation for y and use the result to write an expression in x for the height of the poster.

(d) Write a function $A(x)$ for the area of the poster as a function of the width x.

(e) Produce the graph of A and use the *Minimum* feature to estimate the width of the poster for which the amount of poster material will be a minimum. Round your answer to the nearest inch.

78. Recycling Bin An open bin for recycled newspaper has a square base and a volume of 125 cubic yards. The base of the bin costs \$8 per square yard to manufacture, and the sides cost \$3 per square yard.

(a) Write an equation to model the condition that the volume of the bin is 125 cubic yards.

(b) Solve the equation for h to obtain an expression in x for the height of the bin.

(c) Represent the cost of the base of the bin. Represent the total cost of the four sides of the bin.

(d) Use the results in part (c) to write a function $C(x)$ for the total cost of the bin.

(e) Produce the graph of C and use the *Minimum* feature to estimate the width of the base for which the cost of the bin will be a minimum. Round your answer to the nearest tenth of a yard.

79. Average Cost The total cost C of manufacturing x refrigerators per week is given by the following function:

$$C(x) = 0.02x^3 - 0.2x^2 + x + 20$$

The average cost $A(x)$ per refrigerator is
$$A(x) = \frac{C(x)}{x}.$$

Produce the graph of A and use the *Minimum* feature to estimate the number of refrigerators that should be manufactured per week to minimize the average cost. Round your result to the nearest whole number.

3.4   *Complex Numbers*

Imaginary Numbers   ■   *Operations with Complex Numbers*   ■
Complex Conjugates   ■   *Quadratic Equations*

Imaginary Numbers

In the real number system, the equation $x^2 = -1$ has no solution. In order to solve such equations, the number system must be expanded. We begin by defining an *imaginary number i*.

> ### Definition of the Imaginary Number i
> The symbol *i* represents an **imaginary number** with the properties that $i^2 = -1$ and $i = \sqrt{-1}$.

The expanded number system, which includes *i*, is called the **complex number system.**

■ *NOTE   If a = 0 and b ≠ 0, then a + bi is called a pure imaginary number.*

> ### Definition of a Complex Number
> For real numbers *a* and *b*, $a + bi$ is the *standard form* of a **complex number.** The number *a* is called the **real part,** and *b* is called the **imaginary part.** If $b \neq 0$, then $a + bi$ is an **imaginary number.** If $b = 0$, then $a + bi$ is simply the real number *a*.

The complex number system is the union of all real numbers and all imaginary numbers. A complex number is either real or imaginary but not both. Figure 3.29 shows the structure of the complex number system.

FIGURE 3.29

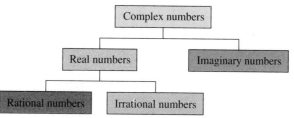

Using the number *i*, we can define the square root of *any* real number.

> **Definition of $\sqrt{-n}$**
> For any positive real number n, $\sqrt{-n} = i\sqrt{n}$.

EXAMPLE 1 **Writing $\sqrt{-n}$ as $i\sqrt{n}$**

■ *NOTE As illustrated in Example 1, we write $i\sqrt{10}$ rather than $\sqrt{10}i$ so that the expression is not confused with $\sqrt{10i}$.*

Write each square root in terms of i.

(a) $\sqrt{-81} = i\sqrt{81} = 9i$

(b) $\sqrt{-10} = i\sqrt{10}$

Two complex numbers are equal if their real parts are equal and their imaginary parts are equal.

> **Definition of Equality of Complex Numbers**
> For two complex numbers $a + bi$ and $c + di$, $a + bi = c + di$ if and only if $a = c$ and $b = d$.

Operations with Complex Numbers

■ *NOTE Many of the familiar properties of real numbers are not valid for complex numbers. For example, $\sqrt{-9} \cdot \sqrt{-4} \neq \sqrt{36}$ because the Product Rule for Radicals does not hold in the complex number system. Instead,*

$$\sqrt{-9} \cdot \sqrt{-4} = (3i)(2i) = 6i^2$$
$$= 6(-1) = -6$$

The complex number system has a commutative property, an associative property, and a distributive property. Thus we can perform the arithmetic operations of addition, subtraction, multiplication, and division with complex numbers as if they were binomials.

Keep in mind that i is not a variable. However, the procedure for adding or subtracting complex numbers is the same as that used for adding or subtracting binomials.

EXAMPLE 2 **Sums and Differences of Complex Numbers**

(a) $(3 - 5i) + (2 + i) = (3 + 2) + (-5 + 1)i = 5 - 4i$

(b) $7i + (9 - 2i) = 9 + 5i$

(c) $(-1 + 6i) - (1 - 5i) = (-1 - 1) + (6 + 5)i = -2 + 11i$

(d) $(8 - 7i) - (8 - 6i) = -i$

Along with the fact that $i^2 = -1$, we can use the Distributive Property to multiply two complex numbers in the same way that we multiply binomials.

EXAMPLE **3**   **Products of Complex Numbers**

(a) $2i(7 - 3i) = 14i - 6i^2 = 14i - 6(-1) = 6 + 14i$

(b) $(2 + 5i)(3 - 2i) = 6 - 4i + 15i - 10i^2$
$= 6 + 11i - 10(-1)$
$= 16 + 11i$

(c) $(3 + i)^2 = 3^2 + 2(3i) + i^2$
$= 9 + 6i - 1$
$= 8 + 6i$

(d) $(1 + 2i)(1 - 2i) = 1 - 4i^2$
$= 1 + 4$
$= 5$

The following table shows that the powers of i follow a cyclical pattern.

i^1	i^2	i^3	i^4		i^5	i^6	i^7	i^8		i^9	i^{10}	...
i	-1	$-i$	1		i	-1	$-i$	1		i	-1	...

Example 4 shows how we can take advantage of this pattern to evaluate any power of i.

EXAMPLE **4**   **Evaluating Powers of i**

METHOD 1

By writing powers of i as $(i^2)^n$ or $(i^2)^n \cdot i$, we can reduce the computation of powers of i to raising -1 to an integer power.

(a) $i^{17} = (i^2)^8 \cdot i = (-1)^8 \cdot i = 1 \cdot i = i$

(b) $i^{22} = (i^2)^{11} = (-1)^{11} = -1$

METHOD 2

By writing powers of i as a product with one factor being $(i^4)^n$, we can take advantage of the fact that $i^4 = 1$.

(c) $i^{11} = (i^4)^2 \cdot i^3 = (1)^2 \cdot -i = -i$ $i^3 = -i$

(d) $i^{30} = (i^4)^7 \cdot i^2 = (1)^7 \cdot (-1) = -1$ $i^2 = -1$

■ *NOTE   The advantage of Method 1 is that you need to know only that $i^2 = -1$. For Method 2, you need to know the values of i^r for r = 1, 2, 3, and 4.*

In effect, if we divide the exponent n by 4 and the remainder is r, then $i^n = i^r$.

Complex Conjugates

In Example 3(d), the product of $1 + 2i$ and $1 - 2i$ was the real number 5. The two complex numbers differ only in the sign of the imaginary part. Such numbers are called **complex conjugates,** and their product is a real number.

> ### Definition of Complex Conjugates
> Let a and b be real numbers. The complex numbers $a + bi$ and $a - bi$ are called **complex conjugates,** and their product is
> $$(a + bi)(a - bi) = a^2 + b^2$$

EXAMPLE 5 **Products of Complex Conjugates**

	Number	*Conjugate*	*Product*
(a)	$-5 + 2i$	$-5 - 2i$	$(-5)^2 + 2^2 = 25 + 4 = 29$
(b)	$2 - i$	$2 + i$	$2^2 + (-1)^2 = 4 + 1 = 5$
(c)	$4i = 0 + 4i$	$-4i = 0 - 4i$	$0^2 + 4^2 = 0 + 16 = 16$

To divide a complex number by a real number c, $c \neq 0$, we divide the real part and the imaginary part by c.

> ### Dividing a Complex Number by a Real Number
> For real numbers a, b, and c, $c \neq 0$,
> $$\frac{a + bi}{c} = \frac{a}{c} + \frac{b}{c}i$$

To divide by an imaginary number, we multiply the numerator and the denominator by the conjugate of the denominator.

EXAMPLE 6 **Dividing by an Imaginary Number**

Write the quotient in standard form.

$$\frac{3 - i}{1 - i} = \frac{3 - i}{1 - i} \cdot \frac{1 + i}{1 + i} = \frac{3 + 3i - i + 1}{1 + 1} = \frac{4 + 2i}{2} = 2 + i$$

FIGURE **3.30**

```
(3+5i)+(-2-9i)
                1-4i
(4-i)(-3+2i)
              -10+11i
(3-i)/(1-i)
                 2+i
```

Mode By selecting the complex number mode, we can use a calculator to perform operations with complex numbers.

Figure 3.30 illustrates how we can perform addition and multiplication of complex numbers as well as the division in Example 6.

Quadratic Equations

If we include complex numbers and allow for double roots, then all quadratic equations have two solutions. For instance, the equation $x^2 + 1 = 0$ has solutions $\pm i$.

Recall that when we use the Quadratic Formula to solve a quadratic equation, a negative discriminant indicates that the equation has no real solutions. However, the formula can be used to determine complex solutions.

EXAMPLE 7 **Complex Solutions of Quadratic Equations**

Solve $9x^2 - 6x + 5 = 0$.

SOLUTION

$$x = \frac{-b \pm \sqrt{b^2 - 4ac}}{2a}$$

Use the Quadratic Formula to determine the solutions.

$$= \frac{-(-6) \pm \sqrt{(-6)^2 - 4(9)(5)}}{2(9)}$$

$a = 9, b = -6, c = 5$

$$= \frac{6 \pm \sqrt{-144}}{18}$$

The discriminant is negative, so the solutions are imaginary numbers.

$$= \frac{6 \pm 12i}{18}$$

Divide out the common factor of 6.

$$= \frac{1}{3} \pm \frac{2}{3}i$$

Observe that the two solutions are complex conjugates.

3.4 Quick Reference

Imaginary Numbers
■ The symbol i represents an **imaginary number** with the properties that $i^2 = -1$ and $i = \sqrt{-1}$.

■ If a and b are real numbers, $a + bi$ is the *standard form* of a **complex number.** For $a + bi$, a is called the **real part** and b is called the **imaginary part.**

■ For any positive real number n, $\sqrt{-n} = i\sqrt{n}$.

■ Imaginary numbers and real numbers are also complex numbers.

Operations with Complex Numbers
■ The arithmetic operations of addition, subtraction, and multiplication with complex numbers are performed as if the numbers were binomials.

Complex Conjugates
■ The complex numbers $a + bi$ and $a - bi$ are called **complex conjugates.**

■ The product of complex conjugates is a real number.

$$(a + bi)(a - bi) = a^2 + b^2$$

■ If c is a nonzero real number, then $\dfrac{a + bi}{c} = \dfrac{a}{c} + \dfrac{b}{c}i$.

■ To find the quotient of $a + bi$ and $c + di$, multiply the numerator and the denominator by the conjugate of the denominator.

Quadratic Equations ■ The Quadratic Formula can be used to determine complex solutions of a quadratic equation.

3.4 Speaking the Language

1. The number $3i$ is both a(n) ▨▨▨▨ number and a(n) ▨▨▨▨ number.

2. The numbers $3 + i$ and $3 - i$ are examples of ▨▨▨▨.

3. For the number $5 + 2i$, 5 is called the ▨▨▨▨ part and 2 is called the ▨▨▨▨ part.

4. To add, subtract, and multiply complex numbers, we can treat the numbers as though they were ▨▨▨▨.

3.4 Exercises

Concepts and Skills

1. In the definition of $\sqrt{-n}$, how can we be certain that the radicand is a negative number?

2. What kind of number, real or imaginary, is $a + bi$ under the following conditions?
   (i) $a = 0, b \neq 0$
   (ii) $a \neq 0, b = 0$
   (iii) $a = 0, b = 0$
   (iv) $a \neq 0, b \neq 0$

In Exercises 3–6, write the expression in the complex number form $a + bi$.

3. $\sqrt{-36}$ 4. $\sqrt{-18}$

5. $-3 - \sqrt{-75}$ 6. $\sqrt{-4} - \sqrt{4}$

In Exercises 7–24, perform the indicated operations.

7. $(2 + 3i) + (1 - 4i)$ 8. $(2i - 3) - (4 - i)$

9. $(6i - 4) - (7 + 2i)$ 10. $(9 + 5i) + (-3 - 4i)$

11. $5i - (-3 + 2i)$ 12. $4 + (1 - 6i)$

13. $3i(2i - 1) - 3(2 + i)$ 14. $2(4i - 3) - i(1 - 3i)$

15. $(1 + 3i)(2 - i)$ 16. $(4 - i)(9 + 4i)$

17. $(5 - 4i)(4 - 5i)$ 18. $(6 + 7i)(3 + i)$

19. $(3 + 5i)(3 - 5i)$ 20. $(2 + 7i)(2 - 7i)$

21. $\left(\sqrt{3} - 2i\right)\left(\sqrt{3} + 2i\right)$

22. $\left(\sqrt{7} + i\sqrt{5}\right)\left(\sqrt{7} - i\sqrt{5}\right)$

23. $(1 + 3i)^2$

24. $(7 - 3i)^2$

In Exercises 25–30, use either method to perform the indicated operation.

25. i^9 26. i^{12}

27. i^{31} 28. i^{25}

29. $(i^3)^5$ 30. $i^3 \cdot i^7$

In Exercises 31–38, write the complex number in standard form.

31. $\sqrt{-9} \cdot \sqrt{-1}$ 32. $\sqrt{-3} \cdot \sqrt{-12}$

33. $\sqrt{-2}\left(\sqrt{-8} + \sqrt{2}\right)$ 34. $\sqrt{-6}\left(\sqrt{2} - \sqrt{-3}\right)$

35. $\left(2 + \sqrt{-3}\right)\left(2 - \sqrt{-3}\right)$ 36. $\left(1 + \sqrt{-5}\right)^2$

37. $\dfrac{-4 + \sqrt{-12}}{2}$ 38. $\dfrac{10 - \sqrt{-50}}{5}$

39. Compare the product $(a + b)(a - b)$ with the product $(a + bi)(a - bi)$.

40. How is dividing by a complex number similar to the process of rationalizing a denominator?

In Exercises 41–44, write the conjugate of the given number. Then determine the product of the given number and its conjugate.

41. $3 + 2i$

42. $1 - 4i$

43. $i\sqrt{5} - 4$

44. $7 + i\sqrt{3}$

In Exercises 45–54, perform the indicated division.

45. $\dfrac{1 + i}{2 - i}$

46. $\dfrac{3 - 2i}{3 + 2i}$

47. $\dfrac{2 + i}{1 + 3i}$

48. $\dfrac{1 + i}{1 - 5i}$

49. $\dfrac{3 + 2i}{i}$

50. $\dfrac{4 - i}{2i}$

51. $\dfrac{5i}{1 + 3i}$

52. $\dfrac{2}{5 - i}$

53. $\dfrac{\sqrt{3} - i\sqrt{2}}{\sqrt{2} + i\sqrt{3}}$

54. $\dfrac{5 - i\sqrt{7}}{5 + i\sqrt{7}}$

55. Suppose that you have memorized the values of i^n for $n = 1, 2, 3,$ and 4. Describe a method for evaluating i^n for any value of n.

56. Suppose that the only power of i that you know is $i^2 = -1$. Describe a method for evaluating i^n for any value of n.

In Exercises 57–66, determine all complex number solutions of the given quadratic equation.

57. $x^2 + 25 = 0$

58. $2x^2 + 15 = 0$

59. $x^2 - 6x + 13 = 0$

60. $x^2 - 4x + 5 = 0$

61. $16x = 16x^2 + 5$

62. $5x^2 + x = -1$

63. $x^2 = 8x - 18$

64. $2x(x + 2) = -3$

65. $5 = 4x(1 - x)$

66. $3x^2 + 1 = x$

67. Evaluate $i^4 + i^8 + i^{12} + i^{16}$.

68. Show that each of the following statements is true.

(a) The product of a complex number and its conjugate is a real number.

(b) The sum of a complex number and its conjugate is a real number.

(c) The difference of a complex number and its conjugate is a pure imaginary number.

Concept Extension

In Exercises 69–72, determine a and b.

69. $(a + 3i) - (5 - bi) = 4i$

70. $\left(a + \sqrt{-16}\right) + (1 + bi) = 1$

71. $\dfrac{a + bi}{2i} = 3 - \dfrac{3}{2}i$

72. $(a + bi)(4 + 7i) = 18 - i$

In Exercises 73–76, evaluate the given expression for the given complex number.

73. $x^2 - 2x + 5;\quad 1 - 2i$

74. $x^2 + x + 2;\quad -\dfrac{1}{2} + \dfrac{\sqrt{7}}{2}i$

75. $x^2 + 4x + 9;\quad -2 + i\sqrt{5}$

76. $x^2 - 6x + 10;\quad 3 - i$

In Exercises 77–80, determine the product.

77. $(x + 2i)(x - 2i)$

78. $(2x + 5i)(2x - 5i)$

79. $[x - (3 + i)][x - (3 - i)]$

80. $[x - (2 + 5i)][x - (2 - 5i)]$

81. For a nonzero complex number $a + bi$, write an expression in standard form for the reciprocal of the number.

82. Let z represent a complex number $a + bi$, and let $\overline{z}$ represent the conjugate of z. Show that $\overline{\overline{z}} = z$.

In Exercises 83–86, solve the equation.

83. $ix + 3 - 4i = i$

84. $3ix + 4 - i = 1 + i$

85. $(5 + 2i)x + i = 2i + (1 + i)x$

86. $3 - (1 - 2i)x = (i - 1)x$

87. For a nonzero complex number $a + bi$ and a positive integer n, we define $(a + bi)^{-n}$ as $\dfrac{1}{(a + bi)^n}$. Use this definition to evaluate each of the following.

(a) i^{-3}

(b) i^{-2}

(c) $(1 + 2i)^{-1}$

(d) $(2 - 3i)^{-2}$

88. The number 16 has two square roots. Show that 16 has four fourth roots.

3.5 *Fundamental Theorem of Algebra*

Factors and Zeros ■ *Conjugate Zeros*

Factors and Zeros

We have seen that the graph of a polynomial function may or may not intersect the x-axis, and so the function may or may not have a real zero. However, an important theorem guarantees that all polynomial functions (other than constant functions) have at least one complex zero.

> ### Fundamental Theorem of Algebra
> A polynomial function of degree 1 or higher has at least one complex number zero.

One consequence of the Fundamental Theorem of Algebra is that a polynomial has at least one linear factor. By extension, we can show that a polynomial of degree $n \geq 1$ has n linear factors.

DEVELOPING THE CONCEPT

Factoring a Polynomial

Consider a polynomial function $P(x)$ of degree n, where $n \geq 1$. The Fundamental Theorem of Algebra guarantees that P has at least one complex number zero; that is, there is a complex number c_1 such that $P(c_1) = 0$.

By the Factor Theorem, an equivalent statement is that $x - c_1$ is a factor of $P(x)$, and so $P(x)$ can be written in the factored form

$$P(x) = (x - c_1) \cdot Q(x)$$

Now the Fundamental Theorem of Algebra and the Factor Theorem can be applied again to factor $Q(x)$. Because $P(x)$ has degree n, we can repeat the process n times to write

$$P(x) = a_n(x - c_1)(x - c_2) \cdots (x - c_n)$$

where a_n is the leading coefficient of $P(x)$.

These results lead to the following theorem.

> **Linear Factor Theorem**
>
> If $P(x)$ is a polynomial function of degree $n \geq 1$, then the polynomial can be written as a product of n linear factors, where c_i is a complex number.
>
> $$P(x) = a_n(x - c_1)(x - c_2) \cdots (x - c_n)$$

Note that each factor leads to a zero of the polynomial. However, any factor that is repeated k times leads to a zero with multiplicity k. For example, consider the following polynomial function:

$$P(x) = (x + 1)^2(x - 2)^3$$

Function P has two *distinct* zeros, -1 (multiplicity 2) and 2 (multiplicity 3). However, taking multiplicity into account, the function has a total of five zeros. Thus a polynomial of degree n has at most n distinct zeros, but if we consider multiplicity, the polynomial has exactly n zeros.

> **Zeros of a Polynomial Function**
>
> A function defined by a polynomial of degree n has exactly n complex zeros if a zero of multiplicity k is counted k times.

EXAMPLE **1** **Zeros of a Polynomial Function**

For $P(x) = x^3(x^2 + 4)(x - 5)^4$, determine the zeros and multiplicity.

SOLUTION

Because the degree of the polynomial is 9, taking multiplicity into account, the number of zeros is 9. Observe that $x^2 + 4 = 0$ has two solutions, $2i$ and $-2i$.

Zero	0	5	$2i$	$-2i$
Multiplicity	3	4	1	1

EXAMPLE **2** **Determining Zeros of a Function**

Determine all zeros of the following function:

$$P(x) = 2x^7 - 7x^6 - 13x^5 + 68x^4 - 30x^3$$

SOLUTION

DETERMINING A ZERO OF THE FUNCTION

Because its degree is 7, function P has seven zeros. To determine them, we begin by factoring out x^3.

$$P(x) = x^3(2x^4 - 7x^3 - 13x^2 + 68x - 30)$$

Because x^3 is a factor, we now know that $P(x)$ has a zero 0 of multiplicity 3.

VISUALIZING THE X-INTERCEPTS

To factor $2x^4 - 7x^3 - 13x^2 + 68x - 30$, we use a graph to estimate the x-intercepts. (See Figure 3.31.)

One x-intercept appears to be $(-3, 0)$. Figure 3.31 also shows the use of the *Zero* feature to locate the other x-intercept $(0.5, 0)$.

DETERMINING THE OTHER ZEROS

Because x-intercepts represent the zeros of a function, we conclude that -3 and $\frac{1}{2}$ are zeros of function P. We use synthetic division to verify these zeros and to write $Q(x)$ in factored form.

FIGURE 3.31

$$Q(x) = 2x^4 - 7x^3 - 13x^2 + 68x - 30$$

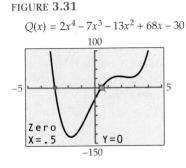

$$
\begin{array}{r|rrrrr}
-3 & 2 & -7 & -13 & 68 & -30 \\
& & -6 & 39 & -78 & 30 \\
\hline
\frac{1}{2} & 2 & -13 & 26 & -10 & 0 \\
& & 1 & -6 & 10 & \\
\hline
& 2 & -12 & 20 & 0 &
\end{array}
$$

The remainder is 0, so -3 is a zero.

The remainder is 0, so $\frac{1}{2}$ is a zero.

Because -3 and $\frac{1}{2}$ are zeros of the function, $x + 3$ and $x - \frac{1}{2}$ are factors of $Q(x)$. We can use the result of the division to write the following factorization of $Q(x)$.

$$Q(x) = (x + 3)\left(x - \frac{1}{2}\right)(2x^2 - 12x + 20)$$

$$= (x + 3)\left(x - \frac{1}{2}\right) \cdot 2 \cdot (x^2 - 6x + 10) \qquad \text{Factor out the 2.}$$

$$= (x + 3)(2x - 1)(x^2 - 6x + 10) \qquad 2\left(x - \tfrac{1}{2}\right) = 2x - 1$$

Thus, $Q(x) = (x + 3)(2x - 1)(x^2 - 6x + 10)$.

Finally, we use the Quadratic Formula to obtain the zeros of the function $S(x) = x^2 - 6x + 10$.

$$x = \frac{-b \pm \sqrt{b^2 - 4ac}}{2a} = \frac{-(-6) \pm \sqrt{(-6)^2 - 4(1)(10)}}{2} = 3 \pm i$$

We can now write the factored form of $P(x)$ as follows:

$$P(x) = x^3(x + 3)(2x - 1)(x - 3 + i)(x - 3 - i)$$

The seven zeros of P are 0 (multiplicity 3), $-3, \frac{1}{2}, 3 + i$, and $3 - i$.

Conjugate Zeros

We have observed that the zeros of quadratic functions occur in pairs. For example, we can use the Quadratic Formula to determine the zeros of the function $f(x) = x^2 - 4x + 1$.

$$x = \frac{4 \pm \sqrt{(-4)^2 - 4(1)(1)}}{2} = 2 \pm \sqrt{3}$$

The pair of zeros, $2 + \sqrt{3}$ and $2 - \sqrt{3}$, are square root conjugates.

In Examples 1 and 2, observe that the two imaginary zeros are complex conjugates. These examples illustrate the following theorem.

■ *NOTE   Because imaginary zeros come in pairs, the number of imaginary zeros must be even.*

> ### Conjugate Zero Theorem
> 1. If $a + b\sqrt{c}$ is a zero of a polynomial $P(x)$ with rational coefficients, then its conjugate $a - b\sqrt{c}$ is also a zero.
> 2. If a complex number $a + bi$ is a zero of a polynomial $P(x)$ with real coefficients, then its conjugate $a - bi$ is also a zero.

An advantage of knowing that certain types of zeros come in pairs is that knowing one zero immediately allows us to know another zero.

EXAMPLE 3   **Determining the Zeros of a Polynomial Function**

Find all zeros of $x^4 - 2x^3 + 14x^2 - 18x + 45$ given that $1 - 2i$ is a zero.

SOLUTION

Because the polynomial has real coefficients and $1 - 2i$ is a zero, $1 + 2i$ is also a zero. Using synthetic division verifies these two zeros and produces the remaining factor.

$1 - 2i$	1	-2	14	-18	45
		$1 - 2i$	-5	$9 - 18i$	-45
	1	$-1 - 2i$	9	$-9 - 18i$	0

$1 + 2i$	1	$-1 - 2i$	9	$-9 - 18i$
		$1 + 2i$	0	$9 + 18i$
	1	0	9	0

The resulting polynomial $x^2 + 9$ has two complex zeros, $\pm 3i$. Thus the original polynomial has four zeros: $1 \pm 2i$ and $\pm 3i$. The factored form of the polynomial is as follows:

$$P(x) = (x - 1 + 2i)(x - 1 - 2i)(x + 3i)(x - 3i)$$

EXAMPLE 4 **Writing a Polynomial Function**

Write a polynomial function of lowest possible degree having a zero -3 of multiplicity 2, a zero 0 of multiplicity 4, a zero $1 + \sqrt{2}$ of multiplicity 1, and a zero $3i$ of multiplicity 1. Assume that the leading coefficient is 1 and that all other coefficients are integers.

SOLUTION

Because $1 + \sqrt{2}$ and $3i$ are zeros, their conjugates $1 - \sqrt{2}$ and $-3i$ are also zeros. Taking multiplicity into account, the lowest possible degree of the function is 10. The polynomial function can be written as follows:

$$P(x) = x^4(x + 3)^2\left(x - 1 - \sqrt{2}\right)\left(x - 1 + \sqrt{2}\right)(x + 3i)(x - 3i)$$

Multiplying $\left(x - 1 - \sqrt{2}\right)\left(x - 1 + \sqrt{2}\right)$ and $(x + 3i)(x - 3i)$ leads to the following form:

$$P(x) = x^4(x + 3)^2(x^2 - 2x - 1)(x^2 + 9)$$

3.5 Quick Reference

Factors and Zeros
■ The Fundamental Theorem of Algebra states that a polynomial function of degree 1 or higher has at least one complex zero.

■ The Linear Factor Theorem states that a polynomial function P of degree $n \geq 1$ can be written as a product of n linear factors $x - c_i$, where c_i is a complex number.

$$P(x) = a_n(x - c_1)(x - c_2) \ldots (x - c_n)$$

■ If a factor $(x - c)$ is repeated k times, then c is a zero with multiplicity k.

■ A polynomial function of degree n has exactly n complex zeros if a zero of multiplicity k is counted k times.

Conjugate Zeros
■ If $a + b\sqrt{c}$ is a zero of a polynomial $P(x)$ with rational coefficients, then its conjugate $a - b\sqrt{c}$ is also a zero.

■ If $a + bi$ is a zero of a polynomial $P(x)$ with real coefficients, then its conjugate $a - bi$ is also a zero.

3.5 Speaking the Language

1. If $P(x)$ is a polynomial function of degree $n \geq 1$, then the ▨▨▨▨▨ states that P has at least one complex number zero.

2. If $P(x) = x^3 + 2x - 1$, then the ▨▨▨▨▨ Theorem states that P can be written as $P(x) = a_3(x - c_1)(x - c_2)(x - c_3)$, where c_1, c_2, and c_3 are complex numbers.

3. If a complex number is a zero of a polynomial function with real number coefficients, then its ▨▨▨▨▨▨ is also a zero.

4. Although $P(x) = (x - 1)^2(x + 3)^4$ has only two distinct zeros, the total number of zeros equals the degree of the polynomial if ▨▨▨▨▨▨ is taken into account.

3.5 Exercises

Concepts and Skills

1. Suppose that a polynomial function with rational coefficients has no real number zeros. Can the degree of the function be odd? Why?

2. Suppose that a fifth-degree polynomial function P with rational coefficients has at least two real zeros. What is the possible number of x-intercepts that the graph of P may have? Explain.

In Exercises 3–6, write the polynomial function of lowest possible degree that has the given zeros. Assume that the leading coefficient is 1, and that all other coefficients are integers.

3. $6, 1 + \sqrt{2}, 1 - \sqrt{2}$ 4. $-4, -i\sqrt{5}, i\sqrt{5}$

5. $-\sqrt{3}, \sqrt{3}, -2i, 2i$

6. $3 + \sqrt{2}, 3 - \sqrt{2}, 2 + 5i, 2 - 5i$

In Exercises 7–12, write the polynomial function of lowest possible degree that has the given zeros. Assume that the leading coefficient is 1, and that all other coefficients are integers.

7. $3, 2 - i$ 8. $-4, 3 + i$

9. $1, -2, i$ 10. $3 + \sqrt{2}, 1 - i$

11. $1 - \sqrt{2}, i\sqrt{3}$ 12. $\sqrt{6}, 4i$

13. Suppose that a polynomial function with rational coefficients has an imaginary zero of multiplicity 3. What is the lowest possible degree of the function? Why?

14. Suppose that the only zeros of a polynomial function P with rational coefficients are 1 and -2. Does this imply that $P(x) = x^2 + x - 2$? Why?

In Exercises 15–26, a polynomial function is given along with one of its zeros. Determine all the other zeros.

15. $Q(x) = x^4 - 2x^3 + 8x^2 - 18x - 9;\quad 1 + \sqrt{2}$

16. $P(x) = x^4 + 4x^3 + 3x^2 + 16x - 4;\quad -2i$

17. $f(x) = x^4 - 6x^3 + 8x^2 + 30x - 65;\quad 3 + 2i$

18. $g(x) = x^4 - 8x^3 + 18x^2 + 56x - 175;\quad \sqrt{7}$

19. $P(x) = x^4 + 6x^3 + 11x^2 + 10x + 2;\quad -1 - i$

20. $F(x) = x^4 - 16x^3 + 92x^2 - 204x + 87;\quad 3 - \sqrt{6}$

21. $q(x) = x^4 - 3x^2 + 10x - 6;\quad -1 + \sqrt{3}$

22. $p(x) = x^4 - 6x^3 - 11x^2 + 32x + 138;\quad -2 - i\sqrt{2}$

23. $f(x) = x^4 + 4x^3 + 22x^2 + 36x + 117;\quad 3i$

24. $g(x) = x^4 - 2x^3 + 6x^2 - 8x + 8;\quad -2i$

25. $h(x) = x^4 + 2x^3 - x^2 - 2x + 10;\quad -2 + i$

26. $f(x) = x^4 + 2x^3 + 5x^2 + 4x + 40;\quad 1 - 2i$

27. One zero of $P(x) = x - i$ is i. Is $-i$ also a zero? Explain why this situation does not violate the Conjugate Zero Theorem.

28. When we say that complex zeros of a polynomial function occur in pairs, what assumption are we making about the function?

In Exercises 29–34, determine all zeros of the given function.

29. $P(x) = 2x^3 + 9x^2 + 2x - 30$

30. $Q(x) = 2x^3 - 15x^2 + 32x + 20$

31. $f(x) = x^4 - 6x^3 + 10x^2 + 42x + 25$

32. $h(x) = 9x^4 + 24x^3 + 28x^2 - 80x + 32$

33. $q(x) = 4x^6 - 4x^5 - 9x^4 + 8x^3 + 6x^2 - 4x - 1$

34. $p(x) = 3x^5 + 5x^4 - 44x^3 - 76x^2 + 57x - 9$

35. The Rational Zero Theorem states that the only possible rational zeros of $P(x) = x^4 - x^2 - 2$ are ± 1 and ± 2. Are any of these numbers zeros? If not, should we conclude that P has four complex zeros? Test your conjecture by using the factoring method to solve $x^4 - x^2 - 2 = 0$.

36. (a) Produce the graph of the following function:

$$P(x) = x^5 + 3x^4 + 4x^3 + 4x^2 + 3x + 1$$

Use the graph to estimate a real zero c_1 of P and test the zero with synthetic division. Does the graph prove that the remaining zeros are imaginary numbers?

(b) Write P in the form $P(x) = Q(x) \cdot (x - c_1)$. Now produce the graph of $Q(x)$ and repeat part (a) for function Q.

(c) Continue the steps above until you have determined all zeros of function P.

In Exercises 37–42, determine all solutions of the given equation.

37. $x^3 - 10x^2 + 35x - 44 = 0$

38. $x^3 + 5x^2 + 14x + 24 = 0$

39. $2x^4 - 7x^3 - 9x^2 + 2x + 2 = 0$

40. $3x^4 + 16x^3 + 22x^2 - 4x - 16 = 0$

41. $x^5 - 4x^4 + 7x^3 - 4x^2 - 8x + 8 = 0$

42. $8x^5 + 4x^4 - 98x^3 + 251x^2 - 732x + 315 = 0$

43. Suppose that $P(x) = x^3 - x + 2$. Compare the zeros of P to the solutions of the equation $x^3 - x + 2 = 0$.

44. Suppose that you divide a third-degree polynomial P by $2 + i$, and the remainder is 0. If P has real coefficients, how many real zeros does the polynomial function have? Why?

In Exercises 45–50, one factor of the function is given. Write the function as a product of linear factors.

45. $F(x) = x^4 - 8x^3 + 15x^2 - 2x - 4$; $\quad x - 3 + \sqrt{5}$

46. $H(x) = x^4 - 8x^3 + 19x^2 - 12x + 2$; $\quad x - 2 - \sqrt{2}$

47. $R(x) = x^4 + 6x^3 + 26x^2 + 32x - 40$; $\quad x + 2 - 4i$

48. $P(x) = x^4 - 8x^3 + 22x^2 - 16x - 11$; $\quad x - 3 - i\sqrt{2}$

49. $G(x) = x^4 - 6x^3 + 27x^2 - 58x + 90$; $\quad x - 2 + i\sqrt{5}$

50. $W(x) = x^4 + 2x^3 + x^2 + 6x + 20$; $\quad x + 2 + i$

In Exercises 51–56, write the given function as a product of linear factors.

51. $f(x) = x^3 - 3x^2 - 11x + 5$

52. $g(x) = 2x^3 + 7x^2 - 6x + 1$

53. $p(x) = 6x^4 + 13x^3 + 13x^2 - 2$

54. $r(x) = 3x^4 - 11x^3 + 15x^2 + 15x - 14$

55. $q(x) = 4x^4 + 16x^3 - x^2 - x + 12$

56. $f(x) = 9x^4 - 57x^3 + 119x^2 - 97x + 30$

Concept Extension

The Quadratic Formula applies to quadratic equations with complex coefficients. In Exercises 57 and 58, determine the complex solutions to the given quadratic equation.

57. $x^2 - 5ix - 6 = 0$

58. $2x^2 - ix + 3 = 0$

Chapter Review Exercises

Section 3.1

In Exercises 1 and 2, produce the graphs of both functions in the same coordinate system. Describe the graph of the second function relative to the graph of the first function.

1. $y = x^3, f(x) = x^3 + 4$

2. $y = x^4, g(x) = (x - 2)^4 - 3$

In Exercises 3 and 4, identify the term that determines the end behavior of the graph of the function. Predict the end behavior.

3. $g(x) = 2x^2 - x^4$ **4.** $h(x) = x^5 + 3x^2 - 7$

In Exercises 5 and 6, determine the zeros of the function and state the multiplicity of each.

5. $P(x) = (2x - 1)(x + 1)^3(x - 2)^2$

6. $Q(x) = 12x^6 + 60x^5 + 75x^4$

7. Suppose that a function has a zero c of multiplicity n. Explain the relationship between the multiplicity and the behavior of the graph of the function at the x-intercept $(c, 0)$.

8. Determine the x-intercepts of the graph of the function $f(x) = x^5 - 3x^4$. Predict whether the graph crosses the x-axis or is tangent to the x-axis at each intercept.

9. Use a calculator to determine a third-degree polynomial function P that models the given data.

x	1	3	4	6
y	-6	-12	0	84

10. Beginning one hour after taking a medication for a certain viral infection, the patient's temperature for the next four hours can be modeled by the following function:

$$T(x) = -0.58x^4 + 7.2x^3 - 31x^2 + 54x + 70$$

Predict the patient's highest temperature.

In Exercises 11 and 12, estimate the relative extrema of the given function.

11. $f(x) = x^3 - 2x^2 - 5x + 5$

12. $g(x) = -x^4 + 7x^3 - 10x^2 - 6x + 2$

Section 3.2

13. Let $P(x) = 4x^3 + 6x^2 + 2x + 5$ and $D(x) = 2x - 1$. Use long division to write $P(x)$ as $D(x) \cdot Q(x) + R$.

14. Use synthetic division to determine the quotient and remainder when $3x^4 - 9x^2 - 10$ is divided by $x - 2$.

In Exercises 15 and 16, determine all rational zeros of the given function.

15. $P(x) = 6x^3 - 25x^2 + 2x + 8$

16. $Q(x) = x^4 + 4x^3 - 2x^2 - 12x + 9$

In Exercises 17 and 18, determine all real zeros of the given function.

17. $f(x) = x^4 + 15x^3 + 70x^2 + 84x - 72$

18. $h(x) = 2x^4 - 7x^3 - 2x^2 + 12x$

In Exercises 19 and 20, write the polynomial function in completely factored form.

19. $H(x) = 4x^6 - 12x^5 - 15x^4 - 4x^3$

20. $F(x) = x^4 - 6x^3 + 11x^2 - 18x + 24$

21. Determine all real number solutions of the following equation:

$$5x^3 + 10x + 6 = 2x^4 + x^2$$

22. Explain how to show that $x^3 + x^2 - 2x + 3 = 0$ has no rational number solution.

In Exercises 23 and 24, solve the given inequality.

23. $x^4 + 8x^3 + 15x^2 \geq 0$

24. $-x^4 + 6x^3 + x^2 - 30x < 0$

Section 3.3

In Exercises 25–28, determine the intercepts and asymptotes of the given function.

25. $h(x) = \dfrac{x}{x^2 + 1}$

26. $r(x) = \dfrac{4 - x}{x + 5}$

27. $f(x) = \dfrac{x^2 - 2x - 3}{x^2 + 2x}$

28. $g(x) = \dfrac{-x^2 + 3x - 2}{x}$

In Exercises 29–32, sketch the graph of the given function.

29. $f(x) = \dfrac{-2}{x + 3}$

30. $w(x) = \dfrac{x^2}{x^2 - 4}$

31. $g(x) = \dfrac{x}{x^2 + 6}$

32. $f(x) = \dfrac{x^2 + 1}{x}$

In Exercises 33 and 34, solve the given inequality.

33. $\dfrac{x - 2}{x + 3} \geq -4$

34. $\dfrac{x - 2}{x - 4} < \dfrac{x}{x + 2}$

35. Explain why the line $x = 2$ is not a vertical asymptote of the graph of $y = \dfrac{x^2 - 2x}{x - 2}$.

36. The cost of removing p percent of the chemical spilled into a stream following a tanker truck accident is modeled by the rational function $C(p) = \dfrac{5000p}{100 - p}$.

(a) What is the cost of removing 60% of the chemical?

(b) What is the cost of removing 80% of the chemical?

(c) What does the graph indicate about the cost as the amount removed nears 100%?

Section 3.4

37. For the complex number $4 - 7i$, write the

(a) real part.

(b) imaginary part.

(c) conjugate.

38. Explain why a real number is also a complex number.

In Exercises 39–44, perform the indicated operations.

39. $(3i - 2) + (2 + i)$ **40.** $3i - (-2 + 7i)$

41. $3(i - 5) - i(2 + i)$ **42.** $(4 + 2i)(1 - i)$

43. $(3 - 4i)^2$ **44.** i^{11}

45. Divide $\dfrac{2 - i}{1 - 3i}$.

46. Write the complex number $\sqrt{-3}\left(\sqrt{-12} - \sqrt{3}\right)$ in standard form.

In Exercises 47 and 48, determine the complex solutions of the given quadratic equation.

47. $x^2 + 2x + 5 = 0$ **48.** $3x^2 = 2x - 4$

Section 3.5

49. Suppose that you know that one zero of a polynomial with real coefficients is $-4 - 5i$. Explain why you know another zero.

50. Write the following polynomial as the product of linear factors. $\left(\text{Hint: One factor is } 2x - 1 + 2\sqrt{2}.\right)$

$$4x^4 - 12x^3 - 7x^2 + 22x + 14$$

In Exercises 51 and 52, write a polynomial function of lowest degree with rational coefficients with the given zeros.

51. $1 \pm \sqrt{2}, \pm 5i$ **52.** $2, -3 + 2i$

In Exercises 53 and 54, one zero of the polynomial function is given. Find all other zeros.

53. $F(x) = x^4 - 4x^3 + 6x^2 - 28x - 7; \quad i\sqrt{7}$

54. $P(x) = x^4 + 4x^3 + 8x^2 + 8x - 5; \quad -1 - \sqrt{2}$

In Exercises 55 and 56, determine all zeros of the given function.

55. $f(x) = 3x^3 + 7x^2 - 2x - 30$

56. $g(x) = x^4 + 6x^3 + 9x^2 - 6x - 10$

In Exercises 57 and 58, write the given function as a product of linear factors.

57. $f(x) = 2x^4 - 3x^3 - 4x^2 + 14x - 12$

58. $h(x) = x^4 - 3x^3 - 4x^2 + 10x - 4$

In Exercises 59 and 60, determine all solutions of the given equation.

59. $2x^3 + 50x = 5x^2 + 125$

60. $2x^4 + x^3 + 5x^2 + 24x + 18 = 0$

Chapter 4

Exponential and Logarithmic Functions

How do we learn, and what motivates us to want to learn? Some researchers believe that we learn more quickly if we are punished for erroneous behavior. Others have concluded that being rewarded for correct behavior is more effective.

Psychologists have developed learning theories based on the behavior of lab animals. For example, a classic study involves determining the number of trials a mouse needs before it learns how to move through a maze with no errors.

The scatterplot might depict the results for a particular maze experiment. To model the data, a new kind of a function called an exponential function might be appropriate.

In this chapter, we discuss operations that can be performed with functions, and we introduce the concept of an inverse function. Then we consider two new types of functions, exponential and logarithmic functions, and their properties. Finally, we discuss methods for solving equations that involve exponential and logarithmic expressions, and we apply these skills to solving real-life applications and modeling problems.

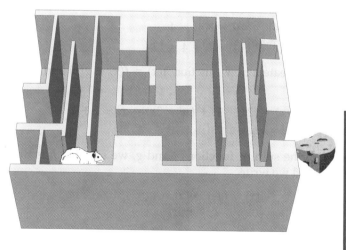

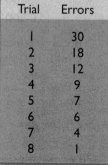

Trial	Errors
1	30
2	18
3	12
4	9
5	7
6	6
7	4
8	1

$y = 45.39(0.67)^x$

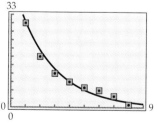

The following definition gives the precise meaning of the composition of functions.

> **Definition of the Composition of Functions**
>
> The **composite function** or **composition** of functions f and g is a function $f \circ g$ defined by $(f \circ g)(x) = f(g(x))$.
>
> The domain of $f \circ g$ is the set of numbers x in the domain of g for which $g(x)$ is in the domain of f.

The general relationship among functions f, g, and $f \circ g$ is shown in Figure 4.4.

FIGURE **4.4**

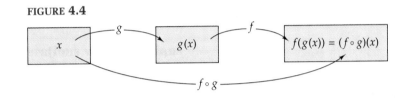

>
>
> **Compose**   You can use your calculator to form the composition of two functions. If functions f and g are entered on the **Y** screen as Y_1 and Y_2, respectively, then the form of $(f \circ g)(x)$ is $Y_1(Y_2)$.
>
> **Activate**   If a function that is entered on the **Y** screen is "active," its graph can be displayed. We can deactivate functions on the **Y** screen so that their graphs are not displayed.

EXAMPLE **4**   **Composition of Two Functions**

Let $f(x) = 2x^2 + 3$ and $g(x) = x - 3$. Determine each of the following:

(a) $(f \circ g)(2)$
(b) $(g \circ f)(1)$
(c) $(f \circ g)(x)$
(d) $(g \circ f)(x)$

FIGURE **4.5**

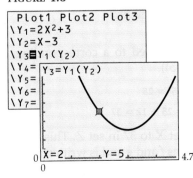

SOLUTION

(a) VISUALIZING THE VALUE OF THE COMPOSED FUNCTION

Figure 4.5 shows the entries of f, g, and $f \circ g$ on the **Y** screen. Note that only $f \circ g$ is active.

Figure 4.5 also shows the graph of $f \circ g$ with the tracing cursor at $x = 2$. We see that $(f \circ g)(2) = 5$. This result can be verified by evaluating $f(g(2))$.

DETERMINING THE VALUE OF THE COMPOSED FUNCTION

$$(f \circ g)(2) = f(g(2))$$
$$= f(-1) \qquad g(2) = 2 - 3 = -1$$
$$= 5 \qquad f(-1) = 2(-1)^2 + 3 = 5$$

(b) $(g \circ f)(1) = g(f(1))$
$$= g(5) \qquad f(1) = 2(1)^2 + 3 = 5$$
$$= 2 \qquad g(5) = 5 - 3 = 2$$

(c) $(f \circ g)(x) = f(g(x))$
$$= f(x - 3) \qquad \text{Replace } g(x) \text{ with } x - 3.$$
$$= 2(x - 3)^2 + 3 \qquad \text{Replace } x \text{ in function } f \text{ with } x - 3.$$

(d) $(g \circ f)(x) = g(f(x))$
$$= g(2x^2 + 3) \qquad \text{Replace } f(x) \text{ with } 2x^2 + 3.$$
$$= (2x^2 + 3) - 3 \qquad \text{Replace } x \text{ in function } g \text{ with } 2x^2 + 3.$$
$$= 2x^2$$

Parts (c) and (d) of Example 4 demonstrate that, in general, $f \circ g$ is not equal to $g \circ f$.

In Example 4, the domains of f and g are **R,** so the domains of $f \circ g$ and $g \circ f$ are also **R.** As illustrated in the next example, the domain of the composition of two functions may not be the same as the domain of either of the two functions.

EXAMPLE 5   **Domain of the Composition of Functions**

Let $f(x) = x - 4$ and $g(x) = \sqrt{x}$. Determine the domain of (a) $f \circ g$ and (b) $g \circ f$.

SOLUTION

Note that the domain of f is **R** and the domain of g is $\{x \mid x \geq 0\}$.

(a) VISUALIZING THE DOMAIN

Figure 4.6 shows the graph of $f \circ g$. The domain appears to be $\{x \mid x \geq 0\}$.

FIGURE **4.6**

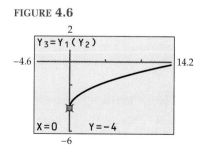

DETERMINING THE DOMAIN

Because the domain of function f is **R**, we can evaluate f for any value of $g(x)$. However, $g(x)$ is a real number only if $x \geq 0$. Thus the domain of $f \circ g$ is the same as the domain of g: $\{x \mid x \geq 0\}$.

(b) VISUALIZING THE DOMAIN

Figure 4.7 shows the graph of $g \circ f$. The domain appears to be $\{x \mid x \geq 4\}$.

FIGURE **4.7**

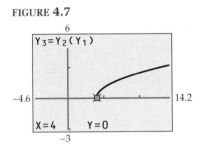

DETERMINING THE DOMAIN

Because only nonnegative values are permitted when we evaluate g, only values of x such that $f(x) \geq 0$ are acceptable.

$$x - 4 \geq 0 \qquad f(x) \geq 0$$
$$x \geq 4 \qquad \textbf{Solve for } x.$$

The domain of $g \circ f$ is $\{x \mid x \geq 4\}$.

Sometimes we want to write a known function as a composition of simpler functions. We call this process *decomposing* a function.

EXAMPLE **6**   **Decomposing a Function**

Write $h(x) = \sqrt{2x + 1}$ as the composition of two functions.

SOLUTION

Let $f(x) = \sqrt{x}$ and $g(x) = 2x + 1$.

Then $h(x) = (f \circ g)(x) = f(g(x)) = f(2x + 1) = \sqrt{2x + 1}$.

In Example 6, the rules for functions f and g are not unique. Other rules can be found for these functions such that $h(x) = (f \circ g)(x)$. For instance, composing $f(x) = \sqrt{x + 1}$ and $g(x) = 2x$ as $(f \circ g)(x)$ also results in $h(x)$.

$$(f \circ g)(x) = f(g(x)) = f(2x) = \sqrt{2x + 1} = h(x)$$

4.1 Quick Reference

Basic Operations ■ Definitions of $f + g, f - g, fg$, and $\dfrac{f}{g}$:

Suppose that f and g are functions whose domains have at least one element in common.

The **sum $f + g$:** $(f + g)(x) = f(x) + g(x)$.
The **difference $f - g$:** $(f - g)(x) = f(x) - g(x)$.
The **product $f \cdot g$:** $(f \cdot g)(x) = f(x) \cdot g(x)$.
The **quotient $\dfrac{f}{g}$:** $\left(\dfrac{f}{g}\right)(x) = \dfrac{f(x)}{g(x)}$, where $g(x) \neq 0$.

■ The domains of these functions are the sets of numbers that are common to the domains of f and g, except that $g(x) \neq 0$ in the case of $\dfrac{f}{g}$.

Composition of Functions ■ The **composite function** or **composition** of functions f and g is a function $f \circ g$ defined by $(f \circ g)(x) = f(g(x))$.

■ The domain of $f \circ g$ is the set of numbers x in the domain of g for which $g(x)$ is in the domain of f.

■ *Decomposing* a function is the process of writing a known function as a composition of simpler functions.

4.1 Speaking the Language

1. If $f(x) = \sqrt{x}$ and $g(x) = x$, then the ▨▨▨▨ function is defined by $(f + g)(x) = \sqrt{x} + x$.

2. The domain of the function $(fg)(x)$ is the ▨▨▨▨ of the domains of functions f and g.

3. The function $(f \circ g)(x)$ is called the ▨▨▨▨ of functions f and g.

4. The domain of $f \circ g$ is the set of numbers x in the domain of g for which ▨▨▨▨.

4.1 Exercises

Concepts and Skills

1. Suppose the function $f(x) = x^2 + 2x - 5$ and the function $g(x) = x^3 - 1$. Describe two ways of determining $(fg)(-2)$. Which method is easier?

2. Suppose that $f(x) = x + 1$ and $g(x) = x - 1$. For which of the functions $f + g$, $f - g$, fg, and $\dfrac{f}{g}$ is the domain of the resulting function *not* **R**? Why?

In Exercises 3–6, for the given functions, evaluate each of the following.

(a) $(f + g)(1)$ (b) $(f - g)(2)$

(c) $(fg)(-1)$ (d) $\left(\dfrac{f}{g}\right)(0)$

3. $f(x) = 3x, g(x) = 1 - x^2$

4. $f(x) = \sqrt[3]{x - 1}, g(x) = 2 - |x|$

5. $f(x) = 2^x, g(x) = \dfrac{1}{x - 3}$

6. $f(x) = \dfrac{4}{x^3 + 3}, g(x) = -2$

In Exercises 7 and 8, for the given functions, evaluate the combined function.

7. $f(x) = \sqrt{1 - 2x}, g(x) = |2x| - 1$

   (a) $(f + g)(0)$ (b) $(f - g)(0.5)$

   (c) $\left(\dfrac{f}{g}\right)(-4)$ (d) $(ff)(-2)$

8. $f(x) = \sqrt[3]{x - 2}, g(x) = x^3$

   (a) $(g - f)(-6)$ (b) $(f + g)(3)$

   (c) $(gg)(-2)$ (d) $\left(\dfrac{g}{f}\right)(1)$

In Exercises 9–12, for the given functions, evaluate and simplify $(f + g)(x)$, $(f - g)(x)$, $(fg)(x)$, and $\left(\dfrac{f}{g}\right)(x)$. Give the domain of $\dfrac{f}{g}$.

9. $f(x) = x + 1, g(x) = x^2 + 2x + 1$

10. $f(x) = 3x^2 + 6x, g(x) = x^2 - 2x - 8$

11. $f(x) = \dfrac{x}{x - 1}, g(x) = \dfrac{1 - x}{x + 2}$

12. $f(x) = \dfrac{2}{x}, g(x) = \dfrac{1}{x + 4}$

13. Suppose that for the functions f and g, $(fg)(-1)$ is not defined. Two of the following pairs are possible definitions of f and g. Identify them and explain why the third pair is not possible.

   (i) $f(x) = \sqrt{1 - x}$

      $g(x) = \dfrac{1}{x - 1}$

   (ii) $f(x) = x + 1$

      $g(x) = \sqrt{x - 1}$

   (iii) $f(x) = \dfrac{1}{x^2 - 1}$

      $g(x) = x^2 + 1$

14. Suppose that $f(x) = \sqrt{x}$ and $g(x) = \sqrt{-x - 1}$. Is $(f + g)(x) = \sqrt{x} + \sqrt{-x - 1}$? Why?

In Exercises 15–20, determine the domain of the combined functions.

   (a) $(f + g)(x)$ (b) $(f - g)(x)$

   (c) $(fg)(x)$ (d) $\left(\dfrac{f}{g}\right)(x)$

15. $f(x) = x^2 - 4, g(x) = x^2 + 4$

16. $f(x) = 2x + 5, g(x) = 3x - 2$

17. $f(x) = \dfrac{1}{x - 3}, g(x) = \dfrac{x + 2}{x}$

18. $f(x) = \sqrt{x}, g(x) = \dfrac{1}{x - 4}$

19. $f(x) = \sqrt{x}, g(x) = \sqrt{x - 3}$

20. $f(x) = \sqrt{5 - x}, g(x) = \sqrt{x + 2}$

21. If $f(x) = \sqrt{x}$ and $g(x) = x - 1$, then the composite function $(f \circ g)(x) = \sqrt{x - 1}$. According to the definition of the composition of functions, why is 0 not in the domain of $f \circ g$ even though 0 is in the domain of g?

22. Suppose that $h(x) = (f \circ g)(x) = 2x + 6$. Which of the following are valid definitions of f and g?

   (i) $f(x) = 2x$ (ii) $f(x) = x - 1$

      $g(x) = x + 3$ $g(x) = 2x + 7$

In Exercises 23–26, evaluate the composite function.

23. $f(x) = 2x - 3, g(x) = x^2 - 1$

   (a) $(f \circ g)(1)$ (b) $(g \circ f)(2)$

   (c) $(f \circ g)(-3)$ (d) $(g \circ f)(0)$

24. $f(x) = x^2 - x, g(x) = \sqrt{x + 1}$

   (a) $(f \circ g)(3)$ (b) $(g \circ f)(1)$

   (c) $(f \circ g)(6)$ (d) $(g \circ f)(2)$

25. $f(x) = \sqrt[3]{3 - x}, g(x) = x + 2$

   (a) $(f \circ g)(0)$ (b) $(g \circ f)(-5)$

   (c) $(f \circ f)(4)$ (d) $(g \circ g)(-7)$

26. $f(x) = x - 3|x|, g(x) = \dfrac{1}{x}$

   (a) $(f \circ g)(-\frac{1}{4})$ (b) $(g \circ f)(-2)$

   (c) $(g \circ g)(\frac{2}{3})$ (d) $(f \circ f)(-1)$

In Exercises 27–36, determine $(f \circ g)(x)$, $(g \circ f)(x)$, and $(f \circ f)(x)$.

27. $f(x) = x^2, g(x) = 3 - x$

28. $f(x) = \sqrt{x + 2}, g(x) = x^2 - 2$

29. $f(x) = x^3 - 2, g(x) = \sqrt[3]{x + 1}$

30. $f(x) = 4 - 3x, g(x) = 2x + 1$

31. $f(x) = \dfrac{1}{x + 1}, g(x) = 3 - 2x$

32. $f(x) = |x|, g(x) = x + 5$

33. $f(x) = x^4, g(x) = x^{3/2}$

34. $f(x) = \dfrac{1}{2}x + 2, g(x) = 2x - 1$

35. $f(x) = x + 3, g(x) = x^3 + 2x^2 - x - 1$

36. $f(x) = 2x - 1, g(x) = x^3 - x^2 - x + 4$

In Exercises 37–42, show that $(f \circ g)(x) = x$ and that $(g \circ f)(x) = x$.

37. $f(x) = 2x - 6, g(x) = \dfrac{1}{2}x + 3$

38. $f(x) = \dfrac{x + 7}{2}, g(x) = 2(x - 3.5)$

39. $f(x) = \dfrac{x}{1 - x}, g(x) = \dfrac{x}{x + 1}$

40. $f(x) = \dfrac{2x + 3}{x - 1}, g(x) = \dfrac{x + 3}{x - 2}$

41. $f(x) = \sqrt[5]{x} + 1, g(x) = (x - 1)^5$

42. $f(x) = x^3 + 5, g(x) = \sqrt[3]{x - 5}$

In Exercises 43–46, what is the domain of (a) $(f \circ g)(x)$ and (b) $(g \circ f)(x)$?

43. $f(x) = 3x + 1, g(x) = \dfrac{x}{3}$

44. $f(x) = \dfrac{1}{x}, g(x) = x^2 - 1$

45. $f(x) = 3 - x, g(x) = \sqrt{x}$

46. $f(x) = \dfrac{2}{x^2 - 4}, g(x) = |x|$

In Exercises 47–56, find f and g such that $h = f \circ g$. (There are several possibilities. In the *Answers to Exercises*, one possibility is given.)

47. $h(x) = \sqrt{2x + 1}$

48. $h(x) = (x - 2)^2 - 4(x - 2)$

49. $h(x) = \dfrac{2}{x^2 - x}$ **50.** $h(x) = 3|x^2 - x|$

51. $h(x) = \dfrac{x^2 - 4}{x^2 + 4}$ **52.** $h(x) = \sqrt{\sqrt{x + 3} + 3}$

53. $h(x) = (x - 1)^3 + 3(x - 1)^2 + 3(x - 1) + 5$

54. $h(x) = (3 - x)^{2/3} - (3 - x)^{4/3}$

55. $h(x) = |2 - |2 - x||$ **56.** $h(x) = \sqrt[3]{\sqrt{x}}$

Concept Extension

In Exercises 57 and 58, for the given functions, evaluate and simplify each combined function.

57. $f(x) = x^2 - 2x, g(x) = x - 4$

(a) $(f + g)(2t)$ (b) $(f - g)(t + 1)$

(c) $\left(\dfrac{g}{f}\right)(a^2)$ (d) $(gf)(1 - b)$

58. $f(x) = \dfrac{1}{x}, g(x) = \dfrac{1}{x + 1}$

(a) $(g + f)(w - 1)$ (b) $(f - g)\left(\dfrac{1}{t}\right)$

(c) $\left(\dfrac{f}{g}\right)\left(\dfrac{t}{2}\right)$ (d) $(fg)(2s - 1)$

In Exercises 59–64, the figure shows the graphs of functions f and g. Use the figure to answer each part.

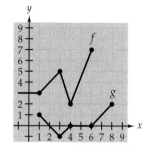

59. (a) Draw the graph of $f + g$.
   (b) What is the domain of $f + g$?

60. (a) Draw the graph of $f - g$.
   (b) What is the domain of $f - g$?

61. (a) Evaluate $\left(\dfrac{f}{g}\right)(1)$.

   (b) What is the domain of $\dfrac{f}{g}$?

62. (a) Evaluate $(fg)(5)$.   (b) Evaluate $(fg)(3)$.

63. (a) Evaluate $(f \circ g)(4)$. (b) Evaluate $(g \circ f)(3)$.

64. (a) Evaluate $(f \circ g)(5)$. (b) Evaluate $(g \circ f)(0)$.

65. If f is an even function and g is an odd function, what kind of function is fg?

66. Produce the graph of $g(x) = x^2 - 9$. Compare this graph to the graph of $(f \circ g)(x)$, where $f(x) = |x|$. What is your conjecture about the graph of any function of the form $y = |h(x)|$?

Applications

67. Revenue and Costs   A small company with x sales representatives finds that its revenue can be modeled by $R(x) = 100{,}000\sqrt{x}$, and its costs can be modeled by $C(x) = 30{,}000x$.

(a) Suppose that profit P is defined as the difference of revenue and cost. Write function P as a combination of functions R and C.

(b) Determine $R(4) - C(4)$.

(c) Compare the value of $P(4)$ to your result in part (b).

68. Distance and Speed Suppose that the speed limit on a long stretch of interstate is 70 mph.

(a) Write a function f to describe the distance traveled in t hours if the driver maintains a constant speed that exceeds the speed limit by 10 mph.

(b) Write a function s to describe the distance traveled in t hours if the driver maintains a constant speed at the speed limit.

(c) Evaluate and interpret $(f - s)(2)$.

69. Fuel Consumption Suppose that a driver starts on a trip with a full 18-gallon gas tank. The driver averages 50 mph for the duration of the trip, and the gas mileage is 20 miles per gallon.

(a) Write a function h to describe the number of hours needed for a trip of d miles.

(b) Write a function f to describe the number of gallons of gasoline remaining after h hours of driving.

(c) Write and interpret the function $(f \circ h)(d)$.

(d) Evaluate and interpret $(f \circ h)(200)$.

70. Distance and Time Two police officers started walking from an intersection. One headed north at a rate of 6 mph, and the other headed east. Both officers walked for exactly 2 minutes.

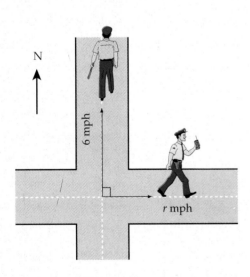

(a) If the eastbound officer walked at r mph, write a function $d(r)$ that describes the distance walked.

(b) Write a function $f(d)$ that describes the distance between the officers after 2 minutes. (*Hint*: Use the Pythagorean Theorem.)

(c) Write and interpret the function $(f \circ d)(r)$.

(d) Evaluate and interpret $(f \circ d)(5)$. (Round your result to the nearest hundredth.)

71. Pond Ripple When a stone is dropped into a pond, the radius of the circular ripple increases at a rate of 1.5 feet per second.

(a) Write a function r to describe the length of the radius at time t (in seconds).

(b) Write a function A that describes the area enclosed by the ripple in terms of the radius r.

(c) Write a composite function f that gives the area of the ripple as a function of time t.

(d) To the nearest tenth, what is the area of the ripple after 2.5 seconds?

72. Water Pressure The pressure (in pounds per square foot) at the base of a water tank is found by multiplying the depth (in feet) of the water by 62.4. Suppose an intake valve is opened to fill the tank, and the water level increases at a rate of 3 inches per minute.

(a) Write a function that describes the depth d (in feet) of the water at time t (in minutes).

(b) Write a function that describes the water pressure at the base of the tank when the depth of the water is *d*.

(c) Write a composite function that gives the water pressure at time *t*.

(d) What is the water pressure after filling the tank for one hour?

73. Home Health-Care Agencies The table shows that the number of home health-care agencies that were certified and not certified for Medicare increased dramatically from 1989 to 1995. (*Source*: National Association of Home Care.)

Year	Certified	Not certified
1989	5676	4824
1995	9120	7897

With *x* representing the number of years since 1989, the number *C* of certified agencies can be modeled by $C(x) = 574x + 5676$, and the number *N* of noncertified agencies can be modeled by $N(x) = 512x + 4824$.

(a) Write and interpret the function $(C + N)(x)$.

(b) Produce the graphs of *C*, *N*, and $C + N$. Trace the graphs to determine the relationship among $C(4)$, $N(4)$, and $(C + N)(4)$.

74. Long Distance Calls In the section on linear regression models we were given data on the number of long distance telephone calls made on Mother's Day and Father's Day during the years 1993 to 1996. With *n* representing the number of years since 1993, the number of calls (in millions) can be modeled by the following functions:

Mother's Day:
$$M(n) = -1.175n^2 + 11.075n + 107.425$$

Father's Day:
$$F(n) = 7.46n + 90.36.$$

(a) Write and interpret the function $(M - F)(n)$.

(b) Produce the graphs of *M*, *F*, and $M - F$. Trace the graphs to determine the relationship among $M(3)$, $F(3)$, and $(M - F)(3)$.

4.2 *Inverse Functions*

Inverse Relations ■ *One-to-One Functions* ■
Composition and Restricted Domains

Inverse Relations

■ *NOTE In the symbol R^{-1}, the -1 is not an exponent. The symbol means "the inverse of relation R" or, more simply, "R inverse."*

Recall that a relation is any set of ordered pairs. By reversing the coordinates of the ordered pairs of a relation *R*, we obtain another relation called the **inverse relation,** denoted by R^{-1}.

EXPLORATION

The Inverse of a Relation

The relation $R = \{(-2, -4), (3, 1), (4, -3)\}$ contains three ordered pairs.

(a) Plot the points that represent the ordered pairs in *R*.

(b) By reversing the coordinates of the ordered pairs in *R*, write the inverse relation R^{-1}.

(c) Compare the domains and ranges of *R* and R^{-1}.

(d) In the same coordinate system that you used for part (a), plot the points that represent the ordered pairs in R^{-1}.

(e) Draw a line about which the points of R and R^{-1} appear to be symmetric. What is your conjecture about the equation of that line?

DEVELOPING THE CONCEPT

The Inverse of a Relation

Consider the following relation R and its inverse R^{-1}:

$$R = \{(-4, -7)(-1, 3), (2, 4), (5, -1)\}$$
$$R^{-1} = \{(-7, -4)(3, -1), (4, 2), (-1, 5)\}$$

Because we reverse the coordinates of the ordered pairs in R to obtain the ordered pairs in R^{-1}, the domains and ranges of the two relations are also reversed.

Relation	Domain	Range
R	$\{-4, -1, 2, 5\}$	$\{-7, 3, 4, -1\}$
R^{-1}	$\{-7, 3, 4, -1\}$	$\{-4, -1, 2, 5\}$

Figure 4.8 shows the graphs of R and R^{-1}. Observe that the corresponding points are symmetric with respect to the line $y = x$.

FIGURE **4.8**

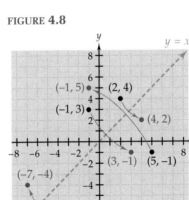

The following is a formal definition of an inverse relation.

Definition of an Inverse Relation

If R is a relation containing ordered pairs of the form (a, b), then the **inverse relation** R^{-1} is a relation containing ordered pairs of the form (b, a).

EXAMPLE **1**   **Writing an Inverse Relation**

Suppose that R is a relation in which the first coordinate x is an integer and $-1 \leq x \leq 2$. The second coordinate y of each ordered pair is 3 more than the first coordinate.

(a) Write relations R and R^{-1} as sets of ordered pairs.
(b) Write the domains and ranges of R and R^{-1}.
(c) Describe R^{-1} in words.

SOLUTION

(a) $R = \{(-1, 2), (0, 3), (1, 4), (2, 5)\}$
$R^{-1} = \{(2, -1), (3, 0), (4, 1), (5, 2)\}$
(b) Domain of R = range of $R^{-1} = \{-1, 0, 1, 2\}$
Domain of R^{-1} = range of $R = \{2, 3, 4, 5\}$
(c) R^{-1} is a relation in which the first coordinate x is an integer and $2 \leq x \leq 5$. The second coordinate y of each ordered pair is 3 less than the first coordinate.

We can extend the concept of an inverse relation to a relation R that is defined by an equation.

DEVELOPING THE CONCEPT

Relations Defined by Equations

When we say that the relation R is defined by the equation $2x + y = 4$, we mean that $R = \{(x, y) \mid 2x + y = 4\}$. In other words, R contains the solutions of the equation.

By the definition of an inverse relation, R^{-1} contains ordered pairs of the form (y, x); that is, the coordinates of the ordered pairs of R are reversed. This implies that the equation that defines R^{-1} is determined by reversing the variables x and y in the equation that defines R. Thus, the equation that defines R^{-1} is $2y + x = 4$.

Consider the following tables of values for R and R^{-1}.

Relation: $2x + y = 4$

x	1	-3	5	-1	0
y	2	10	-6	6	4

Inverse Relation: $2y + x = 4$

x	2	10	-6	6	4
y	1	-3	5	-1	0

We can use the pairs in the table to graph the relation R and its inverse R^{-1}. (See Figure 4.9.) Note again that the two graphs are symmetric with respect to the line $y = x$.

FIGURE 4.9

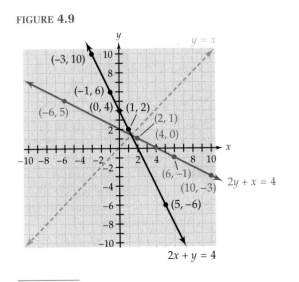

$2x + y = 4$

▪ **Draw Inverse**   You can use a calculator to draw the graph of an inverse relation. Enter the equation that defines the relation on the **Y** screen. Then the calculator reverses the coordinates of the solutions of the equation to draw the inverse relation.

Figure 4.10 shows the entry of $y = -2x + 4$ (the equation that defines relation R) and the graphs of R and R^{-1}.

FIGURE 4.10

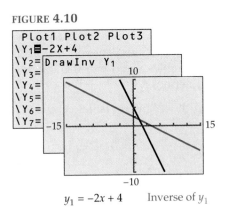

$y_1 = -2x + 4$ Inverse of y_1

One-to-One Functions

Recall that a relation is a function if its graph passes the Vertical Line Test. The same test is applied to determine whether the graph of an inverse relation represents a function.

DEVELOPING THE CONCEPT

Inverse Functions

Figure 4.11 shows the graphs of four functions and their inverse relations.

FIGURE **4.11**

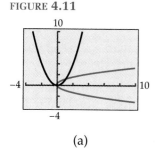 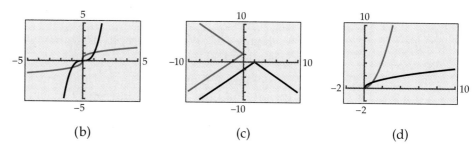

(a)	(b)	(c)	(d)

We see that the graphs of the inverse relations in (a) and (c) do not pass the Vertical Line Test, and so those inverse relations are not functions. However, the graphs of the inverse relations in (b) and (d) do represent functions because each one passes the Vertical Line Test. Thus we call each of those inverse relations an **inverse function.**

As shown in parts (b) and (d) of Figure 4.11, the graph of an inverse relation passes the Vertical Line Test only if no two points of the graph have the same *first* coordinate. This means that the graph of the function itself cannot have two points with the same *second* coordinate. When this is the case, we call the function a **one-to-one function.**

■ *NOTE   A function f is a one-to-one function if and only if f(a) = f(b) implies that a = b. This also implies that if a ≠ b, then f(a) ≠ f(b).*

Definition of a One-to-One Function

A **one-to-one function** is a function for which no two different ordered pairs have the same second coordinate.

We can use a test that is similar to the Vertical Line Test to judge whether the graph of a function represents a one-to-one function.

Horizontal Line Test

A function is a one-to-one function if and only if no horizontal line intersects the graph of the function at more than one point.

EXAMPLE **2**   **Using the Horizontal Line Test**

Use the Horizontal Line Test to determine whether the given functions are one-to-one.

(a) $f(x) = \sqrt{x}$ (b) $g(x) = |x|$ (c) $h(x) = \dfrac{1}{x}$

SOLUTION

Figure 4.12 shows the graphs of the three functions with a representative horizontal line.

Because the graph in Figure 4.12(b) does not pass the Horizontal Line Test, g is not a one-to-one function. Figure 4.12(a) and (c) shows that functions f and h do pass the Horizontal Line Test, and so each is one-to-one.

FIGURE 4.12

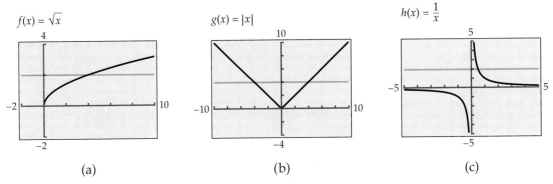

(a)   (b)   (c)

If a function $f(x)$ is one-to-one, then the inverse relation is a function that we name $f^{-1}(x)$ (read "f inverse of x"). Note again that the domain of f is the same as the range of f^{-1}, and the range of f is the same as the domain of f^{-1}.

If a one-to-one function is described by an equation in x and y, we can write an equation for the inverse function by interchanging the variables. Sometimes we can solve the equation that defines the inverse for y and write a rule for $f^{-1}(x)$.

Writing the Inverse of a Function

The inverse of a one-to-one function f can often be determined as follows:

1. Replace $f(x)$ with y.
2. Interchange the variables x and y.
3. Solve the resulting equation for y.
4. Replace y with $f^{-1}(x)$.

EXAMPLE **3**   **Determining an Inverse Function**

Let $f(x) = 3 - 5x$. Determine a rule for f^{-1}.

SOLUTION

The graph of the linear function f indicates that f is one-to-one.

$$f(x) = 3 - 5x \qquad \text{The given function}$$
$$y = 3 - 5x \qquad \text{Replace } f(x) \text{ with } y.$$
$$x = 3 - 5y \qquad \text{Interchange } x \text{ and } y.$$
$$x - 3 = -5y \qquad \text{Solve for } y.$$
$$y = \frac{x - 3}{-5}$$
$$f^{-1}(x) = \frac{x - 3}{-5} \qquad \text{Replace } y \text{ with } f^{-1}(x).$$

FIGURE **4.13**

$f(x) = 3 - 5x$

$f^{-1}(x) = \dfrac{x - 3}{-5}$

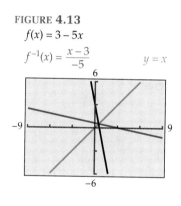

Figure 4.13 shows the graphs of f and f^{-1} in Example 3. The graph of f^{-1} is the reflection of the graph of f across the line $y = x$.

In general, for each point (a, b) of the graph of f, the point (b, a) is a point of the graph of f^{-1}. Thus the graph of f^{-1} is the reflection of the graph of f across the line $y = x$.

> **Graphs of f and f^{-1}**
>
> For a one-to-one function f, the graph of f^{-1} is the reflection of the graph of f across the line $y = x$.

Composition and Restricted Domains

An inverse function f^{-1} reverses (or "undoes") the operations indicated by f. For instance, if $f(x) = 4x$, then $f^{-1}(x) = \dfrac{x}{4}$; if $g(x) = x^3$, then $g^{-1}(x) = \sqrt[3]{x}$.

Observe the results of composing these functions and their inverses.

$$f(f^{-1}(x)) = f\left(\frac{x}{4}\right) = 4\left(\frac{x}{4}\right) = x$$

$$f^{-1}(f(x)) = f^{-1}(4x) = \frac{4x}{4} = x$$

$$g(g^{-1}(x)) = g\left(\sqrt[3]{x}\right) = \left(\sqrt[3]{x}\right)^3 = x$$
$$g^{-1}(g(x)) = g^{-1}(x^3) = \sqrt[3]{x^3} = x$$

VISUALIZING THE INVERSE

Figure 4.15 shows the graphs of f and f^{-1}.

FIGURE 4.15

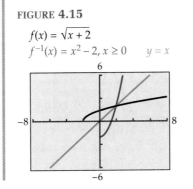

$f(x) = \sqrt{x+2}$
$f^{-1}(x) = x^2 - 2, x \geq 0 \qquad y = x$

Sometimes a function is not one-to-one unless we restrict the domain in some way. For instance, the function $g(x) = (x-4)^2$ is not one-to-one. However, if we restrict the domain to $\{x \mid x \geq 4\}$, as in the following example, then the function is one-to-one and has an inverse function.

EXAMPLE 6   Restricting the Domain

Let $g(x) = (x-4)^2$, $x \geq 4$. Write a rule for g^{-1}.

SOLUTION

DETERMINING THE INVERSE

$y = (x-4)^2, x \geq 4$ Replace $g(x)$ with y. Note the restricted domain.

$x = (y-4)^2, y \geq 4$ The range of g^{-1} is the same as the domain of g.

$\sqrt{x} = y - 4$ Solve for y.

$y = \sqrt{x} + 4$

$g^{-1}(x) = \sqrt{x} + 4$ Replace y with $g^{-1}(x)$.

VISUALIZING THE INVERSE

Figure 4.16 shows the graph of g (with the restricted domain) and the graph of g^{-1}.

FIGURE 4.16

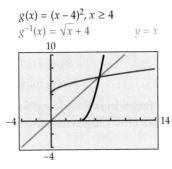

$g(x) = (x-4)^2, x \geq 4$
$g^{-1}(x) = \sqrt{x} + 4 \qquad y = x$

4.2 Quick Reference

Inverse Relations ■ If R is a relation containing ordered pairs of the form (a, b), then the **inverse relation** R^{-1} is a relation containing ordered pairs of the form (b, a).

■ If a relation R is given as a set of ordered pairs, then the inverse relation R^{-1} can be written by interchanging the x- and y-coordinates of the ordered pairs.

■ If a relation R is given as an equation in x and y, then we can obtain an equation for the inverse relation R^{-1} by interchanging the x- and y-variables.

1. Replace $f(x)$ with y.
2. Interchange the variables x and y.
3. Solve the resulting equation for y.
4. Replace y with $f^{-1}(x)$.

■ The graphs of a relation R and the inverse relation R^{-1} are symmetric with respect to the line $y = x$.

One-to-One Functions ■ If the inverse of a function f is a function, then we call the inverse relation the **inverse function,** denoted by f^{-1}.

■ A **one-to-one function** is a function for which no two different ordered pairs have the same second coordinate.

■ A function is a one-to-one function if and only if no horizontal line intersects the graph of the function at more than one point. This test is called the *Horizontal Line Test.*

■ If a function f is one-to-one, then the inverse of f is an inverse function.

■ The graph of f^{-1} is the reflection of the graph of f across the line $y = x$.

Composition and Restricted Domains ■ If f is a one-to-one function, then f^{-1} exists.

1. $(f \circ f^{-1})(x) = x$ for each x in the domain of f^{-1}, and
2. $(f^{-1} \circ f)(x) = x$ for each x in the domain of f.

■ For one-to-one functions f and g, if $(f \circ g)(x) = x$ for each x in the domain of g and $(g \circ f)(x) = x$ for each x in the domain of f, then we can conclude that f and g are inverse functions.

■ Sometimes we must restrict the domain of a function in order for the function to be one-to-one.

4.2 Speaking the Language

1. If a relation contains ordered pairs of the form (x, y), then the ▨▨▨▨ relation contains ordered pairs of the form (y, x).

2. If f is a function such that no two different ordered pairs have the same second coordinate, then f is a(n) ▨▨▨▨ function.

3. The ▨▨▨▨ Test can be used on the graph of a function to determine whether the function is a one-to-one function.

4. The graph of f^{-1} is the ▨▨▨▨ of the graph of f across the line $y = x$.

4.2 Exercises

Concepts and Skills

1. Why must a one-to-one function pass both the Vertical Line Test and the Horizontal Line Test?

2. Suppose that no two different ordered pairs of a function f have the same second coordinate. What does this imply about the inverse of f?

In Exercises 3 and 4, write the inverse relation.

3. $\{(-3, 4), (0, -1), (5, 9)\}$
4. $\{(5, -2), (3, 4), (-6, 2), (1, 0)\}$

In Exercises 5–8, write an equation for the inverse relation.

5. $x = \sqrt{9 - y^2}$ 6. $x^2 + 2y = 4x$
7. $xy + y = 3$ 8. $y + 2x^2 = 3 + x$

9. Suppose that the graph of function f is in Quadrant II. Where is the graph of its inverse? Why?

10. If you know the y-intercept of the graph of function f, what point of the graph of f^{-1} do you know?

In Exercises 11–16, for each function, use the *Draw Inverse* feature of your calculator to produce the graph of the inverse relation. Based on the graph, determine whether the inverse relation is a function.

11. $y = |x + 3|$ 12. $y = 4 - x$
13. $y = 2^x$ 14. $y = 3^{1-x}$
15. $y = \dfrac{-5}{x}$ 16. $y = x^3 - 4x$

In Exercises 17–24, use the *Draw Inverse* feature of your calculator to produce the graphs of the function and its inverse. Use the graphs to estimate the domain and range of (a) the function and (b) its inverse.

17. $f(x) = \dfrac{2x}{x - 4}$ 18. $f(x) = \dfrac{1}{2 - x}$
19. $f(x) = \sqrt{3 - x}$ 20. $f(x) = 5 - \sqrt{x}$
21. $f(x) = |x - 2|, x \geq 2$ 22. $f(x) = 1 - x^2, x \leq 0$
23. $f(x) = \sqrt[3]{x - 5}$ 24. $f(x) = (1 - 2x)^3$

In Exercises 25–28, determine whether the function whose graph is given is a one-to-one function.

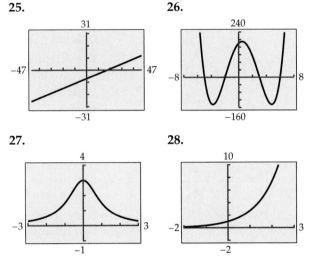

25.

26.

27.

28.

In Exercises 29–44, produce a graph to decide whether the given function is one-to-one. If the function is one-to-one, determine the inverse function.

29. $f(x) = 5 - 2x$ 30. $f(x) = \dfrac{1}{2}x + 3$
31. $f(x) = 3 - x^2$ 32. $g(x) = \left|\dfrac{1}{2}x - 1\right|$
33. $g(x) = 8x^3$ 34. $h(x) = x^5 + 2$
35. $f(x) = \dfrac{1}{x^2 + 1}$ 36. $h(x) = x^4 - x^3$
37. $p(x) = \sqrt{x + 3}$ 38. $r(x) = 2\sqrt{x} - 5$
39. $f(x) = \dfrac{x + 3}{x - 1}$ 40. $f(x) = \dfrac{2x + 1}{x}$
41. $h(x) = x\sqrt{9 - x^2}$ 42. $g(x) = x^2 - 3\sqrt{x}$
43. $f(x) = x - |x|$
44. $g(x) = |x - 1| + |x - 3|$

In Exercises 45–48, use composition to verify that the given functions are inverse functions.

45. $f(x) = \dfrac{2x - 1}{3}, f^{-1}(x) = \dfrac{3x + 1}{2}$
46. $f(x) = -2 + \sqrt[3]{x}, f^{-1}(x) = (x + 2)^3$
47. $f(x) = \dfrac{x + 1}{x}, f^{-1}(x) = \dfrac{1}{x - 1}$

48. $f(x) = \dfrac{-4x}{3}, f^{-1}(x) = -\dfrac{3}{4}x$

In Exercises 49–56, determine $f^{-1}(x)$.

49. $f(x) = \dfrac{x+2}{5}$

50. $f(x) = 4 - \dfrac{2}{3}x$

51. $f(x) = \sqrt[3]{x} - 7$

52. $f(x) = x^3 + 4$

53. $f(x) = -\dfrac{3}{x+5}$

54. $f(x) = \dfrac{x}{x-1}$

55. $f(x) = 2 - \sqrt{x}$

56. $f(x) = \sqrt{2x+5}$

In Exercises 57–62, determine the inverse function and state its domain.

57. $f(x) = (x-1)^2, x \geq 1$

58. $g(x) = \sqrt{25 - x^2}, 0 \leq x \leq 5$

59. $g(x) = 4 - x^2, x \geq 0$

60. $f(x) = |x + 5|, x \geq -5$

61. $h(x) = (x+2)^2, x \leq -2$

62. $h(x) = \sqrt{x^2 - 9}, x \leq -3$

63. Explain why the inverse of a constant function cannot be a function.

64. Suppose that $f(x_1) = f^{-1}(x_1)$. What do you know about the point $P(x_1, y_1)$?

Concept Extension

65. Show that $f(x) = \dfrac{x}{x-1}$ is its own inverse.

66. Show that two of the following basic functions are their own inverse.

(i) $f(x) = |x|$

(ii) $f(x) = x$

(iii) $f(x) = \dfrac{1}{x}$

(iv) $f(x) = x^2$

In Exercises 67–70, the graph of f is given. Sketch the graph of the inverse relation. Is the inverse relation a function?

67.

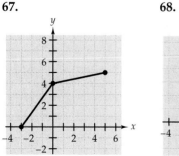

68.

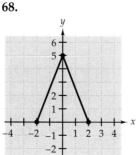

69.

70.

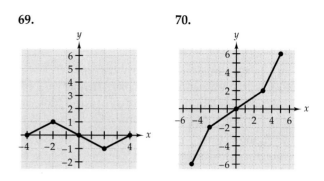

In Exercises 71–74, trace the graphs of $f \circ g$ and $g \circ f$ to decide whether g is the inverse of f.

71. $f(x) = x^5, g(x) = x^{1/5}$

72. $f(x) = x + 3, g(x) = \dfrac{1}{x+3}$

73. $f(x) = x^2 - 1, g(x) = x^2 + 1$

74. $f(x) = -2(3 - x), g(x) = 3 + \dfrac{x}{2}$

75. Is the function $f(x) = \begin{cases} -x^2, x \leq 0 \\ x^2, x > 0 \end{cases}$ a one-to-one function? If so, write $f^{-1}(x)$.

76. For functions of the form $f(x) = x^n$, for what values of n does f^{-1} exist? What is $f^{-1}(x)$ for those values of n?

77. Show that the inverse relation of an even function cannot be a function.

78. Show whether f^{-1} always exists when f is an odd function.

Applications

79. Telephone Keypad Look at the numbers 2 through 9 on a telephone keypad and observe the association of numbers and letters. (See the figure.)

Referring to the figure on the previous page, let R be a relation whose ordered pairs are of the form (number, letter).

(a) Is R a function? Why or why not?
(b) Is R^{-1} a function? Why or why not?
(c) Can a relation that is *not* a function have an inverse that *is* a function?

80. Exchange Rate Suppose that the exchange rate for Mexican pesos is 8.38 pesos per U.S. dollar.

(a) Write a function p that describes the value of d dollars in pesos.
(b) Write p^{-1} and describe what it does.

81. Professional Seminar A professional association conducts a seminar for which tickets cost $5.00 for members and $8.00 for nonmembers. The seminar is attended by 100 people.

(a) Write a function r to describe the total revenue if m of the attenders were members.
(b) For this application, what is the domain of function r?
(c) Write r^{-1} and describe what it does.

82. Temperature Scales The function $F(t) = \frac{9}{5}t + 32$ converts Celsius temperatures t to Fahrenheit temperatures. Determine the inverse function and describe what it does.

83. House Painters A painter requires 3 days to paint a house. A second painter, working alone, would require x days to paint the house, where $x \geq 1$.

(a) Write a function f that describes the time needed to paint the house if both painters work together.
(b) Evaluate and interpret $f(4.5)$.
(c) Produce the graph of f to help you decide whether f is a one-to-one function. If it is, write f^{-1}.
(d) Evaluate and interpret $f^{-1}(2)$.

84. Dual Investment An employee invests her year-end bonus of $1000 in two funds. She places x dollars in a money market that pays 3% interest and the remainder in a bond fund that pays 7%.

(a) Write a function I to describe the total annual interest from the two investments. (Assume simple interest.)
(b) Write I^{-1} and describe what it does.
(c) Evaluate and interpret $I^{-1}(58)$.

85. Smart Cards "Smart cards" are predicted to replace credit cards in the near future. The graph shows the number of smart cards used worldwide in the 1990s. (*Source*: Meridien Research.)

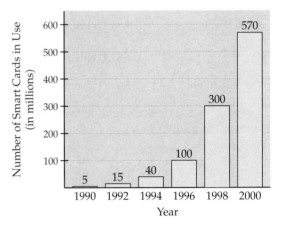

(a) Letting x represent the number of years since 1988, produce a cubic regression function f to model the data. Round all coefficients to the nearest hundredth.
(b) Evaluate and interpret $f(7)$.
(c) Let c represent your answer in part (b). Evaluate $f^{-1}(c)$.
(d) Using only the time period given in the bar graph, what is the domain of $f^{-1}(x)$?

86. Alcohol-Related Traffic Fatalities The number of traffic deaths in which alcohol was a factor decreased from 22,716 in 1985 to 17,274 in 1995. (*Source*: National Highway Traffic Safety Administration.)

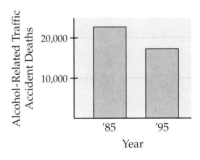

(a) Letting x represent the number of years since 1980, write a linear function f to model the data.
(b) Write and interpret f^{-1}.
(c) Evaluate and interpret $f^{-1}(0)$.

4.3 *Exponential Functions*

Definition and Graphs   ■   *The Natural Exponential Function*   ■
Applications and Modeling

The sum total of all human knowledge is said to be growing "exponentially." This means that not only is knowledge increasing, but in fact the rate of increase is increasing. This is particularly true, for example, in the area of technology.

In this section, we consider a new function called an *exponential function.* This function is often an appropriate model for data that are rapidly increasing or decreasing over time. Understanding the properties and uses of exponential functions will greatly enhance your ability to solve application problems in business and science.

Definition and Graphs

Consider the function $f(x) = 5^x$, in which the exponent is a variable. Having defined rational exponents, we know that the function is defined for all rational numbers. The expression 5^x can also be defined for irrational number exponents. (The details are beyond the scope of this text.) For example, your calculator should give you the following approximation: $5^{\sqrt{2}} \approx 9.738517742$.

In short, the expression 5^x is defined for all real numbers. All properties pertaining to rational exponents are also valid for real number exponents.

> **Definition of an Exponential Function**
>
> An **exponential function** is a function that can be written as $f(x) = b^x$, where $b > 0$, $b \neq 1$, and x is any real number.

Note that the definition places certain limitations on the base b.

1. The definition excludes a base of 1 because $f(x) = 1^x$ would simply be the constant function $f(x) = 1$.
2. The base b must be positive to avoid complex numbers in the range of the function. For example, if $b = -9$ and $x = \frac{1}{2}$, then evaluating the function would lead to $(-9)^{1/2} = \sqrt{-9}$, which is not a real number.

─────── **EXAMPLE 1**   **Evaluating an Exponential Function**

(a) If $f(x) = 5^x$, then $f(-2) = 5^{-2} = \dfrac{1}{5^2} = \dfrac{1}{25}$.

(b) If $g(x) = 9^x$, then $g\left(\dfrac{1}{2}\right) = 9^{1/2} = \sqrt{9} = 3$.

FIGURE **4.17**

```
(0.7)^(-1.3)
                1.59
2^π
                8.82
```

(c) If $r(x) = \left(\dfrac{1}{8}\right)^x$, then $r\left(-\dfrac{2}{3}\right) = \left(\dfrac{1}{8}\right)^{-2/3} = 8^{2/3} = \left(\sqrt[3]{8}\right)^2 = 4$.

(d) If $h(x) = (0.7)^x$, then $h(-1.3) = (0.7)^{-1.3} \approx 1.59$.   See Figure 4.17.

(e) If $s(x) = 2^x$, then $s(\pi) = 2^\pi \approx 8.82$.   See Figure 4.17.

Figure 4.17 shows the results for parts (d) and (e) rounded to two decimal places.

We can learn much about the properties and behavior of exponential functions by observing the characteristics of their graphs.

As always, we can sketch the graph of $f(x) = b^x$ by selecting values of x, evaluating $f(x)$, and plotting the points $(x, f(x))$. This process is easier if we know the basic shape of the graph in advance.

EXPLORATION

Graphs of Exponential Functions

Consider functions of the form $y = b^x$.

(a) Select several values of b where $b > 1$ and produce the graphs of the functions.

(b) What characteristics of the graphs do you observe?

(c) Select several values of b where $0 < b < 1$ and repeat parts (a) and (b).

DEVELOPING THE CONCEPT

Graphs of Exponential Functions

Figure 4.18 shows the graphs of the functions in each of the following groups:

(a) $f(x) = b^x$, for $b = 2, 4,$ and 7

(b) $f(x) = b^x$, for $b = \dfrac{1}{2}, \dfrac{1}{3},$ and $\dfrac{1}{5}$

FIGURE **4.18**

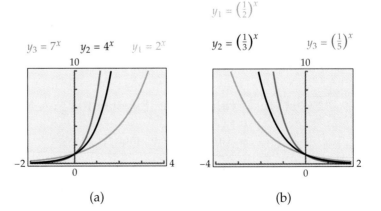

(a)   (b)

The primary difference in the two groups is the value of the base b. If $b > 1$, as in Figure 4.18(a), then f is an *increasing* function; that is, the graph rises from left to right. However, if $0 < b < 1$, as in Figure 4.18(b), then f is a *decreasing* function; that is, the graph falls from left to right.

The value of b^x can never be 0, and so the graph of $f(x) = b^x$ has no x-intercept. Because the graph approaches the x-axis but never intersects it, the x-axis is a horizontal asymptote. Specifically, if $b > 1$, as in Figure 4.18(a), then as $x \to -\infty, f(x) \to 0$. If $0 < b < 1$, as in Figure 4.18(b), then as $x \to \infty, f(x) \to 0$.

From both sets of graphs, we can conclude that, for all permissible values of b, the y-intercept is (0, 1). The domain of each function is **R**, and the range is $\{y \mid y > 0\}$. Finally, applying the Horizontal Line Test, we conclude that each function is a one-to-one function.

The following is a summary of these observations.

Properties of Exponential Functions

Let $f(x) = b^x, b > 0, b \neq 1$.

1. The domain is **R**, and the range is $\{y \mid y > 0\}$.
2. The y-intercept is (0, 1); there is no x-intercept.
3. The function is a one-to-one function.
4. For $b > 1$, the function is increasing, and the x-axis is a horizontal asymptote as $x \to -\infty$.

   For $0 < b < 1$, the function is decreasing, and the x-axis is a horizontal asymptote as $x \to \infty$.

EXAMPLE 2 **Properties of Exponential Functions**

Produce the graph of the given function and use it to determine the domain, range, y-intercept, and horizontal asymptote. Describe the relationship between the graph of the given function and the graph of $y = 2^x$.

(a) $g(x) = 2^{x+1}$
(b) $r(x) = 3 + 2^x$
(c) $h(x) = 8 - 2^x$

FIGURE 4.19

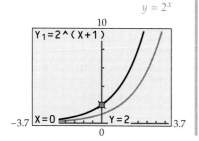

SOLUTION

The domain of each function is **R**.

(a) Figure 4.19 indicates that the x-axis is the horizontal asymptote as $x \to -\infty$, and so the range is $\{y \mid y > 0\}$. Tracing the graph suggests that the y-intercept is (0, 2). We can verify this algebraically:

$$g(0) = 2^{0+1} = 2^1 = 2$$

From our previous experience with transformations, we know that adding a constant to the variable has the effect of shifting the graph horizontally. In this case, the graph of $g(x) = 2^{x+1}$ is the graph of $y = 2^x$ shifted 1 unit to the left.

(b) The graphs of $r(x) = 3 + 2^x$ and $y = 2^x$ are shown in Figure 4.20. We know that adding a constant to a function has the effect of shifting the graph of the function vertically. In particular, the graph of $r(x) = 3 + 2^x$ is simply the graph of $y = 2^x$ shifted upward 3 units. Thus we can analyze function r by simply "translating" what we know about the graph of $y = 2^x$. The horizontal asymptote as $x \to -\infty$ is $y = 3$, and so the range is $\{y \mid y > 3\}$. The y-intercept is $(0, 4)$.

FIGURE 4.20

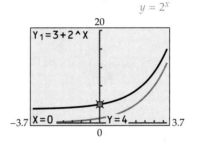

(c) Figure 4.21 shows that the graph of function h is the graph of $y = 2^x$ reflected across the x-axis and shifted up 8 units. Thus, the horizontal asymptote is the line $y = 8$, and the range is $\{y \mid y < 8\}$. The tracing cursor is on the apparent y-intercept $(0, 7)$, which can easily be verified algebraically.

FIGURE 4.21

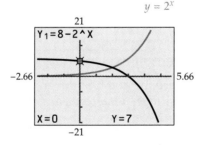

The Natural Exponential Function

Applications in science, engineering, and business frequently involve an exponential function whose base is an irrational number approximately equal to 2.7182818284. Because this number occurs so frequently in mathematics, the letter e is used to represent it, just as the Greek letter π is used to represent the irrational number approximately equal to 3.14159.

The function $f(x) = e^x$ is called the **natural exponential function.** Its graph is shown in Figure 4.22.

FIGURE **4.22**

$y = e^x$

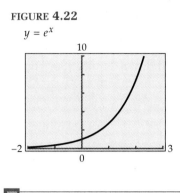

> **Base *e***   Most calculators have special keys for the number *e* and the expression e^x.

EXAMPLE **3**   **Evaluating the Natural Exponential Function**

Evaluate the function $f(x) = e^x$ as indicated.

(a) $f(1)$ (b) $f(-2)$ (c) $f(2.5)$

SOLUTION

Figure 4.23 shows the required values rounded to three decimal places.

FIGURE **4.23**

```
e^1
                    2.718
e^(-2)
                     .135
e^2.5
                  12.182
```

The graph of the natural exponential function can be shifted and/or reflected according to the rules discussed in the section on transformations.

EXAMPLE **4**   **Properties of Natural Exponential Functions**

Describe how the graph of the given function is a reflection or translation of the graph of $y = e^x$.

(a) $f(x) = -e^x$ (b) $g(x) = e^{-x}$ (c) $h(x) = -2 + e^{x+3}$

SOLUTION

(a) The graph of $f(x) = -e^x$ is the reflection of the graph of $y = e^x$ across the x-axis. [See Figure 4.24(a).]
(b) The graph of $g(x) = e^{-x}$ is the reflection of the graph of $y = e^x$ across the y-axis. [See Figure 4.24(b).]
(c) The graph of $h(x) = -2 + e^{x+3}$ is the graph of $y = e^x$ shifted left 3 units and down 2 units. [See Figure 4.24(c).]

FIGURE **4.24**

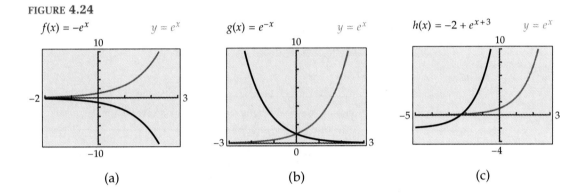

(a)　　　　　　　　　　(b)　　　　　　　　　　(c)

_____

EXAMPLE **5**　**Zeros of Exponential Functions**

Use the graphing method to estimate the zero of $f(x) = 2e^x - 5$.

SOLUTION

Figure 4.25 shows the graph of $f(x) = 2e^x - 5$.

FIGURE **4.25**

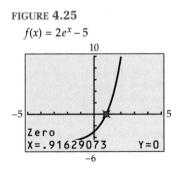

Using the *Zero* feature of the calculator, we estimate that the x-intercept is $(0.92, 0)$, and so the approximate zero of the function is 0.92. (Later in this chapter, we will learn how to determine the exact zero of such exponential functions.)

Applications and Modeling

Financial advisers always recommend that people begin saving and investing at a young age for such eventualities as college expenses and retirement. The amount by which such funds will grow depends on the number of times per year that interest is added to the amount in the fund. *Compound interest* is interest paid on the principal and all interest accumulated to date.

For interest that is compounded just once per year, the value A of the fund after t years is given by the formula $A = P(1 + r)^t$, where P is the original amount of the investment and r is the interest rate. However, most institutions compound more frequently. For n compoundings of interest per year, the formula is adjusted as follows:

$$A = P\left(1 + \frac{r}{n}\right)^{nt}$$

The best plan is *continuous compounding*, which can be regarded as having infinitely many compoundings.

DEVELOPING THE CONCEPT

Continuous Compounding

Suppose that 1 dollar is invested at an interest rate of 6% for one year. For n compoundings, the value A of the investment is

$$A = P\left(1 + \frac{r}{n}\right)^{nt} = 1\left(1 + \frac{0.06}{n}\right)^{n \cdot 1} = \left(1 + \frac{0.06}{n}\right)^{n}$$

The following table shows the value of the investment for certain numbers of compoundings.

Compounding	n	A
Yearly	1	$1.060000
Quarterly	4	1.061364
Monthly	12	1.061678
Daily	365	1.061831
Hourly	8760	1.061836

We see that as the number of compoundings increases, the value of A appears to be approaching a limiting value. To determine what that limiting value is, we will continue to increase n without bound.

We begin with the formula

$$A = \left(1 + \frac{0.06}{n}\right)^{n}$$

and we replace $\frac{0.06}{n}$ with $\frac{1}{x}$. This means that $x = \frac{n}{0.06}$ or $n = 0.06x$.

$$A = \left(1 + \frac{0.06}{n}\right)^{n} = \left(1 + \frac{1}{x}\right)^{0.06x} = \left[\left(1 + \frac{1}{x}\right)^{x}\right]^{0.06}$$

As n becomes larger and larger, x also becomes larger and larger. The following table shows the values of the expression $\left(1 + \frac{1}{x}\right)^{x}$ for increasingly larger values of x.

x	$\left(1 + \dfrac{1}{x}\right)^x$
10	2.59374246
100	2.704813829
1,000	2.716923932
10,000	2.718145927
100,000	2.718268237
1,000,000	2.718280469

As $x \to \infty$, the value of $\left(1 + \dfrac{1}{x}\right)^x$ approaches a number that should look familiar to you. The number is e.

FIGURE **4.26**

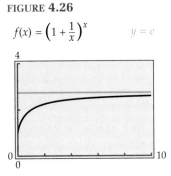

$$f(x) = \left(1 + \frac{1}{x}\right)^x \qquad y = e$$

Figure 4.26 shows the graph of $f(x) = \left(1 + \dfrac{1}{x}\right)^x$ and the line $y = e$, which appears to be a horizontal asymptote for the graph of f.

As suggested by the table and the graph, the value of $\left(1 + \dfrac{1}{x}\right)^x$ approaches e as $x \to \infty$. Thus the value of A approaches $e^{0.06}$. In general, if $A = P\left(1 + \dfrac{r}{n}\right)^{nt}$, then $A \to Pe^{rt}$ as $n \to \infty$.

These observations lead to the following formula for continuously compounded interest.

Continuously Compounded Interest

If interest is compounded continuously, then the value A of an investment of P dollars after t years is

$$A = Pe^{rt}$$

where r is the interest rate.

EXAMPLE **6** **Compound Interest**

When a couple marry at age 25, their family and friends provide them with a starting "nest egg" of $10,000. Rather than spending the money, the couple create an untouchable fund for their retirement at age 65. Assuming a constant interest rate of 8%, what will the value of the fund be with

(a) quarterly compounding?   (b) continuous compounding?

SOLUTION

(a) For quarterly compounding, we use the formula $A = P\left(1 + \dfrac{r}{n}\right)^{nt}$.

$$P = 10{,}000 \qquad \text{The original investment is \$10,000.}$$
$$r = 0.08 \qquad \text{The interest rate is 8\%.}$$
$$n = 4 \qquad \text{Interest is compounded 4 times per year.}$$
$$t = 40 \qquad \text{The life of the investment is 40 years.}$$

$$A = 10{,}000\left(1 + \frac{0.08}{4}\right)^{4 \cdot 40}$$
$$= 10{,}000(1.02)^{160}$$
$$\approx \$237{,}699.07$$

(b) For continuous compounding, we use the formula $A = Pe^{rt}$, with the same values of P, r, and t.

$$A = 10{,}000e^{0.08 \cdot 40}$$
$$\approx \$245{,}325.30$$

EXAMPLE 7   **The Populations of China and India**

From 1997 to 2050, the *increase* in the population of China is projected to be equal to the entire population of the United States in 2050. The population of India is projected to increase 57% in that same period. (*Source:* United Nations.)

With x representing the number of years since 1990, the following functions can be used to model the populations (in billions of people) of China and India:

$$\text{China:} \quad C(x) = 1.20(1.004)^{x}$$
$$\text{India:} \quad I(x) = 0.90(1.009)^{x}$$

(a) What is the projected population of China in 2048?
(b) What is the projected population of India in 2048?
(c) What conclusion can you draw from the results in parts (a) and (b)?
(d) If you were to graph the two model functions, what would be the approximate point of intersection?

SOLUTION

(a) For the year 2048, $x = 58$.

$$C(58) = 1.20(1.004)^{58} \approx 1.51$$

The projected population for China in 2048 is approximately 1.51 billion.
(b) $I(58) = 0.90(1.009)^{58} \approx 1.51$
The projected population for India in 2048 is approximately 1.51 billion.
(c) The model functions indicate that the populations of China and India will be approximately equal in the year 2048.
(d) Because $C(58) \approx I(58)$, the approximate point of intersection of the graphs would be $(58, 1.51)$.

4.3 Quick Reference

Definition and Graphs
■ An **exponential function** is a function that can be written as $f(x) = b^x$, where $b > 0$, $b \neq 1$, and x is any real number.

■ Properties of Exponential Functions
Let $f(x) = b^x$, $b > 0$, $b \neq 1$.
1. The domain is **R,** and the range is $\{y \mid y > 0\}$.
2. The y-intercept is $(0, 1)$; there is no x-intercept.
3. The function is a one-to-one function.
4. For $b > 1$, the function is increasing, and the x-axis is a horizontal asymptote as $x \to -\infty$.
   For $0 < b < 1$, the function is decreasing, and the x-axis is a horizontal asymptote as $x \to \infty$.

The Natural Exponential Function
■ The number e is an irrational number that is approximately equal to 2.7182818284.

■ The function $f(x) = e^x$ is called the **natural exponential function.**

Applications and Modeling
■ If interest is compounded n times per year, then the value A of an initial investment P is given by $A = P\left(1 + \dfrac{r}{n}\right)^{nt}$, where r is the interest rate and t is the number of years.

■ The formula $A = Pe^{rt}$ is used for *continuous compound interest.*

4.3 Speaking the Language

1. A function of the form $f(x) = b^x$ is called a(n) ▩▩▩▩ function.
2. For the function $f(x) = b^x$, the number b cannot be 0 or 1, nor can it be ▩▩▩▩ .
3. If $b > 1$, then the function $f(x) = b^x$ is ▩▩▩▩ , and the x-axis is a horizontal ▩▩▩▩ as $x \to -\infty$.
4. The function $g(x) = e^x$ is called the ▩▩▩▩ function.

4.3 Exercises

Concepts and Skills

1. Explain why $f(x) = 2^x$ and $g(x) = \left(\frac{1}{2}\right)^{-x}$ have the same graph.

2. In Example 1(e), we evaluated 2^π. How could we have known in advance that the value would be between 8 and 16?

In Exercises 3–8, let $f(x) = 5^x$, $g(x) = \pi^x$, and $h(x) = 6.8^x$. Evaluate the function as indicated. (Round answers to the nearest hundredth.)

3. $f(3.2)$
4. $h(4.1)$
5. $g(-\pi)$
6. $f\left(\sqrt{3}\right)$
7. $h\left(-\sqrt{2}\right)$
8. $g(-1.7)$

In Exercises 9–14, evaluate $f(x) = e^x$ as indicated.

9. $f(3)$
10. $f(-7)$
11. $f\left(-\dfrac{3}{4}\right)$
12. $f(2.71)$
13. $f(-0.3)$
14. $f(e)$

15. What names are given to the functions $f(x) = 2x$, $g(x) = 2^x$, and $h(x) = x^2$?

16. Order the expressions 2^x, 3^x, and e^x from smallest to largest on the interval $(-\infty, 0)$. Do the same on the interval $(0, \infty)$.

In Exercises 17–20, determine the missing coordinate so that the ordered pair belongs to the given function. (*Hint:* Trace the graph of the function.)

17. $y = 4^x$; $(a, 1), (b, 16), (3, c), (-1, d)$

18. $y = 2^{-x}$; $(-3, a), (2, b), (c, 2), \left(d, \dfrac{1}{8}\right)$

19. $y = e^x$; $(a, e), (b, 1), (c, e^3), \left(d, \dfrac{1}{e}\right)$

20. $y = \left(\dfrac{1}{3}\right)^{x+1}$; $(-3, a), (1, b), (c, 1), \left(d, \dfrac{1}{3}\right)$

21. Which of the following is a variable or variable expression? Why?

(i) e (ii) e^3 (iii) e^x

22. If the graph of $f(x) = b^x$ has a horizontal asymptote as $x \to \infty$, what do we know about the base b?

In Exercises 23–26, produce the graphs of the given function, function f, and function g. Describe the graphs of f and g relative to the graph of the given function.

23. $y = 5^x$ (a) $f(x) = 5^{x+4}$ (b) $g(x) = 4 + 5^x$
24. $y = 2^x$ (a) $f(x) = 2^{x-3}$ (b) $g(x) = 2^x - 3$
25. $y = 3^x$ (a) $f(x) = 5 + 3^x$ (b) $g(x) = -1 + 3^x$
26. $y = e^x$ (a) $f(x) = e^{x-2}$ (b) $g(x) = -2 + e^x$

27. Suppose that the graph of $f(x) = b^x$ contains the point $\left(-2, \dfrac{1}{9}\right)$. What is b?

28. Which is larger, π^e or e^π?

In Exercises 29–44, produce the graph of the given function. Use the graph to estimate (a) the intercept(s), (b) the asymptote, and (c) the range of the function.

29. $f(x) = 2 + 5^x$ **30.** $T(x) = \left(\dfrac{2}{3}\right)^x - 1$

31. $w(x) = \left(\dfrac{1}{2}\right)^{x-1}$ **32.** $y(x) = 3^{x+4}$

33. $p(x) = 6^{1-x}$ **34.** $h(x) = -3^x$
35. $f(x) = 3^{-x} - 7$ **36.** $h(x) = 5 + 2^{-x}$
37. $r(x) = 1 + e^x$ **38.** $f(x) = e^{x-2}$
39. $p(x) = e^{x+3}$ **40.** $f(x) = 4 - e^x$
41. $f(x) = 3 - e^{x-2}$ **42.** $g(x) = e^{x+3} - 1$
43. $f(x) = 4^{x+1} - 5$ **44.** $g(x) = 2 + 4^{x-4}$

In Exercises 45–50, match the graph to the function.

(A) $y = \left(\dfrac{2}{3}\right)^{|x|}$ (B) $y = |2 - 0.5^x|$

(C) $y = |3^{x^2} - 4|$ (D) $y = 5^x - 5^{-x}$

(E) $y = \dfrac{5}{1 + 2^x}$ (F) $y = \dfrac{2^x}{x}$

45.   **46.**

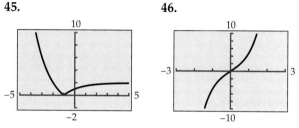

47.   **48.**

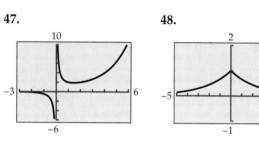

49.   **50.**

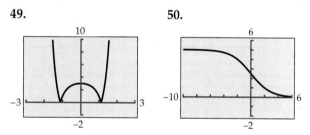

In Exercises 51–56, use the graphing method to estimate the zeros of the function. (Round answers to the nearest hundredth.)

51. $f(x) = 3^x - 5$ **52.** $g(x) = 1 - 2e^x$
53. $h(x) = -2e^{-x} + 7$ **54.** $t(x) = -8 + 3^{x/2}$
55. $y(x) = x - 2^x + 9$ **56.** $f(x) = e^x - x - 8$

57. If $k > 0$, explain how the graphs of the following functions compare to the graph of $f(x) = b^x$.

(i) $g(x) = b^x - k$ (ii) $h(x) = b^{x-k}$

58. Explain why the function $f(x) = e^x$ has no zeros.

Concept Extension

In Exercises 59–62, use a graph of the given function to estimate (a) the relative extrema, (b) the horizontal asymptote, and (c) the range.

59. $f(t) = 5te^{-t}$

60. $g(t) = \dfrac{e^t + e^{-t}}{2}$

61. $h(t) = \dfrac{3}{1 + e^t}$

62. $r(t) = e^{-t^2}$

63. If $f(x) = b^x$, show that the following are true.

(i) $f(m + n) = f(m) \cdot f(n)$

(ii) $f(kx) = [f(x)]^k$

64. If $f(x) = b^x$, show that $\dfrac{f(x + h) - f(x)}{h}$ can be written in the form $k \cdot f(x)$, where k is a constant.

Applications

65. College Trust Fund On the day a child was born, an uncle set up a trust fund worth $10,000 to be used for college expenses. If the constant interest rate was 7%, with interest compounded monthly, what was the value of the fund when the child turned 18 years old?

66. In Exercise 65, how much more would the fund have been worth if the interest had been compounded continuously?

67. Automobile Depreciation Suppose that the initial value P_0 of an automobile was $22,000 and that the depreciated value D after x years is given by $D = P_0(0.8)^x$.

(a) What is the value of the automobile after 3 years?

(b) The formula for the depreciated value has the form $D = P_0 a^x$. If the value of a is decreased, does the rate of depreciation increase or decrease?

68. Bacteria Count A stomach bacterium known as *H. pylori* is sometimes increased in number in patients who undergo chemotherapy. Suppose that, under controlled conditions, the number n of *H. pylori* bacteria found at time t (in minutes) is given by $n(t) = \dfrac{1000}{2^{-t/10}}$.

(a) How many bacteria were present at $t = 0$?

(b) What is the approximate percent increase in bacteria after 15 minutes?

69. Cycling and Jogging Paths Cycling and jogging enthusiasts have organized to raise funds for the construction of paths and walkways. In 1996 alone, $325 million was raised for such projects, compared to a total of only $41 million for the entire period 1973–1991. (*Source*: Bicycle Federation of America.)

With x representing the number of years since 1990, the annual funding F can be modeled by $F(x) = 54(1.36)^x$.

(a) According to the model function, what is the estimated funding for 1998?

(b) What feature of the graph of F makes the model function unreliable for future years?

70. Intercollegiate Sports Over 70% of all boys and girls who participate in organized sports in the eighth grade expect to play intercollegiate sports. The percentages decrease as students move through high school. (*Source*: National Athletic Testing Program.)

With x representing the grade level, the percentage P of male athletes who expect to play intercollegiate sports can be modeled by the function $P(x) = 127(0.94)^x$.

(a) If 80 high school seniors are male athletes, according to the model function, how many of them expect to go on to college athletics?

(b) What value in the model function indicates that the percentages are decreasing?

71. Stock Option Profits An employee took advantage of her company's stock option plan and invested her profits in a mutual fund that paid 8%, compounded continuously. After 5 years, the fund was worth $10,517.36. How much profit did she earn from the stock option plan?

72. Lump Sum Benefit Upon retirement, a person rolled his lump sum benefit into a conservative money market fund that paid 3.5% interest, compounded monthly. After 4 years, the value of the fund was $69,002.37. What was the amount of the lump sum benefit?

73. Property Sale A property owner sold two parcels of land. He sold one parcel for $120,000, which he invested in a fund that paid 6.5%, and he sold the other parcel for $80,000, which he invested in another fund that paid 6.75%. If the interest from both funds was compounded monthly, what was the total combined worth of the two funds after 20 years?

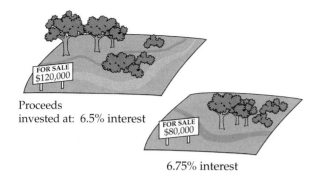

Proceeds
invested at: 6.5% interest

6.75% interest

74. Inheritance Investments One brother invested his inheritance of $20,000 in an income fund that paid 7%, compounded continuously. His younger brother, who inherited the same amount, invested in a growth fund that paid 7.25%, compounded monthly. Compare the values of the two funds after 15 years.

In Exercises 75 and 76, use the graphing method to estimate the answers to the questions.

75. Property Purchase A person has found a 2-acre plot of property on which he wants to build a home. However, the property costs $30,900, and he has only $18,000. If he invests his money in a fund that pays 6%, compounded continuously, in approximately how many years will he be able to afford the property?

76. In Exercise 75, what interest rate must the buyer be able to obtain in order to be able to buy the property in approximately 7 years?

77. Labor Strikes During the period 1970–1996, the number of labor strikes involving 1000 or more workers decreased rapidly. In 1970 there were 381 such strikes, whereas there were only 37 strikes in 1996. (*Source*: National Labor Relations Board.)

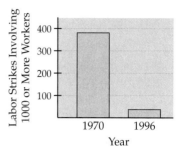

With t representing the number of years since 1970, the number of strikes N can be modeled by the exponential function $N_1(t) = 374(0.91)^t$. Another model is the quadratic function $N_2(t) = 0.60t^2 - 29t + 380$.

Produce the graphs of both model functions. For the years following 1996, what assumptions would you be making about the trend in the number of strikes if you select the exponential model? The quadratic model?

4.4 *Logarithmic Functions*

Definitions and Properties ■ *Evaluating Logarithms* ■
Graphs of Logarithmic Functions ■ *Applications and Modeling*

Definitions and Properties

Because the exponential function $f(x) = b^x$ is a one-to-one function, the function has an inverse f^{-1}. In the section on inverse functions, we wrote inverse functions by interchanging the variables x and y in the equation that defined the function. We then solved the resulting equation for y to obtain the inverse function.

For an exponential function such as $3^x = y$, interchanging x and y leads to the equation $3^y = x$. Note that we cannot solve this equation for y using our current algebraic methods. However, we can at least describe the role of y: *y is the power to which 3 must be raised to obtain x.*

To write this long description more compactly, we use the following notation:

$$y = \log_3 x$$

Note that the symbol $\log_3 x$ (read "logarithm base 3 of x") replaces the phrase "the power to which 3 must be raised to obtain x."

Because the value of y depends on the value of x, $y = \log_3 x$ is a function, which we call a *logarithmic* function. Furthermore, we derived the logarithmic function $y = \log_3 x$ by interchanging the variables in the exponential function $y = 3^x$. Thus the logarithmic function $y = \log_3 x$ is the inverse of the exponential function $y = 3^x$.

This procedure can be extended to any permissible base b. We define the inverse of the exponential function $y = b^x$ to be a logarithmic function $f(x) = \log_b x$.

Definition of a Logarithmic Function

For $b > 0$ and $b \neq 1$, the function $f(x) = \log_b x$ is called the **logarithmic function** with base b. The logarithmic function f is the inverse of the exponential function $y = b^x$.

The following are some details regarding logarithmic functions.

1. In the logarithmic function $f(x) = \log_b x$, the independent variable x is called the *argument*.
2. For the logarithmic function, the base b is limited in the same way as it is for the exponential function; that is, $b > 0$ and $b \neq 1$.
3. Because the range of the exponential function is $\{y \mid y > 0\}$, the domain of the logarithmic function is $\{x \mid x > 0\}$. A logarithmic function is defined only if its argument is positive.
4. Because the domain of the exponential function is **R**, the range of the logarithmic function is **R**.

Although the base b of a logarithmic function can be any positive number other than 1, the most frequently used bases are 10 and e. Such logarithmic functions are given special names.

Common and Natural Logarithmic Functions

The logarithmic function with base 10 is called the **common logarithmic function** and is written $y = \log x$ rather than $y = \log_{10} x$.

The logarithmic function with base e is called the **natural logarithmic function** and is written $y = \ln x$ rather than $y = \log_e x$.

Because exponential functions and logarithmic functions are inverses, the functions are related in the following way:

$$\log_b x = y \text{ if and only if } b^y = x$$

This relationship allows us to convert a logarithmic form to an exponential form or vice versa.

EXAMPLE 1   Converting from Logarithmic Form to Exponential Form

	Logarithmic form	*Exponential form*
(a)	$\log_2 32 = 5$	$2^5 = 32$
(b)	$\log_3 \dfrac{1}{9} = -2$	$3^{-2} = \dfrac{1}{9}$
(c)	$\log 100 = 2$	$10^2 = 100$

EXAMPLE 2   Converting from Exponential Form to Logarithmic Form

	Exponential form	*Logarithmic form*
(a)	$2^{-1} = \dfrac{1}{2}$	$\log_2 \dfrac{1}{2} = -1$
(b)	$16^{1/2} = 4$	$\log_{16} 4 = \dfrac{1}{2}$
(c)	$\left(\dfrac{2}{3}\right)^{-3} = \dfrac{27}{8}$	$\log_{2/3} \dfrac{27}{8} = -3$

Evaluating Logarithms

The expression $\log_2 x$ is a variable expression, whereas $\log_2 8$ is a constant. We refer to both of these expressions simply as "logarithms." For example, in the logarithmic *function* $y = \log_3 x$, the *logarithm* is $\log_3 x$.

To evaluate logarithms, we sometimes begin by writing the expression in the more familiar exponential form. Then we can use the fact that if $b^m = b^n$, then $m = n$.

EXAMPLE 3   Evaluating Logarithms

Evaluate the given logarithms.

(a) $\log_7 49$ (b) $\log_9 27$ (c) $\log_{1/2} 16$

SOLUTION

In each part, we begin by writing the logarithmic form in the equivalent exponential form.

(a) $\log_7 49 = y$

$\qquad 7^y = 49$

$\qquad 7^y = 7^2$

$\qquad\quad y = 2$ If $b^m = b^n$, then $m = n$.

Therefore, $\log_7 49 = 2$.

(b) $\log_9 27 = y$

$\qquad 9^y = 27$

$\qquad (3^2)^y = 3^3 \qquad$ Write both sides with the same base.

$\qquad 3^{2y} = 3^3$

$\qquad 2y = 3 \qquad$ If $b^m = b^n$, then $m = n$.

$\qquad y = \dfrac{3}{2} \qquad$ Solve for y.

Therefore, $\log_9 27 = \dfrac{3}{2}$.

(c) $\log_{1/2} 16 = y$

$\qquad \left(\dfrac{1}{2}\right)^y = 16$

$\qquad 2^{-y} = 2^4 \qquad$ Write both sides with the same base.

$\qquad -y = 4 \qquad$ If $b^m = b^n$, then $m = n$.

$\qquad y = -4$

Therefore, $\log_{1/2} 16 = -4$.

■ *NOTE The method illustrated in Example 3 works because we can easily write both sides of the equation with the same base. When this is not possible, other methods for evaluating a logarithm are needed.*

 LOG, LN Calculators have special keys for common and natural logarithms.

EXAMPLE **4** **Evaluating Logarithms with a Calculator**

Use a calculator to evaluate each logarithm.

(a) (i) $\log 5$ (ii) $\log 1000$ (iii) $\log 0.001$

(b) (i) $\ln 8$ (ii) $\ln e$ (iii) $\ln 0.2$

SOLUTION

Figure 4.27 shows the reported values for parts (a) and (b), respectively, rounded to the nearest thousandth.

FIGURE **4.27**

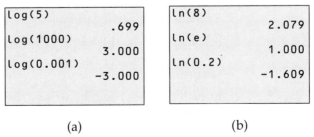

      (a)   (b)

To use a calculator to evaluate logarithms with bases other than 10 or e, we express the logarithms in terms of common or natural logarithms.

> **Change of Base Formula**
>
> For positive real numbers x, a, and b, where $a \neq 1$ and $b \neq 1$,
>
> $$log_b x = \frac{\log_a x}{\log_a b}$$

With the Change of Base Formula, we can change the base of a logarithm to any permissible base a. However, our choice for the base a is usually 10 or e to enable us to use a calculator.

EXAMPLE 5 **Evaluating Logarithms with the Change of Base Formula**

Figure 4.28 shows the following calculations.

(a) $\log_4 7 = \dfrac{\log 7}{\log 4} \approx 1.40$

(b) $\log_{3.5} 20 = \dfrac{\ln 20}{\ln 3.5} \approx 2.39$

FIGURE **4.28**

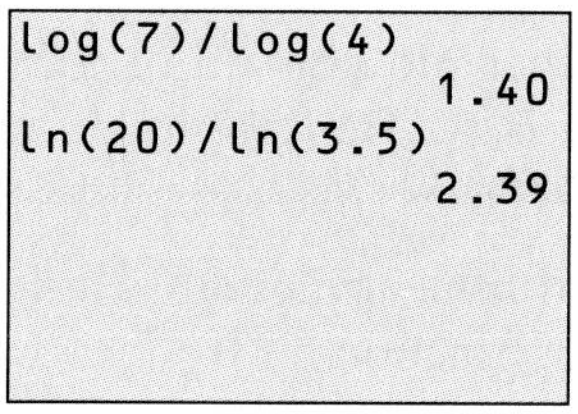

```
Log(7)/Log(4)
                1.40
Ln(20)/Ln(3.5)
                2.39
```

Graphs of Logarithmic Functions

By using our knowledge of exponential functions, we can determine properties of logarithmic functions and their graphs.

EXPLORATION

Graphs of Logarithmic Functions

Consider functions of the form $y = \log_b x$.

(a) Select several values of b where $b > 1$ and produce the graphs of the functions. One way to produce the graph of $y = \log_b x$ is to use the Change of Base Formula to write the function in terms of common logarithms: $y = \log_b x = \dfrac{\log x}{\log b}$.

(b) What characteristics of the graphs do you observe?

(c) Select several values of b where $0 < b < 1$ and repeat parts (a) and (b).

DEVELOPING THE CONCEPT

Graphs of Logarithmic Functions

Consider the exponential functions $f(x) = 3^x$ and $g(x) = \left(\frac{1}{3}\right)^x$ and their inverses $f^{-1}(x) = \log_3 x$ and $g^{-1}(x) = \log_{1/3} x$. Figure 4.29 shows the graphs of f and g along with their inverses.

FIGURE **4.29**

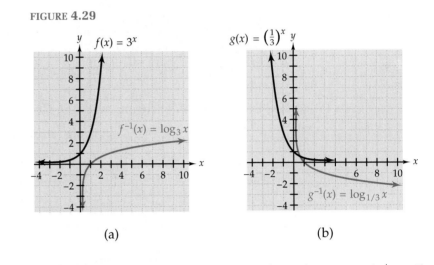

(a)   (b)

Observe that the domain of each logarithmic function is $\{x \mid x > 0\}$ and the range is **R**. The graphs do not have a y-intercept, but each has an x-intercept $(1, 0)$.

The logarithmic function is an increasing function for $b > 1$ and a decreasing function for $0 < b < 1$.

As x approaches 0 from the right, $\log_b x \to -\infty$ when $b > 1$, and $\log_b x \to +\infty$ when $0 < b < 1$. Thus the y-axis is a vertical asymptote.

The following is a summary of our observations.

Properties of the Logarithmic Function and Its Graph

Let $f(x) = \log_b x$, $b > 0$, $b \neq 1$.

1. The domain of f is $\{x \mid x > 0\}$, and the range is **R**.
2. The x-intercept is $(1, 0)$, and there is no y-intercept.
3. For $b > 1$, the function is increasing and the y-axis is a vertical asymptote as $x \to 0$ from the right.

   For $0 < b < 1$, the function is decreasing and the y-axis is a vertical asymptote as $x \to 0$ from the right.

EXAMPLE **6**   **Domains of Logarithmic Functions**

What is the domain of each function?

(a) $f(x) = \ln (x + 3)$
(b) $g(x) = 2 - \log_2 (1 - x)$
(c) $h(x) = \log x^2$

SOLUTION

In each part, the argument must be positive.

(a) $x + 3 > 0$
$$x > -3$$

The domain is $(-3, \infty)$. [See Figure 4.30(a).]

(b) $1 - x > 0$
$$x < 1$$

The domain is $(-\infty, 1)$. [See Figure 4.30(b).]

(c) Because $x^2 > 0$ for all x except 0, the domain is $(-\infty, 0) \cup (0, \infty)$. [See Figure 4.30(c).]

FIGURE **4.30**

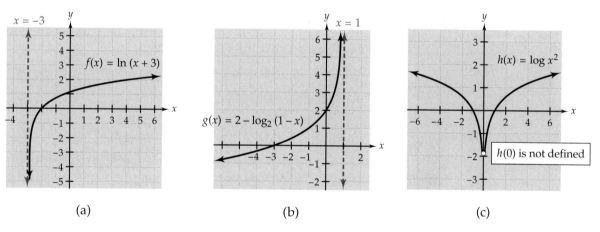

(a)   (b)   (c)

The transformation rules for shifting and/or reflecting graphs apply to the graphs of logarithmic functions.

EXAMPLE **7**   **Graphs of Logarithmic Functions**

Describe the graph of each given function relative to the graph of $y = \log_2 x$.

(a) $f(x) = \log_2 (x - 4)$
(b) $s(x) = \log_2 x - 4$
(c) $h(x) = \log_2 (-x)$

SOLUTION

Note that in part (a) the argument is $x - 4$, whereas in part (b) the argument is x. The Change of Base Formula is used to produce all the graphs.

(a) To produce the graph, we write

$$f(x) = \log_2 (x - 4) = \frac{\log (x - 4)}{\log 2}$$

Subtracting 4 from the variable x has the effect of shifting the graph of $y = \log_2 x$ to the right 4 units. (See Figure 4.31.)

FIGURE **4.31**

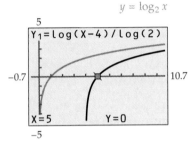

The graph indicates that the line $x = 4$ is the vertical asymptote. The domain is $\{x \mid x > 4\}$. By tracing, we estimate that the x-intercept is $(5, 0)$. We can verify this algebraically.

$$f(5) = \log_2 (5 - 4) = \log_2 1 = 0$$

(b) We use common logarithms to rewrite $s(x)$.

$$s(x) = \log_2 x - 4 = \frac{\log x}{\log 2} - 4$$

This time we are subtracting 4 from the function rather than from the variable. The effect is that the graph of $y = \log_2 x$ is shifted downward 4 units. (See Figure 4.32.)

FIGURE **4.32**

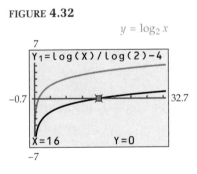

The domain is $\{x \mid x > 0\}$, and the y-axis is the vertical asymptote. Because $s(16) = \log_2 16 - 4 = 4 - 4 = 0$, the x-intercept is $(16, 0)$.

(c) We rewrite $h(x)$ as $h(x) = \log_2 (-x) = \dfrac{\log (-x)}{\log 2}$.

The graph is the reflection of the graph of $y = \log_2 x$ across the y-axis. (See Figure 4.33.)

FIGURE **4.33**

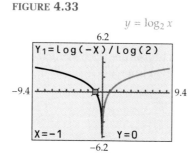

$y = \log_2 x$

Y₁=log(-X)/log(2)

6.2

−9.4   9.4

X=−1   Y=0

−6.2

From the graph, we see that the y-axis is the vertical asymptote, and the domain is $\{x \mid x < 0\}$. The x-intercept is $(-1, 0)$.

EXAMPLE **8** **Determining the Inverse of a Function**

Determine the inverse of the function $f(x) = 5 - e^{x-1}$. Then produce the graphs of f and f^{-1} and observe their symmetry about the line $y = x$.

SOLUTION

DETERMINING THE INVERSE FUNCTION

$$y = 5 - e^{x-1}$$
$$x = 5 - e^{y-1} \qquad \text{Interchange } x \text{ and } y.$$
$$e^{y-1} = 5 - x \qquad \text{Isolate the exponential expression.}$$
$$y - 1 = \ln (5 - x) \qquad \text{Write in natural logarithmic form.}$$
$$y = 1 + \ln (5 - x) \qquad \text{Solve for } y.$$
$$f^{-1}(x) = 1 + \ln (5 - x) \qquad \text{Write the inverse function.}$$

VISUALIZING THE INVERSE FUNCTION

Figure 4.34 shows the symmetry of the graphs of f and f^{-1} about the line $y = x$.

FIGURE **4.34**

$f(x) = 5 - e^{x-1}$
$f^{-1}(x) = 1 + \ln (5 - x)$ $y = x$

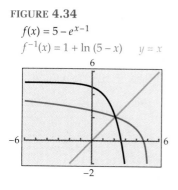

6

−6   6

−2

Applications and Modeling

Two applications of logarithmic models are the Richter scale, which is used to measure the intensity of earthquakes, and the decibel scale, which is used to measure the intensity of sound.

The two models have a similar appearance because both involve the ratio of intensity I to a base-level intensity I_0. In the case of earthquakes, I_0 is the lowest intensity that a seismograph can record. For sound, I_0 is the lowest intensity that a healthy human ear can hear.

EXAMPLE **9**   **The Intensity of an Earthquake**

On the Richter scale, the magnitude R of an earthquake of intensity I is given by $R = \log \dfrac{I}{I_0}$. The Mexico City earthquake in 1985 registered 7.8 on the Richter scale. How much more intense was this earthquake than the base-level intensity?

Solution

$$R = \log \frac{I}{I_0}$$

$$7.8 = \log \frac{I}{I_0} \qquad \text{The Richter scale reading was 7.8.}$$

$$10^{7.8} = \frac{I}{I_0} \qquad \text{Convert to exponential form.}$$

$$I = 10^{7.8}I_0 \qquad \text{Solve for } I.$$

The intensity of the Mexico City earthquake was $10^{7.8}$ times the base-level intensity.

EXAMPLE **10**   **The Intensity of Sound**

On the decibel scale, the loudness D (in decibels) of a sound of intensity I is given by $D = 10 \log \dfrac{I}{I_0}$. If the sound of an airplane taking off is 10^{14} times as intense as the sound of a barely audible whisper, what is the loudness in decibels?

Solution

$$D = 10 \log \frac{I}{I_0}$$

$$D = 10 \log \frac{10^{14}I_0}{I_0} \qquad \text{The intensity } I \text{ is } 10^{14} \text{ times the base-level intensity } I_0.$$

$$D = 10 \log 10^{14} \qquad \text{Divide out the common factor } I_0.$$

$$\frac{D}{10} = \log 10^{14} \qquad \text{Isolate the logarithm.}$$

$$10^{D/10} = 10^{14} \qquad \text{Convert to exponential form.}$$

$$\frac{D}{10} = 14 \qquad \text{If } b^x = b^y, \text{ then } x = y.$$

$$D = 140$$

The loudness of the airplane is 140 decibels.

4.4 Quick Reference

Definitions and Properties

- For $b > 0$ and $b \neq 1$, the function $f(x) = \log_b x$ is called the **logarithmic function** with base b. The logarithmic function f is the inverse of the exponential function $y = b^x$.
- In the logarithmic function $f(x) = \log_b x$, the independent variable x is called the *argument*.
- The domain of the logarithmic function is $\{x \mid x > 0\}$, and the range is **R.**
- The logarithmic function with base 10 is called the **common logarithmic function** and is written $y = \log x$ rather than $y = \log_{10} x$.
- The logarithmic function with base e is called the **natural logarithmic function** and is written $y = \ln x$ rather than $y = \log_e x$.
- Logarithmic forms can be converted to exponential forms (and vice versa) with the following relationship:

$$\log_b x = y \text{ if and only if } b^y = x$$

Evaluating Logarithms

- In the logarithmic function $y = \log_b x$, the expression $\log_b x$ is called a *logarithm*.
- To evaluate logarithms, we sometimes begin by writing the expression in the more familiar exponential form. Then we can use the fact that if $b^m = b^n$, then $m = n$.
- The Change of Base Formula allows us to convert a logarithm with base b to a logarithm with base a. If x, a, and b are positive and $a \neq 1$ and $b \neq 1$, then

$$\log_b x = \frac{\log_a x}{\log_a b}$$

Graphs of Logarithmic Functions

- The graph of $f(x) = \log_b x$ has the following properties:
  1. The x-intercept is $(1, 0)$, and there is no y-intercept.
  2. For $b > 1$, the function is increasing and the y-axis is a vertical asymptote as $x \to 0$ from the right.

     For $0 < b < 1$, the function is decreasing and the y-axis is a vertical asymptote as $x \to 0$ from the right.

4.4 Speaking the Language

1. The inverse of the exponential function is called the ________ function.
2. For the function $f(x) = \log(x - 1)$, we call $x - 1$ the ________.
3. The function $y = \log x$ is called the ________ logarithmic function, whereas the function $y = \ln x$ is called the ________ logarithmic function.
4. To evaluate $\log_3 5$ with a calculator, you will need to use the ________ Formula.

4.4 Exercises

Concepts and Skills

1. Suppose that $f(x) = \log x$ and $f(a) \cdot f(b) = 0$. What can you conclude about a and b? Why?

2. Explain why $f(x) = \log(-x^2)$ is undefined.

In Exercises 3–6, write the given logarithmic form in exponential form.

3. $\log_9 81 = 2$

4. $\log_{1/3} 3 = -1$

5. $\log_t x = 3$

6. $\ln a = x$

In Exercises 7–10, write the given exponential form in logarithmic form.

7. $7^{-2} = \dfrac{1}{49}$

8. $8^{2/3} = 4$

9. $P = e^{rt}$

10. $e^{-r} = y$

In Exercises 11–22, evaluate each logarithm.

11. $\log_3 81$

12. $\log_2 8$

13. $\log_4 \dfrac{1}{16}$

14. $\log 0.01$

15. $\log_{1/2} \dfrac{1}{16}$

16. $\log_{1/5} 125$

17. $\ln e^4$

18. $\ln e^{-2}$

19. $\log_7 \sqrt[3]{49}$

20. $\log_3 \sqrt{27}$

21. $\log_8 0.25$

22. $\log_{16} \dfrac{1}{64}$

In Exercises 23–32, evaluate each logarithm. Use the Change of Base Formula as needed. (Round answers to the nearest hundredth.)

23. $\ln 2 + \ln \sqrt{2}$

24. $\log\left(3 + \sqrt{5}\right)$

25. $\dfrac{\ln \sqrt{2}}{\ln 5}$

26. $\sqrt{\log 7}$

27. $\log_7 4$

28. $\log_3 25$

29. $\log_5 \sqrt[3]{324}$

30. $(\log_{12} 6)^2$

31. $\log_4 3^5$

32. $\log_9 \dfrac{\sqrt{15}}{3}$

In Exercises 33–38, produce the graph of the given function. Use the graph to estimate (a) the intercept(s), (b) the asymptote, and (c) the domain of the function.

33. $y = -5 + \log_4 x$

34. $y = \log_{1/3}(x + 1)$

35. $y = -\log_{1/2} x$

36. $y = \log(2 - x)$

37. $y = 3 + \ln(x + 1)$

38. $y = 1 - \ln x$

39. Compare the vertical asymptotes of the graphs of $y_1 = \log_3 x$, $y_2 = -\log_3 x$, $y_3 = \log_3(-x)$, and $y_4 = \log_3 x - 4$.

40. Explain how you can determine the vertical asymptote for the graph of $f(x) = \log_a(x - c)$, where $c > 0$.

In Exercises 41–44, produce the graphs of functions f and g and the given related function $y = \log_b x$. Describe the graphs of f and g relative to the graph of $y = \log_b x$.

41. $y = \log_3 x$
   (a) $f(x) = \log_3(x + 2)$
   (b) $g(x) = \log_3(4 - x)$

42. $y = \log_5 x$
   (a) $f(x) = 2 + \log_5 x$
   (b) $g(x) = 1 - \log_5 x$

43. $y = \log x$
   (a) $f(x) = 2 + \log(1 - x)$
   (b) $g(x) = 4 - \log(x + 1)$

44. $y = \ln x$
   (a) $f(x) = -4 - \ln(x - 2)$
   (b) $g(x) = 3 - \ln(x + 1)$

In Exercises 45–50, match the graph to the function.

(A) $y = \log x^2$

(B) $y = |\log x|$

(C) $y = \dfrac{\log |x|}{x}$

(D) $y = x \log |x|$

(E) $y = \log(9 - x^2)$

(F) $y = \log(x^2 - 9)$

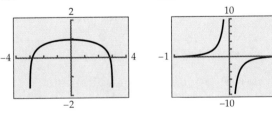

45.

46.

47.

48.

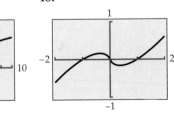

49.

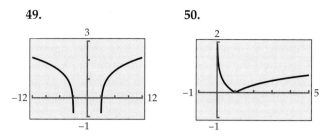

50.

74. Solve the equation $\log_2 [\log_3 (\log_4 x)] = 0$.

In Exercises 51–58, determine the inverse of the given function.

51. $f(x) = 5^x$

52. $h(x) = \left(\dfrac{2}{3}\right)^x$

53. $g(x) = \log_{1/2} (x + 1)$

54. $f(x) = 3 + \log x$

55. $f(x) = 4^{x-2}$

56. $g(x) = 3^{x+1}$

57. $r(x) = 1 - \ln x$

58. $p(x) = \ln (x - 3)$

59. Explain why the graph of $f(x) = \dfrac{\log_4 x}{\log_4 2}$ is the same as the graph of $g(x) = \log_2 x$.

60. How can you determine whether the function $f(x) = \log (-x)$ is an increasing or a decreasing function without producing the graph of f?

In Exercises 61–64, determine the domain of the given function.

61. $g(x) = \log_3 (2x - 7)$

62. $h(x) = \log_5 (x^2 + 1)$

63. $f(x) = \log_2 |x - 2|$

64. $g(x) = \log (4 - x^2)$

Concept Extension

65. Estimate the point of intersection of the graphs of $y = e^{-x}$ and $y = \ln x$. Do the graphs of $y = e^x$ and $y = \ln x$ intersect?

66. Determine the interval of x-values for which $\ln x > \log x$.

67. Are the graphs of $y = \ln e^x$ and $y = e^{\ln x}$ the same? Why?

68. What is the point of intersection of $y = \log_a x$ and $y = \log_b x$, where $a \neq b$?

In Exercises 69–72, use a graph of the given function to estimate (a) the relative extrema and (b) the intercepts.

69. $f(x) = \ln (1 - x^2)$

70. $h(x) = \ln \dfrac{x}{x^2 + 1}$

71. $g(x) = x \ln (x + 5)$

72. $r(x) = e^{-x} \ln x$

73. If $f(x) = \log (\log x)$, what is f^{-1}?

Applications

75. **Altitude and Pressure** Altitude (in meters) is related to atmospheric pressure h (in mm Hg) by $A(h) = 22{,}860 \ln \dfrac{h_0}{h}$, where h_0 is the atmospheric pressure at sea level (760 mm Hg).

(a) Determine the altitude (to the nearest meter) for the given atmospheric pressure.
(i) 680 mm Hg   (ii) 650 mm Hg

(b) Determine the atmospheric pressure (to the nearest hundredth) at the given altitudes.
(i) 3500 meters   (ii) 5000 meters

76. **pH Values** The acidity of a substance depends on the hydrogen ion concentration H^+ (moles/liter) and is measured by the pH value, which is found by the formula $pH = -\log H^+$. (Note that low pH values represent high acidity.)

(a) Determine the pH value for the food whose approximate hydrogen ion concentration is given.

Food	Hydrogen ion concentration
(i) Dill pickles	$4.0 \cdot 10^{-4}$
(ii) Peas	$7.9 \cdot 10^{-7}$
(iii) Shrimp	$1.3 \cdot 10^{-7}$

(b) Determine the hydrogen ion concentration for each food whose pH value is given.

Food	pH value
(i) Milk	6.5
(ii) Oranges	3.5
(iii) Carrots	5.1

77. **Richter Scale** A Richter scale value is given by $R = \log \dfrac{I}{I_0}$, where I_0 is the intensity of the smallest earthquake that can be recorded.

(a) Determine the Richter scale value of each earthquake whose intensity is given.

Year	Location	Deaths	Intensity
(i) 1920	China	100,000	$(4.0 \cdot 10^8)I_0$
(ii) 1960	Morocco	12,000	$(6.3 \cdot 10^5)I_0$
(iii) 1970	Turkey	1,100	$(2.5 \cdot 10^7)I_0$

(b) Determine the intensity of each earthquake whose Richter scale value is given.

Year	Location	Deaths	Richter scale value
(i) 1933	California	115	6.2
(ii) 1964	Alaska	131	8.4
(iii) 1995	Russia	2000	7.6

78. Decibel Scale   The decibel scale D of a sound of intensity I is given by

$$D = 10 \log \frac{I}{I_0}$$

where I_0 is the intensity of a very faint sound.

(a) Determine the decibel scale value of each sound whose intensity is given.

Sound	Intensity
(i) Automobile horn	$(1 \cdot 10^{12})I_0$
(ii) Vacuum cleaner	$(1 \cdot 10^7)I_0$
(iii) Helicopter	$(1 \cdot 10^{10})I_0$

(b) Determine the intensity of each sound whose decibel scale value is given.

Sound	Decibel scale value
(i) Refrigerator	45
(ii) Snowmobile	110
(iii) Noisy restaurant	80

79. Foreign Language Proficiency   At the beginning of a study abroad program, a group of students were tested for achievement in a foreign language. The students were tested weekly during their stay in the country. The average score can be modeled as a function of the number of weeks t after the program began.

$$S(t) = 58 + 16 \ln (t + 1) \qquad 0 \le t \le 10$$

(a) What was the initial average score?
(b) What was the score after 4 weeks?
(c) After how many weeks did the score exceed 90?

80. Advertising Promotes Demand   An advertising agency is conducting a media blitz to promote a new brand of cereal. The agency predicts that the demand d (in number of units) for the cereal t months after the ad campaign begins is modeled by the following function:

$$d(t) = 6800 - 1500 \ln t \qquad t \ge 1$$

(a) What is the demand one month after the initial advertising?
(b) What is the predicted demand after 3 months?
(c) After how many months is demand predicted to fall below 4000 units?

81. Business Revenue   A person began a tool rental business and determined that annual revenue R (in thousands of dollars) grew according to the model $R(t) = 142 + 111 \ln t$, where t is the number of years after the business opened.

(a) What was the first-year revenue?
(b) After how many years did the first-year revenue more than double?
(c) What is the anticipated revenue after 10 years?

82. Short-Term Memory   A psychologist tests short-term memory by presenting a set of items to a subject and asking the subject to remember as many of the items as possible. Each hour thereafter, the subject is tested and the percentage of items remembered is recorded. The psychologist finds that the percentage $P(t)$ can be modeled by

$$P(t) = 85 - 24 \ln t$$

where t is the number of hours after the initial familiarization with the items and $t \ge 1$.

(a) After how many hours is the percentage less than 50%?
(b) Calculate and interpret $P(1)$.

4.5 *Properties of Logarithms*

Basic Properties   ■   *Product, Quotient, and Power Rules*
for Logarithms   ■   *Combining Properties*

Basic Properties

Because logarithmic forms can be written in equivalent exponential forms, the following basic properties of logarithms are easy to verify.

Basic Properties of Logarithms		
Logarithmic form	*Exponential form*	*Example*
1. $\log_b 1 = 0$	$b^0 = 1$	$\log_4 1 = 0$
2. $\log_b b = 1$	$b^1 = b$	$\log 10 = 1$

Because $f(x) = \log_b x$ and $g(x) = b^x$ are inverse functions, their compositions result in x.

$$\text{For all } x, (f \circ g)(x) = f(g(x)) = \log_b g(x) = \log_b b^x = x$$
$$\text{For } x > 0, (g \circ f)(x) = g(f(x)) = b^{f(x)} = b^{\log_b x} = x$$

These results lead to the following inverse properties.

> *Inverse Properties of Logarithms*
> For $b > 0, b \neq 1$,
> 1. $\log_b b^x = x$, for any real number x.
> 2. $b^{\log_b x} = x$, for $x > 0$.

In the remainder of this section, we assume that all logarithms are defined.

EXAMPLE **1**   **Using the Inverse Properties of Logarithms**

(a) $\log_3 3^{2t} = 2t$

(b) $\ln e^{x^2} = x^2$

(c) $10^{\log (1 - x)} = 1 - x$

(d) $7^{\log_7 (y/5)} = \dfrac{y}{5}$

Product, Quotient, and Power Rules for Logarithms

Because logarithms are defined in terms of exponents, their properties are closely related to the properties of exponents.

EXPLORATION

The Product Rule for Logarithms

Select several pairs of positive numbers a and b. For each pair of numbers, evaluate log (ab) and log a + log b. How do the expressions log (ab) and log a + log b appear to be related?

DEVELOPING THE CONCEPT

The Product Rule for Logarithms

Figure 4.35(a) shows the calculations of log $(4 \cdot 7)$ and log 4 + log 7. Observe that the results are the same.

FIGURE **4.35**

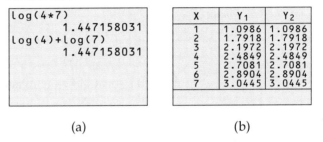

(a)   (b)

Figure 4.35(b) shows a table of values for the functions $y_1 = \ln (3x)$ and $y_2 = \ln 3 + \ln x$. The table suggests that $\ln (3x) = \ln 3 + \ln x$. Producing the graphs of f and g would lead to the same conclusion.

These results suggest that the logarithm of a product of numbers is the sum of the logarithms of the numbers.

> ***Product Rule for Logarithms***
> For $b > 0$, $b \neq 1$, and positive real numbers M and N,
> $$\log_b MN = \log_b M + \log_b N$$

We can justify the Product Rule for Logarithms by using the corresponding property of exponents.

■ *NOTE There is no property for the logarithm of the sum or difference of two numbers. Specifically, $\log_b (M + N)$ is not equal to $\log_b M + \log_b N$, and $\log_b (M - N)$ is not equal to $\log_b M - \log_b N$.*

Let $\log_b M = r$ and $\log_b N = s$.

Then $b^r = M$ and $b^s = N$.   Convert each logarithmic form to exponential form.

$MN = b^r \cdot b^s$ Write the product of M and N.

$MN = b^{r+s}$ Product Rule for Exponents

$\log_b MN = r + s$ Convert to logarithmic form.

$\log_b MN = \log_b M + \log_b N$ $r = \log_b M$ and $s = \log_b N$

Writing log 21 as log 3 + log 7 is called *expanding* the logarithm. Conversely, writing log 4 + log z as log ($4z$) is called *combining* the logarithm.

EXAMPLE 2   **Using the Product Rule to Expand Logarithms**

(a) $\log_3 5y = \log_3 5 + \log_3 y$

(b) $\log xy(2 + z) = \log x + \log y + \log (2 + z)$

EXAMPLE 3   **Using the Product Rule to Combine Logarithms**

(a) $\log_7 y + \log_7 3 = \log_7 (3y)$

(b) $\ln 4 + \ln a + \ln b^3 = \ln (4ab^3)$

(c) $\log_4 x + \log_4 (x + 2) = \log_4 x(x + 2)$

The following rule can be proven with methods similar to those used for the Product Rule for Logarithms. The proof is left as an exercise.

> **Quotient Rule for Logarithms**
> For $b > 0, b \neq 1$, and positive real numbers M and N,
> $$\log_b \frac{M}{N} = \log_b M - \log_b N$$

The Quotient Rule for Logarithms allows us to expand the logarithm of a quotient of two numbers into the difference of the logarithms of the two numbers.

EXAMPLE 4   **Using the Quotient Rule to Expand Logarithms**

(a) $\log \dfrac{t}{7} = \log t - \log 7$

(b) $\log_5 \dfrac{2x + 1}{x - 1} = \log_5 (2x + 1) - \log_5 (x - 1)$

Conversely, we can use the Quotient Rule for Logarithms to combine a difference of the logarithms of two numbers into the logarithm of the quotient of the two numbers.

EXAMPLE 5   **Using the Quotient Rule to Combine Logarithms**

(a) $\log_2 5 - \log_2 y = \log_2 \dfrac{5}{y}$

(b) $\ln(x + 1) - \ln(x^2 + 1) = \ln \dfrac{x + 1}{x^2 + 1}$

(c) $\log a^5 - \log a^3 = \log \dfrac{a^5}{a^3} = \log a^2$

Our last rule deals with the logarithm of a number raised to a power. Again, the proof of this rule is left as an exercise.

> **Power Rule for Logarithms**
>
> For $b > 0, b \neq 1$, any positive real number M, and any real number c,
>
> $$\log_b M^c = c \log_b M$$

EXAMPLE 6   **Using the Power Rule for Logarithms**

(a) $\ln a^3 = 3 \ln a$

(b) $\log_2 \sqrt[3]{x} = \log_2 x^{1/3} = \dfrac{1}{3} \log_2 x$

(c) $\log \dfrac{1}{5} = \log 5^{-1} = -\log 5$

EXAMPLE 7   **Using the Power Rule for Logarithms**

(a) $2 \ln y = \ln y^2$

(b) $-3 \log_7 z = \log_7 z^{-3} = \log_7 \dfrac{1}{z^3}$

(c) $\dfrac{\log_6 x}{2} = \dfrac{1}{2} \log_6 x = \log_6 x^{1/2} = \log_6 \sqrt{x}$

(d) $\dfrac{2}{3} \log t = \log t^{2/3} = \log \sqrt[3]{t^2}$

Combining Properties

Expanding or combining logarithms sometimes requires more than one rule.

EXAMPLE **8** **Expanding Logarithmic Expressions**

(a) $\log_7 \dfrac{a^2}{bc}$

$= \log_7 a^2 - \log_7 bc$ — Quotient Rule for Logarithms

$= \log_7 a^2 - (\log_7 b + \log_7 c)$ — Product Rule for Logarithms

$= \log_7 a^2 - \log_7 b - \log_7 c$ — Distributive Property

$= 2 \log_7 a - \log_7 b - \log_7 c$ — Power Rule for Logarithms

(b) $\log \sqrt{\dfrac{x^3 y}{z^5}}$

$= \log \left(\dfrac{x^3 y}{z^5} \right)^{1/2}$ — Definition of rational exponents

$= \dfrac{1}{2} \log \dfrac{x^3 y}{z^5}$ — Power Rule for Logarithms

$= \dfrac{1}{2} (\log x^3 + \log y - \log z^5)$ — Quotient and Product Rules for Logarithms

$= \dfrac{1}{2} (3 \log x + \log y - 5 \log z)$ — Power Rule for Logarithms

$= \dfrac{3}{2} \log x + \dfrac{1}{2} \log y - \dfrac{5}{2} \log z$ — Distributive Property

EXAMPLE **9** **Combining Logarithmic Expressions**

(a) $2 \log (x + 1) + \dfrac{1}{2} \log x$

$= \log (x + 1)^2 + \log x^{1/2}$ — Power Rule for Logarithms

$= \log (x + 1)^2 + \log \sqrt{x}$ — Definition of rational exponents

$= \log (x + 1)^2 \sqrt{x}$ — Product Rule for Logarithms

(b) $3 \ln a + 3 \ln (a - 2) - 6 \ln b$

$= 3[\ln a + \ln (a - 2) - 2 \ln b]$ — Distributive Property

$= 3[\ln a + \ln (a - 2) - \ln b^2]$ — Power Rule for Logarithms

$= 3[\ln a(a - 2) - \ln b^2]$ — Product Rule for Logarithms

$= 3 \ln \dfrac{a(a - 2)}{b^2}$ — Quotient Rule for Logarithms

$= \ln \left[\dfrac{a(a - 2)}{b^2} \right]^3$ — Power Rule for Logarithms

$= \ln \dfrac{a^3 (a - 2)^3}{b^6}$ — Power Rule for Exponents

(c) $\dfrac{1}{2} \log_3 t + \dfrac{3}{2} \log_3 (2 - t) - 2 \log_3 (t - 5)$

$= \dfrac{1}{2}[\log_3 t + 3 \log_3 (2 - t) - 4 \log_3 (t - 5)]$ Distributive Property

$= \dfrac{1}{2}[\log_3 t + \log_3 (2 - t)^3 - \log_3 (t - 5)^4]$ Power Rule for Logarithms

$= \dfrac{1}{2} \log_3 \dfrac{t(2 - t)^3}{(t - 5)^4}$ Product and Quotient Rules for Logarithms

$= \log_3 \left[\dfrac{t(2 - t)^3}{(t - 5)^4}\right]^{1/2}$ Power Rule for Logarithms

$= \log_3 \sqrt{\dfrac{t(2 - t)^3}{(t - 5)^4}}$ Definition of rational exponents

4.5 Quick Reference

Basic Properties   ■ Basic Properties of Logarithms

1. $\log_b 1 = 0$
2. $\log_b b = 1$

■ Inverse Properties of Logarithms

For $b > 0, b \neq 1$,

1. $\log_b b^x = x$, for any real number x.
2. $b^{\log_b x} = x$, for $x > 0$.

Product, Quotient, and   ■ For $b > 0, b \neq 1$, and positive real numbers M and N,
Power Rules for Logarithms

1. $\log_b MN = \log_b M + \log_b N$ Product Rule for Logarithms
2. $\log_b \dfrac{M}{N} = \log_b M - \log_b N$ Quotient Rule for Logarithms
3. $\log_b M^c = c \log_b M$ Power Rule for Logarithms

■ Writing log MN as log M + log N is called *expanding* log MN. The reverse process is called *combining* the logarithms.

Combining Properties   ■ Expanding or combining logarithms sometimes requires more than one rule.

4.5 Speaking the Language

1. The value of log $(2 \cdot 3)$ is equal to the ▭▭▭▭ of log 2 and log 3.

2. Writing $\ln \dfrac{M}{N}$ as $\ln M - \ln N$ is called ▭▭▭▭ $\ln \dfrac{M}{N}$, whereas writing log M + log N as log MN is called ▭▭▭▭ the logarithms.

3. Many properties of logarithms can be proved by writing the logarithms in ▭▭▭▭ form.

4. According to the ▭▭▭▭ Rule for Logarithms, log x^3 = 3 log x.

4.5 Exercises

Concepts and Skills

1. Explain why $\dfrac{\log M}{\log N} \neq \log M - \log N$. Use a counterexample to illustrate.

2. Explain why $(\log M)(\log N) \neq \log M + \log N$. Use a counterexample to illustrate.

In Exercises 3–10, evaluate the given expression.

3. $10^{-\log 3}$
4. $\ln e^{\ln e}$
5. $8^{-\log_2 5}$
6. $49^{\log_7 6}$
7. $e^{\ln 6 - \ln 2}$
8. $7^{\log_7 2 + \log_7 3}$
9. $\log_7 (\log_3 3)$
10. $\log_4 (\log_2 16)$

In Exercises 11–14, simplify the given expression.

11. $\log_5 5^{x-2}$
12. $\log 10^{|3-x|}$
13. $e^{\ln (3x)}$
14. $7^{\log_7 z^5}$

In Exercises 15–18, use the Product Rule for Logarithms to expand or combine the logarithms.

15. $\ln 3(x-1)$
16. $\log_4 5x(y+2)$
17. $\log x + \log \sqrt{x+3}$
18. $\log_2 (a+1) + \log_2 (a-1) + \log_2 6$

In Exercises 19–22, use the Quotient Rule for Logarithms to expand or combine the logarithms.

19. $\log_5 \dfrac{t}{t-2}$
20. $\log_4 \dfrac{x^2+1}{x-2}$
21. $\ln (t^2-1) - \ln \sqrt{t}$
22. $\log_7 (xy) - \log_7 z^2$

In Exercises 23–26, use the Power Rule for Logarithms to rewrite the given logarithm.

23. $\ln a^{-3}$
24. $\log_2 \sqrt[3]{t}$
25. $-\log_3 w$
26. $\dfrac{3}{2} \log_9 (x+2)$

In Exercises 27–40, expand the given logarithm.

27. $\log_9 \dfrac{xy^2}{z}$
28. $\log_3 \dfrac{2x}{yz^3}$
29. $\ln x(y+3)^2$
30. $\log a^3(b-1)^2 c$
31. $\log \sqrt[3]{\dfrac{a}{2b}}$
32. $\log_2 \dfrac{\sqrt{3x}}{y}$
33. $\log_4 \dfrac{\sqrt{z+2}}{x^3 y}$
34. $\ln \sqrt[3]{\dfrac{a^2}{b^2 c}}$
35. $\log_7 \dfrac{x^2-4}{x^3}$
36. $\log_6 \dfrac{a+1}{a^2-9}$
37. $\ln (x^2+6x+9)$
38. $\log (2x^2-x-3)$
39. $\log (100x - 10x^2)$
40. $\ln Ae^{rt}$

In Exercises 41–54, write the given expression as a single logarithm with a coefficient of 1.

41. $2 \log_6 y + 3 \log_6 x$
42. $4 \log_5 t - 2 \log_5 (t-1)$
43. $\dfrac{1}{2} \log_3 a - \log_3 b$
44. $\dfrac{1}{3}[\log_8 x + \log_8 (x-3)]$
45. $3 \log x - \dfrac{1}{2} \log y + \log z$
46. $\dfrac{1}{2} \ln t - \dfrac{3}{4} \ln w$
47. $\dfrac{2}{3} \log (x+1) + \dfrac{1}{3} \log (x-2)$
48. $\dfrac{1}{2} \log z - \dfrac{3}{2} \log (z+2)$
49. $\ln (a+1) + 2 \ln b - 3 \ln (c-3)$
50. $\log x - \log (x+4) - \log (x-4)$
51. $3 + 2 \log_5 y$
52. $1 - 3 \log_7 x$
53. $6x + \ln y$
54. $rt + \ln P$

55. Using the Power Rule for Logarithms, we can write $\ln x^2 = 2 \ln x$. Produce graphs of $y = \ln x^2$ and $y = 2 \ln x$. Why are the graphs not the same?

56. Explain the difference between the domains of $\log x^2$ and $(\log x)^2$.

In Exercises 57–64, determine whether the given statement is true or false.

57. $\log_{1/b} b = -1$
58. $\log_3 10 = \dfrac{1}{\log 3}$
59. $\ln \dfrac{7}{3} = \dfrac{\ln 7}{\ln 3}$
60. $\log x^2 = (\log x)^2$

61. $(\log x)(\log y) = \log (xy)$

62. $\dfrac{\log 12}{\log 9} = \log 12 - \log 9$

63. $\dfrac{\log 6}{\log 15} = \dfrac{\ln 6}{\ln 15}$

64. $\log (9.3 \cdot 10^7) = 7 + \log 9.3$

In Exercises 65–68, evaluate the given expression by using the properties of logarithms.

65. $\log_4 24 - \log_4 3$

66. $\log_6 24 + \log_6 3 - \log_6 2$

67. $\log \sqrt{10} + 10^{\log 3}$

68. $\log_9 12 - 2 \log_9 6$

In Exercises 69–72, let $\log_b 2 = A$, $\log_b 3 = B$, and $\log_b 5 = C$. Write the expression in terms of A, B, and/or C.

69. $\log_b \left(\dfrac{2}{15} \right)$

70. $\log_b 6\sqrt[3]{25}$

71. $\log_b \dfrac{\sqrt{b}}{60}$

72. $\log_b 18b$

In Exercises 73–76, let $\ln x = a$, $\ln y = b$, and $\ln z = c$. Write the given expression in terms of a, b, and/or c.

73. $\ln x\sqrt{yz}$

74. $\ln \sqrt[3]{xy^2}$

75. $\ln (exz) - \ln \left(\dfrac{x}{e} \right)$

76. $\ln \left(\dfrac{xy}{e} \right) - \ln y^2$

Concept Extension

77. Show that $\log_B \dfrac{1}{A} = \log_{1/B} A$.

78. Evaluate the following expression (a) with a calculator and (b) by using properties of logarithms.

$$\ln 7 + \ln \left(\dfrac{\sqrt{2}}{7} \right) + \dfrac{1}{2} \ln \left(\dfrac{5}{2} \right) + \ln \dfrac{e}{\sqrt{5}}$$

79. Produce a table of values of $\log x$ for $x = 2.3$, 23, 230, 2300. Explain the result. (*Hint*: Write the x-values in scientific notation and use the Product Rule for Logarithms.)

80. Show that $\log_b a = \dfrac{1}{\log_a b}$. Use your result to evaluate the following expressions.

(a) $(\log_3 7)(\log_7 3)$

(b) $\log_6 12 + \dfrac{1}{\log_3 6}$

81. Using steps similar to those used to justify the Product Rule for Logarithms, prove the Quotient Rule for Logarithms.

82. Prove the Power Rule for Logarithms.

Applications

In Exercises 83 and 84, the loudness D (in decibels) of a sound is given by

$$D = 10 \log \dfrac{I}{I_0}$$

where I is the intensity of the sound and I_0 is the intensity of a barely audible sound.

83. Rock Music The "threshold of pain" is 140 decibels. How much more intense is a danger-level sound than the sound at a rock concert where the loudness has been measured at 130 decibels?

84. Sound Intensity The difference in the loudness of a normal conversation and that of a screaming child is 30 decibels. How much more intense is the sound of a screaming child than the sound of a normal conversation?

4.6 *Exponential and Logarithmic Equations*

Definitions and Rules ■ *Exponential Equations* ■
Logarithmic Equations

Definitions and Rules

An equation such as $32^x = 4^{1+x}$ or $2e^x - 7 = 41$ is called an **exponential equation.** Similarly, an equation such as $\log 3x = 2$ or $\ln(x+2) = \ln(1-x)$ is called a **logarithmic equation.**

Because exponential and logarithmic functions are one-to-one functions, we have the following properties, which are the basis for solving exponential and logarithmic equations.

> ### *Properties of Exponential and Logarithmic Equations*
> Assume that $b > 0$, $b \neq 1$, and that all expressions are defined.
>
> 1. If $x = y$, then $b^x = b^y$. Equating exponential expressions
> 2. If $b^x = b^y$, then $x = y$. Equating exponents
> 3. If $x = y$, then $\log_b x = \log_b y$. Equating logarithmic expressions
> 4. If $\log_b x = \log_b y$, then $x = y$. Equating arguments of logarithms

---

Exponential Equations

We can use the graphing method to estimate the solutions of an exponential equation.

If both sides of an exponential equation can be written as exponential expressions with the same base, we can solve the resulting equation algebraically by equating the exponents.

EXAMPLE 1 **Solving Exponential Equations**

Use the graphing method to estimate the solution of each equation. Then solve the equation algebraically by equating exponents.

(a) $32^x = 4^{1+x}$ (b) $\left(\dfrac{9}{4}\right)^{3x} = \left(\dfrac{2}{3}\right)^{2-3x}$ (c) $1 = e^{x^2 - 2x - 3}$

SOLUTION

(a) VISUALIZING THE SOLUTION

Figure 4.36 shows the graphs of the two sides of the equation. Using the *Intersect* feature of the calculator, we estimate the solution to be $0.\overline{6}$ or $\frac{2}{3}$.

FIGURE 4.36

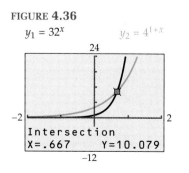

$y_1 = 32^x$ $y_2 = 4^{1+x}$

Intersection
X=.667   Y=10.079

DETERMINING THE SOLUTION

The equation can be solved algebraically by writing both sides with the same base and equating exponents.

$$32^x = 4^{1+x}$$

$(2^5)^x = (2^2)^{1+x}$ Write each expression with base 2.

$2^{5x} = 2^{2+2x}$ Power to a Power Rule.

$5x = 2 + 2x$ Equate exponents.

$x = \dfrac{2}{3}$ Solve for x.

FIGURE 4.37

$y_1 = \left(\dfrac{9}{4}\right)^{3x}$ $y_2 = \left(\dfrac{2}{3}\right)^{2-3x}$

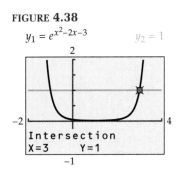

(b) VISUALIZING THE SOLUTION

From Figure 4.37, we estimate the solution to be $-0.\overline{6}$ or $-\dfrac{2}{3}$.

DETERMINING THE SOLUTION

To solve the equation algebraically, we write both sides with the same base and equate exponents.

$$\left(\dfrac{9}{4}\right)^{3x} = \left(\dfrac{2}{3}\right)^{2-3x}$$

$\left(\dfrac{4}{9}\right)^{-3x} = \left(\dfrac{2}{3}\right)^{2-3x}$ $\left(\dfrac{a}{b}\right)^n = \left(\dfrac{b}{a}\right)^{-n}$

$\left(\dfrac{2}{3}\right)^{-6x} = \left(\dfrac{2}{3}\right)^{2-3x}$ Write each expression with base $\frac{2}{3}$.

$-6x = 2 - 3x$ Equate the exponents.

$x = -\dfrac{2}{3}$ Solve for x.

FIGURE 4.38

$y_1 = e^{x^2-2x-3}$ $y_2 = 1$

(c) VISUALIZING THE SOLUTION

Figure 4.38 indicates that the equation has two solutions. The solution shown is 3. We can also trace to the other point of intersection, which represents the solution −1.

DETERMINING THE SOLUTION

We begin the algebraic solution by replacing 1 with e^0.

$$1 = e^{x^2-2x-3}$$

$e^0 = e^{x^2-2x-3}$ Write both expressions with base e.

$0 = x^2 - 2x - 3$ Equate exponents.

$0 = (x-3)(x+1)$ Factor.

$x - 3 = 0$ or $x + 1 = 0$ Zero Factor Property.

$x = 3$ or $x = -1$ Solve for x.

For exponential equations for which a common base is not obvious, we can use the method of equating the logarithms of both sides of the equation. The process is illustrated in the next example.

EXAMPLE **2**   **Solving an Exponential Equation**

Use the graphing method to estimate the solution of each equation. Then solve the equation algebraically by equating exponents.

(a) $10^x = 25$ (b) $2e^x - 7 = 41$

SOLUTION

(a) VISUALIZING THE SOLUTION

With the graphing method, we estimate the solution to be about 1.4. (See Figure 4.39.)

DETERMINING THE SOLUTION

To solve the equation algebraically, we note that the base is 10, and so we equate the common logarithms.

$$10^x = 25$$
$$\log 10^x = \log 25 \qquad \text{Equate the common logarithms of both sides.}$$
$$x = \log 25 \qquad \text{Recall that } \log_b b^x = x.$$
$$x \approx 1.40$$

FIGURE **4.39**

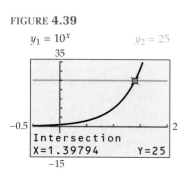

$y_1 = 10^x$ $y_2 = 25$

(b) VISUALIZING THE SOLUTION

From Figure 4.40, we see that the estimated solution is 3.18.

DETERMINING THE SOLUTION

Because the base is e, equating the natural logarithms is convenient.

$$2e^x - 7 = 41$$
$$2e^x = 48 \qquad \text{Isolate } e^x.$$
$$e^x = 24$$
$$\ln e^x = \ln 24 \qquad \text{Equate the natural logarithms of both sides.}$$
$$x = \ln 24 \qquad \log_b b^x = x$$
$$x \approx 3.18$$

FIGURE **4.40**

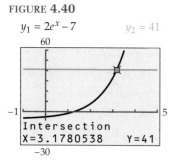

$y_1 = 2e^x - 7$ $y_2 = 41$

■ *NOTE   When you use the graphing method to estimate solutions, you may prefer to write the equation with 0 on one side and graph just the other side. Then the x-intercepts represent the solutions.*

For bases other than 10 or e, we usually equate either the common logarithms or the natural logarithms of both sides. As the next example shows, the choice is arbitrary.

EXAMPLE **3**   **Solving an Exponential Equation**

Estimate the solutions of the following equations graphically. Then solve the equations algebraically.

(a) $2^{3x} = 25$ (b) $5^{2x+1} = 3^{2-x}$

SOLUTION

(a) VISUALIZING THE SOLUTION

We can use the graph in Figure 4.41 to estimate that the solution is about 1.55.

FIGURE **4.41**

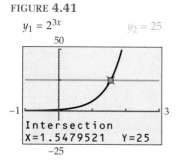

$y_1 = 2^{3x}$ $y_2 = 25$

DETERMINING THE SOLUTION

To solve the equation algebraically, we choose to equate the natural logarithms of both sides.

$$2^{3x} = 25$$

$\ln 2^{3x} = \ln 25$ Equate the natural logarithms of both sides.

$3x \cdot \ln 2 = \ln 25$ Power Rule for Logarithms

$$x = \frac{\ln 25}{3 \ln 2}$$ Solve for x.

$$x \approx 1.55$$

(b) VISUALIZING THE SOLUTION

From the graph, the estimated solution is 0.14. (See Figure 4.42.)

DETERMINING THE SOLUTION

To solve the equation algebraically, we choose to equate the common logarithms of both sides.

$$5^{2x+1} = 3^{2-x}$$

$\log 5^{2x+1} = \log 3^{2-x}$ Equate the common logarithms of both sides.

$(2x + 1)\log 5 = (2 - x)\log 3$ Power Rule for Logarithms

$2x \log 5 + \log 5 = 2 \log 3 - x \log 3$ Distributive Property

$2x \log 5 + x \log 3 = 2 \log 3 - \log 5$ Isolate the variable terms.

$x(2 \log 5 + \log 3) = 2 \log 3 - \log 5$ Distributive Property

$$x = \frac{2 \log 3 - \log 5}{2 \log 5 + \log 3}$$

$$x \approx 0.14$$

FIGURE **4.42**

$y_1 = 5^{2x+1}$ $y_2 = 3^{2-x}$

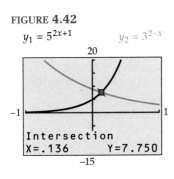

Intersection
X=.136   Y=7.750

Some exponential equations cannot be solved by equating exponents or logarithms.

EXAMPLE **4**   **Estimating Solutions of Exponential Equations**

Use the graphing method to estimate the solution of $e^{1-x} = 5 - 2x$.

SOLUTION

By writing the equation as $e^{1-x} - 5 + 2x = 0$, we can graph the function $y = e^{1-x} - 5 + 2x$ and use the *Zero* feature of the calculator to estimate the x-intercepts. (See Figure 4.43.)

The figure shows that one solution is approximately 2.37. The other solution is approximately -0.92.

FIGURE **4.43**

$y = e^{1-x} - 5 + 2x$

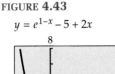

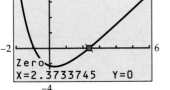

Zero
X=2.3733745   Y=0

In our final example of solving exponential equations, we illustrate the technique of using a substitution to write the equation in quadratic form.

EXAMPLE 5 **Exponential Equations That Are Quadratic in Form**

Solve $2e^{2x} - 7e^x + 3 = 0$.

SOLUTION

VISUALIZING THE SOLUTION

Figure 4.44 shows one of the estimated solutions, 1.10.

DETERMINING THE SOLUTION

The equation can be solved algebraically by recognizing that e^{2x} in the first term is the square of e^x in the second term. Thus the equation is quadratic in form.

FIGURE **4.44**

$y_1 = 2e^{2x} - 7e^x + 3$

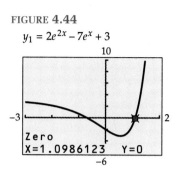

Zero
X=1.0986123   Y=0

$$2e^{2x} - 7e^x + 3 = 0$$

$$2(e^x)^2 - 7e^x + 3 = 0 \qquad e^{2x} = (e^x)^2$$

$$2u^2 - 7u + 3 = 0 \qquad \text{Let } u = e^x.$$

$$(2u - 1)(u - 3) = 0 \qquad \text{Factor.}$$

$2u - 1 = 0 \qquad \text{or} \quad u - 3 = 0 \qquad$ Zero Factor Property

$u = \dfrac{1}{2} \qquad \text{or} \qquad u = 3 \qquad$ Solve for u.

$e^x = \dfrac{1}{2} \qquad \text{or} \qquad e^x = 3 \qquad$ Replace u with e^x.

$x = \ln\dfrac{1}{2} \qquad \text{or} \qquad x = \ln 3 \qquad$ Solve for x.

$x \approx -0.69 \quad \text{or} \qquad x \approx 1.10$

Logarithmic Equations

When we solve logarithmic equations, we must be aware that the argument of a logarithm must be positive. An apparent solution that violates this requirement is called an *extraneous solution* and must be disqualified. Combining graphing and algebraic methods can be helpful in identifying extraneous solutions.

As Example 6 shows, some logarithmic equations can be solved by converting from logarithmic form to exponential form or by equating arguments.

EXAMPLE 6 **Solving Logarithmic Equations**

FIGURE **4.45**

$y_1 = 7 - \ln(x + 2) \qquad y_2 = 9$

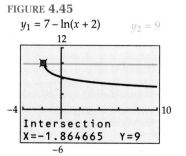

Intersection
X=-1.864665   Y=9

Use the graphing method to estimate the solution of each given logarithmic equation. Then solve the equation algebraically.

(a) $7 - \ln(x + 2) = 9$
(b) $\log(x + 11) = \log(2x + 5)$

SOLUTION

(a) VISUALIZING THE SOLUTION

Figure 4.45 shows the estimated solution, -1.86.

DETERMINING THE SOLUTION

We solve the equation algebraically by isolating the logarithm and converting to exponential form.

$7 - \ln(x + 2) = 9$	Because $x + 2 > 0$, $x > -2$.
$-\ln(x + 2) = 2$	Isolate the logarithm.
$\ln(x + 2) = -2$	
$x + 2 = e^{-2}$	Write in exponential form.
$x = -2 + e^{-2}$	Solve for x.
$x \approx -1.86$	Because $-1.86 > -2$, the solution is permissible.

(b) VISUALIZING THE SOLUTION

Figure 4.46 shows the estimated solution, 6.

DETERMINING THE SOLUTION

When two logarithms are equal, their arguments are equal.

$\log(x + 11) = \log(2x + 5)$	Because $x + 11 > 0$ and $2x + 5 > 0$, $x > -\frac{5}{2}$.
$x + 11 = 2x + 5$	Equate the arguments.
$x = 6$	Because $6 > -\frac{5}{2}$, the solution is permissible.

FIGURE 4.46

$y_1 = \log(x + 11)$
$y_2 = \log(2x + 5)$

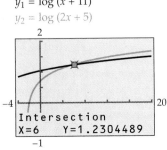

Intersection
X=6   Y=1.2304489

We sometimes use the properties of logarithms to simplify the expressions.

EXAMPLE 7   **Solving Logarithmic Equations**

Solve the following equation:

$$\log_5 x = \log_5(x + 8) - \log_5(x + 3)$$

SOLUTION

VISUALIZING THE SOLUTION

Figure 4.47 shows the graphs of the functions defined by each side of the equation. (Recall that the Change of Base Formula is needed to produce the graphs.) The equation appears to have one solution, 2.

DETERMINING THE SOLUTION

To solve the equation algebraically, we use the Quotient Rule for Logarithms to write the right side of the equation as a single logarithm.

FIGURE 4.47

$y_1 = \log_5 x = \dfrac{\log x}{\log 5}$

$y_2 = \log_5(x + 8) - \log_5(x + 3)$

$= \dfrac{\log(x + 8)}{\log 5} - \dfrac{\log(x + 3)}{\log 5}$

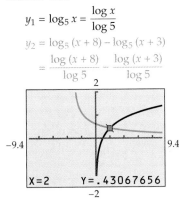

X=2   Y=.43067656

$\log_5 x = \log_5(x + 8) - \log_5(x + 3)$	Note that x must be positive.
$\log_5 x = \log_5 \dfrac{x + 8}{x + 3}$	Quotient Rule for Logarithms
$x = \dfrac{x + 8}{x + 3}$	Equate the arguments of the logarithms.
$x^2 + 3x = x + 8$	Multiply both sides by $x + 3$.
$x^2 + 2x - 8 = 0$	Write the equation in standard form.

$$(x + 4)(x - 2) = 0 \qquad \text{Factor.}$$

$$x + 4 = 0 \quad \text{or} \quad x - 2 = 0 \qquad \textbf{Zero Factor Property}$$

$$x = -4 \quad \text{or} \qquad x = 2$$

Because x must be positive, -4 is an extraneous solution. As suggested by the graph, the only solution of the equation is 2.

EXAMPLE **8**   **Solving Logarithmic Equations**

Solve the following equation:

$$\log (2x + 1) + \log (x - 2) = 1$$

SOLUTION

VISUALIZING THE SOLUTION

From Figure 4.48, the estimated solution is 3.31.

FIGURE **4.48**

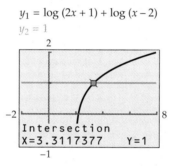

$y_1 = \log (2x + 1) + \log (x - 2)$

$y_2 = 1$

DETERMINING THE SOLUTION

To solve the equation algebraically, we use the Product Rule for Logarithms to write the left side of the equation as a single logarithm.

$\log (2x + 1) + \log (x - 2) = 1$	**Note that $x > 2$.**
$\log (2x + 1)(x - 2) = 1$	**Product Rule for Logarithms**
$\log (2x^2 - 3x - 2) = 1$	**Distributive Property**
$2x^2 - 3x - 2 = 10^1$	**Write in exponential form.**
$2x^2 - 3x - 12 = 0$	**Write the quadratic equation in standard form.**

Using the Quadratic Formula, we obtain two solutions, 3.31 and -1.81. Because $x > 2$, -1.81 is an extraneous solution. As indicated by the graph, the only solution of the equation is approximately 3.31.

As with exponential equations, not all logarithmic equations can be solved with our current algebraic methods. However, the graphing method can provide estimated solutions.

EXAMPLE **9**   **Estimating Solutions of Logarithmic Equations**

Estimate the solution of $\ln x = x^2 - 5x + 4$.

SOLUTION

Figure 4.49 shows the graphs and the estimated points of intersection.

FIGURE **4.49**

$y_1 = \ln x$
$y_2 = x^2 - 5x + 4$

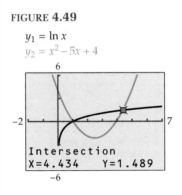

The graph suggests that 1 is a solution, which can easily be verified by substitution. The other solution, as shown in the figure, is approximately 4.43.

4.6 Quick Reference

Definitions and Rules

■ An **exponential equation** is an equation that contains at least one term of the form $a(b^x)$.

■ A **logarithmic equation** is an equation that contains at least one term of the form $a \log_b x$.

■ The following properties are used to solve exponential and logarithmic equations:

1. If $x = y$, then $b^x = b^y$.
2. If $b^x = b^y$, then $x = y$.
3. If $x = y$, then $\log_b x = \log_b y$.
4. If $\log_b x = \log_b y$, then $x = y$.

Exponential Equations

■ If both sides of an exponential equation can be written as exponential expressions with the same base, we can solve the resulting equation algebraically by equating the exponents.

■ For exponential equations for which a common base is not obvious, we can use the method of equating the logarithms of both sides of the equation.

■ For bases other than 10 or e, we usually equate either the common logarithms or the natural logarithms of both sides.

■ When algebraic methods are not available, the solution(s) of an exponential equation can be estimated graphically.

Logarithmic Equations   ■ The argument of a logarithm must be positive. An apparent solution that violates this requirement is called an *extraneous solution* and must be disqualified.

■ Some logarithmic equations can be solved by converting from logarithmic form to exponential form or by equating arguments.

■ Sometimes the properties of logarithms are used to simplify one or both sides of a logarithmic equation.

■ When algebraic methods are not available, the solution(s) of a logarithmic equation can be estimated graphically.

4.6 Speaking the Language

1. If both sides of an exponential equation can be written as exponential expressions with the same base, then we can solve the equation by ▨▨▨▨▨ exponents.

2. The equation $\log (2x + 3) = 5$ is an example of a(n) ▨▨▨▨ equation.

3. The logarithmic equation $\ln (2 - x) = \ln (x + 3)$ can be solved by equating ▨▨▨▨.

4. If an equation contains the term $\log (x - 3)$ and solving the equation results in $x = 1$, we say that 1 is a(n) ▨▨▨▨ solution because $x - 3$ must be positive.

4.6 Exercises

Concepts and Skills

1. The equation $10^x = 100^{2x}$ can be solved by writing both sides with base 10 and equating exponents. Describe an alternative method.

2. To solve $3 = e^{2x - 1}$ by equating exponents, we would need to write 3 as e to a power. Describe how to do that.

In Exercises 3–14, solve the exponential equation.

3. $3^{2x} = 81$

4. $\dfrac{1}{125} = 5^{1 - x}$

5. $4^{-3x} = 32^{1 - x}$

6. $27^{2x} = 9^{2x - 3}$

7. $\dfrac{1}{25^{x+1}} = 125^{2 - x}$

8. $\dfrac{1}{8} = 2^{2x - 1} \cdot 4^{2x}$

9. $8 \cdot 2^{x^2} = 4^{-2x}$

10. $5^{x^2 + 3x} = \dfrac{1}{25}$

11. $9^{x^2} = 3^x$

12. $3^{|x + 1|} = 9$

13. $\left(\dfrac{4}{3}\right)^x = \dfrac{9}{16}$

14. $\left(\dfrac{8}{125}\right)^x = \left(\dfrac{25}{4}\right)^{x+3}$

In Exercises 15–22, solve the exponential equation.

15. $3^{2x} = 16$

16. $6^{1 - x} = 27$

17. $3^x = 4^{x + 1}$

18. $5^{x + 2} = 2^{1 - x}$

19. $5 \cdot 3^{1 - 2x} = 14$

20. $500e^{-0.09t} = 200$

21. $3e^x + 7 = 12$

22. $10 - 3^x = 2$

In Exercises 23–36, solve the logarithmic equation.

23. $3 + \log_5 x = 5$

24. $2 \log_4 x = -1$

25. $-2 + 5 \ln (3x) = 0$

26. $\log_2 (x - 2) + 7 = 6$

27. $\log_7 49^x = 3$

28. $\log_3 x^2 = 4$

29. $\log_2 \sqrt[3]{x} = 4$

30. $\log_8 (x^2 - 3x) = 2$

31. $\ln x(x + 3) = 0$

32. $\log_2 (x + 1)(x - 2) = 1$

33. $\ln (2x - 5) = \ln (x + 1)$

34. $\log_8 (x^2 + 6) = \log_8 (5x)$

35. $\dfrac{1}{2} \log_2 (2x - 1) = \log_2 3$

36. $\dfrac{1}{3} \ln (4 - x) = \ln 3$

37. Suppose that you want to solve $4^{3x} = 5$ by equating logarithms. Describe an advantage and a disadvantage of writing $\log_4 4^{3x} = \log_4 5$.

38. Explain why the equation $\log x^{\ln x} = \ln x^{\log x}$ is an identity for $x > 0$.

In Exercises 39–50, solve the equation.

39. $\log_3 x + \log_3 (x + 2) = 1$

40. $\log_2 (x + 4) + \log_2 (x - 4) = 3$

41. $\log_2 (x + 3) - \log_2 (x - 3) = 2$

42. $\log_3 x - \log_3 (x + 2) = -1$

43. $1 - \log_6 \dfrac{x}{3} = \log_6 (x + 3)$

44. $\log (x - 1) = \log (x + 2) - 1$

45. $\ln (4 - x) - \ln (x + 1) = \ln 2$

46. $\ln (7x - 2) = 2 \ln x + \ln 3$

47. $\log x = \log (x + 6) - \log (x + 2)$

48. $\log (x - 2) + \log (2x) = \log (x - 3)$

49. $\log x + \log (x + 3) - \log (x + 1) = \log (x - 3)$

50. $\ln x - \ln (x + 1) = \ln (x + 2) - \ln (x + 4)$

51. Solving both of the following equations leads to the equation $x^2 + 2x - 15 = 0$.

(i) $e^{x^2 + 2x - 15} = 1$
(ii) $\log x + \log (x + 3) = \log (x + 15)$

How do the solution sets of the two equations differ? Why?

52. Explain how to determine k so that $f(x) = (1.2)^x$ can be written as $y = e^{kx}$.

In Exercises 53–74, solve.

53. $\log_4 |x| = 2$

54. $|3 - \log_7 (x + 2)| = 4$

55. $(\log x)^2 - \log x^2 - 8 = 0$

56. $(\log_3 x)^2 = \log_3 x^2$

57. $x^{\log x} = 1000x^2$ **58.** $\ln x^{\ln x} = \ln x + 2$

59. $\log (\ln x) = 0$ **60.** $\ln (\ln x) = 1$

61. $-2 + \log x = \sqrt{-2 + \log x}$

62. $\ln x + 3 = 4\sqrt{\ln x}$

63. $\ln [\ln (\ln x)] = 1$

64. $\log_2 [\log_3 (\log_4 x)] = 1$

65. $7^{|x - 1|} = 6$ **66.** $|3 - 5^x| = 2$

67. $e^x + 3e^{-x} = 4$ **68.** $e^x + 2 = 3e^{-x}$

69. $e^{2 \ln x^3 - 3 \ln x} = \ln e^8$ **70.** $\dfrac{e^{-t}}{2 - e^{-t}} = \dfrac{1}{2}$

71. $\dfrac{e^x + e^{-x}}{2} = 1$ **72.** $\dfrac{e^x + e^{-x}}{e^x - e^{-x}} = 5$

73. $6^{\log_6 (3x)} = 2$ **74.** $\log 10^{x^2} = 9$

Concept Extension

75. Compare the solutions of the following equations.

(i) $\log x^2 = \log x$
(ii) $(\log x)^2 = \log x$
(iii) $2 \log x = \log x$

76. Compare the solutions of the following equations.

(i) $\ln x = \dfrac{\log x}{\log e}$ (ii) $\ln x = \dfrac{\log e}{\log x}$

77. Produce a table of values of $\log x$ for $x = 0.47$, 0.047, 0.0047, 0.00047. Explain the result. (*Hint:* Write the numbers in scientific notation and use the Product Rule for Logarithms.)

78. Compare the solutions of the following equations.

(i) $\log (5x^2) = 2 \log \sqrt{5}x$
(ii) $\log (5x^2) = 2 \log (5x)$

In Exercises 79–82, use the graphing method to estimate the solution(s) of the given equation.

79. $xe^x = 5 - x^2$ **80.** $2^x = x^3 + 2x^2$

81. $\log x = \dfrac{1}{2}x - 1$ **82.** $\ln (3 - x) = x$

In Exercises 83–86, use the graphing method to estimate the solutions of the given inequality.

83. $3^{-x} \le 2 - x$ **84.** $\dfrac{10x}{2^x} \ge 3$

85. $x \log (x + 3) \le 1$ **86.** $\ln x \ge 2 - x$

87. Consider the function $f(x) = x^x$.

(a) Is the function defined on $(0, \infty)$? Why?
(b) Explain why the function is defined for some but not all values in $(-\infty, 0)$.
(c) Use the decimal setting and the *Dot* mode to produce the graph of the function. Explain the result on the interval $(-\infty, 0)$.

88. Consider the equation $\dfrac{e^x - e^{-x}}{2} = 1$.

(a) Clear the fraction and multiply both sides by e^x.
(b) Write the equation in quadratic form.
(c) Use the Quadratic Formula to solve the equation.

4.7 *Exponential and Logarithmic Models*

Exponential Growth and Decay Models ▪ *Logarithmic Models*
▪ *Logistic Models* ▪ *Regression Equations*

Many real-life applications can be modeled by exponential or logarithmic functions and equations. Exponential models are often appropriate when data increase at an increasing rate or decrease at a decreasing rate. When data increase or decrease but then begin to level off, we can often model the conditions with a logarithmic equation or function.

In some applications, a known exponential or logarithmic formula can be used. Typically, such formulas contain constants that must be determined as a first step in the solving process. In other cases, we must make a judgment as to the type of model that is most appropriate for modeling the given data.

In this section, we illustrate the use of exponential and logarithmic functions as well as a variety of regression equations for solving application problems.

Exponential Growth and Decay Models

Exponential growth or *decay* refers to the increase or decrease in a quantity over a period of time. If P_0 is the original amount of the quantity (at time $t = 0$), then the formula $P = P_0 e^{kt}$ might be used to model the quantity P after time t has elapsed. The constant k is the *growth* $(k > 0)$ or *decay* $(k < 0)$ *constant*.

The *half-life* of a substance is the time required for half the original amount of the substance to disintegrate. Organic matter contains carbon 14, which has a half-life of 5770 years. Archeologists sometimes estimate the age of an object by measuring the amount of carbon 14 it contains.

$\overline{\text{EXAMPLE } 1}$ **Carbon Dating**

Suppose that an archeologist discovered a wooden tool that had lost 18% of its original carbon 14. What was the approximate age of the tool?

SOLUTION

$$P = P_0 e^{kt} \qquad \text{The model for exponential decay.}$$

$$\frac{1}{2}P_0 = P_0 e^{k(5770)} \qquad \text{After 5770 years, the amount of carbon 14 is half the original amount: } t = 5770, P = \tfrac{1}{2}P_0.$$

$$\frac{1}{2} = e^{k(5770)} \qquad \text{Divide both sides by } P_0.$$

$$\ln\left(\frac{1}{2}\right) = 5770k \qquad \text{Convert to logarithmic form.}$$

$$k = \frac{\ln\left(\frac{1}{2}\right)}{5770} \approx -0.00012 \qquad \text{The decay constant is negative.}$$

The wood has lost 18% of its original carbon 14, so its contains 82% of the original carbon 14. Use the exponential decay model again, this time with the known decay constant.

$$P = P_0 e^{kt} \qquad \text{The model for exponential decay.}$$

$$0.82 P_0 \approx P_0 e^{-0.00012t} \qquad P = 0.82 P_0 \text{ and } k \approx -0.00012.$$

$$0.82 \approx e^{-0.00012t} \qquad \text{Divide both sides by } P_0.$$

$$\ln 0.82 \approx -0.00012t \qquad \text{Convert to logarithmic form.}$$

$$t \approx 1654 \qquad \text{Solve for } t.$$

The tool was approximately 1654 years old.

Example 1 illustrates exponential decay. The next example illustrates exponential growth.

EXAMPLE 2   **The Mass of a Bacteria Colony**

The mass M of a bacteria colony at time t is given by

$$M = M_0 (2)^{kt}$$

where M_0 is the mass at $t = 0$. If the mass of a particular colony doubles in 2.5 hours, in how many hours will the colony triple in mass?

SOLUTION
The first step is to determine the value of the constant k.

$$M = M_0 2^{kt} \qquad \text{The model of exponential growth}$$

$$2^{kt} = \frac{M}{M_0} \qquad \text{Divide both sides by } M_0.$$

$$2^{2.5k} = 2 \qquad \text{When } t = 2.5, \ \frac{M}{M_0} = 2.$$

$$\log_2 2^{2.5k} = \log_2 2 \qquad \text{For convenience, we use logarithms of base 2.}$$

$$2.5k = 1 \qquad \log_b b^x = x$$

$$k = 0.4$$

Now we can rewrite the model with the known value of k.

$$M = M_0 (2)^{0.4t}$$

$$2^{0.4t} = \frac{M}{M_0} \qquad \text{Divide both sides by } M_0.$$

$$2^{0.4t} = 3 \qquad \text{At some time } t, \ \frac{M}{M_0} = 3.$$

$$\log 2^{0.4t} = \log 3 \qquad \text{This time we use common logarithms.}$$

$$0.4t \log 2 = \log 3 \qquad \text{Power Rule for Logarithms}$$

$$t = \frac{\log 3}{0.4 \log 2} \approx 3.96$$

The colony triples in mass in approximately 4 hours.

Logarithmic Models

Applications may involve formulas or equations that contain logarithmic expressions.

Newton's Law of Cooling can be written $\ln (T - R) = kt + C$, where T is the temperature of an object at time t, R is the constant temperature of the surrounding environment, and k and C are constants. Typically, we calculate C for $t = 0$ and k for some known time t.

EXAMPLE **3**   **The Law of Cooling**

A chef needs to cool melted chocolate to 90°F before adding it to other ingredients. At a room temperature of 70°F, the chocolate cools from 180°F to 150°F in 12 minutes. How many minutes are required to cool the chocolate from 180°F to a usable temperature?

SOLUTION
We begin by calculating the constant C.

$$\ln (T - R) = kt + C \qquad \text{The Law of Cooling}$$
$$\ln (180 - 70) = k(0) + C \qquad \text{At } t = 0, T = 180 \text{ and } R = 70.$$
$$C = \ln 110 \approx 4.7$$
$$\ln (T - R) = kt + 4.7 \qquad \text{The Law of Cooling with } C = 4.7$$

Now we calculate the constant k.

$$\ln (150 - 70) = k(12) + 4.7 \qquad \text{At } t = 12, T = 150 \text{ and } R = 70.$$
$$\ln 80 = 12k + 4.7$$
$$k = \frac{\ln 80 - 4.7}{12} \approx -0.026$$
$$\ln (T - R) = -0.026t + 4.7 \qquad \text{The Law of Cooling with } k = -0.026$$

Finally, we calculate the value of t for which $T = 90$.

$$\ln (90 - 70) = -0.026t + 4.7$$
$$\ln 20 = -0.026t + 4.7$$
$$t = \frac{\ln 20 - 4.7}{-0.026} \approx 65.5$$

Cooling the chocolate from 180°F to 90°F requires approximately 65.5 minutes.

Logistic Models

In some applications, data increase but then begin to level off near an apparent limiting value. Such data can sometimes be modeled with a *logistic* equation of the form $y = \dfrac{c}{1 + ae^{-bx}}$.

EXAMPLE **4**   **Spreading of Food Poison Bacteria**

An outbreak of food poisoning occurred on a cruise ship carrying 2000 passengers. The number y of passengers infected with the bacteria after x hours can be modeled by $y = \dfrac{1400}{1 + 18e^{-1.5x}}$.

(a) Produce a table of values to determine the number of passengers infected after $1, 2, 4, 7, 15,$ and 24 hours.

(b) Produce a graph of the model function. Based on the graph and the table of values in part (a), how many passengers were not infected?

SOLUTION

(a) The Y column of the table in Figure 4.50 shows the number (to the nearest integer) of affected passengers after the indicated number of hours.

FIGURE **4.50**

X	Y₁	
1	279	
2	738	
4	1340	
7	1399	
15	1400	
24	1400	

Y₁∎1400/(1+18e^...

(b) Figure 4.51 shows the graph of the model function. The table of values and the graph indicate that the number of passengers infected approaches a limiting value of 1400. So approximately 600 passengers would not be infected by the bacteria.

FIGURE **4.51**

$$y_1 = \frac{1400}{1 + 18e^{-1.5x}} \qquad y_2 = 1400$$

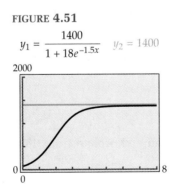

Regression Equations

Several regression models involve exponential or logarithmic functions. The choice of a model depends on the nature of the data. Each part of Figure 4.52 shows a scatterplot of certain data along with the graph of a model function that might be used to model the data.

FIGURE **4.52**

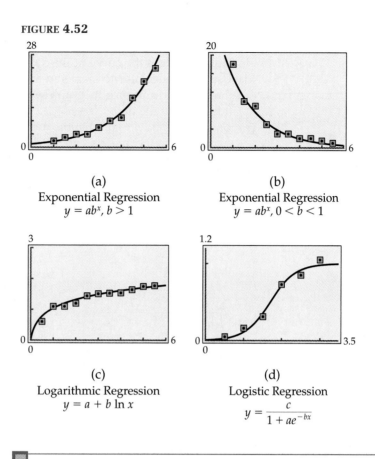

(a)
Exponential Regression
$y = ab^x, b > 1$

(b)
Exponential Regression
$y = ab^x, 0 < b < 1$

(c)
Logarithmic Regression
$y = a + b \ln x$

(d)
Logistic Regression
$$y = \frac{c}{1 + ae^{-bx}}$$

Regression   The typical menu selections for exponential, logarithmic, and logistic regression equations are *ExpReg*, *LnReg*, and *Logistic*, respectively.

EXAMPLE **5**   **Industrial Accidents**

In 1996, an industrywide program was instituted to reduce the number of on-the-job accidents. The table shows the success of the program for the period 1996–2000.

Year	Accidents
1996	50
1997	29
1998	18
1999	11
2000	7

Suppose that we model the data with $P = P_0 e^{-0.5t}$.

(a) If t is the number of years since 1996, what is an appropriate value for P_0?

(b) What is the value of k, and what does its sign imply about the trend in accidents?

(c) Produce a scatterplot of the data. Using your value of P_0 from part (a), produce the graph of $y = P_0 e^{-0.5t}$ in the same coordinate system. How well does the curve fit the data?

(d) Use your calculator to create an exponential regression equation for the data and produce its graph with the scatterplot. How well does this curve fit the data?

(e) Use either model to predict the number of accidents in the year 2002. According to the models, can the number of accidents ever reach 0?

SOLUTION

(a) From the table, when $t = 0$, $P_0 = 50$.

(b) Because $k = -0.5$, the number of accidents is decreasing.

(c) Figure 4.53(a) shows a scatterplot of the data along with the graph of the equation $P = 50e^{-0.5x}$. The curve appears to fit the data well.

FIGURE **4.53**

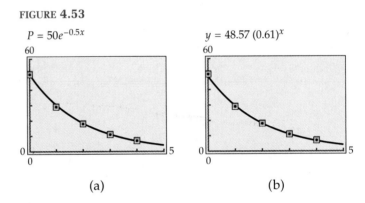

(a)   (b)

■ *NOTE In Example 5, the regression equation that we obtained was $y = 48.57(0.61)^x$, which appears to be quite different from the formula $P = 50e^{-0.5x}$. However, we can show that the two equations are approximately equivalent.*

$y = 48.57(0.61)^x =$
$48.57e^{(\ln 0.61)x} \approx 48.57e^{-0.49x}$

In general, exponential regression equations can always be written with base e by applying the following formula:

$$ab^x = ae^{(\ln b)x}$$

(d) The approximate regression equation is $y = 48.57(0.61)^x$. The graph of this equation is shown with the scatterplot in Figure 4.53(b). Again, the regression equation models the data well.

(e) Both models predict about 2.5 accidents in 2002. Because their graphs have no x-intercept, neither model can project 0 accidents.

A logarithmic function may be an appropriate model for certain data. In the following example, we take advantage of the calculator's logarithmic regression equation option.

EXAMPLE **6**   **Trends in College Enrollment**

The table shows the total public and private college enrollments (in thousands) for selected years. The enrollment for 2007 is projected. (*Source*: Council for Aid to Education.)

Year	Enrollment
1970	8,094
1980	11,750
1985	12,242
1990	13,087
1994	15,020
1997	15,100
2007	15,600

(a) Letting x represent the number of years since 1965, produce a natural logarithmic regression equation to model the data. Then produce the graph of the equation along with a scatterplot of the data.

(b) For projected data, statisticians frequently use "high" and "low" models. Based on the projected data for 2007, would you say that the regression equation is a higher or lower model than the one used for the table?

FIGURE 4.54

$y = 1940 + 3664 \ln x$

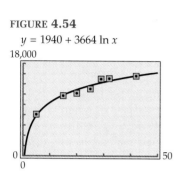

SOLUTION

(a) The natural logarithmic regression equation is reported in the form $y = a + b \ln x$, where a and b are constants. Figure 4.54 shows the scatterplot and the graph of the approximate regression equation $y = 1940 + 3664 \ln x$.

(b) The regression equation projects a 2007 enrollment of approximately 15,635,000. Thus the regression equation appears to be a slightly higher model than the one used for the table.

4.7 Exercises

Exponential Models

Bacteria Growth *Escherichia coli* (usually called *E. coli*) is a type of bacteria that is found in the human digestive system. *E. coli* colonies grow exponentially. The mass M of a colony after t hours is given by $M = M_0 e^{rt}$, where M_0 is the initial mass of the colony and r is the growth rate.

1. Suppose that the initial mass of an *E. coli* colony is $1.2 \cdot 10^{-8}$ gram. After 3 hours, the mass is $3.6 \cdot 10^{-7}$ gram. Approximate the growth rate of the colony.

2. Suppose that the initial mass of an *E. coli* colony is $1.2 \cdot 10^{-8}$ gram and the growth rate is 1.08. After approximately how many hours will the mass of the colony reach $4.8 \cdot 10^{-8}$ gram?

3. After approximately how many minutes will the mass of an *E. coli* colony triple if the rate of growth is 1.04?

4. If the rate of growth of an *E. coli* colony is 1, what is the ratio of mass to initial mass after 1 hour?

Radioactive Decay Radioactive material decays exponentially according to the formula $A = A_0(2)^{-t/T}$, where A is the amount of a radioactive substance at time t, A_0 is the amount of the substance at $t = 0$, and T is the half-life of the substance. The following table lists some radioactive materials and the half-life of each.

The Half-Life of Selected Isotopes

Isotope	Half-life
Carbon 14	5770 years
Argon 41	1.83 hours
Plutonium 236	2.85 years
Uranium 229	58 minutes
Radium 221	30 seconds
Krypton 74	15 minutes
Iodine 131	8.065 days
Neon 17	0.107 seconds
Sodium 24	15 hours
Magnesium 27	9.5 minutes
Titanium 45	3.08 hours

5. To diagnose certain diseases, a small amount of radioactive material is injected into patients and traced. Doses can be used only after the radioactive substance has decayed to a certain level. Suppose that a patient is to receive an injection of iodine 131 when the radioactivity level is 100. If the radioactivity level is 300 on June 1, on approximately what date should the injection be made?

6. A small piece of cloth discovered by archaeologists contained 77% of its original amount of carbon 14. How old was the cloth?

7. To conduct an experiment, a physicist required 1 milligram of radioactive krypton 74. At 9:00 A.M., the available krypton sample was tested and was found to contain 3 milligrams of radioactive substance. At what time could the experiment begin?

8. One ton of radioactive plutonium 236 was stored in an underground nuclear waste facility. If the radioactive level of the facility will not be safe until only 100 pounds of the substance remain, how long will it be before the facility is safe?

9. **Interest Rate** If an initial amount of $10,000 is invested with continuously compounded interest, what interest rate must be earned for the investment to double in 9 years?

10. **Investment Period** With interest compounded continuously at 8.5%, how many years are required for an investment to double in value? to triple in value?

11. **Heart Transplants** The first heart transplant occurred in 1967. In 1987, 1418 heart transplant operations were performed. Assuming exponential growth, the number N of heart transplants can be modeled by $N = N_0 e^{kt}$, where t is the number of years since the first operation. In what year were 40,000 heart transplants predicted?

12. **Cost of a Bus Ride** The cost of riding a city bus in 1955 was 10 cents. By 1995, the cost had risen to $1.00. If the cost C can be modeled by $C = C_0 e^{kt}$, where t is the number of years since 1955, in what year will the cost reach $5.00?

13. **The Population of San Diego** The population of San Diego increased from 875,000 in 1980 to 1.11 million in 1990. Assuming exponential growth, in what year is the population projected to be 1.3 million?

14. **Rabbit Population** If a population of rabbits that grows exponentially triples in 30 days, in how many days would the population double?

15. **Air Resistance** The velocity of an object (in feet per second) is decreased by the resistance of air. Suppose that the velocity v of an object at time t is $v(t) = 40(1 - 2e^{-t})$. What is the velocity after 1 second? after 2 seconds? What does the function indicate about the velocity as time increases?

16. **Depreciation** The owner of a construction company depreciates the value V of a front-end loader by using $V(t) = V_0(1 - r)^t$, where r is the depreciation rate, V_0 is the original value of the equipment, and t is the time in years. If the depreciation rate is 20%, after how many years is the value of the front-end loader one-third of its original value?

Logarithmic Models

Newton's Law of Cooling The Law of Cooling can be written $\ln (T - R) = kt + C$, where T is the temperature of an object at time t, R is the constant temperature of the surrounding environment, and k and C are constants. Typically, we calculate C for $t = 0$ and k for the ending time t.

17. The temperature of a fresh pot of tea was 203°F when it was poured into a cup. After 1 minute, the tea had cooled to 194°F. If the temperature of the room was 70°F, approximately how long did the tea take to cool to 149°F?

18. Suppose that the temperature of freshly brewed coffee is 200°F. To the nearest degree, what is the room temperature if $C = 5$?

19. A small ingot of aluminum was heated to 600°C, and after 20 minutes, the ingot had cooled to 400°C. If the temperature of the facility was 40°C, how much longer did the ingot take to cool another 100°C?

20. Solve Newton's cooling formula for T.

The Power Law of Practice In psychology, the "power law of practice" states that the logarithm of the reaction time needed to perform a certain task is a linear function of the logarithm of the number of practice trials taken. The specific formula can be written $\log t = k \log n + C$, where t is the reaction time, n is the number of practice trials, and k and C are constants. Note that when $n = 1$, $C = \log t$.

21. In an experiment involving motor-perceptual skills, a subject needed 5 seconds to succeed on the first trial. After 10 practice trials, the subject succeeded in 2 seconds. After approximately how many trials would the subject's reaction time be reduced to 1 second?

22. Using the information in Exercise 21, what would the subject's approximate reaction time be after 30 trials?

23. Letting $C = \log c$, solve the power law of practice formula for t. Referring to your formula, explain why the reaction time t can never be 0.

24. A new driver ($n = 1$) requires 1.5 seconds to move from the accelerator to the brake pedal after hearing an alarm. However, the driver is able to reduce the reaction time to 1 second after only 10 trials. How many trials are needed for the driver to achieve the $\frac{3}{8}$-second reaction time of an experienced driver?

The Species/Area Curve   Ecologists have studied the number of species N in relation to the land area A in which the species are found. A model of this relationship is $\ln N = k \ln A + 2.834$, where k is a constant between 0.15 and 0.40, depending on the geographical location of the study.

25. Use the fact that $\ln e^c = c$ to write the constant 2.834 as a natural logarithm. Then combine logarithms on the right side and solve the formula for N.

26. Using the revised formula in Exercise 25, determine the approximate value of k when 22 species are found in an area of 4 square miles.

27. For the Galapagos Islands, researchers have arrived at the formula $\ln N = 0.28 \ln A + 2.834$. What is the predicted number of species on an island of 0.5 square mile?

28. In Exercise 27, what is the predicted land area on which 16 species were found?

29. **The Sound of Septuplets**   The intensities of sounds are additive. For example, the total intensity of seven crying babies is 7 times the intensity of one crying baby. Loudness D (in decibels) is given by the logarithmic formula $D = 10 \log\left(\dfrac{I}{10^{-12}}\right)$, where I is the intensity. If the loudness of one crying baby is 30 decibels, what is the loudness of crying septuplets?

30. **Laundry Noise**   The loudness of a washing machine is 65 decibels. The loudness of a clothes dryer is 60 decibels. Use the formula in Exercise 29 and the fact that the intensities of sounds are additive to determine the loudness in your laundry when both machines are running.

Brightness of Stars   The magnitude M of a star of brightness B is $M = -2.5 \log \dfrac{B}{B_0}$, where B_0 is the brightness of a star with $M = 0$.

31. The magnitude of the sun is -26.8, and the magnitude of the star Sirius is -1.5. How many times brighter is the sun than Sirius?

32. What is the difference in magnitude of two stars if the brightness of one is one million times the brightness of the other?

Logistic Models

33. **Black Bear Population**   Suppose that the model $B(t) = \dfrac{580}{1 + 8.3e^{-0.12t}}$ describes the black bear population in a certain area in Alaska. In this model, t is the number of years after logging ended in the area. What was the initial population? In how many years will the initial population double?

34. **Personal Computers**   Suppose that the function $P(t) = \dfrac{0.93}{1 + 4.1e^{-0.3t}}$ models the fraction of homes that have personal computers, where t is the number of years after 1998. In what year does the model predict that half of all homes will have computers?

Limited Population Growth   Under certain conditions and circumstances, population growth is limited. If M represents the maximum population that can be sustained, the population P at time t is modeled by $P(t) = \dfrac{MP_0}{P_0 + (M - P_0)e^{-kt}}$, where P_0 is the initial population.

35. Suppose that an area can sustain a maximum of 200 bald eagles. If 20 eagles were initially introduced into the area and 30 eagles were counted 2 years later, after how many years will the population reach 100?

36. A biologist studied mosquitos in a laboratory system that could support at most 3000 mosquitos. If an initial population of 200 mosquitos climbed to 300 in 2 days, in how many days did the population triple?

37. Futurist organizations, such as the Delta Group, have concluded that the maximum population that can be sustained on Earth is 10 billion. In 1994, the world population was approximately 5.643 billion and the annual growth rate was 1.5%. Using the limited growth model, in approximately what year will the world population reach 9 billion?

38. Compare your result in Exercise 37 to the result obtained from using the standard exponential growth model $P(t) = P_0 e^{rt}$.

Exponential Regression

39. Mouse in a Maze Psychologists have studied learning behavior by observing mice in a maze. A passage through the maze is considered a success if the mouse makes no wrong turns in arriving at the cheese placed at the end. The following table is a record of the number of errors made by a mouse during the first 7 trials.

Trial number	Errors
1	54
2	40
3	29
4	20
5	13
6	7
7	3

(a) Letting x represent the trial number, produce an exponential regression equation that models the data.

(b) Produce a scatterplot of the data and the graph of the regression equation.

(c) Assuming that the mouse will eventually learn the maze so well that it makes no errors, in what respect does the model equation fail?

40. Half-Life Radioactive substances decay as a result of nuclear emissions of radiation. The *half-life* of a substance is the elapsed time at which half of the original amount of substance remains. If we start with 5 grams of actinium 225, the table shows the amount remaining after a given number of days.

Time elapsed (days)	Amount remaining (grams)
2	4.35
5	3.54
8	2.87
10	2.50
20	1.25
30	0.63

(a) What is the half-life of actinium 225?

(b) The data can be modeled by the equation $A = 5\left(\frac{1}{2}\right)^{t/10}$, where A is the amount remaining and t is the number of days. Interpret the constants 5 and 10.

(c) Letting x represent the number of days, produce a scatterplot of the data and the graph of the model equation. Trace the graph to estimate the amount remaining after 15 days.

(d) Produce an exponential regression equation and verify that its graph fits the data well.

41. In-Flight Entertainment The table shows the amounts (in billions of dollars) spent annually by airlines on in-flight entertainment. (*Source*: World Airline Entertainment Association.)

Year	Expenditures
1992	0.4
1993	0.6
1994	0.8
1995	0.9
1996	1.2
1997	1.4

(a) Letting x represent the number of years since 1990, produce a scatterplot of the data. Then produce and graph both a linear regression equation and an exponential regression equation with the scatterplot.

(b) Which appears to be the better model of the 1992–1997 data? How does the table support your conclusion?

(c) Which model would you regard as the more realistic for years following 1997? Why?

42. Net Stock Fund Assets The bull market of the 1990s produced a rapid increase in stock fund assets. The table below shows the trend in net assets (in trillions of dollars) for the period 1989–1997. (*Source*: Investment Company Institute.)

Year	Net assets
1989	0.3
1991	0.4
1993	0.8
1995	1.3
1997	2.2

(a) Letting *x* represent the number of years since 1987, produce an exponential regression equation to model the data. Then produce its graph along with a scatterplot to verify that the equation models the data well.

(b) Use the model to estimate the net stock fund assets in 1996.

43. The Mormon Church From meager beginnings, the Church of the Latter-Day Saints has amassed $30 billion in assets and enjoys an annual income of nearly $6 billion. Its spectacular increase in membership is shown in the accompanying graph. (*Source*: Global Media Guide.)

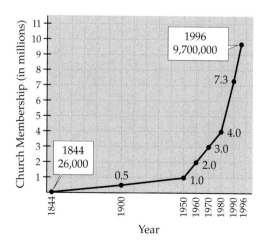

Let *x* represent the number of years since 1844.

(a) Enter the given membership data in your calculator and produce an exponential regression equation.

(b) If the population of the United States were to remain constant at 260 million, in what year does the model project that 10% of the population would belong to the Mormon church?

(c) For the period 1950–1980, what type of model would be more appropriate? Why?

44. Online Shopping The commercialization of the Internet is marked by massive increases in advertising and online shopping. The accompanying table illustrates this trend. (*Source*: Jupiter Communications.)

Year	Online sales of recorded music (in millions of dollars)
1996	18
1997	47
1998	110
1999	240

(a) From an inspection of the table, would you select an exponential or a logarithmic equation to model the data? Why?

(b) Based on your decision in part (a), and letting *x* represent the number of years since 1990, produce a regression equation to model the data. Use the model to predict the online sales of recorded music for the year 2000.

Logarithmic Regression

45. Sport-Utility Vehicles Of all vehicles that were priced at $28,000 or more, the percentage that were sport-utility vehicles rose from 15% in 1990 to 46% in 1997. (*Source*: Mercedes-Benz.)

Year	Percentage
1990	15%
1991	19%
1992	24%
1993	33%
1994	34%
1995	38%
1996	42%
1997	46%

(a) Use a calculator to find a natural logarithmic regression model that is in the form $P(t) = a + b \ln t$, where t is the number of years since 1989. Round coefficients to the nearest tenth.

(b) In what year does the model predict that half the vehicles in this category will be sport-utility vehicles?

46. Viral Load *Viral load* refers to the quantity of a virus that is in the blood. For persons with HIV, measurements of viral load are used to determine when to start antiretroviral therapy. Changes in viral load are often reported as logarithmic changes. For example, if the baseline load were 20,000, then a 1 log increase means a 10-fold increase to 200,000. (*Source*: HIV/AIDS Treatment Information Service.)

(a) Assuming a baseline load of 15,000, what would the viral load be after a 2 log increase? a 1 log decrease?

(b) Any load change of less than one-half log is not considered significant. If the baseline load is 18,000, what range of viral loads would be considered to represent an insignificant change when the next blood sample is tested?

Logistic Regression

In some applications, data increase but then begin to level off near an apparent limiting value. Such data can sometimes be modeled with a *logistic* regression equation of the form $y = \dfrac{c}{1 + ae^{-bx}}$.

47. Cable Television For households that have television sets, the table shows the percent that had cable television in the period 1970–1994. (*Source*: Television Bureau of Advertising, Inc.)

Year	Percent with cable television
1970	6.7
1980	19.9
1985	42.8
1988	49.4
1989	52.8
1990	56.4
1991	58.9
1992	60.2
1993	61.4
1994	62.4

(a) Letting x represent the number of years since 1960, produce a scatterplot of the data. Use the scatterplot to describe the trends in the 1980s and 1990s.

(b) Produce a logistic regression equation and graph it with the scatterplot. Trace the graph to predict the percentages for the years 2000 and 2010.

(c) Look at your regression equation and its graph and approximate the limiting value of y as $x \to \infty$. Interpret the result.

48. Automatic Dishwashers The bar graph shows the percentage of households with automatic dishwashers in the period 1978–1993. (*Source*: Energy Information Administration.)

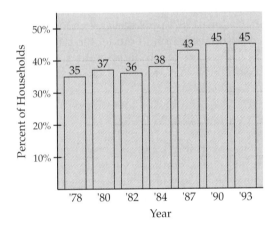

Letting x represent the number of years since 1970, produce both a linear regression equation and a logistic regression equation.

(a) Graph the linear regression equation along with a scatterplot of the data. What does this model predict about the trend in percentages in future years?

(b) Deactivate the linear function and graph the logistic regression equation along with a scatterplot of the data. What does this model predict about the trend in percentages in future years?

(c) Use a large enough window setting to see where the two graphs begin to part. Estimate the year in which this occurs.

49. Internet Investing The accompanying bar graph shows the actual and projected number of brokerage accounts that use the Internet for investing. (*Source*: Forrester Research, Inc.)

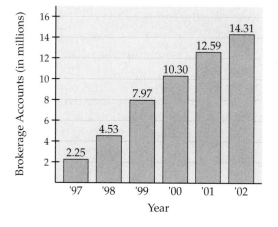

Let x represent the number of years since 1990.

(a) Use your calculator to write a logistic model for the data in the bar graph.

(b) In what year does the logistic model predict that the number of accounts will reach 15 million?

(c) Use your calculator to write a natural logarithmic model for the data.

(d) In what year does the natural logarithmic model predict that the number of accounts will reach 15 million?

(e) What is the long-term trend for each model?

50. Corporations and Charities The table shows the amount (in millions of dollars) that companies paid to charities to use their names in marketing. (*Source*: IEG Sponsorship Report.)

Year	Amount
1994	340
1995	423
1996	485
1997	535

Let x represent the number of years since 1990.

(a) Use your calculator to write a logistic model for the data in the table.

(b) In what year does the logistic model predict that the amount will be twice the 1994 level?

(c) Use your calculator to write a natural logarithmic model for the data in the table.

(d) In what year does the natural logarithmic model predict that the amount will be twice the 1994 level?

(e) What is the long-term trend for each model?

Chapter Review Exercises

Section 4.1

In Exercises 1–3, let $f(x) = -3x$ and $g(x) = x^2 - x + 1$. Evaluate each function as indicated.

1. (a) $(f + g)(-2)$ (b) $(f - g)(x)$

2. (a) $\left(\dfrac{g}{f}\right)(3)$ (b) $(fg)(x)$

3. (a) $(f \circ g)(1)$ (b) $(g \circ f)(x)$

In Exercises 4 and 5, let $f(x) = \dfrac{4}{x - 1}$ and $g(x) = \dfrac{x + 3}{1 - x}$. Determine the indicated function and give the domain.

4. $(g - f)(x)$ **5.** $\left(\dfrac{f}{g}\right)(x)$

6. Let $f(x) = 4 - 3x$ and $g(x) = -6x^2 + 11x - 4$. Determine the indicated function and give the domain.

(a) $\left(\dfrac{f}{g}\right)(x)$ (b) $\left(\dfrac{g}{f}\right)(x)$

7. Let $f(x) = |x + 1|$ and $g(x) = \dfrac{1}{x - 1}$. Evaluate each of the following.

(a) $(g \circ g)\left(\dfrac{3}{2}\right)$ (b) $(f \circ f)(-4)$

In Exercises 8 and 9, determine the composite functions $(f \circ g)(x)$ and $(g \circ f)(x)$.

8. $f(x) = \dfrac{x}{x + 2}, g(x) = \dfrac{1}{x}$ **9.** $f(x) = \dfrac{2}{3}x, g(x) = \dfrac{3}{2}x$

10. Let $f(x) = \sqrt{x + 1}$ and $g(x) = x - 3$. Determine the domain of each of the following.

(a) $(f \circ g)(x)$ (b) $(g \circ f)(x)$

In Exercises 11 and 12, determine functions f and g so that $H(x) = (f \circ g)(x)$.

11. $H(x) = \dfrac{2}{x^2 + x}$

12. $H(x) = (x^2 + 1)^3 + 8(x^2 + 1)^4$

Section 4.2

13. The equation $3xy - y^2 = 7$ describes a relation. Write an equation for the inverse relation.

In Exercises 14 and 15, use a graph to determine whether the inverse relation is a function.

14. $x^2 + y = 1 + 2x$ **15.** $y = -2^{x+3}$

In Exercises 16 and 17, use the graphs of the function and its inverse to estimate the domain and range of (a) the function and (b) its inverse.

16. $f(x) = \sqrt{x - 5}$ **17.** $g(x) = \dfrac{3x}{x + 2}$

18. Explain how to use a graph of a function to determine whether the function is a one-to-one function.

In Exercises 19–21, use a graph to decide whether the function is one-to-one. If the function is one-to-one, determine the inverse function.

19. $f(x) = \sqrt{2 - x}$ **20.** $g(x) = 3 - |x|$

21. $h(x) = \dfrac{x + 1}{x}$

22. The function $C(t) = \frac{5}{9}(t - 32)$ converts Fahrenheit temperature to Celsius temperature. Determine $C^{-1}(t)$ and describe what it does.

23. For $f(x) = x^2 - 1$, $x \geq 0$, determine $f^{-1}(x)$.

24. Use composition to verify that $f(x) = \frac{2}{3}x + 4$ and and $g(x) = \dfrac{3x - 12}{2}$ are inverse functions.

Section 4.3

25. Evaluate $f(x) = e^x$ as indicated.

(a) $f(2.5)$ (b) $f(-1)$

26. How are the graphs of $f(x) = \left(\frac{3}{4}\right)^x$ and $g(x) = \left(\frac{4}{3}\right)^x$ related? Why?

27. For $k > 0$, explain how the graphs of the functions $f(x) = b^x + k$ and $g(x) = b^x - k$ are related.

In Exercises 28 and 29, produce the graphs of f, g, and the given related function $y = b^x$. Describe the graphs of f and g relative to the graph of $y = b^x$.

28. $y = 3^x$

(a) $f(x) = 5 + 3^x$ (b) $g(x) = -1 + 3^x$

29. $y = 7^x$

(a) $f(x) = -7^x$ (b) $g(x) = 1 - 7^x$

In Exercises 30 and 31, produce the graph of the given function. Use the graph to estimate (a) the intercept(s), (b) the asymptote, and (c) the range of the function.

30. $g(x) = 2 + e^{-x}$ **31.** $h(x) = 3^{x+1} - 5$

32. On a child's fourth birthday, the child's grandparents invested \$15,000 in a college fund at 9% interest compounded continuously. What will be the value of the fund when the child is 18 years old?

33. Use the graphing method to estimate the zeros of the function $f(x) = e^x + x - 3$.

34. Use a graph to estimate the relative extrema and horizontal and vertical asymptotes of the function $g(x) = \dfrac{3^x}{x^2}$.

Section 4.4

35. Write each expression in an equivalent logarithmic or exponential form.

(a) $\log_{1/4} 16 = -2$ (b) $e^{-rt} = A$

In Exercises 36 and 37, evaluate the given logarithm.

36. (a) $\log_5 125$ (b) $\ln e$

37. (a) $\log_{2/3} \dfrac{27}{8}$ (b) $\log_6 \sqrt[3]{36}$

38. Approximate the value of each logarithm.

(a) $\ln\left(\dfrac{\sqrt{3}}{2}\right)$ (b) $\log_6 20$

39. How are the graphs of $f(x) = e^x$ and $g(x) = \ln x$ related? Why?

In Exercises 40 and 41, produce the graphs of f, g, and the given related function $y = \log_b x$. Describe the graphs of functions f and g relative to the graph of $y = \log_b x$.

40. $y = \log_2 x$

  (a) $f(x) = \log_2 (1 - x)$
  (b) $g(x) = \log_2 (x + 5)$

41. $y = \log_4 x$

  (a) $f(x) = -1 - \log_4 x$
  (b) $g(x) = 5 + \log_4 x$

42. Produce the graph of $f(x) = 4 - \ln (x + 3)$. Use the graph to estimate (a) the intercepts, (b) the asymptote, and (c) the domain of the function.

43. The decibel scale D of a sound of intensity I is given by $D = 10 \log \left(\dfrac{I}{I_0} \right)$, where I_0 is the intensity of a very faint sound.

  (a) Determine the decibel level for a jackhammer that has intensity $10^{10} \cdot I_0$.
  (b) Determine the intensity of a dishwasher that has a loudness of 65 decibels.

In Exercises 44 and 45, determine the inverse of the given function.

44. $f(x) = 3 - e^x$ **45.** $g(x) = \log (x + 5)$

46. Determine the domain of $f(x) = \log_5 (7 - x)$.

Section 4.5

47. Evaluate the given expression.

  (a) $e^{-2 \ln 3}$ (b) $\log_3 3^7$

48. Simplify the given expression.

  (a) $\ln e^{x + 1}$ (b) $5^{\log_5 t^4}$

49. Use the Product Rule for Logarithms to expand or combine the given expression.

  (a) $\log_5 (2ab)$ (b) $\log t + \log (t + 2)$

50. Use the Quotient Rule for Logarithms to expand or combine the given expression.

  (a) $\ln \left(\dfrac{y}{3 - y} \right)$ (b) $\log_7 w - \log_7 3$

51. Use the Power Rule for Logarithms to rewrite the given logarithm.

  (a) $\log_2 z^{-5}$ (b) $\dfrac{1}{3} \ln t$

52. Explain how to rewrite the logarithm $\log \dfrac{1}{w}$ using (a) the Quotient Rule for Logarithms and (b) the Power Rule for Logarithms.

In Exercises 53–55, use the properties of logarithms to expand the given expression.

53. $\log_7 \dfrac{5a^2}{b}$ **54.** $\log \sqrt[3]{\dfrac{x}{y^2 z}}$

55. $\ln 2x(x + 4)^3$

In Exercises 56–58, use the properties of logarithms to combine the given expression.

56. $\dfrac{1}{2} \ln (x - 2) - \ln x$

57. $2 \log_9 t + 3 \log_9 (t^2 + 1)$

58. $\log (y + 3) - \log y - \log (y - 4)$

Section 4.6

In Exercises 59–61, solve the exponential equation.

59. $8^{x - 2} = 16^{x + 1}$ **60.** $\left(\dfrac{1}{9} \right)^x = 27^{x^2}$

61. $e^{1 - x} = 10^x$

62. Explain why $\log_3 x^2 = 4$ has two solutions, whereas $2 \log_3 x = 4$ has only one solution, although $\log_3 x^2 = 2 \log_3 x$.

In Exercises 63–67, solve the logarithmic equation.

63. $\log_2 (x^2 - 2x) - 1 = 2$

64. $\dfrac{1}{2} \log_5 (4 - 5x) = \log_5 3$

65. $\log_2 (x - 1) - \log_2 (2x + 5) = -2$

66. $\ln x + \ln (x - 3) - \ln 18 = 0$

67. $(\log_5 x)^2 + 3 = \log_5 x^4$

68. Use the graphing method to estimate the solution of $e^x = 2 - x$.

Section 4.7

69. While excavating, an archeologist discovered the skeletal remains of an animal. The bones contained 90% of the original carbon 14. The half-life of carbon 14 is 5770 years. How old was the skeleton?

70. If interest is compounded continuously, what interest rate is required for an investment to triple in value in 12 years?

71. A medical team began working in a remote area to combat an epidemic of influenza. The table shows the number of new cases treated each week after the team began working.

Week	Cases
1	650
2	520
3	450
4	370
5	300

(a) Determine an exponential regression equation to model the data.

(b) How many new cases are predicted during the eighth week?

72. The percentage of customers in an area that a new home improvement store can expect t months after opening is modeled by the logistic equation $P(t) = \dfrac{50}{1 + 6.7e^{-1.5t}}$.

(a) What percentage of the customer base does the store have after 2 months?

(b) Estimate and interpret the horizontal asymptote of the graph of the function.

Chapter 5

Systems of Equations and Inequalities

Following the end of the Cold War, and as a result of budget cutbacks, all branches of the Armed Forces have been downsized dramatically. The table shows the trends in personnel reductions.

To model the data in the table, we would need one equation for each service branch. Any two of those equations, considered simultaneously, form a system of equations. The graph shows linear models for the Navy and the Air Force. As you will see in this chapter, the point of intersection represents the solution of the system of equations and indicates the year when the projected number of Navy and Air Force personnel will be equal.

Chapter 5 focuses on systems of equations and various methods for solving them. In particular, we introduce the concept of a matrix as a numerical method for solving systems, and we discuss the operations that can be performed with matrices. We conclude with systems of inequalities and their application to solving linear programming problems.

TOTAL NUMBER OF ARMED FORCES PERSONNEL			
Year	Army	Navy	Air Force
1993	590,000	512,000	444,000
1995	521,000	464,000	400,000
1997	487,000	398,000	379,000

Navy: $y_1 = -28.5x + 515$
Air Force: $y_2 = -16.25x + 440$

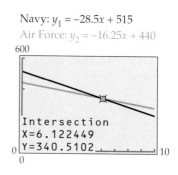

```
Intersection
X=6.122449
Y=340.5102
```

5.1 *Graphing and Algebraic Methods*

Systems of Equations ■ *The Graphing Method* ■
The Substitution Method ■ *The Addition Method* ■
Applications and Modeling

Application problems often involve two or more unknown quantities. Although you can sometimes solve such problems with a single variable, the use of two or more variables is often more convenient. However, if you use two variables, for example, then you will need to be able to write two equations to model the conditions of the problem. If you use three variables, then you will need three equations, and so on.

The new methods and techniques introduced in this section will expand your ability and give you greater flexibility in modeling and solving real-life applications.

Systems of Equations

Suppose that the sum of two numbers is 9 and that the difference of the numbers is 3. If we let x represent the larger number and y the smaller number, then we can describe the given conditions with the following two linear equations in two variables:

$$x + y = 9$$
$$x - y = 3$$

Taken together, all of the equations that describe the conditions of a problem are called a **system of equations.** Because we must find values of the variables for which each equation is true, we say that we consider the equations *simultaneously.*

> **Definition of a System of Equations**
> Two or more equations considered simultaneously form a **system of equations.**

The following are other examples of systems of equations.

$$
\begin{aligned}
2x + 3y - z &= 4 \\
x - 2y + 5z &= -1 \\
-x + y - 3z &= 0
\end{aligned}
\qquad
\begin{aligned}
x^2 + y^2 &= 25 \\
x^2 - 2y^2 &= -6
\end{aligned}
$$

Although a system of equations may consist of several equations in several variables, in this section we will limit our discussion to systems of two linear equations in two variables.

The Graphing Method

We know that the graph of a linear equation in two variables is a straight line and that each point of the line represents a solution of the equation. If the graphs of the two equations of a system intersect, then the point of intersection represents a solution of both equations simultaneously; that is, the point of intersection represents the solution of the system of equations.

EXPLORATION

The Graphing Method

Consider again the following system of equations:

$$x + y = 9$$
$$x - y = 3$$

(a) Solve each equation for y and use the standard window setting to produce the graphs of the equations.
(b) What is true about every point of the graph of $x + y = 9$? What is true about every point of the graph of $x - y = 3$?
(c) Use the *Intersect* feature to determine the coordinates of the point of intersection of the graphs. What is the significance of this point?
(d) By substitution, verify your conclusion in part (c).

DEVELOPING THE CONCEPT

The Graphing Method

Consider the following system of two linear equations in two variables:

$$x - 2y = 6$$
$$x + \ y = 3$$

FIGURE 5.1

$y_1 = -x + 3$ $\quad$ $y_2 = \frac{1}{2}x - 3$

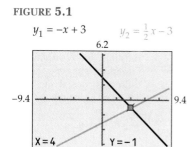

Solving each equation for y, we obtain $y = \frac{1}{2}x - 3$ and $y = -x + 3$. Figure 5.1 shows the graphs of the two equations in the same coordinate system.

Each point of the graph of $y = \frac{1}{2}x - 3$ has coordinates that satisfy that equation. Similarly, each point of the graph of $y = -x + 3$ has coordinates that satisfy that equation. Thus, the point of intersection of the two lines represents a solution of *both* equations.

In Figure 5.1, the coordinates of the point of intersection appear to be $(4, -1)$. We can verify by substitution that both equations of the system are true when $x = 4$ and $y = -1$.

For the preceding system of equations, we say that the ordered pair $(4, -1)$ is the **solution** of the system.

> **Definition of a Solution of a System of Equations in Two Variables**
>
> A **solution** of a system of equations in two variables is an ordered pair (x, y) that satisfies both equations of the system.

If both equations of a system are linear equations, then, as we have seen, the two lines may intersect at exactly one point, in which case the system has a unique solution. However, two other possibilities exist: The lines may coincide, or the lines may be parallel.

When the lines coincide, every point of the line is a point of intersection and each point represents a solution. This means that the system has **infinitely many solutions.** We say that the equations of such a system are **dependent.**

When the lines are parallel, there is no point of intersection, and so the system has no solution. The solution set is ∅. We say that such systems are **inconsistent.**

■ *NOTE Equations are dependent or independent; systems are consistent or inconsistent.*

If the lines do not coincide, then the equations of the system are **independent.** If the lines are not parallel, the system is **consistent,** which means that it has at least one solution.

Figure 5.2 summarizes the possible outcomes for a system of two linear equations in two variables.

FIGURE 5.2

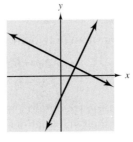
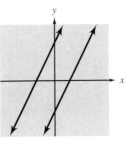

Solution is unique.
Equations are independent.
System is consistent.
The lines intersect.

(a)

No solution.
Equations are independent.
System is inconsistent.
The lines are parallel.

(b)

Infinitely many solutions.
Equations are dependent.
System is consistent.
The lines coincide.

(c)

The Substitution Method

Although the graphing method is useful in visualizing the solution(s), if any, of a system of equations, algebraic methods are needed to determine exact solutions. Consider the following system of equations:

$$x + y = 21$$
$$y = x - 3$$

Because the value of y must be the same in both equations, we can replace y in the first equation with the expression for y in the second equation:

$$x + y = x + (x - 3) = 21$$

We can now begin the process of solving the system by solving the resulting equation for x.

This method, known as the *substitution method*, is illustrated in Example 1.

EXAMPLE **1**   **Graphing and Substitution Methods**

Use the graphing method to estimate the solution of the system of equations. Then solve the system algebraically with the substitution method.

$$3y - 4x = 18$$
$$2x + y = -4$$

FIGURE 5.3

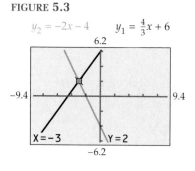

$y_2 = -2x - 4$ $y_1 = \frac{4}{3}x + 6$

6.2

-9.4   9.4

X = -3   Y = 2

-6.2

SOLUTION

VISUALIZING THE SOLUTION

To produce the graphs of the two equations, we solve each one for y.

$$3y - 4x = 18 \qquad\qquad 2x + y = -4$$
$$3y = 4x + 18 \qquad\qquad y = -2x - 4$$
$$y = \frac{4}{3}x + 6$$

Figure 5.3 shows the graphs of the equations and the point of intersection. The solution of the system appears to be $(-3, 2)$.

DETERMINING THE SOLUTION

To use the substitution method, we solve one of the equations for one of the variables. Because the second equation is easy to solve for y, we begin with that equation.

$$2x + y = -4 \qquad\text{The second equation}$$
$$y = -2x - 4 \qquad\text{Solve for } y.$$

Now we replace y in the first equation with $-2x - 4$.

$$\begin{aligned} 3y - 4x &= 18 & &\text{The first equation}\\ 3(-2x - 4) - 4x &= 18 & &\text{Replace } y \text{ with } -2x - 4.\\ -6x - 12 - 4x &= 18 & &\text{Solve the resulting equation for } x.\\ -10x - 12 &= 18 \\ -10x &= 30 \\ x &= -3 \end{aligned}$$

We now know that the solution has the form $(-3, y)$. To determine y, we replace x (in either equation) with -3.

$$y = -2x - 4 \qquad \text{The second equation}$$
$$y = -2(-3) - 4 \qquad \text{Replace } x \text{ with } -3.$$
$$y = 6 - 4$$
$$y = 2$$

As suggested by the graph, the solution is $(-3, 2)$.

The substitution method will also detect the special cases of inconsistent systems, which have no solution, and systems of dependent equations, which have infinitely many solutions.

EXAMPLE 2   Special Cases

Solve the following systems of equations.

(a) $6x = 4y + 20$ (b) $x - 2y = 6$
    $3x - 2y = 10$ $6y - 3x = 30$

SOLUTION

(a) **VISUALIZING THE SOLUTION**

To produce the graphs of the equations, we solve each one for y.

$$6x = 4y + 20 \qquad\qquad 3x - 2y = 10$$
$$6x - 20 = 4y \qquad\qquad -2y = -3x + 10$$
$$\frac{3}{2}x - 5 = y \qquad\qquad y = \frac{3}{2}x - 5$$

FIGURE 5.4

$$y_1 = \frac{3}{2}x - 5$$

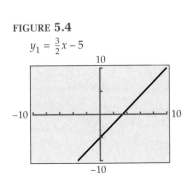

Observe that the two equations are identical and so, as Figure 5.4 shows, the lines coincide.

We conclude that the equations of the system are dependent and that the system has infinitely many solutions.

DETERMINING THE SOLUTION

$$6x = 4y + 20 \qquad \text{Solve the first equation for } x.$$
$$x = \frac{2}{3}y + \frac{10}{3}$$
$$3\left(\frac{2}{3}y + \frac{10}{3}\right) - 2y = 10 \qquad \text{Substitute the expression for } x \text{ in the second equation.}$$
$$2y + 10 - 2y = 10 \qquad \text{Distributive Property}$$
$$10 = 10 \qquad \text{True}$$

The resulting equation has no variables and is an identity. This indicates that the equations of the system are dependent, and so the system has infinitely many solutions.

Earlier, we found that solving either equation of the system for y resulted in $y = \frac{3}{2}x - 5$. Thus the coordinates of the points of the coinciding lines have the form $\left(x, \frac{3}{2}x - 5\right)$, and the solution set can be written $\left\{\left(x, \frac{3}{2}x - 5\right)\right\}$.

Alternatively, we can solve either equation for x to obtain $x = \frac{2}{3}y + \frac{10}{3}$, and we can write the solution set as $\left\{\left(\frac{2}{3}y + \frac{10}{3}, y\right)\right\}$. This is the format that we will use to describe the solution sets for systems of dependent equations.

Note that when the equations of a system are dependent, the system has infinitely many solutions. However, the solution set is *not* **R**, the set of all real numbers. Rather, the solution set consists of all ordered pairs represented by the points of the line.

FIGURE **5.5**

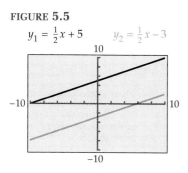

$y_1 = \frac{1}{2}x + 5$ $y_2 = \frac{1}{2}x - 3$

(b) VISUALIZING THE SOLUTION

Solving the equations for y, we obtain the following:

$$
\begin{array}{ll}
x - 2y = 6 & 6y - 3x = 30 \\
-2y = -x + 6 & 6y = 3x + 30 \\
y = \dfrac{1}{2}x - 3 & y = \dfrac{1}{2}x + 5
\end{array}
$$

Note that the slopes of the lines are the same, but the y-intercepts are different. Thus, as shown in Figure 5.5, the lines are parallel.

The system is inconsistent, and the solution set is ∅.

DETERMINING THE SOLUTION

$$
\begin{array}{ll}
x - 2y = 6 & \text{Solve the first equation for } x. \\
x = 2y + 6 & \\
6y - 3(2y + 6) = 30 & \text{Substitute the expression for } x \text{ in the second equation.} \\
6y - 6y - 18 = 30 & \text{Distributive Property} \\
-18 = 30 & \text{False}
\end{array}
$$

The resulting equation has no variables and is a contradiction. This indicates that the system is inconsistent. The solution set is ∅.

The Addition Method

With the substitution method, we eliminated a variable by making a substitution. With the addition method, we eliminate a variable by adding the equations.

Adding two equations results in the elimination of a variable when that variable has coefficients that are opposites. Sometimes we use the Multiplication Property of Equations to write an equivalent system so that the coefficients of one of the variables are opposites. (*Equivalent systems* are systems with the same solution set.)

EXAMPLE 3 **Solving a System with the Addition Method**

■ *NOTE For convenience, when we use the addition method to solve a system, we generally write the equations in standard form.*

Use the graphing method to estimate the solution of the system of equations. Then solve the system algebraically with the addition method.

$$3x + 2y = 8$$
$$-2x + y = -10$$

SOLUTION

VISUALIZING THE SOLUTION

Solving the equations for y, we obtain $y = -\frac{3}{2}x + 4$ and $y = 2x - 10$. From the graphs in Figure 5.6, we estimate the solution of the system to be $(4, -2)$.

FIGURE 5.6

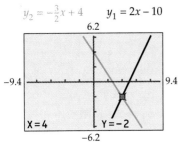

$y_2 = -\frac{3}{2}x + 4$ $y_1 = 2x - 10$

DETERMINING THE SOLUTION

To solve the system with the addition method, we must write an equivalent system in which the coefficients of one of the variables are opposites. To do this, we choose to adjust the second equation.

$3x + 2y = 8$	$\rightarrow$	$3x + 2y = 8$	The first equation
$-2(-2x + y) = -2(-10)$	$\rightarrow$	$4x - 2y = 20$	Multiply the second equation by -2 and add the equations.
		$7x = 28$	The coefficients of y are opposites, so y is eliminated.
		$x = 4$	Solve for x.

$$-2x + y = -10 \qquad \text{The second equation}$$
$$-2(4) + y = -10 \qquad \text{Replace } x \text{ with 4.}$$
$$-8 + y = -10$$
$$y = -2 \qquad \text{Solve for } y.$$

The solution is $(4, -2)$, which is consistent with the solution indicated by the graph.

In Example 3, we chose to eliminate y because, to do so, we had to multiply only one equation by a constant. If we had chosen to eliminate x, we would have had to multiply both equations by constants in order to make the coefficients of x opposites. Sometimes we need to adjust both equations regardless of the variable we choose to eliminate.

EXAMPLE 4 **Solving a System with the Addition Method**

Use the graphing method to estimate the solution of the system of equations. Then solve the system algebraically with the addition method.

$$3x - 5y = -10$$
$$2x + 3y = 6$$

FIGURE 5.7

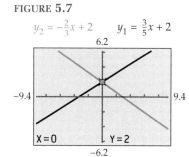

$y_2 = -\frac{2}{3}x + 2$ $y_1 = \frac{3}{5}x + 2$

SOLUTION

VISUALIZING THE SOLUTION

Figure 5.7 shows the graphs of $y = \frac{3}{5}x + 2$ and $y = -\frac{2}{3}x + 2$, which we obtain by solving the given equations for y.

The solution of the system appears to be $(0, 2)$. Note that we can be confident about this solution because we can see from the equations that the lines have the same y-intercept.

DETERMINING THE SOLUTION

To solve the system with the addition method, we choose to eliminate x. However, you can choose to eliminate either variable.

$2(3x - 5y) = 2(-10)$	$\rightarrow$ $6x - 10y = -20$	Multiply the first equation by 2.
$-3(2x + 3y) = -3(6)$	$\rightarrow$ $\underline{-6x - 9y = -18}$	Multiply the second equation by -3 and add the equations.
	$-19y = -38$	The coefficients of x are opposites, so x is eliminated.
	$y = \quad 2$	Solve for y.

We now substitute 2 for y in the second equation. Again, the substitution can be made in either equation.

$2x + 3y = 6$	**The second equation**
$2x + 3(2) = 6$	**Substitute 2 for y.**
$2x = 0$	**Solve for x.**
$x = 0$	

As the graph indicates, the solution is $(0, 2)$.

Like the substitution method, the addition method will detect systems for which the lines are parallel (inconsistent system) or coincide (dependent equations).

EXAMPLE 5   **Special Cases**

Use the addition method to solve each system of equations.

(a) $2x = y + 5$ (b) $12x - 4y = 8$
    $3y - 6x = 6$ $y - 3x = -2$

SOLUTION

(a) VISUALIZING THE SOLUTION

    After solving each equation for y, we produce the graphs shown in Figure 5.8.

    Note that the slopes of the lines are the same and that the y-intercepts are different. Thus the lines are parallel, and the system has no solution.

FIGURE 5.8

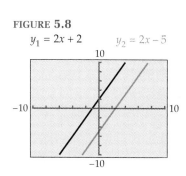

$y_1 = 2x + 2$ $y_2 = 2x - 5$

DETERMINING THE SOLUTION

We write both equations in standard form and then multiply both sides of the first equation by 3.

$$2x = y + 5 \quad \rightarrow \quad 2x - y = 5 \quad \rightarrow \quad 6x - 3y = 15$$
$$3y - 6x = 6 \quad \rightarrow \quad -6x + 3y = 6 \quad \rightarrow \quad \underline{-6x + 3y = 6}$$
$$0 = 21$$

Because the resulting equation is false, the system is inconsistent. The solution set is Ø.

(b) After writing the second equation in standard form, we multiply both sides by 4.

$$12x - 4y = 8 \quad \rightarrow \quad 12x - 4y = 8 \quad \rightarrow \quad 12x - 4y = 8$$
$$y - 3x = -2 \quad \rightarrow \quad -3x + y = -2 \quad \rightarrow \quad \underline{-12x + 4y = -8}$$
$$0 = 0$$

Because the resulting equation is true, the equations of the system are dependent, and their graphs coincide. Solving either equation for x, we obtain $x = \frac{1}{3}y + \frac{2}{3}$. Thus the solution set can be written as $\left(\frac{1}{3}y + \frac{2}{3}, y\right)$.

Applications and Modeling

As you analyze the information given in a problem, one of your first considerations should be the number of unknown quantities and the number of variables that you will use to represent them. Keep in mind that you must be able to write as many equations as you have variables.

EXAMPLE 6   Remodeling Costs

A builder has a contract to remodel two homes, one on Wood Drive and the other on Oak Street. Of the total cost of the Wood Drive project, 45% will go to labor and 55% will be for materials. Of the total cost of the Oak Street project, 30% will go to labor and 70% will be for materials. For the two projects combined, the total cost of labor is $30,000 and the total cost of materials is $42,000. What is the total cost of each project?

SOLUTION

Let x = the total cost of the Wood Drive project
y = the total cost of the Oak Street project

Then $0.45x$ = the cost of labor at Wood Drive
$0.30y$ = the cost of labor at Oak Street
$0.55x$ = the cost of materials at Wood Drive
$0.70y$ = the cost of materials at Oak Street

$0.45x + 0.30y = 30,000$	(1)	The total cost of labor is $30,000.
$0.55x + 0.70y = 42,000$	(2)	The total cost of materials is $42,000.

We solve this system of equations with the addition method.

$$3.15x + 2.10y = 210{,}000 \qquad \text{Multiply both sides of (1) by 7.}$$
$$\underline{-1.65x - 2.10y = -126{,}000} \qquad \text{Multiply both sides of (2) by } -3.$$
$$1.50x = 84{,}000 \qquad \text{Add the equations to eliminate } y.$$
$$x = 56{,}000$$

Now we replace x in equation (1) with 56,000 and solve for y.

$$0.45(56{,}000) + 0.30y = 30{,}000$$
$$25{,}200 + 0.30y = 30{,}000$$
$$0.30y = 4{,}800$$
$$y = 16{,}000$$

The total cost of the Wood Drive project is $56,000, and the total cost of the Oak Street project is $16,000.

5.1 Quick Reference

Systems of Equations
- Two or more equations considered simultaneously form a **system of equations.**
- A **solution** of a system of equations in two variables is an ordered pair (x, y) that satisfies both equations of the system.

The Graphing Method
- A point of intersection of two lines represents a solution of a system of linear equations in two variables.
  1. If the lines intersect at one point, the solution of the system is unique.
  2. If the lines are parallel, the system is **inconsistent** and has no solution.
  3. If the lines coincide, the equations are **dependent,** and the system has infinitely many solutions.
- If the lines do not coincide, then the equations of the system are **independent.** If the lines are not parallel, the system is **consistent** and has at least one solution.

The Substitution Method
- To solve a system with the *substitution method*, we can solve either equation for either variable and substitute the resulting expression into the other equation.
- If both variables are eliminated when the substitution method is used, then
  1. if the resulting equation is an identity (true), the system has infinitely many solutions.
  2. if the resulting equation is a contradiction (false), the system has no solution.

The Addition Method
- The *addition method* involves adding equations to eliminate a variable whose coefficients are opposites.
- Multiplying one or both equations of the system may be necessary in order to obtain a system in which the coefficients of one of the variables are opposites.

■ The special cases of inconsistent systems and dependent equations are detected by the addition method in the same way that they are detected by the substitution method.

5.1 Speaking the Language

1. Two or more equations considered simultaneously form a(n) ▭.
2. A system of equations that has no solution is called a(n) ▭ system.
3. Two algebraic methods for solving a system of linear equations are the ▭ method and the ▭ method.
4. If we add two equations in order to eliminate a variable, we will succeed only if the ▭ of that variable are opposites.

5.1 Exercises

Concepts and Skills

1. How can you tell that the solution of the following system is $(0, -3)$ without using graphing or algebraic methods?

$$y = 5.7x - 3$$
$$y = -0.4x - 3$$

2. After you have solved a system of equations, you should check your result. In how many of the equations of the system should you check the solution? Why?

In Exercises 3–8, match the system of equations to the graph and use the graph to estimate the solution of the system. Verify your estimate by substitution.

3. $y + 10 = 2x$
   $2y + 3x = 8$

4. $2y - 3x - 14 = 0$
   $y + 3 = 1.5x$

5. $5y + 3x = 5$
   $0.6x + y = 1$

6. $x + y + 2 = 0$
   $5y = 2 - x$

7. $2x - y + 10 = 0$
   $x + 8 = 2$

8. $2y = 6$
   $x + y = 5$

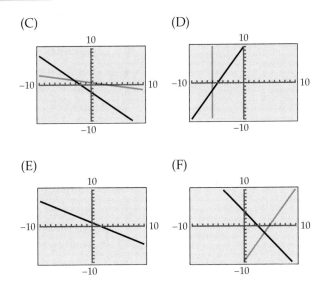

(C)   (D)

(E)   (F)

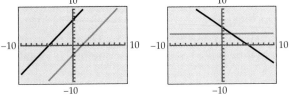

(A)   (B)

In Exercises 9–16, estimate the solution by graphing. Then use the substitution method to solve the system. For special cases, write the solution set.

9. $y = 4x - 13$
   $3x = 29 - 2y$

10. $y = \dfrac{1}{3}x - 2$
    $x + 3y = 6$

11. $3 - x = 0$
    $3x + 5y = -1$

12. $y + x - 3 = 0$
    $2y = 10$

13. $2x + 3y + 8 = 0$
    $y + 2x = 16$

14. $x + 3y = 11$
    $2x - 4y = 2$

15. $y = -2x - 5$
    $y + 2x = 6$

16. $2x + 3y = 5$
    $6y = -4x + 10$

17. Suppose that you are solving a system of two linear equations in two variables. What is the maximum finite number of solutions that the system can have? Why?

18. What is the possible number of solutions of a consistent system of equations?

In Exercises 19–30, use the addition method to solve the given system of equations. For special cases, write the solution set.

19. $x - 3y = 9$
  $2x + y = 11$

20. $5x + 2y = -12$
  $4x + y = -9$

21. $4x + 3y + 2 = 0$
  $2x = y - 1$

22. $x = -3 + 6y$
  $5x - 9y = -1$

23. $2x - 3y + 5 = 0$
  $6y - 4x = 10$

24. $15y - 3x = 3$
  $x - 5y = -1$

25. $6x + 5y = 11$
  $4x + 3y = 5$

26. $7x - 4y = 34$
  $3x + 10y = -44$

27. $5y = 2(3 - x)$
  $5y + 2x - 15 = 0$

28. $3x = 7y + 22$
  $40 + 2x = -9y$

29. $\frac{2}{3}x - \frac{1}{2}y = \frac{3}{2}$
  $\frac{1}{5}x + y = \frac{8}{5}$

30. $1.5y = x + 0.75$

  $2y - \frac{4}{3}x = 0$

In Exercises 31–40, use any method to solve the system of equations. For special cases, write the solution set. Classify the systems as consistent or inconsistent, and classify the equations as dependent or independent.

31. $x + 2y = 1$
  $y + 7 = 0$

32. $5 - x = 0$
  $3x - y = 4$

33. $7y = 2(4 - 3x)$
  $3x + 3.5y = -1$

34. $5y = 2(x + 5)$
  $5(y - 2) = 2x$

35. $3x - 2y = 1$
  $x + 8y = -17$

36. $4x = 9(2 - y)$
  $12 = 6y + 5x$

37. $x = 2y + 6$
  $x + y = -3$

38. $x - 2y = 12$
  $y = 1 - 3x$

39. $x + 0.6y = 3.8$
  $0.3y + 0.5x = 1.9$

40. $\frac{1}{2}x - y = -5$

  $\frac{5}{6}x - \frac{1}{12}y = \frac{7}{6}$

41. The origin represents a solution of both of the following systems:

(i) $3x - 2y = 0$
  $2x - 3y = 0$

(ii) $3x - 2y = 0$
  $-3x + 2y = 0$

In what respect do the two systems differ?

42. Consider the following system:

$$4x + 5y = 11$$
$$x = -3$$

Can this system be solved with the substitution method? with the addition method?

Concept Extension

In Exercises 43 and 44, determine a and b so that the system has the given solution. (*Hint*: A solution of a system of equations satisfies both equations.)

43. $ax - by = 2$
  $bx - (a - b)y = -1$
  $(7, 11)$

44. $ax + 3y = 3b$
  $bx + 5y = -a$
  $(-3, -4)$

Sometimes systems of nonlinear equations can be solved with techniques for linear equations by using an appropriate substitution. In Exercises 45 and 46, replace $\frac{1}{x}$ with u and $\frac{1}{y}$ with v. Solve the resulting system for u and v. Then use those values to determine x and y.

45. $\frac{4}{x} - \frac{1}{y} = 7$

  $-\frac{2}{x} + \frac{3}{y} = 9$

46. $\frac{4}{x} - \frac{1}{y} = 1$

  $\frac{2}{x} + \frac{5}{y} = 6$

In Exercises 47 and 48, solve the given system of three linear equations in two variables.

47. $5x + 3y = -7$
  $x - 2y = -4$
  $-3x - y = 5$

48. $x - 6y = 7$
  $-2x + 3y = -14$
  $-(x + 1) + y = -8$

49. Determine the value of k so that the equations are dependent.

$$2x - y = 5$$
$$kx = 3y + 15$$

50. Determine the value of k so that the system is inconsistent.

$$6x + ky = 1$$
$$y = 3x - 2$$

Applications

51. Shoe Sale   A shoe store advertised one pair of a certain athletic shoe at the regular price of $80, with a coupon for a second pair at $20 off the regular price. The store sold 50 pairs of shoes for a total of $3640. How many coupons were redeemed?

52. Party Admissions The admission to a community centennial party was $15 per person or $25 per couple. If 116 people attended the party and the total receipts were $1525, how many couples attended?

In Exercises 53 and 54, recall that if a bottle contains 40 ounces of a solution that includes 8 ounces of acid, then $\frac{8}{40}$ or 20% of the solution is acid. We say that the bottle contains a 20% acid solution. Also, if a bottle contains x ounces of a 20% acid solution, then the amount of acid in the bottle is $0.20x$.

53. Antiseptic Solution A hospital pharmacist needs to mix a 20% antiseptic solution and a 60% antiseptic solution to obtain 50 milliliters of a 36% antiseptic solution. How many milliliters of each solution are required?

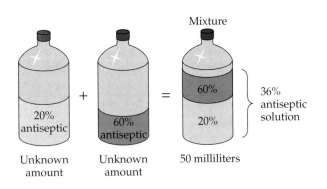

54. Etching Solution A metal etching plant has mixed a 30% acid solution with a 45% acid solution to obtain 60 gallons of a 35% solution. How many gallons of each solution were used?

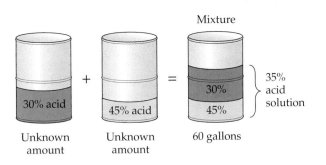

55. Auto Rental An auto rental company charges a daily rate plus mileage. A one-day trip of 235 miles costs $39.45, and a one-day trip of 148 miles costs $33.36. What is the daily rate and the mileage rate?

56. Fast Food Prices The cost of three hamburgers and two orders of fries is $10. One hamburger costs 40 cents more than two orders of fries. What is the price of an order of fries?

57. Computer Disk Packaging CompuDisk uses cartons of two different sizes to ship small and large boxes of floppy disks. The smaller carton, whose volume is 576 cubic inches, holds 10 small boxes of disks and 6 large boxes. The larger carton, whose volume is 896 cubic inches, holds 14 small boxes of disks and 10 large boxes. What are the volumes of the boxes of disks?

58. Fruit Boxes Pittman & Davis is a citrus fruit distributor that offers holiday boxes of oranges and grapefruits. The half-bushel box weighs 18 pounds and contains 16 oranges and 9 grapefruits. The one-bushel box weighs 35.125 pounds and contains 34 oranges and 16 grapefruits. What is the average weight (in ounces) of an orange and of a grapefruit?

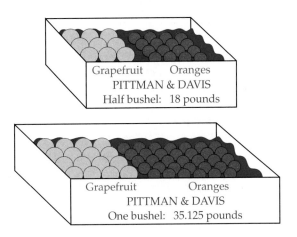

59. Hourly Wages A heating and air-conditioning company employs electricians and sheet metal workers to install industrial equipment. The company's payroll cost for an 8-hour day is $1152 for 1 electrician and 6 metal workers. If 2 additional electricians and 1 fewer metal worker are on the job, then the company's daily payroll cost is increased by $198. What are the hourly wages of the electricians and the metal workers?

60. Restaurant Payroll   A fast food restaurant pays its workers $6.25 per hour, and it pays its shift managers $11.50 per hour. The 8-hour day shift and the 8-hour evening shift have the same number of workers and managers, and each shift has 1 manager for every 4 workers. If the total payroll cost for a 16-hour business day is $1168, how many workers and managers work each shift?

61. Course Enrollment   At a college bookstore, textbooks for Mathematics 101 cost $54 and textbooks for English 109 cost $58. In the fall semester, there were 30 more students enrolled in Mathematics 101 than in English 109. If the total sales of books for these two courses were $11,700, how many students were enrolled in each course?

62. Christmas Tree Sales   A nursery purchased 115 Fraser firs to sell as Christmas trees. The 9-foot trees sold for $70, and the 6-foot trees sold for $45. If the nursery sold all the trees for a total revenue of $5925, how many of each type of tree were sold?

63. Photograph Frames   PrintPlus advertised decorative photograph frames for $11 each or two frames for $20. If PrintPlus sold a total of 121 frames for a total revenue of $1259, how many customers took advantage of the two-for-$20 offer?

64. Freelance Writer   A freelance writer contracted to produce a solutions manual for a mathematics textbook. She received $8 for each page that contained only text and $11 for each page that contained graphics. Suppose that her earnings for a 552-page manual were $5127. How many pages in the manual contained graphics?

65. Bird Seed   A bird lover, knowing that large birds prefer sunflower seed, wants to add sunflower seed to a commercial mix that contains only 15% sunflower seed. If the resulting 50-pound mixture contained 32% sunflower seed, how many pounds of sunflower seed were added to the commercial mix?

66. Fat Content   One cup of Cheerios contains 1 gram of fat, and one cup of Oat Squares contains 2.5 grams of fat. How much of each cereal would be needed to obtain one cup of a mixed cereal with 1.6 grams of fat?

67. Heat Loss   The length of a bedroom is 3 feet greater than its width, and its ceiling is 9 feet high. The longer wall is insulated with R-25 fiberglass insulation, which loses 6 Btu (British thermal units) of heat per square foot. The shorter wall is insulated with R-19 insulation, which loses 9 Btu per square foot. Neither wall has a window or door. If the total heat loss for the two walls is 1782 Btu, what are the dimensions of the bedroom?

68. Fencing Cost   The covenants for a lakeside development permit only split-rail fencing along the front of residential properties. Any type of fencing is permitted along the other sides of the properties. A homeowner decided to fence the front and the two sides of his rectangular lot, which has a perimeter of 720 feet. The split-rail fencing cost $18 per foot, and the chain-link fencing for the sides cost $10 per foot. If the total cost of the fencing was $6880, what are the dimensions of the lot?

69. Cycling   A cyclist leaves her home at noon and heads down the highway at an average speed of 20 mph. At 1:12 P.M., her brother notices that she has forgotten her wallet and drives at an average speed of 50 mph to catch her. How far from home are the two people when they meet?

70. Walking and Running   At 8:30 A.M., a student left his dormitory for his 9:00 class. He ran at an average speed of 7 mph for part of the trip and then walked the rest of the way at an average speed of 4 mph. If he arrived right on time and the distance from the dormitory to the classroom building was 2.5 miles, how far did he walk?

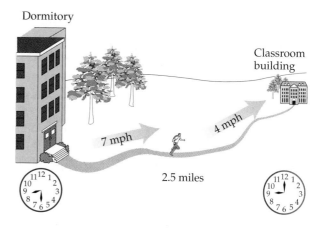

71. **Foundation Award** A foundation presented its Excellence in Education award of $67,500 to the 51 faculty and staff members of an elementary school. Each faculty member received $1500, and each staff member received $500. How many faculty members and how many staff members are employed at the school?

72. **Faculty Credentials** The 40 faculty members at a high school have either bachelor's degrees, which represent 4 years of higher education, or master's degrees, which represent 6 years of higher education. If the faculty has a total of 194 years of higher education, how many faculty members have master's degrees?

73. **Electrical Wiring** An electrician and an apprentice can complete $\frac{7}{24}$ of a residential wiring job in one hour. If a second apprentice is added to the crew, then $\frac{10}{24}$ of the job can be completed in one hour. How long would the electrician take to complete the job if he had no help?

74. **Catering** The time is 6:00 P.M., and a caterer needs to set up tables and chairs for a dinner meeting. If she is helped only by her young son, the job will be only $\frac{3}{4}$ done by her 7:00 P.M. deadline. However, if her daughter also helps, the job can be completed just in time. If the son and daughter work at the same rate, how long would the caterer have needed to do the job by herself?

75. **SAT Scores** Since 1982, SAT scores in mathematics have increased, whereas verbal SAT scores have remained relatively constant. The following table shows data for two selected years. (*Source*: College Board.)

Year	Verbal	Math
1982	504	493
1997	505	511

(a) Assuming that the trends were linear, how can you tell from the table that the verbal and math scores were equal for some year between 1982 and 1997?

(b) Letting x represent the number of years since 1980, produce linear regression equations to model the verbal and math scores.

(c) Solve the system of equations in part (b) and interpret the solution.

76. **College Athletes** The following table compares the numbers of male and female athletes in college sports in 1982 and 1995. (*Source*: National Collegiate Athletic Association.)

Year	Male	Female
1982	180,200	80,000
1995	188,400	116,300

(a) Letting x represent the number of years since 1980, produce linear regression equations to model the data for male and female athletes. Round all constants to whole numbers.

(b) Although the numbers of female athletes are considerably lower than the numbers of male athletes, how can you tell from the regression equations that the graphs will eventually intersect?

(c) In what year do the model equations project that the numbers of male and female athletes will be equal?

5.2 *Systems of Equations in Three Variables*

Definitions and Graphs ■ *The Row-Echelon Form* ■
Special Cases ■ *Applications and Modeling*

Just as certain application problems can be solved with two variables and a system of two equations, other problems may be solved most conveniently with three variables. However, this convenience is somewhat offset by the fact that a system of three equations is needed to model the conditions of the problem.

In this section, the emphasis is on algebraic methods for solving systems of three equations in three variables. Once again, your knowledge and skill with these methods will extend your capabilities so that you can solve a wider range of real-life application problems.

Definitions and Graphs

Linear equations can be defined for more than two variables.

> ### Definition of a Linear Equation in Three Variables
>
> A **linear equation in three variables** is an equation that can be written in the *standard form* $Ax + By + Cz = D$, where A, B, C, and D are real numbers and A, B, and C are not all zero.

Definitions for linear equations in more than three variables are similar.

A **solution** of a linear equation in three variables is an **ordered triple** (x, y, z) of numbers that satisfy the equation. For example, we can verify that the ordered triple $(3, -1, 2)$ is a solution of $2x + 3y - z = 1$ by substitution.

$$2x + 3y - z = 1$$
$$2(3) + 3(-1) - (2) = 1 \qquad \text{Replace } x \text{ with 3, } y \text{ with } -1, \text{ and } z \text{ with 2.}$$
$$6 - 3 - 2 = 1$$
$$1 = 1 \qquad \text{True}$$

Just as an ordered pair (x, y) is associated with a point in a two-dimensional coordinate system, an ordered triple is associated with a point in a three-dimensional coordinate system called a **coordinate space**. Figure 5.9 shows a coordinate space with three perpendicular axes. Figure 5.10 shows the point $P(4, 2, 3)$ plotted in a coordinate space.

We have seen that the graph of a linear equation in two variables is a line drawn in a two-dimensional coordinate system. The graph of a linear equation in three variables is a *plane* drawn in a coordinate space. Figure 5.11

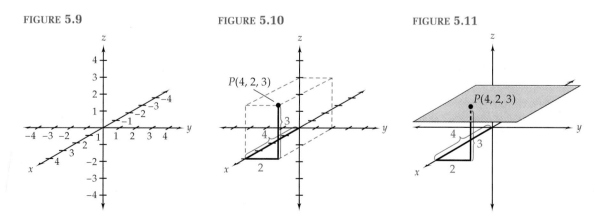

FIGURE **5.9**   FIGURE **5.10**   FIGURE **5.11**

shows one of the many planes that contains the point $P(4, 2, 3)$. Of necessity, planes are drawn as enclosed figures, but like lines, planes extend without bound.

The solution set of a system of three linear equations in three variables is represented by the intersection of the graphs of the three equations. Any point belonging to the intersection of all three planes corresponds to a solution of the system. There are three possible results.

First, the solution could be unique, in which case the planes intersect at exactly one point P. (See Figure 5.12.)

FIGURE 5.12

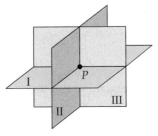

Second, the system could have no solution. This results when at least two of the planes are parallel (see Figure 5.13) or when one plane is parallel to the line L of intersection of the other two planes (see Figure 5.14). Such systems have no solution because in neither case is there a point that is common to all three planes.

FIGURE 5.13   **FIGURE 5.14**

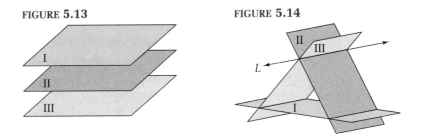

Finally, the system could have infinitely many solutions. This can occur when the planes coincide (see Figure 5.15) or when the planes intersect in a common line L (see Figure 5.16).

FIGURE 5.15   **FIGURE 5.16**

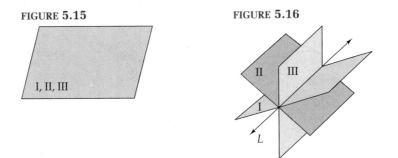

Visualizing the ways in which planes can be oriented provides a geometric interpretation of a system and its solutions. However, actually sketching the planes is frequently difficult. Thus we will emphasize algebraic methods for solving systems of equations.

The Row-Echelon Form

The following systems of equations are written in a special form:

$$
\begin{aligned}
x - y &= 1 \\
y &= 4
\end{aligned}
\qquad
\begin{aligned}
x + y - 2z &= 4 \\
y + z &= 2 \\
z &= -1
\end{aligned}
$$

Observe that each row contains one fewer variable than the previous row and that the coefficient of the first variable in each row is 1. Because of their visual appearance, we say that such systems are in *triangular form*. Another name, the one that we will use in this discussion, is *row-echelon form*.

When a system of equations is in the row-echelon form, the system can be solved from the bottom up by substituting the known values in one row into the equation in the row above it.

To write a system of equations in the row-echelon form, you will usually need to perform certain operations on the equations. The following summary lists the operations that will result in an equivalent system.

> **Allowable Operations on the Equations of a System**
>
> The following operations on the equations of a system will result in an equivalent system.
>
> 1. Any two equations can be interchanged.
> 2. Both sides of any equation can be multiplied by a nonzero constant.
> 3. Any equation can be replaced by the sum of itself and a multiple of another equation.

The following examples illustrate performing these operations to write a system in row-echelon form.

EXAMPLE 1   **Solving a System in Row-Echelon Form**

Write the following system of equations in row-echelon form. Then solve the system.

$$
\begin{aligned}
x + y - 2z &= 4 \quad (1) \\
2x + 3y - 3z &= 10 \quad (2) \\
3x + 4y - 4z &= 13 \quad (3)
\end{aligned}
$$

SOLUTION

Use equation (1) to eliminate x in equations (2) and (3).

$$x + y - 2z = 4 \quad (1)$$

$$y + z = 2 \quad (4)$$ **Multiply equation (1) by −2 and add the result to equation (2).**

$$y + 2z = 1 \quad (5)$$ **Multiply equation (1) by −3 and add the result to equation (3).**

Next, use equation (4) to eliminate y in equation (5).

$$x + y - 2z = 4 \quad (1)$$

$$y + z = 2 \quad (4)$$

$$z = -1 \quad (6)$$ **Multiply equation (4) by −1 and add the result to equation (5).**

Note that the system of equations is now in row-echelon form. From equation (6), we know that the value of z is -1. To determine y, substitute the value of z into equation (4).

$$y + z = 2 \quad (4)$$

$$y + (-1) = 2$$ **Replace z with -1.**

$$y = 3$$ **Solve for y.**

Now substitute the values of y and z into equation 1 to determine x.

$$x + y - 2z = 4 \quad (1)$$

$$x + 3 - 2(-1) = 4$$ **Replace y with 3 and z with -1.**

$$x = -1$$ **Solve for x.**

The solution is $(-1, 3, -1)$.

In Example 1, the technique of adding multiples of one equation to another equation not only eliminated variables but also resulted in equations with leading coefficients of 1. Example 2 shows the steps we need to take when the leading coefficients are not 1.

EXAMPLE 2   **Solving a System in Row-Echelon Form**

Write the following system of equations in row-echelon form. Then solve the system.

$$x + 2y - z = -12 \quad (1)$$

$$x + z = 6 \quad (2)$$

$$2y - z = -14 \quad (3)$$

SOLUTION

Use equation (1) to eliminate x in equation (2).

$$x + 2y - z = -12 \quad (1)$$

$$-2y + 2z = 18 \quad (4)$$ **Multiply equation (1) by −1 and add the result to equation (2).**

$$2y - z = -14 \quad (3)$$

Use equation (4) to eliminate y in equation (3).

$$
\begin{aligned}
x + 2y - z &= -12 \quad (1)\\
-2y + 2z &= 18 \quad (4)\\
z &= 4 \quad (5)
\end{aligned}
$$

Add equations (4) and (3).

To obtain a leading coefficient of 1 in equation (4), divide both sides by -2.

$$
\begin{aligned}
x + 2y - z &= -12 \quad (1)\\
y - z &= -9 \quad (6)\\
z &= 4 \quad (5)
\end{aligned}
$$

The system of equations is now in row-echelon form. Substituting from the bottom of the system to the top, we obtain the solution $(2, -5, 4)$.

Special Cases

Recall that the graph of a linear equation in three variables is a plane. A system of three linear equations in three variables has a unique solution if the three planes have a single point in common. In all other situations, the system either has infinitely many solutions or has no solution.

EXAMPLE 3   **A System with Infinitely Many Solutions**

Solve the following system of equations:

$$
\begin{aligned}
2x + y + 3z &= 2 \quad (1)\\
x + 3y - z &= 1 \quad (2)\\
-x - 4y + 2z &= -1 \quad (3)
\end{aligned}
$$

SOLUTION

$$
\begin{aligned}
x + 3y - z &= 1 \quad (2)\\
2x + y + 3z &= 2 \quad (1)\\
-x - 4y + 2z &= -1 \quad (3)
\end{aligned}
$$

Exchange equations (1) and (2) so that the leading coefficient of x is 1.

We now eliminate x from equations (1) and (3).

$$
\begin{aligned}
x + 3y - z &= 1 \quad (2)\\
-5y + 5z &= 0 \quad (4)\\
-y + z &= 0 \quad (5)
\end{aligned}
$$

Multiply equation (2) by -2 and add the result to equation (1).

Add equations (2) and (3).

$$
\begin{aligned}
x + 3y - z &= 1 \quad (2)\\
y - z &= 0 \quad (6)\\
0 &= 0 \quad (7)
\end{aligned}
$$

Divide both sides of equation (4) by -5.

Add equations (5) and (6).

The true equation $0 = 0$ indicates that the system has infinitely many solutions. To write the solution set, we express x and y in terms of z.

$$y - z = 0 \qquad (6)$$
$$y = z \qquad \text{Write } y \text{ in terms of } z.$$
$$x + 3y - z = 1 \qquad (2)$$
$$x + 3z - z = 1 \qquad \text{Replace } y \text{ with } z.$$
$$x = 1 - 2z \qquad \text{Write } x \text{ in terms of } z.$$

■ *NOTE In Example 3, we describe the solutions as ordered triples $(1 - 2z, z, z)$, where z is any real number. We also could have chosen to describe the solutions in terms of x or y, but the row-echelon form makes describing the solutions in terms of z convenient.*

Now we can write the coordinates of all the ordered triples in terms of z. The solution set is $\{(1 - 2z, z, z)\}$.

EXAMPLE **4** **A System with No Solution**

Solve the following system of equations:

$$\begin{array}{rcl} x + y - 2z &=& 0 \quad (1) \\ 2x + y - z &=& 1 \quad (2) \\ 3x + 5y - 12z &=& -4 \quad (3) \end{array}$$

SOLUTION

$$x + y - 2z = 0 \quad (1)$$
$$-y + 3z = 1 \quad (4) \qquad \text{Multiply equation (1) by } -2 \text{ and add the result to equation (2).}$$
$$2y - 6z = -4 \quad (5) \qquad \text{Multiply equation (1) by } -3 \text{ and add the result to equation (3).}$$

$$x + y - 2z = 0 \quad (1)$$
$$y - 3z = -1 \quad (6) \qquad \text{Multiply equation (4) by } -1.$$
$$0 = -2 \quad (7) \qquad \text{Multiply equation (4) by 2 and add the result to equation (5).}$$

The false equation $0 = -2$ indicates that the system has no solution.

Applications and Modeling

If three noncollinear points are given, a curve that contains the points is the graph of a quadratic function $f(x) = ax^2 + bx + c$. We can determine the function with the methods discussed in the section on quadratic regression models. Alternatively, we can write a system of equations in a, b, and c, and use the solution of the system to write the function. Example 5 illustrates this method along with the regression method.

EXAMPLE **5** **Using a System to Write a Quadratic Function**

Given the points $(1, 3)$, $(-2, -9)$, and $(3, 21)$, determine the values of a, b, and c so that the graph of $f(x) = ax^2 + bx + c$ contains the points. Support your result by producing a quadratic regression equation with your calculator.

SOLUTION

$$f(x) = ax^2 + bx + c$$
$$f(1) = a(1)^2 + b(1) + c = 3 \qquad \text{Because (1, 3) is a point, } f(1) = 3.$$
$$f(-2) = a(-2)^2 + b(-2) + c = -9 \qquad \text{Because (-2, -9) is a point, } f(-2) = -9.$$
$$f(3) = a(3)^2 + b(3) + c = 21 \qquad \text{Because (3, 21) is a point, } f(3) = 21.$$

Now we simplify each equation and write a system of three equations in a, b, and c.

FIGURE **5.17**

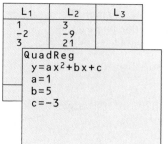

$$
\begin{array}{rl}
a + b + c = 3 & (1) \\
4a - 2b + c = -9 & (2) \\
9a + 3b + c = 21 & (3)
\end{array}
$$

$$
\begin{array}{rl}
a + b + c = 3 & (1) \\
-6b - 3c = -21 & (4) \qquad \text{Multiply equation (1) by } -4 \text{ and add the result to equation (2).} \\
-6b - 8c = -6 & (5) \qquad \text{Multiply equation (1) by } -9 \text{ and add the result to equation (3).}
\end{array}
$$

$$
\begin{array}{rl}
a + b + c = 3 & (1) \\
-6b - 3c = -21 & (4) \\
-5c = 15 & (6) \qquad \text{Multiply equation (4) by } -1 \text{ and add the result to equation (5).}
\end{array}
$$

FIGURE **5.18**

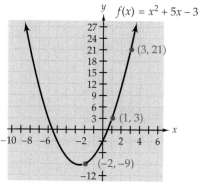

From equation (6), we see that $c = -3$. Substituting back through the system, we find that $b = 5$ and $a = 1$. Thus the required quadratic function is $f(x) = x^2 + 5x - 3$. Figure 5.17 shows that the graph of f contains the given points.

To use the regression method, we enter the given ordered pairs in the statistics list and obtain the quadratic regression equation shown in Figure 5.18. Note that the reported coefficients are the same as those that we derived algebraically.

When a problem involves three unknown quantities, we may be able to use a system of three equations in three variables to describe the conditions of the problem.

EXAMPLE 6   L.L. Bean Shipping Charges

The following table shows L.L. Bean's shipping charges for mail orders. (*Source*: L.L. Bean.)

Order	Shipping charges
$30 or less	$3.95
$30.01–$70.00	4.95
Over $70	5.95

For 640 orders, the shipping charges totaled $3208. If the number of orders of $30.01–$70.00 was the same as the total number of orders in the other two categories, how many orders did L.L. Bean receive in each category?

SOLUTION

Let x = number of orders of \$30 or less
 $\quad y$ = number of orders of \$30.01–\$70.00
 $\quad z$ = number of orders over \$70

Then $3.95x$ = total shipping charges for orders of \$30 or less
 $\quad 4.95y$ = total shipping charges for orders of \$30.01–\$70.00
 $\quad 5.95z$ = total shipping charges for orders over \$70

Using these representations, we write a system of three equations in x, y, and z.

$x + y + z = 640$ There were 640 orders in all.

$y = x + z$ The number of orders of \$30.01–\$70.00 was the same as the total number of orders in the other two categories.

$3.95x + 4.95y + 5.95z = 3208$ The shipping charges totaled \$3208.

$$
\begin{aligned}
x + \quad y + \quad z &= \ 640 \quad (1) \\
x - \quad y + \quad z &= \quad 0 \quad (2) \\
3.95x + 4.95y + 5.95z &= 3208 \quad (3)
\end{aligned}
$$
    Write the equations in standard form.

$$
\begin{aligned}
x + \quad y + \ z &= \ 640 \quad (1) \\
-2y \quad\quad &= -640 \quad (4) \\
y + 2z &= \ 680 \quad (5)
\end{aligned}
$$

(4) Multiply equation (1) by -1 and add the result to equation (2).

(5) Multiply equation (1) by -3.95 and add the result to equation (3).

From equation (4), we see that $y = 320$. By substitution, we can determine that $z = 180$ and $x = 140$.

L.L. Bean received 140 orders of \$30 or less, 320 orders between \$30.01 and \$70.00, and 180 orders for more than \$70.

5.2 Quick Reference

Definitions and Graphs   ■ A **linear equation in three variables** is an equation that can be written in the *standard form* $Ax + By + Cz = D$, where A, B, C, and D are real numbers and A, B, and C are not all zero.

■ A **solution** of a linear equation in three variables is an **ordered triple** (x, y, z) of numbers that satisfy the equation.

■ Ordered triples are associated with points in a three-dimensional coordinate system called a **coordinate space.**

■ The graph of a linear equation in three variables is a *plane* drawn in a coordinate space.

■ The solution set of a system of three linear equations in three variables is represented by the intersection of the graphs of the equations.

1. If the planes intersect at exactly one point, then the solution is unique.
2. If at least two of the planes are parallel, or if one plane is parallel to the line of intersection of the other two planes, then the system has no solution.
3. If the planes coincide, or if the planes intersect in a common line, then the system has infinitely many solutions.

The Row-Echelon Form   ■ The following is an example of a system of equations written in the *row-echelon form* (or *triangular form*).

$$x + 2y - z = 3$$
$$y + z = 2$$
$$z = 5$$

■ The following operations on the equations of a system will result in an equivalent system.

1. Any two equations can be interchanged.
2. Both sides of any equation can be multiplied by a nonzero constant.
3. Any equation can be replaced by the sum of itself and a multiple of another equation.

■ When a system of equations is in the row-echelon form, the system can be solved from the bottom up by substituting the known values in one row into the equation in the row above it.

Special Cases   ■ When you solve a system of three linear equations in three variables, you may find that all the variables are eliminated in at least one of the equations.

1. If the equation is false, then the system has no solution.
2. If the equation is true, then the system has infinitely many solutions.

5.2 Speaking the Language

1. The coordinate system used to plot points that represent ordered triples is called a coordinate ▓▓▓▓▓ .
2. The following system is said to be in triangular or ▓▓▓▓▓ form.

$$x + 3y - z = 4$$
$$y + z = 0$$
$$z = 2$$

3. The graph of a(n) ▓▓▓▓▓ is a plane.
4. A solution of a linear equation in three variables is called a(n) ▓▓▓▓▓ .

5.2 Exercises

Concepts and Skills

1. The following system is not strictly in row-echelon form because the leading coefficient in the second equation is not 1.

$$x + y - z = 3$$
$$2y + z = 1$$
$$z = 5$$

Must we divide both sides of the second equation by 2 before we can solve the system? Why?

2. If we wish to write the following system in the row-echelon form, what is an advantage of exchanging equations (1) and (3)?

$$2x - y + 3z = 4 \quad (1)$$
$$5x + 2y - z = 1 \quad (2)$$
$$x - 3y + 2z = -2 \quad (3)$$

In Exercises 3–12, write the given system of equations in row-echelon form. Then solve the system.

3. $2x + y - 3z = 5$
   $y + 2z = 15$
   $2x - z = 6$

4. $5x + 2y + 4z = -3$
   $4x + z = 1$
   $3y - z = 9$

5. $4x - y - z = 3$
   $2x - 2y + 3z = -5$
   $x + y + z = 2$

6. $4x + 3y + 2z = 1$
   $8x + 9y - 2z = -6$
   $12x + 9y + 4z = 0$

7. $3x + y + z = 5$
   $x - 2y - 5z = -3$
   $5x - 3y - 9z = -1$

8. $x - z = 1$
   $x + 2y + z = 3$
   $x - 2y - 3z = 0$

9. $x - y - z = -1$
   $x + y + z = 3$
   $2x + y - z = 6$

10. $x + 3y - 2z = 0$
    $3x + y + 6z = 4$
    $2x + 2y + z = 3$

11. $5x + z = 5 + 4y$
    $2z + x + 4 = y$
    $5y - 13z = 8x + 29$

12. $y - 9 = z - 3x$
    $4z + 12 = y$
    $x - 5y = 2$

13. For a system of three linear equations in three variables, if the graphs of the equations are parallel planes, what can we conclude about the solution set of the system?

14. Explain why the following system of equations cannot have a unique solution.

$$x - 2y + 3z = 5$$
$$2x + y - z = -3$$

In Exercises 15–22, solve the given system of equations.

15. $y - 3z = 1$
    $x - 2z = 5$
    $x + y = 1$

16. $x - y - z = -1$
    $x + 2y - z = 11$
    $2x + y + 4z = -2$

17. $2x - y + 3z = 4$
    $2x + 7y - 11z = 0$
    $4x + 2y - z = 5$

18. $2x + 6y - 3z = 11$
    $x + 4y + z = 6$
    $x - 9z = 10$

19. $x - z = 2$
    $3x + 4y + 3z = 10$
    $2y + 3z = 2$

20. $3x + 4y - 2z = 2$
    $y + z = 2$
    $x + y - z = 0$

21. $x = 6 + y + 3z$
    $z = 2x + 4y$
    $x = 14 + 5y - 2z$

22. $2x + y = 3(1 + z)$
    $50 + 6y = 5x + z$
    $4x = 18 + 5z$

23. Consider the following system of equations:

$$x + y + z = 1$$
$$x - y + z = 1$$
$$x + 3y + z = 1$$

Explain why the solution set can be written $\{(1 - z, 0, z)\}$ or by $\{(x, 0, 1 - x)\}$. Can the solution set be written in terms of y?

24. Why would it be impossible to write a quadratic function $f(x) = ax^2 + bx + c$ whose graph contains $P(0, 0)$, $Q(1, 1)$, and $R(2, 2)$?

In Exercises 25–28, write and solve a system of three equations to determine a quadratic function $f(x) = ax^2 + bx + c$ whose graph contains the given points.

25. $P(-4, 16)$, $Q(2, -8)$, $R(5, 7)$

26. $R(-4, -3)$, $S(-1, 9)$, $T(4, -11)$

27. $F(-3, -24)$, $G(2, -9)$, $H(3, -18)$

28. $A(-3, 11)$, $B(1, 7)$, $C(4, 25)$

Concept Extension

29. Solve the following system of equations:

$$x - 2y + z = 5$$
$$2x + y - z = 16$$

30. For what value of k is the solution of the following system of equations not unique?

$$x + ky + 2z = 0$$
$$x \quad\ + z = 0$$
$$3x + y + z = 0$$

In Exercises 31 and 32, solve the system of equations.
$\left(\textit{Hint: Let } u = \dfrac{1}{x}, v = \dfrac{1}{y}, \text{ and } w = \dfrac{1}{z}. \text{ Solve the result-}\right.$
ing system for $u, v,$ and w. Then use those values to
determine $x, y,$ and $z.\Big)$

31. $\dfrac{1}{x} - \dfrac{1}{y} + \dfrac{2}{z} = -3$
$\dfrac{3}{x} + \dfrac{1}{y} - \dfrac{1}{z} = 3$
$\dfrac{2}{x} + \dfrac{2}{y} + \dfrac{1}{z} = -2$

32. $\dfrac{2}{x} + \dfrac{1}{y} = \dfrac{1}{z} - 3$
$\dfrac{1}{x} = 1 + \dfrac{3}{y} + \dfrac{2}{z}$
$9 - \dfrac{3}{z} = \dfrac{4}{x} + \dfrac{2}{y}$

In Exercises 33 and 34, solve the given system of
equations.

33. $x + y - 2z + w = -4$
$2x - y + z - 3w = -13$
$x - 2y - 3z - w = -15$
$3x + y - z + 2w = -1$

34. $x - y - 3z + 2w = 8$
$3x + 2y - z + w = 4$
$x - 4y + z - 3w = -4$
$2x + y - z - w = -2$

Applications

35. Envelope Sizes   A mailing service offers three sizes of protective envelopes. The cost of one large envelope is the same as the total cost of one small envelope and one medium envelope. The difference in the cost of a medium and a small envelope is 60 cents. A customer gave the clerk a 10-dollar bill for the purchase of 2 small envelopes, 1 medium envelope, and 1 large envelope, and she received $2.80 in change. What is the cost of each size of envelope?

36. Combined Investments   An investor placed a total of $25,000 into three funds that paid 6%, 7%, and 9%, respectively. The interest earned by the 7% fund was $60 less than 3 times the interest earned by the 6% fund. If the total interest earned by the three funds was $1860, how much was invested in each fund?

37. PIN Number   A bank customer has a four-digit PIN number. The sum of the first and last

digits is equal to the sum of the second and third digits. The fourth digit is the sum of the first two digits. If the sum of the four digits is 14 and the third digit is 4, what is the customer's PIN number?

38. Theater Tickets   A community theater presented a musical at a small auditorium with 187 seats. Tickets cost $2 for children and $8 for adults. Although senior citizens received a $3 discount, there were 5 fewer senior citizens in attendance than children. The total revenue from a sold-out performance was $1151. How many of each type of ticket were sold?

39. Basketball Scoring   A basketball team scored a total of 68 points. The number of points scored from 2-point shots was the same as the total number of points scored from 1-point and 3-point shots. If the number of 2-point shots was 1 less than twice the number of 3-point shots, how many of each type of shot did the team make?

40. Business Loan   The operator of a small business paid $1793 interest on three loans. The interest paid on the 11% loan was $1 more than the interest paid on the 7% loan. The amount of the 11% loan was $3000 more than the amount of the 5% loan. How much was borrowed at each rate?

41. Campus Parking Permits   A campus offers yearly parking permits for $150 for faculty lots, $120 for central student lots, and $70 for perimeter student lots. From selling a total of 5740 permits, the college collects $4800 more from central student lot permits than from the other two types of permits. The difference between the number of perimeter permits and the number of central permits is twice the number of faculty permits. How many permits of each type were sold?

42. Investments   A person invested a total of $17,000 in a money market fund yielding 7%, a bond fund yielding 9%, and a stock fund yielding 10%. The total amount invested in the bond and stock funds was $600 more than the amount invested in the money market fund. Four times the income from the bond fund exceeded the total income from the other two funds by $18. How much was invested in each fund?

43. Photograph Packs The photographer for a shopping mall Santa Claus offers parents three packages of photographs of children. The following table shows the contents and prices of the three packages.

Scrooge Pack: $19

Grandparent Pack: $51

Parent Pack: $32

Package	2×3	3×5	6×8	Price
Scrooge Pack	8	2	1	$19
Parent Pack	10	4	2	32
Grandparent Pack	12	6	4	51

What is the price of each size print?

44. Computer Lab A computer lab has 17 computers that are set up as 1-person, 2-person, and 3-person workstations. The lab accommodates 32 students. If the 2-person stations were converted to 3-person stations and the 3-person stations were converted to 2-person stations, the lab could accommodate 35 students. How many of each type of station are in the lab?

45. U.S. Households The 69 million family households in the U.S. included four times as many female households (husband absent) as male households (wife absent). The number of couples (husband and wife both present) was 6 million less than four times the total number of the other two types. (*Source*: U.S. Bureau of the Census.) Determine the number (in millions) of each type of household.

46. Imported Automobiles In 1997 the United States imported 4 million cars from Canada, Germany, Japan, and other countries. The numbers of imports from Canada and from Japan were the same. The number of imports from Japan was twice the total number of imports from Germany and other countries. The number of imports from Germany was $\frac{1}{5}$ the difference between the number of imports from Canada and from other countries. (*Source*: American Automobile Manufacturers Association.) How many imports (in millions) came from Canada and Germany?

47. Museum Admission A tour group of 42 people paid a total admission of $268 for an art museum and $324 for a natural history museum. The following table shows the admission prices for the two museums.

	Ticket Prices		
	Children	Adults	Senior Citizens
Art	Free	$14	$10
Natural history	$5	$12	$8

How many children, adults, and senior citizens were in the tour group?

48. Computer Shipments A company shipped 700 computer monitors from a distribution center in Raleigh to retail outlets in Atlanta, Baltimore, and Cleveland. The per-unit shipping costs were $18, $15, and $20, respectively. If the total shipping cost was $12,240, and 60 more monitors were shipped to Baltimore than to Cleveland, how many monitors were shipped to each location?

49. School Enrollment Last year the total enrollment at three elementary schools was 1763. This year, the enrollment decreased by 8% at Chapman, increased by 10% at Bascomb, and increased by $\frac{1}{8}$ at Booth. This year's total enrollment at the three schools is 1835, and the enrollment at Bascomb is 53 less than the combined enrollments at Chapman and Booth. What is this year's enrollment at each school?

50. College Expenses Last semester the cost of tuition, books, and parking for a certain student was $1434. This semester the cost of tuition increased by 5%, the cost of books decreased by $\frac{1}{6}$, and the cost of parking increased $10. The total cost this semester was $1471. This semester's tuition cost was $206 more than four times the combined cost of books and parking. What was the cost of tuition, books, and parking this semester?

51. Tennis Players As shown in the accompanying table, the number of tennis players in the United States reached a peak in 1992. (*Source:* Tennis Industry Association.)

Year	Number (millions)
1985	13
1992	23
1994	19

Write and solve a system of equations to determine a function $T(x) = ax^2 + bx + c$ that models the data as a function of the number of years since 1980.

52. Pollution Abatement The accompanying table shows that expenditures for air and water pollution abatement dipped to a low in 1991. (*Source:* U.S. Bureau of Economic Analysis.)

Year	Expenditure (billions of dollars)
1986	30.5
1991	23.0
1993	26.6

Write and solve a system of equations to determine a function $E(t) = at^2 + bt + c$ that models the data as a function of the number of years since 1980.

5.3 *Matrices and Systems of Equations*

Matrices ■ *Row Operations* ■ *Matrix Methods* ■ *Applications*

In the chapter on polynomial and rational functions, we learned how to perform long division with polynomials and how to use synthetic division as a numerical method for quickly performing certain divisions. Recall that synthetic division eliminates the need to write variables repeatedly.

In this section, we consider another numerical method, this time for solving systems of linear equations. Because the solution of a system is dictated by the coefficients and constants of the equations, we can focus on those numbers rather than on the variables. Moreover, because the method is strictly numerical, you can use your calculator to perform the operations quickly and easily.

The methods described in this section apply to systems with any number of variables. Thus you will be able to think more about modeling the conditions of an application problem and less about the process of actually solving the system.

Matrices

A **matrix** is a rectangular array of numbers. The numbers in the array are called the **elements** of the matrix.

$$A = [3 \quad 1 \quad -2] \qquad B = \begin{bmatrix} 4 \\ 1 \end{bmatrix} \qquad C = \begin{bmatrix} 2 & 1 \\ 0 & -3 \end{bmatrix} \qquad D = \begin{bmatrix} 1 & 0 & -1 \\ 2 & 3 & 1 \end{bmatrix}$$

The **rows** of a matrix are read horizontally, and the **columns** are read vertically. Matrix A is an example of a **row matrix** because it has only one row. Matrix B is a **column matrix** because it has only one column. The **dimensions** of a matrix are the number of rows and the number of columns. We say that a matrix with n rows and m columns is an $n \times m$ matrix. For instance, matrix D has 2 rows and 3 columns, so it is a 2×3 matrix. Because the number of rows and columns is the same, matrix C is an example of a **square matrix.**

Consider the following system of equations:

$$3x + 4y = 5$$
$$2x - 7y = -16$$

When the equations of a system are in standard form, the **coefficient matrix** is the matrix whose elements are the coefficients of the variables in the equations. The **constant matrix** is the matrix whose elements are the constants on the right sides of the equations. The **augmented matrix** is obtained by combining the coefficient matrix and the constant matrix.

Coefficient matrix   Constant matrix   Augmented matrix

$$\begin{bmatrix} 3 & 4 \\ 2 & -7 \end{bmatrix} \qquad \begin{bmatrix} 5 \\ -16 \end{bmatrix} \qquad \left[\begin{array}{cc|c} 3 & 4 & 5 \\ 2 & -7 & -16 \end{array} \right]$$

Note that the first column of the coefficient matrix contains the coefficients of x, and the second column contains the coefficients of y. A zero is inserted as the coefficient of any missing term. The number of rows is the same as the number of equations. In the augmented matrix, the vertical line separates the coefficient matrix from the constant matrix.

To solve a system of equations with matrix methods, our goal will be to write the augmented matrix in a form known as the **reduced row-echelon form.** The following are examples of augmented matrices written in the reduced row-echelon form:

$$\left[\begin{array}{cc|c} 1 & 0 & c_1 \\ 0 & 1 & c_2 \end{array} \right] \qquad \left[\begin{array}{ccc|c} 1 & 0 & 0 & c_1 \\ 0 & 1 & 0 & c_2 \\ 0 & 0 & 1 & c_3 \end{array} \right]$$

Note that the coefficient matrix has 1s down its diagonal and 0s elsewhere. (If any row of the coefficient matrix has all 0s, it is placed at the bottom of the matrix.)

Once augmented matrices have been written in reduced row-echelon form, the corresponding systems of equations can be written.

$$1x + 0y = c_1 \qquad\qquad 1x + 0y + 0z = c_1$$
$$0x + 1y = c_2 \qquad\qquad 0x + 1y + 0z = c_2$$
$$\text{Solution: } (c_1, c_2) \qquad\qquad 0x + 0y + 1z = c_3$$
$$\qquad\qquad\qquad\qquad\qquad \text{Solution: } (c_1, c_2, c_3)$$

Row Operations

Recall that when we solve a system of equations, we can interchange equations, multiply both sides of an equation by a nonzero number, add equations, and add a multiple of one equation to another equation. Performing these operations produces an equivalent system of equations.

We can perform similar operations on the rows of a matrix. Matrices obtained by performing these operations are **equivalent matrices.** The permissible row operations are given in the following summary, which also includes the notation that we will use to indicate row operations.

Row Operations on Matrices

Suppose that R_i and R_j are any two rows of a matrix and that c is a nonzero real number. Each of the following row operations produces an equivalent matrix.

1. Any two rows may be interchanged: $R_i \leftrightarrow R_j$.
2. A row may be replaced by a nonzero multiple of itself: $c \cdot R_i \to R_i$.
3. A row may be replaced by the sum of itself and another row: $R_i + R_j \to R_j$.
4. A row may be replaced by the sum of itself and a multiple of another row: $c \cdot R_i + R_j \to R_j$.

One use that we will make of row operations is to write a matrix in reduced row-echelon form.

Matrix Methods

We can use matrix methods to solve a system of equations. Remember that the goal is to write the corresponding augmented matrix in reduced row-echelon form—that is, with 1s down its diagonal and 0s elsewhere. Then we can determine the solution from the numbers in the right-hand column of the matrix.

EXAMPLE 1 **Using Matrices to Solve a System of Two Equations in Two Variables**

Use both the graphing method and matrices to solve the following system of equations:

$$x - 3y = -9$$
$$3x + y = -7$$

SOLUTION

VISUALIZING THE SOLUTION

Solving the equations for y, we obtain $y_1 = \frac{1}{3}x + 3$ and $y_2 = -3x - 7$. As shown in Figure 5.19, the graphs of these equations intersect at $(-3, 2)$, which is the solution of the system.

FIGURE **5.19**

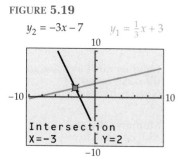

DETERMINING THE SOLUTION

$$\begin{bmatrix} 1 & -3 & | & -9 \\ 3 & 1 & | & -7 \end{bmatrix}$$ Augmented matrix

$-3R_1 + R_2 \rightarrow \begin{bmatrix} 1 & -3 & | & -9 \\ 0 & 10 & | & 20 \end{bmatrix}$ Multiply row 1 by -3 and add the result to row 2. Now we have a 0 in row 2, column 1.

$R_2 \div 10 \rightarrow \begin{bmatrix} 1 & -3 & | & -9 \\ 0 & 1 & | & 2 \end{bmatrix}$ Divide row 2 by 10 to obtain a 1 in row 2, column 2.

$3R_2 + R_1 \rightarrow \begin{bmatrix} 1 & 0 & | & -3 \\ 0 & 1 & | & 2 \end{bmatrix}$ Multiply row 2 by 3 and add the result to row 1. Now we have a 0 in row 1, column 2.

The last matrix is in reduced row-echelon form and represents the following system of equations:

$$1x + 0y = -3$$
$$0x + 1y = 2$$

The solution of the system of equations is $(-3, 2)$, which confirms the result obtained with the graphing method.

Observe that writing the augmented matrix in reduced row-echelon form requires a systematic approach. We begin by using row operations to obtain a 1 in row 1, column 1, and 0s in the rest of column 1. Then we obtain a 1 in row 2, column 2, and 0s in the rest of column 2. We continue with this plan until the matrix is in reduced row-echelon form.

EXAMPLE 2 **Using Matrices to Solve a System of Three Equations in Three Variables**

Use matrices to solve the following system of equations:

$$-3x + 2y + z = 4$$
$$x - y + 2z = -9$$
$$2x + y - z = 3$$

SOLUTION

$$\begin{bmatrix} -3 & 2 & 1 & | & 4 \\ 1 & -1 & 2 & | & -9 \\ 2 & 1 & -1 & | & 3 \end{bmatrix}$$ The augmented matrix

$R_1 \leftrightarrow R_2 \begin{bmatrix} 1 & -1 & 2 & | & -9 \\ -3 & 2 & 1 & | & 4 \\ 2 & 1 & -1 & | & 3 \end{bmatrix}$ Exchange rows 1 and 2 to obtain a 1 in row 1, column 1.

$3R_1 + R_2 \rightarrow \begin{bmatrix} 1 & -1 & 2 & | & -9 \\ 0 & -1 & 7 & | & -23 \\ 2 & 1 & -1 & | & 3 \end{bmatrix}$ Multiply row 1 by 3 and add the result to row 2. Now we have a 0 in row 2, column 1.

$-2R_1 + R_3 \rightarrow \begin{bmatrix} 1 & -1 & 2 & | & -9 \\ 0 & -1 & 7 & | & -23 \\ 0 & 3 & -5 & | & 21 \end{bmatrix}$ Multiply row 1 by -2 and add the result to row 3. Now we have a 0 in row 3, column 1.

$$R_2 \cdot (-1) \quad \rightarrow \quad \begin{bmatrix} 1 & -1 & 2 & | & -9 \\ 0 & 1 & -7 & | & 23 \\ 0 & 3 & -5 & | & 21 \end{bmatrix}$$

Multiply row 2 by −1.
Now we have a 1 in row 2,
column 2.

$$R_2 + R_1 \quad \rightarrow \quad \begin{bmatrix} 1 & 0 & -5 & | & 14 \\ 0 & 1 & -7 & | & 23 \\ 0 & 3 & -5 & | & 21 \end{bmatrix}$$

Add row 2 to row 1.
Now we have a 0 in row 1,
column 2.

$$-3R_2 + R_3 \quad \rightarrow \quad \begin{bmatrix} 1 & 0 & -5 & | & 14 \\ 0 & 1 & -7 & | & 23 \\ 0 & 0 & 16 & | & -48 \end{bmatrix}$$

Multiply row 2 by −3 and
add the result to row 3.
Now we have a 0 in row 3,
column 2.

$$R_3 \div 16 \quad \rightarrow \quad \begin{bmatrix} 1 & 0 & -5 & | & 14 \\ 0 & 1 & -7 & | & 23 \\ 0 & 0 & 1 & | & -3 \end{bmatrix}$$

Divide row 3 by 16.
Now we have a 1 in row 3,
column 3.

$$5R_3 + R_1 \quad \rightarrow \quad \begin{bmatrix} 1 & 0 & 0 & | & -1 \\ 0 & 1 & -7 & | & 23 \\ 0 & 0 & 1 & | & -3 \end{bmatrix}$$

Multiply row 3 by 5 and add
the result to row 1.
Now we have a 0 in row 1,
column 3.

$$7R_3 + R_2 \quad \rightarrow \quad \begin{bmatrix} 1 & 0 & 0 & | & -1 \\ 0 & 1 & 0 & | & 2 \\ 0 & 0 & 1 & | & -3 \end{bmatrix}$$

Multiply row 3 by 7 and add
the result to row 2.
Now we have a 0 in row 2,
column 3.

Now that the matrix is in reduced row-echelon form, we can easily write the solution: $(-1, 2, -3)$.

Example 2 illustrates the many steps required in writing an augmented matrix in reduced row-echelon form. Because the process is strictly numerical, we can use a calculator to perform the steps automatically.

> **RREF**   We can enter an augmented matrix in a calculator and produce the *reduced row-echelon form.*

FIGURE **5.20**

```
[[-3   2   1   4 ]
 [1   -1  2  -9]
 [2    1  -1  3 ]]
rref([A])
     [[1  0  0  -1]
      [0  1  0  2 ]
      [0  0  1  -3]]
```

Figure 5.20 shows the augmented matrix for Example 2 and the reduced row-echelon form.

Matrix and calculator methods will detect the special cases of dependent equations and inconsistent systems.

EXAMPLE **3**   **Dependent Equations**

Use matrices to solve the following system of equations:

$$x - 5y = 2$$
$$3x - 15y = 6$$

SOLUTION

$$\begin{bmatrix} 1 & -5 & | & 2 \\ 3 & -15 & | & 6 \end{bmatrix}$$

The augmented matrix

$$-3R_1 + R_2 \quad \rightarrow \quad \begin{bmatrix} 1 & -5 & | & 2 \\ 0 & 0 & | & 0 \end{bmatrix}$$

Multiply row 1 by −3 and
add the result to row 2.

The last row corresponds to the equation $0x + 0y = 0$, or $0 = 0$, which is true. Thus the equations are dependent. We solve the first equation for x and write the solution set in the form $\{(5y + 2, y)\}$.

EXAMPLE 4   **An Inconsistent System**

Use matrices to solve the following system of equations:

$$x + 2y - z = 5$$
$$x - 2y + z = 2$$
$$2x + 4y - 2z = 7$$

FIGURE 5.21

```
  [[1    2   -1    5]
   [1   -2   1     2]
   [2    4   -2    7]]
rref([A])
  [[1  0  0     0]
   [0  1 -.5   0]
   [0  0  0     1]]
```

SOLUTION

Figure 5.21 shows the augmented matrix along with the reduced row-echelon form.

Because the last row corresponds to the equation $0x + 0y + 0z = 1$, or $0 = 1$, which is false, the system has no solution.

Applications

Modeling the conditions of a problem is the fundamental task in all applications. Once you have done that, you can often use your calculator to quickly arrive at a solution.

EXAMPLE 5   **Pedestrian Traffic Flow**

A small college has studied its pedestrian traffic to and from three junctions in the commons area. Figure 5.22 shows the traffic patterns, where x_1, x_2, and x_3 represent the number of pedestrians along certain paths. Arrows into a junction represent arrivals, and arrows away from a junction represent departures. Note that two of the paths have the same number of pedestrians.

FIGURE 5.22

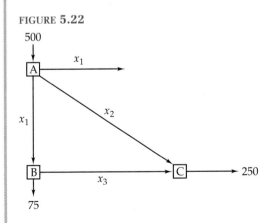

Write and solve a system of equations to determine the traffic flow along each path.

SOLUTION

First, we analyze the traffic flow at each junction, where the number of arrivals equals the number of departures.

Junction	Arrivals	Departures	Equation
A	500	$x_1 + x_2 + x_1$	$500 = 2x_1 + x_2$
B	x_1	$x_3 + 75$	$x_1 = x_3 + 75$
C	$x_2 + x_3$	250	$x_2 + x_3 = 250$

Writing the equations in standard form, we obtain the following system of equations:

$$2x_1 + x_2 \quad\ = 500$$
$$x_1 \quad\ - x_3 = 75$$
$$x_2 + x_3 = 250$$

FIGURE **5.23**

```
rref([A])
   [[1  0  0  175]
    [0  1  0  150]
    [0  0  1  100]]
```

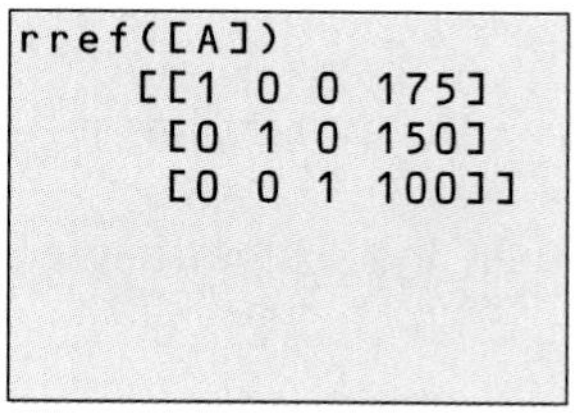

The matrix corresponding to this system is

$$\begin{bmatrix} 2 & 1 & 0 & | & 500 \\ 1 & 0 & -1 & | & 75 \\ 0 & 1 & 1 & | & 250 \end{bmatrix}$$

Figure 5.23 shows this matrix in reduced row-echelon form.

The solutions are $x_1 = 175$, $x_2 = 150$, and $x_3 = 100$.

5.3 Quick Reference

Matrices

■ A **matrix** is a rectangular array of numbers, which are called the **elements** of the matrix.

■ The **rows** of a matrix are read horizontally. A matrix with only one row is a **row matrix.**

■ The **columns** of a matrix are read vertically. A matrix with only one column is a **column matrix.**

■ The **dimensions** of a matrix are the number of rows and the number of columns. A matrix with n rows and m columns is an $n \times m$ matrix.

■ A matrix with the same number of rows and columns is a **square matrix.**

■ If we associate a matrix with a system of equations written in standard form, then

1. the **coefficient matrix** is the matrix whose elements are the coefficients of the variables in the equations.
2. the **constant matrix** is the matrix whose elements are the constants on the right sides of the equations.
3. the **augmented matrix** is obtained by combining the coefficient matrix and the constant matrix.

■ An augmented matrix is in **reduced row-echelon form** if the coefficient matrix has 1s down its diagonal and 0s elsewhere.

Row Operations ■ If R_i and R_j are any two rows of a matrix and c is a nonzero real number, then each of the following row operations produces an equivalent matrix.

1. Any two rows may be interchanged: $R_i \leftrightarrow R_j$.
2. A row may be replaced by a nonzero multiple of itself: $c \cdot R_i \rightarrow R_i$.
3. A row may be replaced by the sum of itself and another row: $R_i + R_j \rightarrow R_j$.
4. A row may be replaced by the sum of itself and a multiple of another row: $c \cdot R_i + R_j \rightarrow R_j$.

Matrix Methods ■ When we use matrix methods to solve a system of equations, the goal is to write the augmented matrix in reduced row-echelon form. Then we can determine the solution from the numbers in the right-hand column of the matrix.

■ The matrix method for solving a system indicates a special case when one row of the coefficient matrix consists of all 0s.

1. If the row corresponds to a true equation, then the system has infinitely many solutions.
2. If the row corresponds to a false equation, then the system has no solution.

5.3 Speaking the Language

1. A matrix with n rows and n columns is called a(n) ▨▨▨▨ matrix.
2. An augmented matrix is formed by combining a(n) ▨▨▨▨ matrix and a(n) ▨▨▨▨ matrix.
3. For an augmented matrix, if the coefficient matrix has 1s down its diagonal and 0s elsewhere, the matrix is said to be in ▨▨▨▨ form.
4. In the matrix $\begin{bmatrix} 1 & 3 & | & 5 \\ 2 & 4 & | & 6 \end{bmatrix}$, the elements 1 and 2 are in a(n) ▨▨▨▨, whereas the elements 1, 3, and 5 are in a(n) ▨▨▨▨.

5.3 Exercises

Concepts and Skills

1. Suppose you write an augmented matrix that corresponds to a system of two linear equations in the form $ax + by = c$. Describe the entries in the coefficient matrix.

2. Suppose you write an augmented matrix that corresponds to a system of equations and that the reduced row-echelon form of the matrix is

$$\begin{bmatrix} 1 & 0 & | & 5 \\ 0 & 1 & | & 2 \end{bmatrix}$$

To what equivalent system of equations does this matrix correspond? How can you determine the solution of the system?

In Exercises 3–6, determine the dimensions of the given matrix.

3. $[3 \quad -5]$

4. $[7]$

5. $\begin{bmatrix} 1 & -3 & 1 \\ 4 & 0 & 2 \end{bmatrix}$

6. $\begin{bmatrix} 4 \\ 2 \\ -7 \end{bmatrix}$

In Exercises 7–10, write the augmented matrix for the given system of equations.

7. $x + 3y = 7$
$2x - 5y = -8$

8. $2y - x = -4$
$3y = 4 - x$

9. $x + 5z = 13$
$2y + z = 4$
$x + y - z = 0$

10. $x - y + z = 3$
$y - x - z = -3$
$2z + x - y = 4$

In Exercises 11–14, write the system of linear equations associated with the given augmented matrix.

11. $\begin{bmatrix} 1 & -1 & | & -2 \\ 3 & 2 & | & 14 \end{bmatrix}$

12. $\begin{bmatrix} 0 & 3 & | & -3 \\ 2 & 1 & | & 3 \end{bmatrix}$

13. $\begin{bmatrix} -2 & 0 & 1 & | & 2 \\ 4 & 1 & -3 & | & -2 \\ 2 & -4 & 1 & | & -10 \end{bmatrix}$

14. $\begin{bmatrix} 1 & 0 & 1 & | & 4 \\ 0 & 1 & 2 & | & 4 \\ 1 & 1 & 1 & | & 2 \end{bmatrix}$

15. In what respect is the row operation $R_i + R_j \rightarrow R_j$ a special case of $c \cdot R_i + R_j \rightarrow R_j$?

16. When we use matrix methods to solve a system of equations, for what purpose do we use row operations?

In Exercises 17–20, an augmented matrix is given. In each step, fill in the missing row that would result from performing the indicated row operation.

17. *Matrix*   *Row operation*

$\begin{bmatrix} 3 & 12 & | & -9 \\ 2 & 9 & | & -8 \end{bmatrix}$ Augmented matrix

$\begin{bmatrix} \blacksquare & \blacksquare & | & \blacksquare \\ 2 & 9 & | & -8 \end{bmatrix}$ (a) $R_1 \div 3$

$\begin{bmatrix} 1 & 4 & | & -3 \\ \blacksquare & \blacksquare & | & \blacksquare \end{bmatrix}$ (b) $-2R_1 + R_2$

$\begin{bmatrix} \blacksquare & \blacksquare & | & \blacksquare \\ 0 & 1 & | & -2 \end{bmatrix}$ (c) $-4R_2 + R_1$

18. *Matrix*   *Row operation*

$\begin{bmatrix} 1 & -2 & | & 7 \\ -5 & 1 & | & -17 \end{bmatrix}$ Augmented matrix

$\begin{bmatrix} 1 & -2 & | & 7 \\ \blacksquare & \blacksquare & | & \blacksquare \end{bmatrix}$ (a) $5R_1 + R_2$

$\begin{bmatrix} 1 & -2 & | & 7 \\ \blacksquare & \blacksquare & | & \blacksquare \end{bmatrix}$ (b) $R_2 \div (-9)$

$\begin{bmatrix} \blacksquare & \blacksquare & | & \blacksquare \\ 0 & 1 & | & -2 \end{bmatrix}$ (c) $2R_2 + R_1$

19. *Matrix*   *Row operation*

$\begin{bmatrix} 1 & -2 & 1 & | & 2 \\ 3 & -5 & 0 & | & 13 \\ -2 & 0 & 11 & | & -35 \end{bmatrix}$ Augmented matrix

$\begin{bmatrix} 1 & -2 & 1 & | & 2 \\ \blacksquare & \blacksquare & \blacksquare & | & \blacksquare \\ \blacksquare & \blacksquare & \blacksquare & | & \blacksquare \end{bmatrix}$ (a) $-3R_1 + R_2$
(b) $2R_1 + R_3$

$\begin{bmatrix} \blacksquare & \blacksquare & \blacksquare & | & \blacksquare \\ 0 & 1 & -3 & | & 7 \\ \blacksquare & \blacksquare & \blacksquare & | & \blacksquare \end{bmatrix}$ (c) $2R_2 + R_1$
(d) $4R_2 + R_3$

$\begin{bmatrix} \blacksquare & \blacksquare & \blacksquare & | & \blacksquare \\ \blacksquare & \blacksquare & \blacksquare & | & \blacksquare \\ 0 & 0 & 1 & | & -3 \end{bmatrix}$ (e) $5R_3 + R_1$
(f) $3R_3 + R_2$

20. *Matrix*   *Row operation*

$\begin{bmatrix} 1 & 3 & 0 & | & 2 \\ -4 & -11 & 6 & | & -32 \\ 2 & 5 & -5 & | & 24 \end{bmatrix}$ Augmented matrix

$\begin{bmatrix} 1 & 3 & 0 & | & 2 \\ \blacksquare & \blacksquare & \blacksquare & | & \blacksquare \\ \blacksquare & \blacksquare & \blacksquare & | & \blacksquare \end{bmatrix}$ (a) $4R_1 + R_2$
(b) $-2R_1 + R_3$

$\begin{bmatrix} \blacksquare & \blacksquare & \blacksquare & | & \blacksquare \\ 0 & 1 & 6 & | & -24 \\ \blacksquare & \blacksquare & \blacksquare & | & \blacksquare \end{bmatrix}$ (c) $-3R_2 + R_1$
(d) $R_2 + R_3$

$\begin{bmatrix} \blacksquare & \blacksquare & \blacksquare & | & \blacksquare \\ \blacksquare & \blacksquare & \blacksquare & | & \blacksquare \\ 0 & 0 & 1 & | & -4 \end{bmatrix}$ (e) $18R_3 + R_1$
(f) $-6R_3 + R_2$

In Exercises 21–24, determine whether the matrix is in reduced row-echelon form. If it is not, perform row operations to write the matrix in reduced row-echelon form.

21. $\begin{bmatrix} 1 & 6 & | & 3 \\ 0 & 1 & | & -1 \end{bmatrix}$

22. $\begin{bmatrix} 3 & 1 & | & 1 \\ 0 & 0 & | & 2 \end{bmatrix}$

23. $\begin{bmatrix} 1 & 1 & 0 & | & 0 \\ 0 & 0 & 1 & | & 3 \end{bmatrix}$

24. $\begin{bmatrix} 1 & 0 & 0 & | & 1 \\ 0 & 1 & 0 & | & 0 \\ 3 & 0 & 1 & | & -2 \end{bmatrix}$

25. Suppose you use matrix methods to solve a system of equations and one row of the reduced row-echelon form is all 0s. How do you interpret this row, and what does it imply about the solution of the system?

26. Give an example of a system of three equations in three variables for which the reduced row-echelon form of the augmented matrix would have two rows of all 0s. Can you give an example for which all three rows would be all 0s?

In Exercises 27–36, use matrix methods to solve the system of equations.

27. $x + 3y = -3$
$2x - y = 8$

28. $x + 2y = 14$
$-3x + y = -7$

29. $x + 3y = 6$
$-2x - 6y = 1$

30. $5x + 2y = 2$
$15x + 6y = 0$

31. $4x = 1 + y$
$20x - 5y = 5$

32. $4x = 9y - 3$
$27y = 9 + 12x$

33. $x \qquad - z = 8$
$3y - 2z = 12$
$2x + y \qquad = 4$

34. $\qquad y - z = 2$
$x + 3y + z = 2$
$2x + y \qquad = -6$

35. $4x - 2y + z = 1$
$x + 3y - 5z = 16$
$x - y + z = 0$

36. $x - 3y + 2z = 11$
$-7y + 5z = 22$
$2x + y - z = 0$

In Exercises 37–54, use a calculator to solve the system of equations.

37. $4x + 10y = 27$
$3x + y = 4$

38. $12x + 15y = 16$
$6x + 9y = 10$

39. $6 + y = x$
$2y = 2x - 12$

40. $y - 3x = 7$
$6x - 2y = 0$

41. $x = y + 2$
$2y = 14 - x$

42. $-x - 6 = 3y$
$-11 = 2y + 3x$

43. $x - y + 3z = 15$
$x + y + z = 5$
$2x - 2y + z = 10$

44. $x + 2y - z = 4$
$2x - y + 3z = -7$
$x - 2y + z = 2$

45. $3x + 4y + z = 4$
$2x - 5y + 6z = 0$
$x - 14y + 11z = 0$

46. $x - 2y + z = 0$
$3x + y + 4z = 6$
$3x + 8y + 5z = 3$

47. $6x - 3y + 3z = 12$
$-4x + 2y - 2z = -8$
$2x - y + z = 4$

48. $9x + 3y - 12z = 45$
$3x + y - 4z = 15$
$12x + 4y - 16z = 60$

49. $y - 2z = -2$
$x - y + z = 1$
$4x + 6y + 2z = 11$

50. $3x - 6y + 3z = -11$
$4x - y = -13$
$y - 3z = -3$

51. $3z + x = 9$
$2y = z$
$18 + 6y = 2x + 9z$

52. $5z + 4 = 2x + 3y$
$z + y - 4 = 2x$
$z + 2 = y$

53. $2x + y = 4z - 1$
$6y = 2x + 12z - 1$
$10 + 2z = 5y + 4x$

54. $5z + 2y + 6x = 27$
$2 - 7z = x - 4y$
$5y = 3x + z$

Concept Extension

In Exercises 55–58, use a calculator to solve the system of equations.

55. $2x + y - z + w = 2$
$x - y + z - 2w = -1$
$3x + 2y + z - w = 0$
$x + 3y - 2z + 4w = 4$

56. $x + 2y \qquad + w = 5$
$x - y + z + 2w = 7$
$3x + 4y - z \qquad = 7$
$y + 3z - w = -6$

57. $x + y = 3$
$y - z = 2$
$z - w = -4$
$x + y + w = 6$

58. $x + y - z = 2$
$x + 2w = 5$
$z - w = -4$
$y + z = -1$

Applications

59. Imports In 1994, the total value of imported coffee, sugar, and wheat was $3.107 billion. The value of imported coffee was $1.433 billion more than the combined value of imported sugar and wheat. The value of imported sugar was $21 million less than twice the value of imported wheat. (*Source*: U.S. Department of Commerce.) What was the value of each of the three imported commodities?

60. Immigrants In the period 1820–1940, a total of 39 million immigrants came to the United States from Europe, Asia, and the Western Hemisphere. The number of immigrants from Europe exceeded the combined total of immigrants from Asia and the Western Hemisphere by 27 million. The number of immigrants from Asia was only one-fifth the number from the Western Hemisphere. (*Source*: U.S. Immigration and Naturalization Service.) How many immigrants came from each region?

61. Population Trends The following table shows the percentage of the population that lived in the Midwest and West in the years 1990 and 1995. (*Source*: Bureau of the Census.)

	1990	*1995*
Midwest	24.0	23.5
West	21.2	21.9

(a) Letting x represent the number of years since 1990, use your calculator to produce linear regression equations to model the data for each region.

(b) Solve the system of equations in part (a) to estimate the year in which the percentage of the population in the two regions is projected to be the same.

62. **House Characteristics** The characteristics of new, privately owned, one-family houses have changed in the last 30 years. For example, the popularity of one-story homes has decreased significantly since 1970, whereas the percentage of homes of two or more stories has increased. (*Source*: U.S. Department of Housing and Urban Development.)

With t representing the number of years since 1970, the following equations model the percentages of new one-story houses and new houses of two stories or more:

One-story houses: $y = -1.21t + 73.1$

Houses with two or more stories: $y = 1.45t + 17.1$

Solve this system of equations to determine the year in which the percentages of one-story and two-story houses are projected to be the same.

63. **Armed Forces Personnel** The number of people serving in the Armed Forces has declined steadily since 1970. The accompanying bar graph shows the number of army and navy enlisted personnel from 1987 to 1995. (*Source*: U.S. Department of Defense.)

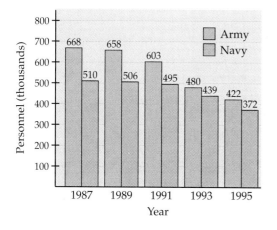

(a) Letting x represent the number of years since 1985, produce regression equations to model the data for the army and the navy.
(b) If you use the equations in part (a) as a system of equations, what conclusions can be drawn from the solution of the system?

64. **Employment Trends** A valuable resource for making career decisions is the Department of Labor's projections of the fastest-growing and fastest-declining occupations. The following table illustrates two occupations with actual and projected employment for 1994 and 2005. (*Source*: U.S. Department of Labor.)

Occupation	Employment (Thousands)	
	1994	2005
Bank tellers	559	391
Home health aides	420	832

(a) Letting t represent the number of years since 1990, produce linear regression equations to model the employment data for the two occupations.
(b) Produce the graphs of the two equations in part (a) and trace to estimate the year in which the employment figures for bank tellers and home health aides were projected to be the same.

65. **Insecticide Solutions** A landscape service worker needs to prepare 50 liters of a 32% insecticide solution from three solutions with concentrations of 20%, 30%, and 50%.

(a) Describe the possible combinations to produce the desired concentration.
(b) What are the least and most amounts of the 50% solution that can be used?
(c) Suppose that the worker must use 20 more liters of the 30% solution than of the 20% solution. How much of each solution should be used?

66. **Water Purification** A city has three water purification facilities that can treat a total of 15,000 gallons in 1 hour. If the largest facility is operating at 50% capacity, then the time needed for the three facilities to process 15,000 gallons increases to 1 hour and 15 minutes.

(a) If the operating capacity of the largest facility were increased by 50%, then the time needed for the three facilities to process 15,000 gallons would be reduced to 50 minutes. How many gallons per hour can the largest facility process?

(b) What are the possible capacities of the other facilities?

(c) If the smallest facility is upgraded to increase its capacity by 75%, the time needed for the three facilities to process 15,000 gallons will decline to 50 minutes. How many gallons per hour can each facility process?

67. **Electrical Network**   The accompanying figure shows a simple electrical network with voltage sources (represented by ⊢), resistors (represented by ‑⌁‑), and currents (represented by i). The currents flow into and out of the two nodes, A and B.

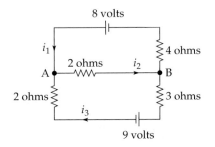

According to *Kirchhoff's Laws*, the following system of equations describes the current in the network:

$$i_1 + i_3 = i_2$$
$$4i_1 + 2i_2 = 8$$
$$2i_2 + 5i_3 = 9$$

Use matrix methods to determine the currents (in amperes) in the network.

68. **Milk Consumption**   In a recent year, the per-capita consumption of plain milk (whole, lowfat, and skim) was 200.2 pounds. The consumption of lowfat milk exceeded three times the consumption of skim milk by 9.2 pounds. The consumption of whole milk was 47 pounds more than that of skim milk. (*Source*: U.S. Department of Agriculture.) What was the per-capita consumption of each type of milk?

5.4 *Matrix Operations*

Equality of Matrices ■ *Addition and Subtraction of Matrices* ■ *Scalar Multiplication* ■ *Matrix Multiplication* ■ *Real-Life Applications*

In addition to using matrices to solve equations, we can use a matrix as a convenient way to organize and classify information. In this section, we will consider the operations that we can perform on such matrices, and we will see how the algebra of matrices can be used to solve application problems.

We can add and subtract matrices, and we can perform two kinds of multiplication with matrices. To understand the methods used to perform such operations, we need to begin with the general matrix notation and the conditions under which two matrices are equal.

Equality of Matrices

Consider the following $m \times n$ matrix, which we denote with the letter A.

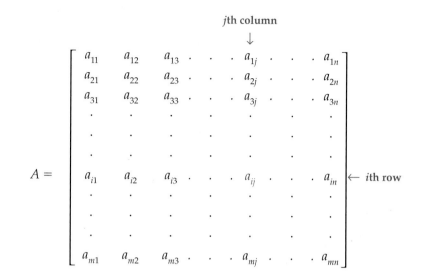

The first subscript of each entry is the row number, and the second subscript is the column number. For example, a_{42} denotes the entry in the 4th row and the 2nd column. In general, a_{ij} represents the entry in row i and column j. We can use this representative entry to denote the matrix:

$$A = [a_{ij}]$$

■ *NOTE   The definition of equality of matrices implies that the matrices must have the same dimensions.*

> **Definition of Equality of Matrices**
>
> Matrices $A = [a_{ij}]$ and $B = [b_{ij}]$ are equal if $a_{ij} = b_{ij}$ for all i and j.
>
> In words, two matrices are equal if their corresponding elements are equal.

EXAMPLE 1   **Equality of Matrices**

Determine a, b, and c so that the matrices are equal.

$$\begin{bmatrix} a & 5 & 2b \\ 1 & c+3 & 0 \end{bmatrix} = \begin{bmatrix} 4 & 5 & -4 \\ 1 & 5 & 0 \end{bmatrix}$$

SOLUTION

For the matrices to be equal, the corresponding elements must be equal: $a = 4$, $2b = -4$, and $c + 3 = 5$. Thus $a = 4$, $b = -2$, and $c = 2$.

Addition and Subtraction of Matrices

We begin our discussion of operations with matrices with addition and subtraction.

> **Definitions of Addition and Subtraction of Matrices**
>
> Suppose that $A = [a_{ij}]$ and $B = [b_{ij}]$ are both $m \times n$ matrices. Then the sum and the difference of the matrices are $m \times n$ matrices given by
>
> $$A + B = [a_{ij} + b_{ij}] \qquad \text{Sum of matrices } A \text{ and } B$$
> $$A - B = [a_{ij} - b_{ij}] \qquad \text{Difference of matrices } A \text{ and } B$$
>
> In words, to add (or subtract) two matrices, add (or subtract) the corresponding elements.

■ *NOTE In order for us to add or subtract two matrices, the dimensions of the matrices must be the same.*

EXAMPLE 2 **Adding and Subtracting Matrices**

(a) $\begin{bmatrix} 1 & -1 & 3 \\ 2 & 5 & -8 \end{bmatrix} + \begin{bmatrix} 6 & 0 & -1 \\ -2 & -1 & 9 \end{bmatrix} = \begin{bmatrix} 7 & -1 & 2 \\ 0 & 4 & 1 \end{bmatrix}$

(b) $\begin{bmatrix} 2 & -5 \\ 4 & 1 \end{bmatrix} + \begin{bmatrix} -2 & 5 \\ -4 & -1 \end{bmatrix} = \begin{bmatrix} 0 & 0 \\ 0 & 0 \end{bmatrix}$

(c) The sum of $\begin{bmatrix} 1 & -5 \\ 2 & 0 \\ -6 & -1 \end{bmatrix}$ and $\begin{bmatrix} 1 & 9 & -5 \\ -4 & 8 & 6 \end{bmatrix}$ is not defined because the dimensions of the matrices are not the same.

(d) $\begin{bmatrix} 2 & -1 \\ 0 & 4 \\ -6 & 5 \end{bmatrix} - \begin{bmatrix} -6 & -1 \\ 3 & 6 \\ 0 & -7 \end{bmatrix} = \begin{bmatrix} 8 & 0 \\ -3 & -2 \\ -6 & 12 \end{bmatrix}$

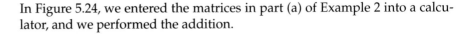

Matrix Sum You can use a calculator to add or subtract matrices.

FIGURE **5.24**

```
[A]+[B]
     [[7  -1   2]
      [0   4   1]]
```

In Figure 5.24, we entered the matrices in part (a) of Example 2 into a calculator, and we performed the addition.

In part (b) of Example 2, the entries in the matrices are opposites, and the sum is a matrix whose entries are all 0s. The $m \times n$ matrix whose entries are all 0s is called the **zero matrix,** which is denoted by **O.**

For matrices, the zero matrix plays the same role as the additive identity in the real numbers. For the real numbers, $a + 0 = a$; for matrices, $A + \mathbf{O} = A$.

> ### *Definition of the Additive Inverse of a Matrix*
>
> For any matrix $A = [a_{ij}]$, the **additive inverse** is $-A = [-a_{ij}]$. In words, each entry of $-A$ is the opposite of the corresponding entry of A.

Just as the sum of a real number and its additive inverse is 0, the sum of a matrix and its additive inverse is **O**. Also, from the definition of $-A$, we can easily show that $A - B = A + (-B)$.

Scalar Multiplication

We can multiply a matrix by a number, called a **scalar,** or by another matrix. To multiply a matrix by a scalar c, we multiply each element by c.

> ### *Definition of Scalar Multiplication*
>
> For an $m \times n$ matrix $A = [a_{ij}]$ and a real number c, $cA = [ca_{ij}]$.

EXAMPLE 3   **Scalar Multiplication of a Matrix**

Given $A = \begin{bmatrix} -2 & 3 \\ 1 & -6 \end{bmatrix}$, find $5A$.

SOLUTION

$$5A = \begin{bmatrix} 5(-2) & 5(3) \\ 5(1) & 5(-6) \end{bmatrix} = \begin{bmatrix} -10 & 15 \\ 5 & -30 \end{bmatrix}$$

FIGURE **5.25**

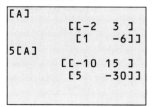

> **Scalar**   You can use a calculator to perform a *scalar* multiplication of a matrix.

Figure 5.25 shows a calculator display of matrices A and $5A$ in Example 3.

Sometimes we need to perform a combination of matrix operations.

EXAMPLE 4   **Scalar Multiplication and Subtraction**

Let $A = \begin{bmatrix} 5 & -3 \\ 4 & 2 \end{bmatrix}$ and $B = \begin{bmatrix} 2 & 5 \\ -1 & 0 \end{bmatrix}$. Find $3A - 2B$.

SOLUTION

$$3A - 2B = 3\begin{bmatrix} 5 & -3 \\ 4 & 2 \end{bmatrix} - 2\begin{bmatrix} 2 & 5 \\ -1 & 0 \end{bmatrix} = \begin{bmatrix} 15 & -9 \\ 12 & 6 \end{bmatrix} + \begin{bmatrix} -4 & -10 \\ 2 & 0 \end{bmatrix} = \begin{bmatrix} 11 & -19 \\ 14 & 6 \end{bmatrix}$$

Figure 5.26 shows a calculator display of the operations after matrices *A* and *B* were entered.

FIGURE **5.26**

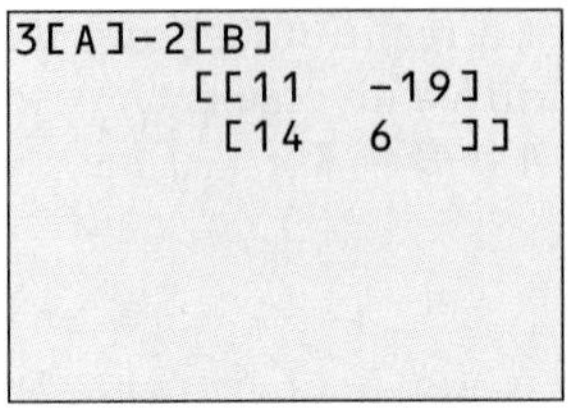

```
3[A]-2[B]
      [[11   -19]
       [14   6  ]]
```

Many familiar properties of real numbers are valid for matrix addition and scalar multiplication. The following summarizes those properties.

Properties of Matrix Addition and Scalar Multiplication

For $m \times n$ matrices *A, B, C,* the zero matrix **O,** and real numbers *c* and *d,*

1. $A + B = B + A$	Commutative Property of Matrix Addition
2. $A + (B + C) = (A + B) + C$	Associative Property of Matrix Addition
3. $(cd)A = c(dA)$	Associative Property of Scalar Multiplication
4. $1A = A$	Multiplicative Scalar Identity Property
5. $A + \mathbf{O} = A$	Additive Identity Property
6. $A + (-A) = \mathbf{O}$	Property of Additive Inverses
7. $c(A + B) = cA + cB$	Distributive Property
8. $(c + d)A = cA + dA$	Distributive Property

Matrix Multiplication

Scalar multiplication involves multiplying a matrix by a real number. *Matrix* multiplication, as the name implies, involves multiplying two matrices.

You might suspect that matrix multiplication would be simply a matter of multiplying the corresponding entries of the matrices. However, that is not the case. We define matrix multiplication in a way that makes the operation useful for certain applications.

Suppose that we are given the following matrices A and B and we wish to determine their product, AB.

$$A = [2 \quad -1 \quad 5] \qquad B = \begin{bmatrix} 3 \\ 6 \\ -4 \end{bmatrix}$$

To perform the multiplication, we multiply the entries in the row of matrix A by the entries in the column of matrix B, and we add the results.

$$AB = [(2)(3) + (-1)(6) + (5)(-4)]$$
$$= [-20]$$

We call this process *row-by-column* multiplication. This example is somewhat simple because matrix A has only one row and matrix B has only one column. More row-by-column multiplications are required when the matrices are larger. Consider the following matrices:

$$A = \begin{bmatrix} 1 & -3 \\ 2 & 5 \end{bmatrix} \qquad B = \begin{bmatrix} 4 & 0 \\ 6 & -1 \end{bmatrix}$$

$$AB = \begin{bmatrix} (1)(4) + (-3)(6) & \\ & \end{bmatrix}$$

Multiply the row 1 entries of A by the column 1 entries of B and add the results.

$$= \begin{bmatrix} -14 & (1)(0) + (-3)(-1) \\ & \end{bmatrix}$$

Multiply the row 1 entries of A by the column 2 entries of B and add the results.

$$= \begin{bmatrix} -14 & 3 \\ (2)(4) + (5)(6) & \end{bmatrix}$$

Multiply the row 2 entries of A by the column 1 entries of B and add the results.

$$= \begin{bmatrix} -14 & 3 \\ 38 & (2)(0) + (5)(-1) \end{bmatrix}$$

Multiply the row 2 entries of A by the column 2 entries of B and add the results.

$$= \begin{bmatrix} -14 & 3 \\ 38 & -5 \end{bmatrix}$$

The following are two important aspects of matrix multiplication.

If A is an $m \times n$ matrix and B is an $n \times p$ matrix, then

1. we can perform matrix multiplication only if the number of columns of A is the same as the number of rows of B.

Matrix A Matrix B

$m \times n$ $n \times p$

These numbers
must be the same.

2. the dimensions of AB are $m \times p$.

Matrix A Matrix B

$m \times n$ $n \times p$

These numbers
are the dimensions of AB.

EXAMPLE 5 **Matrix Multiplication**

Let $A = \begin{bmatrix} 5 & -1 & 3 \\ 2 & 4 & -2 \end{bmatrix}$ and $B = \begin{bmatrix} -4 & 1 \\ 0 & -5 \\ 2 & -3 \end{bmatrix}$

Determine the following matrix products:

(a) AB (b) BA

SOLUTION

(a)

$$AB = \begin{bmatrix} \overbrace{(5)(-4) + (-1)(0) + (3)(2)}^{\text{Row 1 times column 1}} & \overbrace{(5)(1) + (-1)(-5) + (3)(-3)}^{\text{Row 1 times column 2}} \\ \underbrace{(2)(-4) + (4)(0) + (-2)(2)}_{\text{Row 2 times column 1}} & \underbrace{(2)(1) + (4)(-5) + (-2)(-3)}_{\text{Row 2 times column 2}} \end{bmatrix}$$

$$= \begin{bmatrix} -14 & 1 \\ -12 & -12 \end{bmatrix} \qquad \text{Because } A \text{ is } 2 \times 3 \text{ and } B \text{ is } 3 \times 2, AB \text{ is } 2 \times 2.$$

(b)

$$BA = \begin{bmatrix} -18 & 8 & -14 \\ -10 & -20 & 10 \\ 4 & -14 & 12 \end{bmatrix} \qquad \begin{array}{l} \text{Because } B \text{ is } 3 \times 2 \text{ and } A \text{ is } 2 \times 3, \\ BA \text{ is } 3 \times 3. \end{array}$$

Example 5 illustrates the fact that matrix multiplication is not commutative; that is, in general, $AB \neq BA$.

■ **Matrix Product**   Matrix multiplication can be performed with a calculator.

FIGURE 5.27

```
[B][A]
 [[-18   8    -14]
  [-10  -20   10 ]
  [4    -14   12 ]]
```

After we entered matrices A and B in Example 5 into a calculator, we obtained the matrix product BA shown in Figure 5.27.

Although we now know the method for performing matrix multiplication, we have not yet defined the operation formally.

> *Definition of Matrix Multiplication*
>
> If $A = [a_{ij}]$ is an $m \times p$ matrix and $B = [b_{ij}]$ is a $p \times n$ matrix, then the product AB is an $m \times n$ matrix $AB = [c_{ij}]$, where
>
> $$c_{ij} = a_{i1}b_{1j} + a_{i2}b_{2j} + \cdots + a_{in}b_{nj}$$
>
> In words, to determine the element in row i, column j of AB, multiply the elements in row i of matrix A by the elements in column j of matrix B, and add the results.

Matrix multiplication with a calculator is faster and more accurate than applying the definition by hand.

<u>EXAMPLE</u> **6**   Matrix Multiplication

Let $A = \begin{bmatrix} 4 & 5 \\ -2 & 3 \end{bmatrix}$, $B = \begin{bmatrix} 1 & -6 & 0 \\ 5 & 3 & -2 \end{bmatrix}$, and $C = \begin{bmatrix} 5 \\ 2 \end{bmatrix}$. Use your calculator to determine (a) AB and (b) BC.

SOLUTION

(a) The matrix product AB is shown in Figure 5.28.

FIGURE **5.28**

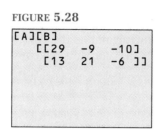

(b)

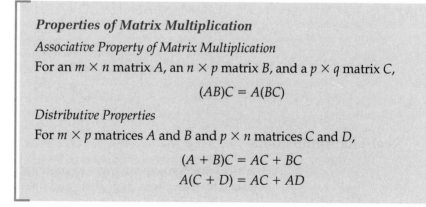

The product BC is not defined. Your calculator may display an error message indicating that the dimensions of the matrices do not permit matrix multiplication.

Many familiar properties of multiplication of real numbers do not apply to matrix multiplication. The following summarizes properties that are valid.

> **Properties of Matrix Multiplication**
>
> *Associative Property of Matrix Multiplication*
> For an $m \times n$ matrix A, an $n \times p$ matrix B, and a $p \times q$ matrix C,
> $$(AB)C = A(BC)$$
> *Distributive Properties*
> For $m \times p$ matrices A and B and $p \times n$ matrices C and D,
> $$(A + B)C = AC + BC$$
> $$A(C + D) = AC + AD$$

One important use of matrix multiplication is in the representation of a system of equations as a matrix equation. In the next section, we will see how such a matrix equation can be used to determine the solution of the system. For now, we will focus just on the representation.

EXAMPLE 7 **Using Matrices to Represent a System of Equations**

Write the system of equations as a matrix equation.

$$\begin{aligned} x + 3y - z &= 6 \\ 2x - y + 5z &= -9 \\ 4x + y - 3z &= 11 \end{aligned}$$

SOLUTION

We can write the left side of the system as the product of the coefficient matrix (which contains the coefficients of the variables) and a column matrix (which contains the variables x, y, and z). The right side is a column matrix that contains the constants.

$$\begin{bmatrix} 1 & 3 & -1 \\ 2 & -1 & 5 \\ 4 & 1 & -3 \end{bmatrix} \begin{bmatrix} x \\ y \\ z \end{bmatrix} = \begin{bmatrix} 6 \\ -9 \\ 11 \end{bmatrix}$$

Just as the number 1 is the multiplicative identity for the real numbers, the $n \times n$ matrix with 1s on the main diagonal and zeros elsewhere is the **identity matrix** I.

$$I = \begin{bmatrix} 1 & 0 & 0 & \cdot & \cdot & \cdot & 0 \\ 0 & 1 & 0 & \cdot & \cdot & \cdot & 0 \\ 0 & 0 & 1 & \cdot & \cdot & \cdot & 0 \\ \cdot & \cdot & \cdot & \cdot & \cdot & \cdot & \cdot \\ 0 & 0 & 0 & \cdot & \cdot & \cdot & 1 \end{bmatrix}$$

The following illustrates the fact that for any $n \times n$ matrix A, $AI = IA = A$.

If $A = \begin{bmatrix} 4 & -2 \\ 3 & 5 \end{bmatrix}$, then

$$AI = \begin{bmatrix} 4 & -2 \\ 3 & 5 \end{bmatrix} \begin{bmatrix} 1 & 0 \\ 0 & 1 \end{bmatrix} = \begin{bmatrix} 4 & -2 \\ 3 & 5 \end{bmatrix} = A$$

and

$$IA = \begin{bmatrix} 1 & 0 \\ 0 & 1 \end{bmatrix} \begin{bmatrix} 4 & -2 \\ 3 & 5 \end{bmatrix} = \begin{bmatrix} 4 & -2 \\ 3 & 5 \end{bmatrix} = A$$

Real-Life Applications

Matrices are convenient ways to organize data. By performing operations on such matrices, we can analyze and interpret the data.

EXAMPLE 8 **Car Rentals**

A car rental company has two locations in a city, one at the airport and another downtown. The tables show the average number of cars rented per day on a weekday and on the weekend.

Weekday	Airport	Downtown
Compact	28	20
Midsize	36	18
Luxury	15	12

Weekend	Airport	Downtown
Compact	25	18
Midsize	30	20
Luxury	10	8

(a) Write 3×2 matrices D and S for the number of cars rented on a weekday and on the weekend.

(b) Use scalar multiplication to write a matrix for the number of cars rented on weekdays. Use scalar multiplication and addition to write a matrix for the total number of cars rented for the entire week.

(c) Suppose that the rental cost per day is \$25 for compact, \$32 for midsize, and \$45 for luxury. Write a row matrix C for the cost.

(d) Determine and interpret the product CD.

SOLUTION

(a) The columns are associated with the location and the rows with the type of car.

$$
\text{Weekday} \qquad\qquad \text{Weekend}
$$

$$
D = \begin{bmatrix} 28 & 20 \\ 36 & 18 \\ 15 & 12 \end{bmatrix} \qquad S = \begin{bmatrix} 25 & 18 \\ 30 & 20 \\ 10 & 8 \end{bmatrix}
$$

(b) The matrix $5D$ represents the number of cars rented on weekdays, and $2S$ represents the number of cars rented on Saturday or Sunday. The matrix $5D + 2S$ represents the total number of cars rented during the week.

$$
5D = 5\begin{bmatrix} 28 & 20 \\ 36 & 18 \\ 15 & 12 \end{bmatrix} = \begin{bmatrix} 140 & 100 \\ 180 & 90 \\ 75 & 60 \end{bmatrix} \qquad 2S = 2\begin{bmatrix} 25 & 18 \\ 30 & 20 \\ 10 & 8 \end{bmatrix} = \begin{bmatrix} 50 & 36 \\ 60 & 40 \\ 20 & 16 \end{bmatrix}
$$

$$
5D + 2S = \begin{bmatrix} 140 & 100 \\ 180 & 90 \\ 75 & 60 \end{bmatrix} + \begin{bmatrix} 50 & 36 \\ 60 & 40 \\ 20 & 16 \end{bmatrix} = \begin{bmatrix} 190 & 136 \\ 240 & 130 \\ 95 & 76 \end{bmatrix}
$$

(c) The elements of matrix C are the respective costs for compact, midsize, and luxury cars.

$$
C = [25 \quad 32 \quad 45]
$$

(d) $CD = [25 \quad 32 \quad 45]\begin{bmatrix} 28 & 20 \\ 36 & 18 \\ 15 & 12 \end{bmatrix} = [2527 \quad 1616]$

The average daily receipts for cars rented on weekdays are \$2527 at the airport and \$1616 at the downtown location.

5.4 Quick Reference

Equality of Matrices

■ A matrix A can be denoted by $[a_{ij}]$, where a_{ij} is the entry in row i and column j.

■ Matrices $A = [a_{ij}]$ and $B = [b_{ij}]$ are equal if $a_{ij} = b_{ij}$ for all i and j.

In words, two matrices are equal if their corresponding elements are equal.

Addition and Subtraction of Matrices

■ Suppose that $A = [a_{ij}]$ and $B = [b_{ij}]$ are both $m \times n$ matrices. Then the sum and the difference of the matrices are $m \times n$ matrices given by

$$A + B = [a_{ij} + b_{ij}] \qquad \text{Sum of matrices } A \text{ and } B$$
$$A - B = [a_{ij} - b_{ij}] \qquad \text{Difference of matrices } A \text{ and } B$$

In words, to add (or subtract) two matrices, add (or subtract) the corresponding elements.

■ The $m \times n$ matrix whose entries are all 0s is called the **zero matrix,** which is denoted by **O.** For a matrix A, $A + \mathbf{O} = A$.

■ For any matrix $A = [a_{ij}]$, the **additive inverse** is $-A = [-a_{ij}]$.

In words, each entry of $-A$ is the opposite of the corresponding entry of A.

Scalar Multiplication

■ For an $m \times n$ matrix $A = [a_{ij}]$ and a real number c, $cA = [ca_{ij}]$.

In words, to multiply a matrix A by a scalar c, multiply each element of A by c.

■ The following summarizes properties of matrix addition and scalar multiplication. For $m \times n$ matrices A, B, C, the zero matrix **O**, and real numbers c and d,

1. $A + B = B + A$ — Commutative Property of Matrix Addition
2. $A + (B + C) = (A + B) + C$ — Associative Property of Matrix Addition
3. $(cd)A = c(dA)$ — Associative Property of Scalar Multiplication
4. $1A = A$ — Multiplicative Scalar Identity Property
5. $A + \mathbf{O} = A$ — Additive Identity Property
6. $A + (-A) = \mathbf{O}$ — Property of Additive Inverses
7. $c(A + B) = cA + cB$ — Distributive Property
8. $(c + d)A = cA + dA$ — Distributive Property

Matrix Multiplication

■ We can multiply matrices A and B only if the number of columns of A is the same as the number of rows of B.

■ If A is an $m \times n$ matrix and B is an $n \times p$ matrix, then AB is an $m \times p$ matrix.

■ If $A = [a_{ij}]$ is an $m \times p$ matrix and $B = [b_{ij}]$ is a $p \times n$ matrix, then the product AB is an $m \times n$ matrix $AB = [c_{ij}]$, where

$$c_{ij} = a_{i1}b_{1j} + a_{i2}b_{2j} + \cdots + a_{in}b_{nj}$$

In words, to determine the element in row i, column j of AB, multiply the elements in row i of matrix A by the elements in column j of matrix B, and add the results.

■ The following are some properties of matrix multiplication:

Associative Property of Matrix Multiplication

For an $m \times n$ matrix A, an $n \times p$ matrix B, and a $p \times q$ matrix C,

$$(AB)C = A(BC)$$

Distributive Properties

For $m \times p$ matrices A and B and $p \times n$ matrices C and D,

$$(A + B)C = AC + BC$$
$$A(C + D) = AC + AD$$

■ In general, matrix multiplication is not commutative: $AB \neq BA$.

■ The $n \times n$ matrix with 1s on the main diagonal and 0s elsewhere is the **identity matrix** I. For any $n \times n$ matrix A, $AI = IA = A$.

5.4 Speaking the Language

1. If each entry in matrix B is the opposite of the corresponding entry in matrix A, then B is the ▨▨▨▨▨ of A.

2. If A is a matrix and c is a real number, then multiplying A by c is called ▨▨▨▨▨ multiplication.

3. If A and B are matrices, then the matrix product AB is defined only if the number of ▨▨▨▨ of A equals the number of ▨▨▨▨ of B.

4. A square matrix with 1s along its diagonal and 0s elsewhere is called a(n) ▨▨▨▨▨ matrix.

5.4 Exercises

Concepts and Skills

1. If A and B are equal matrices, what can you conclude about the entries of A and B?

2. When a matrix is denoted by $[a_{ij}]$, what is the meaning of a_{ij}?

In Exercises 3–6, determine a, b, and c so that the matrices are equal.

3. $\begin{bmatrix} a + 3 \\ 5 \end{bmatrix} = \begin{bmatrix} 0 \\ 3b \end{bmatrix}$

4. $[a \quad 3b \quad 9] = [-1 \quad 0 \quad 1 + 2c]$

5. $\begin{bmatrix} 2a & 3 \\ b - 5 & 0 \end{bmatrix} = \begin{bmatrix} 10 & 3 \\ -2 & c \end{bmatrix}$

6. $\begin{bmatrix} 2 & 3 & -5 \\ 4 & 2 & 0 \\ -1 & 2 & c^3 \end{bmatrix} = \begin{bmatrix} 1 + 2a & 3 & 5 \\ \sqrt{b} & 2 & 0 \\ -1 & 2 & 8 \end{bmatrix}$

In Exercises 7–14, for the given matrices, perform the indicated operation. Assume that the dimensions of I and $\mathbf{O}$ are such that the indicated operation can be performed.

$$A = \begin{bmatrix} 1 & 2 \\ -1 & 5 \end{bmatrix} \qquad B = \begin{bmatrix} 2 & 0 \\ -1 & 3 \end{bmatrix}$$

7. $A + B$

8. $B - 3A$

9. $4B$

10. $-7A$

11. IA

12. $\mathbf{O}B$

13. AB

14. BA

In Exercises 15–22, for the given matrices, perform the indicated operation. Assume that the dimensions of I and $\mathbf{O}$ are such that the indicated operation can be performed.

$$C = \begin{bmatrix} 1 & 0 & 2 \\ 3 & -1 & 1 \\ 1 & 2 & 0 \end{bmatrix} \qquad D = \begin{bmatrix} 4 & 1 & -2 \\ 1 & 2 & 5 \\ -1 & 2 & -1 \end{bmatrix}$$

15. $C - D$　　**16.** $D + C$　　**17.** $2C + D$

18. $D - 3C$　　**19.** $2I + C$　　**20.** $\mathbf{O} - 4C$

21. CD　　**22.** DC

In Exercises 23–26, for the given matrices, determine each of the following, if possible:

(a) $A + B$　(b) $A - 2B$　(c) $5A$　(d) AB　(e) BA

23. $A = \begin{bmatrix} 3 & -2 \\ 1 & 4 \end{bmatrix}$; $B = \begin{bmatrix} 1 & 0 \\ 2 & -1 \end{bmatrix}$

24. $A = \begin{bmatrix} 1 & 2 \\ 0 & 1 \\ -1 & 3 \end{bmatrix}$; $B = \begin{bmatrix} 4 & 3 \\ 2 & 1 \\ 0 & -2 \end{bmatrix}$

25. $A = \begin{bmatrix} 1 \\ 3 \end{bmatrix}$; $B = [2 \quad -6]$

26. $A = \begin{bmatrix} 1 & 2 \\ -1 & 0 \end{bmatrix}$; $B = \begin{bmatrix} 0 \\ 1 \end{bmatrix}$

27. Suppose that A is an $m \times n$ matrix and B is a $p \times q$ matrix. Describe how $m, n, p,$ and q are related if
(a) A and B can be added.
(b) A and B can be multiplied.

28. Explain how to multiply a matrix A by a scalar c.

In Exercises 29–36, determine the indicated product, if possible.

29. $[2 \quad 3 \quad -1] \begin{bmatrix} 1 & 2 \\ -1 & 0 \\ 4 & -1 \end{bmatrix}$

30. $\begin{bmatrix} 1 & 3 & -1 \\ 4 & 2 & -3 \end{bmatrix} \begin{bmatrix} 2 \\ -5 \\ 3 \end{bmatrix}$

31. $\begin{bmatrix} 4 & -1 \\ -3 & 1 \\ 7 & 2 \end{bmatrix} \begin{bmatrix} 3 & 1 \\ -2 & 7 \end{bmatrix}$

32. $\begin{bmatrix} 5 & -1 & 6 \\ 2 & 0 & 1 \\ -4 & 2 & -1 \end{bmatrix} \begin{bmatrix} 1 & 2 \\ 3 & 4 \\ -1 & -3 \end{bmatrix}$

33. $\begin{bmatrix} 2 \\ 4 \end{bmatrix} \begin{bmatrix} 1 & 3 \\ 4 & 2 \\ -1 & 1 \end{bmatrix}$

34. $\begin{bmatrix} 2 & -1 \\ 3 & 2 \\ 1 & 3 \end{bmatrix} \begin{bmatrix} 1 & -1 \\ 3 & 2 \\ -1 & 1 \end{bmatrix}$

35. $\begin{bmatrix} 2 & -1 \\ 3 & -2 \\ 1 & 0 \end{bmatrix} \begin{bmatrix} 4 & 1 & -2 \\ 2 & 5 & -3 \end{bmatrix}$

36. $\begin{bmatrix} 2 \\ 1 \\ -3 \end{bmatrix} [4 \quad -2 \quad 6]$

In Exercises 37–40, write the given matrix equation as a system of equations.

37. $\begin{bmatrix} 1 & 3 \\ 2 & -1 \end{bmatrix} \begin{bmatrix} x \\ y \end{bmatrix} = \begin{bmatrix} 10 \\ -1 \end{bmatrix}$　**38.** $\begin{bmatrix} 2 & -5 \\ 5 & 3 \end{bmatrix} \begin{bmatrix} x \\ y \end{bmatrix} = \begin{bmatrix} -9 \\ -7 \end{bmatrix}$

39. $\begin{bmatrix} 2 & -3 & 1 \\ 1 & 1 & -4 \\ 3 & 2 & -1 \end{bmatrix} \begin{bmatrix} x \\ y \\ z \end{bmatrix} = \begin{bmatrix} -7 \\ 6 \\ 5 \end{bmatrix}$

40. $\begin{bmatrix} 1 & -2 & 0 \\ 0 & 3 & 1 \\ 2 & 1 & -1 \end{bmatrix} \begin{bmatrix} x \\ y \\ z \end{bmatrix} = \begin{bmatrix} -1 \\ 2 \\ 4 \end{bmatrix}$

In Exercises 41–44, write the given system of equations as a matrix equation.

41. $3x - y = 10$
$4x + 5y = 26$

42. $x + y = -1$
$-2x + y = 5$

43. $y + 3z = -4$
$x \quad + z = -1$
$x - 2y \quad = -3$

44. $x - y = 2$
$x + 2y + z = 5$
$y - z = -3$

45. Describe the form and the elements of the identity matrix I.

46. If A is a matrix, describe the elements of $-A$. What matrix is obtained when A and $-A$ are added?

Concept Extension

In Exercises 47 and 48, solve for X, where $A = \begin{bmatrix} 1 & 2 & 0 \\ -1 & 3 & -2 \end{bmatrix}$, $B = \begin{bmatrix} 2 & 0 & -5 \\ 1 & -3 & 1 \end{bmatrix}$, and X is a 2×3 matrix.

47. $2A + X = B$　　**48.** $4A = B - 3X$

49. If a and b are real numbers, then the Zero Factor Property states that if $ab = 0$, then $a = 0$ or $b = 0$. Use the following matrices to determine whether $AB = \mathbf{O}$ implies that $A = \mathbf{O}$ or $B = \mathbf{O}$.

$$A = \begin{bmatrix} 1 & 1 \\ 1 & 1 \end{bmatrix} \qquad B = \begin{bmatrix} 1 & 1 \\ -1 & -1 \end{bmatrix}$$

50. For the matrices in Exercise 49, show that $A(A + B) = A^2 + AB$.

In Exercises 51–54, for the matrices A and B, use your calculator to show that the given pairs of matrix expressions are *not* equal.

$$A = \begin{bmatrix} 2 & -1 & 1 \\ 3 & 4 & -6 \\ 1 & 3 & 0 \end{bmatrix} \qquad B = \begin{bmatrix} -3 & 0 & 5 \\ 2 & -1 & 1 \\ 4 & 3 & 2 \end{bmatrix}$$

51. (a) $A^2 - B^2$ (b) $(A + B)(A - B)$

52. (a) $(AB)^2$ (b) A^2B^2

53. (a) $(A + B)^2$ (b) $A^2 + 2AB + B^2$

54. (a) $(A - B)^2$ (b) $A^2 - 2AB + B^2$

Applications

55. Furniture Production A patio furniture company manufactures tables, chairs, and lounges. The table shows the daily production capabilities at the company's assembly plants in Albany and Canton.

	Albany	Canton
Tables	38	15
Chairs	46	28
Lounges	20	42

(a) Write a 3×2 matrix P to describe the production capabilities. What do the rows and columns represent?
(b) Suppose that the assembly plants operate 5 days per week. What would the matrix $5P$ represent?
(c) After upgrading the equipment, production increased by 20%. Write a matrix S for the amount of increase in production.
(d) Determine the matrix $P + S$. What does it represent?

56. Lunch Menu A restaurant offers four lunch combo specials. The table shows the restaurant's cost (in dollars) per item included in each lunch.

	Combo 1	Combo 2	Combo 3	Combo 4
Sandwich	1.05	1.55	1.30	1.70
Side dish	0.40	0.90	0.80	1.00
Dessert	0.30	1.05	1.00	1.20

(a) Write a 3×4 matrix C for the restaurant's cost.

(b) Suppose that overhead is reduced, and so the cost of all items is decreased by 10%. Write a matrix D for the decrease in cost.
(c) Determine and interpret $C - D$.
(d) What would the matrix $C + 0.25C$ represent?

57. Refer to Exercise 55.

The table gives the production cost and revenue (both in dollars) for the items.

	Tables	Chairs	Lounges
Cost	67	22	58
Revenue	84	35	74

(a) Write a row matrix C for the cost.
(b) Determine and interpret the matrix CP.
(c) Write a row matrix R for the revenue.
(d) Determine and interpret the matrix RP.
(e) Determine and interpret $RP - CP$.
(f) Determine and interpret $R - C$.
(g) Determine and interpret $(R - C)P$.

58. Refer to Exercise 56.
(a) Let $A = [1 \quad 1 \quad 1]$. Determine and interpret AC.
(b) Suppose the restaurant offers the "Double D" option (combo with double desserts). Write a row matrix B to describe the option.
(c) Determine and interpret BC.
(d) The table shows the number of combo meals sold in one day. Write a column matrix S for the number of meals sold.

	Combo 1	Combo 2	Combo 3	Combo 4
Number sold	18	24	15	12

(e) Determine and interpret the product CS.

59. Traffic Count The table shows the average hourly traffic count at a certain bridge.

	Cars	Trucks
Monday–Thursday	64	42
Friday	75	52
Saturday	93	28
Sunday	51	12

(a) Write the traffic count information as a 4×2 matrix C.

(b) On a day with moderate snowfall, the traffic count is half the normal count. What matrix would represent the traffic count on such a day?

(c) What does the matrix $24C$ represent?

(d) Suppose the matrix in part (c) is multiplied by the matrix $M = [4 \ 1 \ 1 \ 1]$. What does the result $M(24C)$ represent?

60. Vehicle Usage The table shows the annual average number of miles (in thousands) driven by cars, buses, and trucks. (*Source*: U.S. Federal Highway Administration.)

	Cars	Buses	Trucks
1980	9.1	11.5	11.9
1985	9.6	8.2	12.7
1990	10.5	9.1	13.8
1995	11.8	9.6	27.0

(a) Write a 4×3 matrix M for the number of miles driven.

(b) Suppose that N is a 3×1 column matrix whose entries represent the average number of cars, buses, and trucks during the period 1980–1995. What would the product MN represent?

61. Refer to Exercise 59.

The table gives the toll charged at the bridge, the percentage of cars from out of town, and the per-vehicle number of units of pollution emitted in the area.

	Toll	Nonlocal Vehicles	Pollutants
Cars	$0.75	34%	15
Trucks	1.25	68%	22

(a) Write the information as a 2×3 matrix D.

(b) Determine and interpret the product CD.

62. Refer to Exercise 60.

Suppose that during the period 1980–1995, the fuel cost (in cents per mile) and the annual operating cost (in dollars per mile) are as shown in the table.

	Fuel Cost	Operating Cost
Cars	0.054	0.412
Buses	0.214	1.121
Trucks	0.152	0.873

(a) Write a 3×2 matrix C for the annual costs.

(b) Determine MC.

(c) What do the elements of MC represent?

63. Class Enrollment The following table gives the enrollment in selected classes at a certain college.

	Math	English	History
Day	258	180	146
Evening	135	110	94

(a) Write the data as a 2×3 matrix E.

(b) Suppose that 45% of day students and 54% of evening students are females. Write the information as a row matrix F.

(c) Determine the product FE. What do the elements represent?

64. College Bookstores The following table shows the number of biology, economics, history, and English books sold during a fall semester at three college bookstores.

	Biology	Economics	History	English
Campus	210	110	150	180
General	140	50	60	110
Bookery	90	70	90	130

(a) Write the data as a 3×4 matrix S.

(b) Suppose that sales declined by 25% in the spring semester. Write a matrix D for the number of units by which sales decreased.

(c) Determine and interpret $S - D$.

(d) What matrix represents the total sales for the fall and spring semesters?

65. Refer to Exercise 63.

(a) Write a row matrix M for the percentages of day and evening students who are male.

(b) What do the elements of ME represent?

(c) Determine and interpret $FE + ME$.

66. Refer to Exercise 64.

Suppose that the wholesale and retail book prices are those given in the table.

	Wholesale	Retail
Biology	$40	$65
Economics	25	45
History	32	58
English	28	50

(a) Write the price information as a 4×2 matrix P.
(b) Determine and interpret the product SP.
(c) What does the product $(2S - D)P$ represent?

5.5 *Inverse of a Matrix*

Definition   ■   *Constructing an Inverse Matrix*   ■
Solving Systems of Equations   ■   *Real-Life Applications*

Recall that every nonzero number a has a *multiplicative inverse* a^{-1} such that $a \cdot a^{-1} = 1$, where 1 is called the *identity element*. We use this fact to solve simple equations such as $3x = 12$. Multiplying both sides by $\frac{1}{3}$ results in the equation $x = 4$.

In this section, we consider a corresponding rule for matrices; that is, we define an *inverse matrix* A^{-1} such that $AA^{-1} = I$, where I is the *identity matrix*. As you will see, we can, under certain conditions, use an inverse matrix to solve a matrix equation. We begin with the definition and some related vocabulary.

Definition

The definition of the inverse of a matrix is similar to the definition of the multiplicative inverse of a real number.

> **Definition of the Inverse of a Matrix**
>
> For an $n \times n$ matrix A, if there is a matrix A^{-1} such that $AA^{-1} = A^{-1}A = I$, then A^{-1} is called the **inverse** of A.

As we will see, not all matrices have an inverse. If a matrix has an inverse, then we say that the matrix is **invertible** or **nonsingular.** Otherwise, we say that the matrix is **singular.** A nonsquare matrix does not have an inverse.

EXAMPLE **1**   **Verifying That Matrices Are Inverses**

Determine whether $B = \begin{bmatrix} 3 & 2 \\ 7 & 5 \end{bmatrix}$ is the inverse of $A = \begin{bmatrix} 5 & -2 \\ -7 & 3 \end{bmatrix}$.

SOLUTION

$$BA = \begin{bmatrix} 3 & 2 \\ 7 & 5 \end{bmatrix}\begin{bmatrix} 5 & -2 \\ -7 & 3 \end{bmatrix} = \begin{bmatrix} 1 & 0 \\ 0 & 1 \end{bmatrix}$$

$$AB = \begin{bmatrix} 5 & -2 \\ -7 & 3 \end{bmatrix}\begin{bmatrix} 3 & 2 \\ 7 & 5 \end{bmatrix} = \begin{bmatrix} 1 & 0 \\ 0 & 1 \end{bmatrix}$$

Because $BA = I$ and $AB = I$, B is the inverse of A, and we write

$$A^{-1} = \begin{bmatrix} 3 & 2 \\ 7 & 5 \end{bmatrix}$$

Constructing an Inverse Matrix

As we can see in Example 1, the elements of A^{-1} are not simply the reciprocals of the elements of A. To construct the inverse of a matrix, we use row operations.

DEVELOPING THE CONCEPT

Constructing an Inverse Matrix

Consider the matrix $A = \begin{bmatrix} 1 & -1 \\ -2 & 3 \end{bmatrix}$. We wish to determine a matrix

$A^{-1} = \begin{bmatrix} a_{11} & a_{12} \\ a_{21} & a_{22} \end{bmatrix}$ such that

$$A \cdot A^{-1} = I$$

$$\begin{bmatrix} 1 & -1 \\ -2 & 3 \end{bmatrix}\begin{bmatrix} a_{11} & a_{12} \\ a_{21} & a_{22} \end{bmatrix} = \begin{bmatrix} 1 & 0 \\ 0 & 1 \end{bmatrix}$$

$$\begin{bmatrix} 1a_{11} - 1a_{21} & 1a_{12} - 1a_{22} \\ -2a_{11} + 3a_{21} & -2a_{12} + 3a_{22} \end{bmatrix} = \begin{bmatrix} 1 & 0 \\ 0 & 1 \end{bmatrix} \quad \begin{array}{l}\text{Multiply the matrices on the}\\ \text{left side of the equation.}\end{array}$$

Because the two matrices are equal, the corresponding elements must be equal. Thus, to determine the elements of the inverse matrix, we need to solve the following two systems of equations:

$$\begin{array}{ll} 1a_{11} - 1a_{21} = 1 & 1a_{12} - 1a_{22} = 0 \\ -2a_{11} + 3a_{21} = 0 & -2a_{12} + 3a_{22} = 1 \end{array}$$

We can solve the systems by writing an augmented matrix for each and performing row operations.

$$\left[\begin{array}{cc|c} 1 & -1 & 1 \\ -2 & 3 & 0 \end{array}\right] \qquad \left[\begin{array}{cc|c} 1 & -1 & 0 \\ -2 & 3 & 1 \end{array}\right]$$

Note that the coefficient matrices are the same. Using previously discussed methods, we obtain the solutions $a_{11} = 3$, $a_{21} = 2$, $a_{12} = 1$, and $a_{22} = 1$. Thus the inverse matrix is

$$A^{-1} = \begin{bmatrix} 3 & 1 \\ 2 & 1 \end{bmatrix}$$

By multiplying matrices A and A^{-1}, we can verify that $AA^{-1} = A^{-1}A = I$.

$$AA^{-1} = \begin{bmatrix} 1 & -1 \\ -2 & 3 \end{bmatrix}\begin{bmatrix} 3 & 1 \\ 2 & 1 \end{bmatrix} = \begin{bmatrix} 1 & 0 \\ 0 & 1 \end{bmatrix}$$

$$A^{-1}A = \begin{bmatrix} 3 & 1 \\ 2 & 1 \end{bmatrix}\begin{bmatrix} 1 & -1 \\ -2 & 3 \end{bmatrix} = \begin{bmatrix} 1 & 0 \\ 0 & 1 \end{bmatrix}$$

Earlier, we observed that the coefficient matrix is the same for both systems of equations. A procedure for determining the inverse of a matrix uses this fact. Rather than solving the systems separately by writing two augmented matrices, we can combine the two matrices and perform the row operations once.

$$\begin{bmatrix} 1 & -1 & \big| & 1 & 0 \\ -2 & 3 & \big| & 0 & 1 \end{bmatrix}$$ Combine matrices A and I in the form $[A\,|\,I]$.

$$\begin{bmatrix} 1 & -1 & \big| & 1 & 0 \\ 0 & 1 & \big| & 2 & 1 \end{bmatrix}$$ Multiply row 1 by 2 and add to row 2. Now we have a 0 in row 2, column 1.

$$\begin{bmatrix} 1 & 0 & \big| & 3 & 1 \\ 0 & 1 & \big| & 2 & 1 \end{bmatrix}$$ Add row 2 to row 1. Now the left side is the identity matrix.

The matrix is now in the form $[I\,|\,A^{-1}]$. In other words, the entries on the right are the elements of A^{-1}.

$$A^{-1} = \begin{bmatrix} 3 & 1 \\ 2 & 1 \end{bmatrix}$$

The following is a summary of the method for determining an inverse matrix.

Constructing an Inverse Matrix

Let A be an $n \times n$ matrix. To construct the inverse matrix, perform the following operations:

1. Write the augmented matrix $[A\,|\,I]$.
2. If possible, perform row operations to convert the left side of the matrix into the identity matrix.
3. If the left side can be converted into the identity matrix, then the augmented matrix has the form $[I\,|\,A^{-1}]$; that is, the right side is the inverse matrix. Otherwise the matrix A is not invertible.

EXAMPLE **2**   **Determining the Inverse of a Matrix**

Determine the inverse of matrix $A = \begin{bmatrix} 1 & 1 \\ 3 & -1 \end{bmatrix}$.

SOLUTION

$$\left[\begin{array}{rr|rr} 1 & 1 & 1 & 0 \\ 3 & -1 & 0 & 1 \end{array}\right]$$

Write the matrix $[A \mid I]$.

$$\left[\begin{array}{rr|rr} 1 & 1 & 1 & 0 \\ 0 & -4 & -3 & 1 \end{array}\right]$$

Multiply row 1 by -3 and add to row 2.
Now we have a 0 in row 2, column 1.

$$\left[\begin{array}{rr|rr} 1 & 1 & 1 & 0 \\ 0 & 1 & \frac{3}{4} & -\frac{1}{4} \end{array}\right]$$

Divide row 2 by -4.
Now we have a 1 in row 2, column 2.

$$\left[\begin{array}{rr|rr} 1 & 0 & \frac{1}{4} & \frac{1}{4} \\ 0 & 1 & \frac{3}{4} & -\frac{1}{4} \end{array}\right]$$

Add -1 times row 2 to row 1.
Now the left side is the identity matrix.

The augmented matrix is now in the form $[I \mid A^{-1}]$, so the matrix on the right side is the inverse matrix.

$$A^{-1} = \left[\begin{array}{rr} \frac{1}{4} & \frac{1}{4} \\ \frac{3}{4} & -\frac{1}{4} \end{array}\right]$$

To verify our result, we multiply A and A^{-1}.

$$\left[\begin{array}{rr} 1 & 1 \\ 3 & -1 \end{array}\right]\left[\begin{array}{rr} \frac{1}{4} & \frac{1}{4} \\ \frac{3}{4} & -\frac{1}{4} \end{array}\right] = \left[\begin{array}{rr} 1 & 0 \\ 0 & 1 \end{array}\right] \text{ and } \left[\begin{array}{rr} \frac{1}{4} & \frac{1}{4} \\ \frac{3}{4} & -\frac{1}{4} \end{array}\right]\left[\begin{array}{rr} 1 & 1 \\ 3 & -1 \end{array}\right] = \left[\begin{array}{rr} 1 & 0 \\ 0 & 1 \end{array}\right]$$

■ **Inverse**   You can use a calculator to determine the inverse of a matrix.

Figure 5.29 shows the results of using a calculator to determine A^{-1} in Example 2.

FIGURE **5.29**

```
[A]
            [[1   1]
             [3  -1]]
[A]⁻¹
        [[.25 .25 ]
         [.75 -.25]]
```

FIGURE **5.30**

```
[A]
            [[1   1 ]
             [3  -1]]
[A]⁻¹▸Frac
        [[1/4   1/4 ]
         [3/4  -1/4]]
```

By using the fraction option, you can display the elements of the inverse matrix as fractions. (See Figure 5.30.)

EXAMPLE **3**   **Determining the Inverse of a Matrix**

Determine the inverse of the matrix $A = \begin{bmatrix} 1 & -1 & 1 \\ 1 & 0 & 2 \\ 0 & 2 & 3 \end{bmatrix}$.

SOLUTION

$$\left[\begin{array}{ccc|ccc} 1 & -1 & 1 & 1 & 0 & 0 \\ 1 & 0 & 2 & 0 & 1 & 0 \\ 0 & 2 & 3 & 0 & 0 & 1 \end{array}\right]$$ Write the matrix $[A \mid I]$.

$$\left[\begin{array}{ccc|ccc} 1 & -1 & 1 & 1 & 0 & 0 \\ 0 & 1 & 1 & -1 & 1 & 0 \\ 0 & 2 & 3 & 0 & 0 & 1 \end{array}\right]$$ Multiply the first row by -1 and add to row 2. The first column now has 0s in rows 2 and 3.

$$\left[\begin{array}{ccc|ccc} 1 & 0 & 2 & 0 & 1 & 0 \\ 0 & 1 & 1 & -1 & 1 & 0 \\ 0 & 0 & 1 & 2 & -2 & 1 \end{array}\right]$$ Add row 2 to row 1. Multiply row 2 by -2 and add to row 3. Now the second column has 0s in rows 1 and 3.

$$\left[\begin{array}{ccc|ccc} 1 & 0 & 0 & -4 & 5 & -2 \\ 0 & 1 & 0 & -3 & 3 & -1 \\ 0 & 0 & 1 & 2 & -2 & 1 \end{array}\right]$$ Multiply row 3 by -2 and add to row 1. Then multiply row 3 by -1 and add to row 2. The left side is the identity matrix.

Now we can use the right side to write the inverse of A.

$$A^{-1} = \begin{bmatrix} -4 & 5 & -2 \\ -3 & 3 & -1 \\ 2 & -2 & 1 \end{bmatrix}$$

Example 4 illustrates the fact that not all matrices have an inverse.

EXAMPLE **4**   **A Singular Matrix**

Determine the inverse of matrix $A = \begin{bmatrix} 1 & -5 \\ -1 & 5 \end{bmatrix}$.

SOLUTION

■ *NOTE   If you use a calculator to attempt to determine the inverse in Example 4, the calculator will usually give an error message or indicate that the matrix is singular.*

$$\left[\begin{array}{cc|cc} 1 & -5 & 1 & 0 \\ -1 & 5 & 0 & 1 \end{array}\right]$$ Write the matrix $[A \mid I]$.

$$\left[\begin{array}{cc|cc} 1 & -5 & 1 & 0 \\ 0 & 0 & 1 & 1 \end{array}\right]$$ Add row 1 to row 2.

Because the last row of the left side contains all 0s, we cannot write the left side as the identity matrix. Thus matrix A is singular (not invertible).

Solving Systems of Equations

We can use inverse matrices to solve certain systems of equations.

DEVELOPING THE CONCEPT

Using Inverse Matrices to Solve a System of Equations

As we saw in the previous section, a system of equations can be written as a product of matrices.

$$4x + 3y = -2 \qquad \begin{bmatrix} 4 & 3 \\ 3 & -2 \end{bmatrix} \begin{bmatrix} x \\ y \end{bmatrix} = \begin{bmatrix} -2 \\ 7 \end{bmatrix}$$
$$3x - 2y = 7$$

When a system is written in the form $AX = B$, we can multiply both sides of the equation by A^{-1} (if it exists) to obtain the solution of the system.

$$AX = B$$
$$A^{-1}AX = A^{-1}B \qquad \text{Multiply both sides of the equation by } A^{-1}.$$
$$IX = A^{-1}B \qquad A^{-1}A = I, \text{ the identity matrix.}$$
$$X = A^{-1}B \qquad \text{Because } I \text{ is the identity matrix, } IX = X.$$

By equating the elements of X with the elements of $A^{-1}B$, we can write the solution of the system.

This procedure gives us a way to solve certain systems of n equations in n variables.

> ### Solving Systems of Equations
>
> If a system of n equations in n variables is written in the form $AX = B$, and if A^{-1} exists, then the system has a unique solution given by $X = A^{-1}B$.

EXAMPLE 5   **Solving a System of Two Equations**

Solve the following system of equations:

$$4x + 3y = -2$$
$$3x - 2y = 7$$

SOLUTION

VISUALIZING THE SOLUTION

Solving the equations for y, we obtain $y_1 = -\frac{4}{3}x - \frac{2}{3}$ and $y_2 = \frac{3}{2}x - \frac{7}{2}$. Figure 5.31 shows that the graphs intersect at $(1, -2)$, which is the solution of the system.

FIGURE 5.31

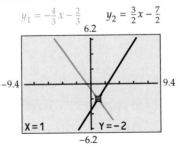

$y_1 = -\frac{4}{3}x - \frac{2}{3} \qquad y_2 = \frac{3}{2}x - \frac{7}{2}$

DETERMINING THE SOLUTION

First, we write the system of equations in the matrix form $AX = B$.

$$\begin{bmatrix} 4 & 3 \\ 3 & -2 \end{bmatrix}\begin{bmatrix} x \\ y \end{bmatrix} = \begin{bmatrix} -2 \\ 7 \end{bmatrix}$$

The coefficient matrix $A = \begin{bmatrix} 4 & 3 \\ 3 & -2 \end{bmatrix}$ has an inverse $A^{-1} = \begin{bmatrix} \frac{2}{17} & \frac{3}{17} \\ \frac{3}{17} & -\frac{4}{17} \end{bmatrix}$.

Multiplying both sides of the matrix equation by A^{-1}, we obtain

$$\begin{bmatrix} \frac{2}{17} & \frac{3}{17} \\ \frac{3}{17} & -\frac{4}{17} \end{bmatrix}\begin{bmatrix} 4 & 3 \\ 3 & -2 \end{bmatrix}\begin{bmatrix} x \\ y \end{bmatrix} = \begin{bmatrix} \frac{2}{17} & \frac{3}{17} \\ \frac{3}{17} & -\frac{4}{17} \end{bmatrix}\begin{bmatrix} -2 \\ 7 \end{bmatrix}$$

Now the equation is in the form $A^{-1}AX = A^{-1}B$, which simplifies to $IX = A^{-1}B$, or $X = A^{-1}B$.

FIGURE 5.32

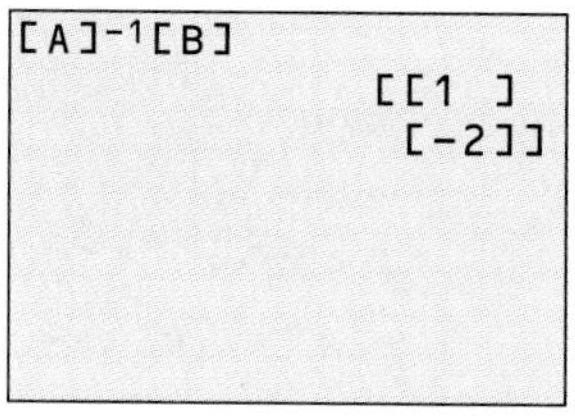

```
[A]⁻¹[B]
        [[1  ]
         [-2]]
```

$$\begin{bmatrix} x \\ y \end{bmatrix} = \begin{bmatrix} 1 \\ -2 \end{bmatrix}$$

Equating the elements of the two matrices, we see that $x = 1$ and $y = -2$. Thus the solution of the system is $(1, -2)$.

After we enter matrices A and B in a calculator, we can obtain the solution by entering $A^{-1}B$, as shown in Figure 5.32.

EXAMPLE 6   **Solving a System of Three Equations**

Solve the following system of equations:

$$\begin{aligned} x - y + z &= -2 \\ x + 2z &= 1 \\ 2y + 3z &= 5 \end{aligned}$$

SOLUTION

$$\begin{bmatrix} 1 & -1 & 1 \\ 1 & 0 & 2 \\ 0 & 2 & 3 \end{bmatrix}\begin{bmatrix} x \\ y \\ z \end{bmatrix} = \begin{bmatrix} -2 \\ 1 \\ 5 \end{bmatrix} \qquad \begin{array}{l}\text{Write the system as a matrix} \\ \text{equation: } AX = B.\end{array}$$

We determined the inverse of matrix A in Example 3:

$$\begin{bmatrix} -4 & 5 & -2 \\ -3 & 3 & -1 \\ 2 & -2 & 1 \end{bmatrix}\begin{bmatrix} 1 & -1 & 1 \\ 1 & 0 & 2 \\ 0 & 2 & 3 \end{bmatrix}\begin{bmatrix} x \\ y \\ z \end{bmatrix} = \begin{bmatrix} -4 & 5 & -2 \\ -3 & 3 & -1 \\ 2 & -2 & 1 \end{bmatrix}\begin{bmatrix} -2 \\ 1 \\ 5 \end{bmatrix} \qquad \begin{array}{l}\textbf{Multiply} \\ \textbf{both sides} \\ \textbf{by } A^{-1}.\end{array}$$

$$\begin{bmatrix} x \\ y \\ z \end{bmatrix} = \begin{bmatrix} 3 \\ 4 \\ -1 \end{bmatrix} \qquad \begin{array}{l}A^{-1}AX = A^{-1}B \text{ simplifies to} \\ X = A^{-1}B.\end{array}$$

Thus $x = 3$, $y = 4$, and $z = -1$. The solution is $(3, 4, -1)$. As before, we would obtain the same result by entering matrices A and B in a calculator and computing $A^{-1}B$.

Real-Life Applications

Matrices are used in situations where secure messages must be transmitted. A message might first be written as a matrix M with 1 for A, 2 for B, and so on. A space is coded with a 0. Then the matrix is multiplied by an *encoding matrix* C to scramble the message and make it more difficult for an unauthorized person to decipher.

The individual who receives the message multiplies the encoded message matrix by a decoding matrix C^{-1} in order to obtain the original message matrix.

In Example 7, we used a calculator to determine inverse matrices and to perform matrix multiplication. We show only the results.

EXAMPLE **7** **Encoding and Decoding a Message**

Write the message LAUNCH AT NOON as a 5×3 matrix. Then use the

encoding matrix $C = \begin{bmatrix} 5 & 3 & -2 \\ -6 & -4 & -1 \\ 3 & 2 & 1 \end{bmatrix}$ to encode the message.

SOLUTION

Using 0 for a space, 1 for A, 2 for B, and so on, we choose a 5×3 matrix M to write the message LAUNCH AT NOON. (Note that the encoding matrix C has 3 rows, so the message matrix must have 3 columns in order for us to multiply M times C.)

$$
\begin{array}{cc}
\text{Message} & \begin{array}{c}\text{Message}\\\text{Matrix}\end{array}\\[4pt]
\begin{bmatrix} L & A & U \\ N & C & H \\ - & A & T \\ - & N & O \\ O & N & - \end{bmatrix} &
M = \begin{bmatrix} 12 & 1 & 21 \\ 14 & 3 & 8 \\ 0 & 1 & 20 \\ 0 & 14 & 15 \\ 15 & 14 & 0 \end{bmatrix}
\end{array}
$$

The sender encodes the message by multiplying the message matrix M by a coding matrix C.

$$
MC = \overset{\begin{array}{c}\text{Message}\\\text{Matrix}\end{array}}{\begin{bmatrix} 12 & 1 & 21 \\ 14 & 3 & 8 \\ 0 & 1 & 20 \\ 0 & 14 & 15 \\ 15 & 14 & 0 \end{bmatrix}} \overset{\begin{array}{c}\text{Coding}\\\text{Matrix}\end{array}}{\begin{bmatrix} 5 & 3 & -2 \\ -6 & -4 & -1 \\ 3 & 2 & 1 \end{bmatrix}} = \overset{\begin{array}{c}\text{Encoded}\\\text{Matrix}\end{array}}{\begin{bmatrix} 117 & 74 & -4 \\ 76 & 46 & -23 \\ 54 & 36 & 19 \\ -39 & -26 & 1 \\ -9 & -11 & -44 \end{bmatrix}}
$$

Upon receipt of the encoded message, the receiver decodes the message by multiplying the encoded matrix by the decoding matrix C^{-1}.

$$
\begin{array}{ccc}
\begin{array}{c}\text{Encoded}\\\text{Matrix}\end{array} & \begin{array}{c}\text{Decoding}\\\text{Matrix}\end{array} & \begin{array}{c}\text{Message}\\\text{Matrix}\end{array}
\end{array}
$$

$$
\begin{bmatrix} 117 & 74 & -4 \\ 76 & 46 & -23 \\ 54 & 36 & 19 \\ -39 & -26 & 1 \\ -9 & -11 & -44 \end{bmatrix}
\begin{bmatrix} 2 & 7 & 11 \\ -3 & -11 & -17 \\ 0 & 1 & 2 \end{bmatrix}
=
\begin{bmatrix} 12 & 1 & 21 \\ 14 & 3 & 8 \\ 0 & 1 & 20 \\ 0 & 14 & 15 \\ 15 & 14 & 0 \end{bmatrix}
$$

Note that the encoded matrix is multiplied by the decoding matrix *on the right*, so that $MCC^{-1} = MI = M$.

Finally, the receiver uses the simple conversion of A for 1, B for 2, and so on, to obtain the original message.

5.5 Quick Reference

Definition ■ For an $n \times n$ matrix A, if there is a matrix A^{-1} such that $AA^{-1} = A^{-1}A = I$, then A^{-1} is called the **inverse** of A.

■ If a matrix has an inverse, then we say that the matrix is **invertible** or **nonsingular.** Otherwise, we say that the matrix is **singular.**

Constructing an Inverse Matrix ■ To construct the inverse of an $n \times n$ matrix A, perform the following operations:

1. Write the augmented matrix $[A \mid I]$.
2. If possible, perform row operations to convert the left side of the matrix into the identity matrix.
3. If the left side can be converted into the identity matrix, then the augmented matrix has the form $[I \mid A^{-1}]$; that is, the right side is the inverse matrix. Otherwise, the matrix A is not invertible.

Solving Systems of Equations ■ If a system of n equations in n variables is written in the form $AX = B$, and if A^{-1} exists, then the system has a unique solution given by $X = A^{-1}B$.

5.5 Speaking the Language

1. If A and B are square matrices such that $AB = BA = I$, where I is the identity matrix, then B is the _____ of A.

2. If A^{-1} exists for a matrix A, then we say that A is invertible or _____ .

3. For a matrix A, to construct A^{-1} we write the augmented matrix $[A \mid I]$ and then perform row operations to convert the left side of the matrix into the _____ matrix.

4. If the steps in Question 3 can be performed, then the right side of the converted matrix is the _____ matrix.

5.5 Exercises

Concepts and Skills

1. How can you tell immediately that the following matrix does not have an inverse?

$$A = \begin{bmatrix} 2 & 5 & -1 \\ 0 & 3 & 4 \end{bmatrix}$$

2. Every nonzero real number a has an inverse a^{-1} such that $a \cdot a^{-1} = 1$. Is it true that every matrix A except $\mathbf{O}$ has an inverse A^{-1} such that $AA^{-1} = I$?

In Exercises 3–10, determine whether the given matrices are inverses.

3. $\begin{bmatrix} 3 & 0 \\ 0 & 2 \end{bmatrix}, \begin{bmatrix} \frac{1}{3} & 0 \\ 0 & \frac{1}{2} \end{bmatrix}$

4. $[3 \quad -2], \begin{bmatrix} \frac{1}{3} \\ \frac{1}{2} \end{bmatrix}$

5. $\begin{bmatrix} 2 & 3 \\ 3 & 4 \end{bmatrix}, \begin{bmatrix} -1 & -3 \\ 1 & 2 \end{bmatrix}$

6. $\begin{bmatrix} -4 & 3 \\ 3 & -2 \end{bmatrix}, \begin{bmatrix} 2 & 3 \\ 3 & 4 \end{bmatrix}$

7. $\begin{bmatrix} 1 & 2 \\ -4 & 3 \\ 2 & 5 \end{bmatrix}, \begin{bmatrix} 1 & \frac{1}{2} \\ -\frac{1}{4} & \frac{1}{3} \\ \frac{1}{2} & \frac{1}{5} \end{bmatrix}$

8. $\begin{bmatrix} -2 & -3 & 5 \\ -3 & -4 & 7 \\ 0 & 0 & 1 \end{bmatrix}, \begin{bmatrix} 4 & -3 & 1 \\ -3 & 2 & 1 \\ 0 & 0 & 1 \end{bmatrix}$

9. $\begin{bmatrix} 1 & 3 & 2 \\ 0 & 1 & 3 \\ 0 & 0 & 1 \end{bmatrix}, \begin{bmatrix} 1 & -3 & 7 \\ 0 & 1 & -3 \\ 0 & 0 & 1 \end{bmatrix}$

10. $\begin{bmatrix} 1 & 2 & 1 \\ -2 & 1 & -1 \end{bmatrix}, \begin{bmatrix} 2 & -1 \\ 1 & 1 \\ -3 & -1 \end{bmatrix}$

In Exercises 11–24, without a calculator, determine the inverse of the given matrix, if it exists.

11. $\begin{bmatrix} 3 & -7 \\ 2 & -5 \end{bmatrix}$

12. $\begin{bmatrix} -2 & 1 \\ 7 & -4 \end{bmatrix}$

13. $\begin{bmatrix} 3 & 2 \\ 5 & 4 \end{bmatrix}$

14. $\begin{bmatrix} 3 & 3 \\ -6 & -5 \end{bmatrix}$

15. $\begin{bmatrix} 15 & -12 \\ 10 & -8 \end{bmatrix}$

16. $\begin{bmatrix} -6 & 9 \\ 4 & -6 \end{bmatrix}$

17. $\begin{bmatrix} 2 & 5 & 6 \\ 1 & -1 & 2 \\ 1 & 2 & 3 \end{bmatrix}$

18. $\begin{bmatrix} 1 & 0 & 1 \\ 2 & 1 & 2 \\ -1 & 1 & 2 \end{bmatrix}$

19. $\begin{bmatrix} 1 & 2 & 1 \\ 0 & 2 & -1 \\ 1 & 0 & 1 \end{bmatrix}$

20. $\begin{bmatrix} -2 & 0 & 1 \\ 3 & 1 & -1 \\ -1 & 1 & 2 \end{bmatrix}$

21. $\begin{bmatrix} 4 & -1 & -3 \\ -1 & 2 & 4 \\ -6 & 5 & 11 \end{bmatrix}$

22. $\begin{bmatrix} 2 & 1 & -1 \\ -1 & 4 & 5 \\ 5 & -2 & -7 \end{bmatrix}$

23. $\begin{bmatrix} -1 & 1 & -2 \\ 1 & -2 & 0 \\ 2 & -1 & 1 \end{bmatrix}$

24. $\begin{bmatrix} 1 & 2 & 3 \\ -1 & -2 & 1 \\ 2 & 1 & 2 \end{bmatrix}$

In Exercises 25–32, use a calculator to find the inverse of the given matrix.

25. $\begin{bmatrix} 10 & -4 \\ 6 & -2 \end{bmatrix}$

26. $\begin{bmatrix} 6 & 5 \\ 3 & 2 \end{bmatrix}$

27. $\begin{bmatrix} 1 & 0 & 0 \\ -2 & 1 & 0 \\ 4 & -2 & 1 \end{bmatrix}$

28. $\begin{bmatrix} -3 & 2 & 4 \\ 1 & 0 & -3 \\ 2 & -3 & -1 \end{bmatrix}$

29. $\begin{bmatrix} 1 & 2 & 3 & 4 \\ 1 & 0 & 1 & 1 \\ -1 & 2 & -2 & 0 \\ 2 & -1 & 3 & 2 \end{bmatrix}$

30. $\begin{bmatrix} -2 & 0 & -3 & 1 \\ 3 & 1 & 2 & -1 \\ 0 & 2 & 3 & 1 \\ 1 & -1 & 2 & 0 \end{bmatrix}$

31. $\begin{bmatrix} 2 & 1 & -3 \\ 1 & 0 & 1 \\ 1 & -1 & 6 \end{bmatrix}$

32. $\begin{bmatrix} 3 & 2 & -1 & 0 \\ 0 & 0 & 2 & 1 \\ 1 & 3 & 1 & -1 \\ 2 & -1 & -6 & -1 \end{bmatrix}$

33. To construct the inverse of matrix A, we can begin by writing the augmented matrix $[A \mid I]$. In what form do we try to write this matrix by performing row operations?

34. If we write a system of equations as a matrix equation of the form $AX = B$, what are the matrices A and B?

In Exercises 35–46, use inverse matrix methods to solve the given system of equations.

35. $3x + 2y = -5$
    $4x + 3y = -6$

36. $2x + y = -2$
    $5x + 2y = -3$

37. $2x + 3y = -1$
    $-6x - 10y = 4$

38. $2x - 3y = 8$
    $-3x + 6y = -15$

39. $x - y + z = -5$
    $2x - y = 0$
    $2y - 2z = 12$

40. $y - z = -2$
    $2x - y + 3z = 16$
    $x + y + 2z = 13$

41.
$$\begin{aligned} 2x - y &= -1 \\ x + 3y + 2z &= 11 \\ -2x + y - z &= -3 \end{aligned}$$

42.
$$\begin{aligned} -6x + y + 5z &= -8 \\ 2x - y - 3z &= 4 \\ -3x + 2y + 6z &= -7 \end{aligned}$$

43.
$$\begin{aligned} 2x + 4y + 6z &= 10 \\ -2x + 2y - z &= -1 \\ x + 3z &= 7 \end{aligned}$$

44.
$$\begin{aligned} 3x - 6y + 4z &= 19 \\ x + 2y - z &= -11 \\ -2x + 6y + z &= -2 \end{aligned}$$

45.
$$\begin{aligned} -3x + 3y + 7z + 5w &= 24 \\ x - 2y - 5z - w &= -7 \\ -x + y + 2z + w &= 6 \\ 3x - 2y - 4z - w &= -11 \end{aligned}$$

46.
$$\begin{aligned} -x + y + 2w &= -8 \\ x + z - w &= 6 \\ 2y - z &= -11 \\ 2x - y + z + w &= 10 \end{aligned}$$

47. Suppose that you are solving a system of two equations in two variables and you write the matrix equation $AX = B$. To solve for X, we multiply both sides by A^{-1}. Why must we write $A^{-1}AX = A^{-1}B$ rather than $AXA^{-1} = BA^{-1}$?

48. Suppose that performing row operations on $[A \mid I]$ results in the matrix $\begin{bmatrix} 3 & 2 & | & -1 & 4 \\ 0 & 0 & | & -2 & 5 \end{bmatrix}$. What can you conclude?

Concept Extension

49. What is the inverse of $\begin{bmatrix} 1 & 0 \\ c & 1 \end{bmatrix}$?

50. What is the inverse of $\begin{bmatrix} 1 & c \\ 0 & 1 \end{bmatrix}$?

51. Show that $(AB)^{-1} = B^{-1}A^{-1}$.

52. What is the inverse of $\begin{bmatrix} a & 0 & 0 \\ 0 & b & 0 \\ 0 & 0 & c \end{bmatrix}$?

53. Suppose that $AB = AC$. Under what condition does $B = C$?

54. Let $A = \begin{bmatrix} 0 & 1 & 0 \\ 0 & 0 & 1 \\ 1 & 0 & 0 \end{bmatrix}$.

(a) What is A^3?

(b) What does your result in part (a) indicate about A^{-1}?

In Exercises 55 and 56, use the following matrix.
$$A = \begin{bmatrix} 1 & 0 & 0 \\ 1 & 1 & 0 \\ 1 & 0 & 1 \end{bmatrix}$$

55. Determine

(a) A^2 (b) A^3 (c) A^n

56. Determine

(a) A^{-1} (b) $(A^{-1})^2$ (c) $(A^{-1})^3$ (d) $(A^{-1})^n$

Applications

In Exercises 57 and 58, use the given coding matrix C to encode the given message. (Use the same coding scheme of 1 for A, 2 for B, and so on, as in Example 7.)

57. $C = \begin{bmatrix} 3 & 2 & 1 \\ -1 & 2 & 0 \\ 2 & 0 & 1 \end{bmatrix}$

Message: MEET ME AT JOES

58. $C = \begin{bmatrix} 2 & 1 & 0 & 1 \\ 0 & -1 & 1 & 2 \\ -1 & 1 & 0 & -1 \\ 3 & 2 & 1 & 0 \end{bmatrix}$

Message: BUY STOCK TUESDAY

In Exercises 59 and 60, use the given coding matrix C to decode the given encoded message.

59. $C = \begin{bmatrix} -5 & -1 & 3 & -1 \\ -6 & -1 & 3 & -1 \\ 4 & 0 & -3 & 1 \\ 9 & 1 & -5 & 2 \end{bmatrix}$

Encoded message matrix:
$$\begin{bmatrix} -33 & -20 & 10 & -2 \\ 108 & 1 & -76 & 30 \\ -49 & -16 & 20 & -2 \\ -213 & -38 & 114 & -38 \end{bmatrix}$$

60. $C = \begin{bmatrix} 3 & 5 & -2 \\ 4 & 6 & 1 \\ 2 & 3 & 1 \end{bmatrix}$

Encoded message matrix:
$$\begin{bmatrix} 84 & 127 & 23 \\ 53 & 81 & 5 \\ 116 & 183 & -20 \\ 22 & 35 & -3 \\ 148 & 234 & -19 \end{bmatrix}$$

5.6 *Systems of Inequalities and Linear Programming*

Linear Inequalities in Two Variables ■ *Systems of Inequalities*
■ *Linear Programming*

Just as *equations* in two variables are often useful models of the conditions of a problem, *inequalities* in two variables may be needed to describe other kinds of conditions. This is particularly true when the unknown quantities have maximum or minimum values.

Typically, application problems that involve inequalities will have more than one condition that must be satisfied, and we might model those conditions with a system of inequalities. As you will see, such systems usually have infinitely many solutions, and the task is to select the optimum (best) solution.

Your knowledge of this topic will give you another valuable tool for analyzing and solving real-life problems.

Linear Inequalities in Two Variables

The inequality $x + 2y < 8$ is an example of a **linear inequality in two variables.**

■ *NOTE   All the inequalities that we discuss in this section are linear inequalities in two variables. For brevity, we will refer to them simply as inequalities.*

> ### Definition of a Linear Inequality in Two Variables
>
> A **linear inequality in two variables** is an inequality that can be written in the form $Ax + By < C$, where A, B, and C are real numbers and A and B are not both 0. (The symbols $\leq$, $>$, and $\geq$ can also be used.)
>
> A **solution** of an inequality in two variables is an ordered pair (a, b) that satisfies the inequality.

EXPLORATION

The Graph of an Inequality

Use the integer setting to produce the graph of $y = x - 8$.

(a) Do any of the points of the line represent solutions of the inequality $y > x - 8$? Why?
(b) Move your general cursor to various points below the line. Do any of these points appear to be solutions of $y > x - 8$?
(c) Move your general cursor to various points above the line. Do any of these points appear to be solutions of $y > x - 8$?
(d) What is your conjecture about the graph of $y > x - 8$?

DEVELOPING THE CONCEPT

The Graph of an Inequality

Consider the inequality $y \leq x + 7$. Figure 5.33(a) shows the graph of the associated equation $y = x + 7$. Because the inequality symbol is $\leq$, the solutions of $y = x + 7$ are also solutions of $y \leq x + 7$. Thus the points of the line represent solutions of the inequality.

FIGURE 5.33

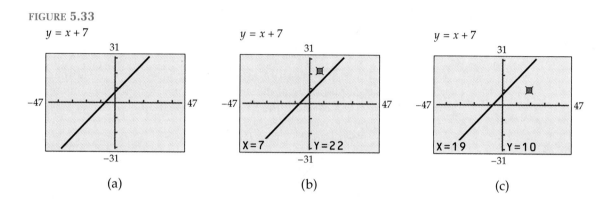

(a)   (b)   (c)

In Figure 5.33(b), the general cursor is on the *test point* (7, 22), which is above the line. We can verify by substitution that (7, 22) is *not* a solution of the inequality.

$$y \leq x + 7$$
$$22 \leq 7 + 7 \qquad \text{Replace } y \text{ with 22 and } x \text{ with 7.}$$
$$22 \leq 14 \qquad \text{False.}$$

By moving the cursor to various other points above the line and testing the coordinates of those points, we find that no point above the line represents a solution.

Figure 5.33(c) shows the cursor on the test point (19, 10), which is below the line. By substitution, we can see that (19, 10) *is* a solution of the inequality.

$$y \leq x + 7$$
$$10 \leq 19 + 7 \qquad \text{Replace } y \text{ with 10 and } x \text{ with 19.}$$
$$10 \leq 26 \qquad \text{True.}$$

In fact, by moving the cursor to various other points below the line and testing their coordinates, we discover that every point below the line represents a solution of the inequality.

In summary, the graph of $y \leq x + 7$ consists of the points of the line $y = x + 7$ and all points below that line. We highlight the points below the line with shading. (See Figure 5.34.)

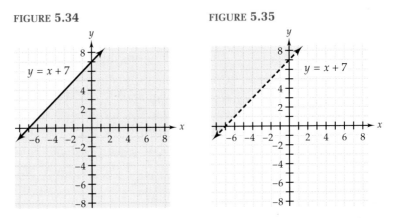

FIGURE **5.34**   FIGURE **5.35**

Figure 5.35 shows the graph of the inequality $y > x + 7$. Note that the graph consists of all points above the line $y = x + 7$ but not points of the line itself. We use a dashed line to indicate that the line $y = x + 7$ is not part of the graph.

For an inequality, the graph of the associated equation is called the **boundary line**. The boundary line divides the plane into two regions called **half-planes**.

The following is an alternative to the test point method for graphing an inequality.

Graphing an Inequality

Follow these steps for graphing an inequality.

1. Solve the inequality for y with y on the left side.
2. Draw the graph (a solid or dashed boundary line) of the associated equation.
3. If the inequality symbol in step 1 is $<$ or $\leq$, shade the half-plane below the line; if the inequality symbol in step 1 is $>$ or $\geq$, shade the half-plane above the line.
4. For a vertical boundary line, shade to the left if the inequality symbol is $<$ or $\leq$; shade to the right if the inequality symbol is $>$ or $\geq$.

EXAMPLE **1**   **Graphing a Linear Inequality**

Graph the given inequalities.

(a) $x - 3y > 12$
(b) $y + 3 \geq 0$
(c) $x - 4 \leq 0$

FIGURE **5.36**

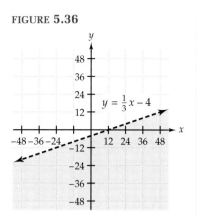

SOLUTION

(a) $x - 3y > 12$ Solve the inequality for y.

$\qquad -3y > -x + 12$ Subtract x from both sides of the inequality.

$\qquad\qquad y < \frac{1}{3}x - 4$ Divide by -3 and reverse the inequality symbol.

The boundary line is the graph of $y = \frac{1}{3}x - 4$. The line is dashed because the inequality symbol is $<$, which also implies that the solutions lie below the boundary line. (See Figure 5.36.)

(b) $y + 3 \geq 0$ Solve the inequality for y.

$\qquad\quad y \geq -3$

Figure 5.37 shows the solid boundary line $y = -3$. We shade all points of the half-plane above the line because the inequality symbol is $\geq$.

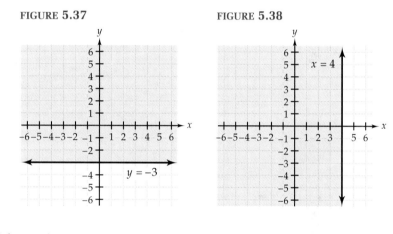

FIGURE **5.37**   FIGURE **5.38**

(c) $x - 4 \leq 0$ Solve the inequality for x.

$\qquad\quad x \leq 4$

FIGURE **5.39**

$y < \frac{1}{3}x - 4$

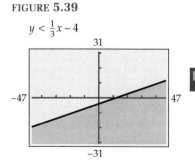

Figure 5.38 shows the solid vertical boundary line $x = 4$ and the points to the left of the boundary line that satisfy the inequality.

Shade   We can use a graphing calculator to produce the graph of a linear inequality.

Figure 5.39 is a typical graphing calculator display of the graph of the inequality in part (a) of Example 1.

Systems of Inequalities

Two or more inequalities considered simultaneously form a **system of inequalities.** The solution set of a system of inequalities is the intersection of the solution sets of the individual inequalities. To graph the solution set of a system of inequalities, we graph the solutions of each inequality and then shade the intersection of the graphs.

EXAMPLE **2** **Graphing a System of Inequalities**

Graph the following system of inequalities:

$$y - x \geq 2$$
$$3x - y > 2$$

SOLUTION

Solving each inequality for y, we determine that the boundary lines have the associated equations $y = x + 2$ (solid line) and $y = 3x - 2$ (dashed line). Figure 5.40(a) shows the graphs of the two inequalities in the same coordinate system. Figure 5.40(b) shows the graph of the system, which is the intersection of the graphs in Figure 5.40(a). Note that the intersection is the overlap of the two half-planes.

FIGURE **5.40**

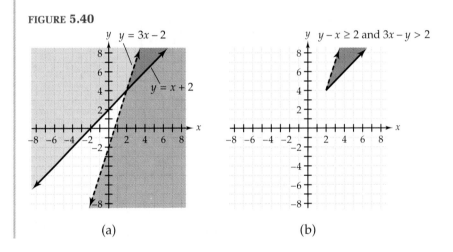

(a)   (b)

EXAMPLE **3** **Graphing a System of Inequalities**

Describe the solution set of the following system of inequalities:

$$2y - x \leq 2$$
$$2y - x \geq 10$$

SOLUTION

When we solve the given inequalities for y, we see that the associated equations are $y = \frac{1}{2}x + 1$ and $y = \frac{1}{2}x + 5$. Figure 5.41 shows the graphs of the inequalities.

FIGURE **5.41**

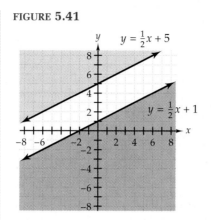

Because the two half-planes have no points in common, the system has no solution. The solution set is Ø.

EXAMPLE **4** **Graphing a System of Inequalities**

Graph the following system of inequalities:

$$2y - x \geq 1$$
$$x + y \geq -4$$
$$y \leq 5$$

SOLUTION

We solve all three inequalities for y and draw their graphs in the same coordinate system.

$$2y - x \geq 1 \quad \rightarrow \quad y \geq \frac{1}{2}x + \frac{1}{2}$$
$$x + y \geq -4 \quad \rightarrow \quad y \geq -x - 4$$
$$y \leq 5 \quad\quad\quad \rightarrow \quad y \leq 5$$

The graph of the system is the region above the line $y = \frac{1}{2}x + \frac{1}{2}$, above the line $y = -x - 4$, and below the line $y = 5$. (See Figure 5.42.)

FIGURE **5.42**

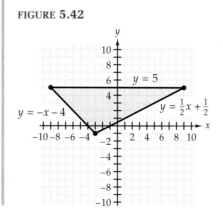

Linear Programming

Many decisions that managers are expected to make are based on the allocation of limited resources. For example, to minimize shipping costs, a company might need to compare the benefits of shipping by railroad with those of shipping by air. Such variable factors as weight, volume, packaging costs, and time might need to be considered.

The manager will usually find that there are certain limitations (*constraints*) that must be considered in making the shipping decisions. For example, perhaps packages cannot exceed a given volume or weight, and the customer may need the product no later than a given date.

Linear programming is a modeling method that the manager might use to take all the constraints into account and to arrive at the best solution—that is, the solution that minimizes the shipping costs.

In a linear programming problem, each **constraint** is described by a linear inequality. Taken together, all the constraints are modeled by a system of linear inequalities. The solutions of the system of inequalities are called **feasible solutions,** and the graph of the system is called the **feasible region.** The boundary lines of the feasible region intersect at corner points called **vertices.**

In addition to graphing constraints to produce a feasible region, linear programming problems involve identifying the quantity that is to be *optimized*—that is, maximized or minimized. For our purposes, we assume that that quantity can be described with a linear function, which we call the **objective function.** The goal is to determine the point in the feasible region whose coordinates optimize the value of the objective function.

EXAMPLE 5 **A Linear Programming Problem**

Maximize the value of $z = 3x + 2y$ subject to the following constraints:

$$-4x + 3y \leq 12$$
$$x + y \leq 11$$
$$6x - y \leq 24$$
$$x \geq 0, y \geq 0$$

SOLUTION

The feasible region is simply the graph of the system of inequalities. (See Figure 5.43.)

The vertices are the points of intersection of the respective pairs of boundary lines. We obtain the coordinates of each vertex by solving the system of two equations whose graphs intersect at that point.

Every point in the feasible region, such as $(2, 3)$, represents a solution of the system of inequalities. However, our goal is to find the solution for which the value of $z = 3x + 2y$ is a maximum. The following table shows the value of z for the point $(2, 3)$ and for the vertices of the feasible region.

FIGURE 5.43

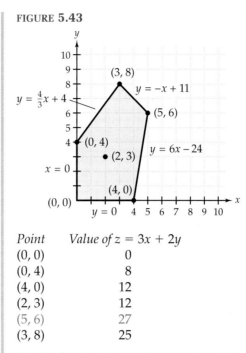

Point	Value of $z = 3x + 2y$
$(0, 0)$	0
$(0, 4)$	8
$(4, 0)$	12
$(2, 3)$	12
$(5, 6)$	27
$(3, 8)$	25

On the basis of only those points shown in the table, we conclude that the solution $(5, 6)$ maximizes the objective function $z = 3x + 2y$.

In Example 5, our conjecture was that the solution occurred at the vertex $(5, 6)$. In fact, for any linear programming problem, the solution (if an optimal solution exists) occurs at a vertex. Thus, to determine the solution, we simply need to evaluate the objective function at each vertex of the feasible region.

Solving a Linear Programming Problem

1. Write all constraints as a system of inequalities.
2. Write the objective function that represents the quantity to be optimized.
3. Graph the feasible region.
4. Determine the vertices of the feasible region.
5. Evaluate the objective function at each vertex. The solution is the ordered pair for which the objective function has an optimum value.

EXAMPLE **6**   **Minimizing the Cost of a Dietary Supplement**

A nutritionist at a hospital needs to prepare a dietary supplement that will contain at least 9 units of vitamin C, at least 24 units of calcium, and at least 200 calories. One ounce of Nutramega costs 10¢ and contains

1 unit of vitamin C, 2 units of calcium, and 30 calories. One ounce of Vitameta costs 12¢ and contains 1 unit of vitamin C, 5 units of calcium, and 20 calories. How many ounces of each supplement should be used to meet the dietary requirements at the least cost?

SOLUTION

Let x = the number of ounces of Nutramega
 y = the number of ounces of Vitameta
 z = the cost of the supplement

The following system of inequalities represents the constraints:

$x + y \geq 9$	One ounce of Nutramega contains 1 unit of vitamin C, and one ounce of Vitameta contains 1 unit of vitamin C. The minimum requirement is 9 units.
$2x + 5y \geq 24$	One ounce of Nutramega contains 2 units of calcium, and one ounce of Vitameta contains 5 units of calcium. The minimum requirement is 24 units.
$30x + 20y \geq 200$	One ounce of Nutramega contains 30 calories, and one ounce of Vitameta contains 20 calories. The minimum requirement is 200 calories.
$x, y \geq 0$	The number of ounces of each product is nonnegative.

The cost function that we are trying to minimize is

$$z = 10x + 12y$$

Nutramega costs 10 cents per ounce, and Vitameta costs 12 cents per ounce.

Figure 5.44 shows the feasible region and the vertices, which we determine by solving the respective systems of equations.

FIGURE 5.44

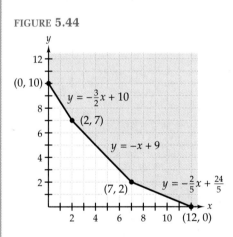

In the following table, we use the coordinates of each vertex to evaluate the objective function.

Vertex	Value of $z = 10x + 12y$
(0, 10)	120
(2, 7)	104
(7, 2)	94
(12, 0)	120

Using 7 ounces of Nutramega and 2 ounces of Vitameta results in a minimum cost of $0.94.

Note that the feasible region in Example 6 is *unbounded*. In such instances, optimizing an objective function is not always possible.

5.6 Quick Reference

Linear Inequalities in Two Variables

■ A **linear inequality in two variables** is an inequality that can be written in the form $Ax + By < C$, where A, B, and C are real numbers and A and B are not both 0. (The symbols $\leq$, $>$, and $\geq$ can also be used.)

■ A **solution** of an inequality in two variables is an ordered pair (a, b) that satisfies the inequality.

■ The graph of the solution set of a linear inequality in two variables is a shaded **half-plane.** The **boundary line** of the half-plane is the graph of the associated equation. The boundary line is solid for $\leq$ and $\geq$ inequalities and dashed for $<$ and $>$ inequalities.

■ If the coordinates of a *test point* satisfy a given inequality, then the point lies in the half-plane to be shaded; otherwise, shade the other half-plane.

■ Alternatively, follow these steps for graphing an inequality.
  1. Solve the inequality for y with y on the left side.
  2. Draw the graph (a solid or dashed boundary line) of the associated equation.
  3. If the inequality symbol in step 1 is $<$ or $\leq$, shade the half-plane below the line; if the inequality symbol in step 1 is $>$ or $\geq$, shade the half-plane above the line.
  4. For a vertical boundary line, shade to the left if the inequality symbol is $<$ or $\leq$, shade to the right if the inequality symbol is $>$ or $\geq$.

Systems of Inequalities

■ Two or more inequalities considered simultaneously form a **system of inequalities.**

■ The solution set of a system of inequalities is the intersection of the solution sets of the individual inequalities.

Linear Programming

■ In a linear programming problem, the **constraints** are described by a system of linear inequalities. The solutions of the system are called **feasible solutions,** and the graph of the system is called the **feasible region.** The boundary lines of the feasible region intersect at corner points called **vertices.**

■ The goal of linear programming is to determine the point in the feasible region whose coordinates optimize the value of the **objective function.**

■ To solve a linear programming problem, use the following steps:
  1. Write all constraints as a system of inequalities.
  2. Write the objective function that represents the quantity to be optimized.
  3. Graph the feasible region.
  4. Determine the vertices of the feasible region.
  5. Evaluate the objective function at each vertex. The solution is the ordered pair for which the objective function has an optimum value.

5.6 Speaking the Language

1. The graph of a linear inequality in two variables is called a(n) ▨▨▨▨▨▨ .

2. To graph $y \le 2x - 3$, we begin by graphing the associated equation $y = 2x - 3$. This graph is called the ▨▨▨▨▨▨ line.

3. In a linear programming problem, the given conditions that are described by inequalities are called ▨▨▨▨▨▨ .

4. In a linear programming problem, the function that is to be optimized is called the ▨▨▨▨▨▨ function.

5.6 Exercises

Concepts and Skills

1. Suppose that the coordinates of a point that lies below the line $Ax + By = C$ do not satisfy the inequality $Ax + By < C$. Describe the graph of the inequality.

2. Suppose that the coordinates of a point that lies below the line $Ax + By = C$ do not satisfy the inequality $Ax + By < C$. Explain how you know that B is negative.

In Exercises 3–6, match the graph with the given inequality.

3. $y \ge -x + 6$

4. $12 \le 2x - 3y$

5. $2x + 5y < -15$

6. $7x - 4y - 8 < 0$

(A)

(B)

(C)

(D)

In Exercises 7–22, sketch the graph of the given inequality.

7. $y \le 4 - x$

8. $2y > x$

9. $x + y < 5$

10. $y \le 2x - 7$

11. $3x + 2y > 8$

12. $5 \ge x - y$

13. $2x - y \ge 0$

14. $x + y > 0$

15. $y < -3$

16. $x > 2$

17. $x - 3y \le 15$

18. $2y - x \le 10$

19. $x \le -4$

20. $y \ge 0$

21. $3y \ge -x - 3$

22. $3y \le 4x + 21$

23. How does the graph of the one-variable inequality $x \le -1$ differ from the graph of the two-variable inequality $x \le -1$?

24. In a linear programming problem, what is the difference between a constraint and an objective function?

In Exercises 25–28, match the graph with the given system of inequalities.

25. $x + y \le 4$
    $y > -4$

26. $3x + 2y \le 8$
    $x \ge 0, y \ge 0$

27. $x < 5$
    $x + y > 3$
    $x \ge 0, y \ge 0$

28. $2y \ge x - 2$
    $x + 2y + 2 \ge 0$
    $y \le 3$

(A)

(B)

(C)

(D)

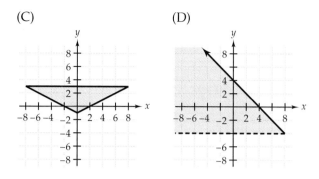

In Exercises 29–44, sketch the graph of the given system of inequalities.

29. $y \le 2x + 5$
  $2y - x > 4$

30. $y < 5$
  $7x - 3y \le 6$

31. $3y - 2x \ge 6$
  $2x - 3y + 18 \ge 0$

32. $x + y \le 8$
  $y \le x + 8$

33. $x + y \ge 4$
  $y \le 9 + x$

34. $4y - 5x < 11$
  $8y > 10x - 20$

35. $x \ge 3$
  $y \le 5$

36. $x < 2$
  $y \ge -2$

37. $2y - 5x > 12$
  $5x - 2y < -4$

38. $y < x$
  $x - y \ge 6$

39. $y \le x + 4$
  $x + y < 8$
  $x \ge 0, y \ge 0$

40. $2x + y < 10$
  $y \ge 4 - 2x$
  $x \ge 0, y \ge 0$

41. $x + y \ge 0$
  $7x - 3y \le 30$
  $9y - x < 30$

42. $2x + 3y > 7$
  $2y - 3x \ge -4$
  $y - 3x < 17$

43. $x + 3y \ge -2$
  $3y - 2x \le 13$
  $-2 \le x \le 4$

44. $y - x \le 4$
  $2x + y < 9$
  $1 < y < 6$

In Exercises 45–52, an objective function z and a list of constraints are given. Determine the maximum and minimum values of the objective function subject to the constraints.

45. $z = y - 2x$
  Constraints:
  $2y - 3x \le 12$
  $2x + y \le 20$
  $x \ge 0, y \ge 0$

46. $z = -3x + 2y$
  Constraints:
  $2x + 5y \le 60$
  $x - y \le 9$
  $x \ge 0, y \ge 0$

47. $z = x + 4y$
  Constraints:
  $x + 2y \le 22$
  $7x - 3y \le 35$
  $x \ge 0, y \ge 0$

48. $z = 6x + 9y$
  Constraints:
  $y - x \le 3$
  $10x - 3y \le 40$
  $x \ge 0, y \ge 0$

49. $z = 3x + 4y$
  Constraints:
  $5x - 3y \ge -12$
  $3x + 7y \le 72$
  $3x - y \le 24$
  $x \ge 0, y \ge 0$

50. $z = x - 2y$
  Constraints:
  $-3x + 5y \le 45$
  $x + 3y \le 41$
  $11x + 6y \le 154$
  $x \ge 0, y \ge 0$

51. $z = 2x + y$
  Constraints:
  $x - y \le 3$
  $4x - 3y \le 16$
  $4x + 5y \le 80$
  $x \ge 0, y \ge 0$

52. $z = 7x - 4y$
  Constraints:
  $5x + 9y \le 126$
  $2x - y \le 9$
  $5x - 3y \le 20$
  $x \ge 0, y \ge 0$

Concept Extension

In Exercises 53–60, sketch the graph of the given inequality in two variables.

53. $-4 < x \le 3$

54. $2 \le y \le 5$

55. $-1 < y < 4$

56. $0 \le x \le 6$

57. $|x| > 4$

58. $|y| \le 7$

59. $y \ge |x|$

60. $y \le |1 - x|$

In Exercises 61 and 62, sketch the graph of the given system of inequalities.

61. $y \le |x|$
  $|y| \ge 3$

62. $y \ge |x + 3|$
  $|x| \le 1$

Applications

63. Financial Planning A financial planner needs to advise a client who wants to invest up to $250,000 in stocks yielding 12% and bonds yielding 8%. The client does not want more than 60% of the total investment to be in stocks. How should the money be allocated to maximize the client's annual income?

64. Landscape Work A worker can plant a shrub in 10 minutes and a tree in 15 minutes. The landscape contract provides that the number of shrubs cannot exceed the number of trees by more than 13. If the worker can work no more than 8 hours per day, what is the maximum total number of shrubs and trees that can be planted in a day?

65. Alaskan Tours In Alaska, Skagway Sightseeing Tours conducts van trips into the Yukon and bus trips of the Klondike Territory. In order to be affiliated with a cruise ship line, the company must be able to offer spaces for at least 240 tourists and must have at least 10 vehicles.

The 12-passenger vans cost $60 per trip to operate, and the 32-passenger buses cost $80 per trip. How many vans and buses should the company operate to minimize cost?

66. **Cleaning Service** A cleaning service charges $5 per window for ground-level windows and $6 per window for second-floor windows. Washing ground-level windows requires 15 minutes each, and washing second-floor windows requires 20 minutes each. The service must be guaranteed at least 8 hours work per day. Suppose that a customer wants to have at least 30 windows washed per day. How many windows of each type should the customer have washed to minimize the cost?

67. **Craft Shop** A small craft shop makes decorative wooden mailboxes and birdhouses. Each item requires three stages of production: cutting, assembly, and painting. A birdhouse sells for $18 and requires 2 hours for cutting, 1 hour to assemble, and 1 hour to paint. A mailbox sells for $30 and requires 1 hour for cutting, 1 hour to assemble, and 2 hours to paint. The artist who does the painting will work at most 30 hours per week. The owner of the shop can devote 18 hours to assembling and employs a helper for at most 30 hours per week to do the cutting. How many of each item should the owner plan to produce to maximize potential revenue?

68. **Petroleum Products** An oil refinery can produce a total of at most 3000 barrels of gasoline and home heating oil. The amount of gasoline produced should be at least two-thirds the amount of heating oil produced. Production cost is $2 per barrel for gasoline and $1 per barrel for heating oil. Management wants to hold production cost to no more than $5000. If the profit on gasoline is $4 per barrel, and the profit on heating oil is $2.50 per barrel, how many barrels of each product should be produced to maximize the profit?

69. **Farm Production** A farmer has 100 acres of land available for planting corn and soybeans. Labor costs $80 per acre for corn and $50 per acre for soybeans. Cost of agricultural products, such as seed and fertilizer, is $40 per acre for corn and $70 per acre for soybeans. The farmer, who has budgeted at most $7550 for labor and at most $6250 for agricultural prod-

ucts, anticipates a profit of $230 per acre of corn and $210 per acre of soybeans. What planting plan would maximize profit?

70. **Crafts Fair** Organizers of a fall crafts fair have 4100 square feet of booth space. The space can accommodate at most 150 booths for craft sales and food sales. Craft booths occupy 30 square feet, and food booths occupy 20 square feet. The organizers charge $120 for craft space and $100 for food space. The setup cost is $10 for each craft booth and $20 for each food booth. If the organizers have a setup budget of $2600, how many of each type of booth are needed in order to maximize the income?

71. **Special Events** A company rents chairs and tents and provides setup for special events. The company has full-time employees who work 40 hours per week at $9 per hour and part-time employees who work 20 hours per week at $6 per hour. For a certain event, the company must have at least 35 employees to provide at least 1000 hours of work for the week. The union contract requires that the ratio of part-time to full-time employees cannot be more than 5 to 1. How many employees should the company have to minimize the labor cost?

72. **Production Planning** One machine produces 8 pounds of crackers and 5 ounces of cracker crumbs per hour and costs $60 per hour to operate. Another machine produces 5 pounds of crackers and 8 ounces of cracker crumbs per hour and costs $40 per hour to operate. The machines can be operated a total of at most 16 hours each day. The company needs at least 86 pounds of crackers and 83 ounces of cracker crumbs each day. What plan would minimize the cost of production?

73. **Parade Seating** The setup cost for seats along a parade route is $1 for bleacher seats and $1.50 for reserved seats. Parade officials need seating for at least 3000 people. The number of reserved seats cannot be more than five times the number of bleacher seats. Parade officials charge $5 for bleacher seating and $15 for reserved seating, and they need to guarantee an income of least $21,000. What seating arrangement will minimize the cost?

74. **Auto Inventory** A car dealer wants at least 200 new and used cars on the lot. The average

daily cost to have a car on the lot is $25 for new cars and $10 for used cars. New cars must represent at least half the total number of cars. A new car requires 2 hours of preparation work, and a used car requires 3 hours of preparation work. To keep the preparation staff busy, the dealer needs at least 450 hours of preparation work. What inventory of cars will minimize the dealer's cost?

75. Golf Shop

(a) A golf shop reconditions and repairs putters and drivers. The shop owner determines that the shop needs to process a total of at least 20 clubs per day. A worker requires 45 minutes to clean and refinish a driver and twice as long to clean and refinish a putter. Replacing grips requires 30 minutes for drivers and 40 minutes for putters. The owner has enough staff to permit 45 hours per day for cleaning and refinishing and 25 hours per day for new grips. What is the maximum number of clubs that the shop can process per day?

(b) Suppose that the profit on a driver is $5 and the profit on a putter is $15. For what number of clubs is the profit maximized?

76. Bakery Products

(a) A bakery operator wants to bake at least 50 pies and sheet cakes per day. A pie requires 20 minutes baking time, and a cake requires 30 minutes. The chef needs 30 minutes to prepare one pie for baking and 15 minutes to prepare one cake for baking. The owner employs enough chefs to have 40 hours of preparation time and has enough ovens for 40 hours of baking time. Suppose that the cost to bake a pie is $5 and the cost to bake a cake is $7. For what total number of pies and cakes is the cost minimized?

(b) Suppose that the cost to bake a pie is $8 and the cost to bake a cake is $4. For what total number of pies and cakes is the cost minimized?

Chapter Review Exercises

Section 5.1

1. Explain why you do not need to solve the following system of equations algebraically or by graphing to know that the system has no solution.

$$y = 3x + 7$$
$$y = 3x - 9$$

2. Explain how knowing the slope and y-intercept of each equation would reveal that a system of two equations in two variables has infinitely many solutions.

In Exercises 3 and 4, use the graphing method to estimate the solution of the system of equations. Then use the substitution method to solve the system.

3. $y = 2x - 11$
$x + y = -2$

4. $3y - 2x = 15$
$12y = 8x - 12$

In Exercises 5 and 6, use the addition method to solve the system of equations.

5. $3x + 2y = 0$
$4x - y = 22$

6. $4y = 6x + 10$
$2y - 3x = 5$

In Exercises 7–10, solve the system of equations. Identify inconsistent systems and dependent equations.

7. $x + 2y = -6$
$\dfrac{y}{2} + \dfrac{x}{4} = 1$

8. $x - y = -3$
$\dfrac{2}{5}x + y = 3$

9. $\dfrac{y + 1}{3} = x$
$3x - y - 1 = 0$

10. $\dfrac{1}{6}x - \dfrac{1}{3}y = 1$
$4 - 2x = y$

11. Cough Medication An infirmary needs 500 milliliters of cough syrup that is 30% cough suppressant. The pharmacist has two cough syrups, one that is 15% cough suppressant and one that is 40% cough suppressant. How many milliliters of each syrup are needed to make the required medication?

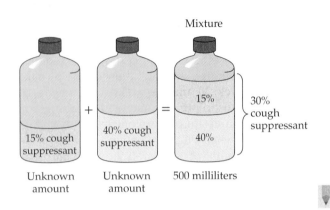

Mixture

15% cough suppressant + 40% cough suppressant = 15% / 40% / 30% cough suppressant

Unknown amount   Unknown amount   500 milliliters

12. **Book Sale** At a used book sale, all paperbacks were the same price and all hardbacks were the same price. If 11 paperbacks and 3 hardbacks cost $9.25, and 8 paperbacks and 5 hardbacks cost $10.25, what was the price of each type of book?

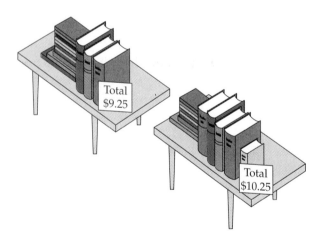

Total $9.25

Total $10.25

Section 5.2

13. Suppose that $(2, -4, 3)$ is the unique solution of a system of three equations in three variables. What is a geometric interpretation of the ordered triple?

14. Determine a, b, and c so that $(1, 6, -3)$ is the solution of the following system of equations.

$$x + by + z = -5$$
$$3x + y + cz = 21$$
$$ax - y - z = -2$$

In Exercises 15–20, solve the system of equations.

15. $x + 2y - z = 6$
  $x \quad\quad + 3z = -10$
  $\quad\quad 4y - 2z = 8$

16. $x + y + 3z = 4$
  $2x - y - z = 10$
  $-3x + 2y + z = -17$

17. $2x + y - 4z = 15$
  $x - 2y + z = -1$
  $x - 7y + 7z = -18$

18. $3x + 2y + z = 4$
  $2y - z = 3$
  $6x - 2y + 5z = 8$

19. $11 + y = x + 3z$
  $5z + 2x = 16$
  $y = 4x - 14$

20. $y = 3x + 2z - 11$
  $y = x + 4z - 17$
  $z = 4x$

21. Explain how to solve the following system of equations without writing the system in row-echelon form.

$$2x + y + z = 3$$
$$3y \quad\quad = -6$$
$$y - z = -6$$

22. Write and solve a system of equations to determine a quadratic function $f(x) = ax^2 + bx + c$ whose graph contains the points $A(-1, -22)$, $B(5, 2)$, and $C(6, -8)$.

23. **School Enrollments** The percentage of public elementary and secondary schools with enrollments of 300 to 699 is 4 less than the percentage of such schools with larger or smaller enrollments. The percentage of schools with enrollments of 300 or more is 10 more than twice the percentage of schools with enrollments of under 300. (*Source*: U.S. Center for Education Statistics.) What percentage of schools fall in the 0–299, 300–699, and over 699 enrollment categories?

24. **Student Loans** A student borrowed money from a relative at 4% simple interest, from the college student loan fund at 6% simple interest, and from a bank at 9% simple interest. The total amount that the student borrowed from the college and the bank was $400 more than 3 times the amount that the student borrowed from the relative. The total amount of interest paid to the college and the relative was $37 more than the interest paid to the bank. If the bank had charged 8% rather than 9%, the interest paid to the bank would have been $12 more than the interest paid to the college. How much did the student borrow from each source?

Section 5.3

25. For a system of two equations in two variables, what are the dimensions of (a) the coefficient matrix and (b) the augmented matrix?

26. Write the augmented matrix for the following system of equations.

$$z = 2x$$
$$3y + 2x = 0$$
$$y + z = 4$$

27. Write the system of linear equations associated with the following augmented matrix.

$$\begin{bmatrix} 2 & -3 & | & -7 \\ -5 & 1 & | & -15 \end{bmatrix}$$

28. The following matrix is not in reduced row-echelon form. Why? What row operation must be performed in order to write the matrix in reduced row-echelon form?

$$\begin{bmatrix} 1 & -1 & | & 3 \\ 0 & 1 & | & 2 \end{bmatrix}$$

In Exercises 29 and 30, use matrix methods to solve the system of equations.

29. $\begin{aligned} x + 3y &= 1 \\ 2x - y &= 9 \end{aligned}$

30. $\begin{aligned} x + 3y - z &= 8 \\ y + 2z &= -5 \\ x + z &= -1 \end{aligned}$

In Exercises 31–34, use a calculator to solve the system of equations.

31. $\begin{aligned} 2x + 7y &= -4 \\ 5x - 6y &= 37 \end{aligned}$

32. $\begin{aligned} y &= 0.4x + 2 \\ 5y - 2x + 10 &= 0 \end{aligned}$

33. $\begin{aligned} 3x - y + 6z &= -6 \\ 2x + 3y - 5z &= 16 \\ 4x - 2y + z &= -3 \end{aligned}$

34. $\begin{aligned} x - y + 4z &= -2 \\ 2x + y + z &= 6 \\ 8x + y + 11z &= 14 \end{aligned}$

35. Carbon Monoxide Emissions The table shows the number of tons (in millions) of carbon monoxide emissions for selected years. (*Source*: U.S. Environmental Protection Agency.)

Year	Emissions
1960	30.5
1970	23.0
1994	26.6

Solve a system of equations to determine a function $E(t) = at^2 + bt + c$ that models the data as a function of the number of years since 1960.

36. CD-ROMs In 1997, 45,900 public schools had CD-ROMs. The number of elementary schools with CD-ROMs exceeded the total number of junior and senior high schools with CD-ROMs by 6900. Half the number of elementary schools with CD-ROMs exceeded the number of high schools with CD-ROMs by 2100. (*Source*: Quality Education Data, Inc.) How many of each type of school had CD-ROMs?

Section 5.4

37. Determine a, b, c, and d so that the matrices are equal.

$$\begin{bmatrix} a+3 & 6 & 9 \\ 0 & -2 & d \end{bmatrix} = \begin{bmatrix} 5 & 6 & 1-2b \\ 0 & 4c & 0 \end{bmatrix}$$

38. What must be true about matrices A and B in order to determine the product AB?

In Exercises 39 and 40, perform the indicated operations with the following matrices.

$$A = \begin{bmatrix} 2 & -1 & 3 \\ 0 & 2 & 1 \\ 1 & -1 & 4 \end{bmatrix} \quad B = \begin{bmatrix} 1 & 3 & 1 \\ 0 & 2 & 4 \\ 0 & 0 & -1 \end{bmatrix} \quad I = \begin{bmatrix} 1 & 0 & 0 \\ 0 & 1 & 0 \\ 0 & 0 & 1 \end{bmatrix}$$

39. (a) $A - B$
(b) $3B + 2I$

40. (a) IB
(b) AB

In Exercises 41–43, perform the following matrix operations, if possible.

(a) $A + B$ (b) $2A - B$ (c) $3A$ (d) AB (e) BA

41. $A = \begin{bmatrix} 1 & -2 \\ 3 & -1 \\ 4 & 3 \end{bmatrix}$, $B = \begin{bmatrix} 1 & 0 & -1 \\ 2 & 1 & 0 \end{bmatrix}$

42. $A = \begin{bmatrix} 2 & 4 \\ -1 & 3 \end{bmatrix}$, $B = \begin{bmatrix} 2 & 0 \\ -1 & 4 \end{bmatrix}$

43. $A = \begin{bmatrix} 1 & 2 & -1 \\ 4 & -5 & 6 \end{bmatrix}$, $B = \begin{bmatrix} 3 & 2 & 0 \\ 4 & -2 & 7 \end{bmatrix}$

44. Suppose that matrix $A + B$ is a 2×3 matrix. Is it possible to determine the dimensions of A and B? Why?

45. Write the following matrix equation as a system of equations.

$$\begin{bmatrix} 2 & 1 \\ 3 & 4 \end{bmatrix}\begin{bmatrix} x \\ y \end{bmatrix} = \begin{bmatrix} 1 \\ -1 \end{bmatrix}$$

46. Write the following system of equations as a matrix equation.

$$x + 2y = -1$$
$$x - 3z = 4$$
$$2x + y - z = 2$$

47. Paper Products A discount store offers four value-packs of household paper products. The table shows the number of rolls of paper towels, the number of boxes of tissues, and the number of packages of table napkins in each pack.

Pack	Towels	Tissue	Napkins
Small	4	8	6
Medium	6	4	4
Large	6	10	8
Mega	10	12	8

Write the data as a 4×3 matrix P.

48. Refer to Exercise 47. Suppose that towels cost 42¢, tissues cost $1.12, and napkins cost 98¢.

(a) Write a column matrix C for the cost of the items.

(b) Determine and interpret PC.

Section 5.5

In Exercises 49 and 50, determine whether the matrices are inverses.

49. $\begin{bmatrix} 3 & 4 \\ -4 & -5 \end{bmatrix}, \begin{bmatrix} -5 & -4 \\ 4 & 3 \end{bmatrix}$

50. $\begin{bmatrix} 2 & 1 \\ -1 & 3 \end{bmatrix}, \begin{bmatrix} 1 & 1 \\ -1 & -2 \end{bmatrix}$

In Exercises 51 and 54, determine the inverse of the matrix, if it exists.

51. $\begin{bmatrix} 3 & 2 \\ 5 & 3 \end{bmatrix}$

52. $\begin{bmatrix} 2 & -1 \\ -6 & 3 \end{bmatrix}$

53. $\begin{bmatrix} -1 & 2 & -1 \\ -2 & 2 & -1 \\ 1 & -1 & 1 \end{bmatrix}$

54. $\begin{bmatrix} 1 & 2 & -1 \\ 0 & 1 & 2 \\ 1 & 2 & 1 \end{bmatrix}$

In Exercises 55 and 56, use inverse matrix methods to solve the given system of equations.

55. $6x + 8y = 5$
$\quad\ 4x + 6y = 3$

56. $7x - 2y = 4$
$\quad\ x + 3y = 17$

57. $3x + 2y + z = 1$
$\quad\ x + y + 2z = 7$
$\quad\ x + 2y + z = 3$

58. $3x = 9 + y$
$\quad\ 0 = z + 4y + 12$
$\quad\ 6z - 1 = x + y$

59. $2x - 3y + 4z - 5w = 27$
$\quad\ x \quad\quad - 3z + 3w = -14$
$\quad\ 7x + y - 2z \quad\quad = 0$
$\quad\ -x + 2y \quad\quad - w = -1$

60. Suppose that you perform row operations to determine the inverse of a matrix. How would you recognize that the matrix is not invertible?

Section 5.6

In Exercises 61 and 62, sketch the graph of the given inequality.

61. $3y \le 2x - 24$

62. $x + 2y > 10$

In Exercises 63–66, sketch the graph of the given system of inequalities.

63. $y < 4x - 17$
$\quad\ 5y - 2x \ge 5$

64. $x + 2 \ge 0$
$\quad\ y \le 5$

65. $4y - 3x < 24$
$\quad\ 4 > 2x - y$
$\quad\ x + 2y > -18$

66. $y + x \ge 0$
$\quad\ y > 3 - x$

67. Explain why the solution set of the system of inequalities in (i) is Ø but the solution set of the system in (ii) is not Ø.

(i) $y > x + 2$
$\quad\ y < x + 2$

(ii) $y \ge x + 2$
$\quad\ y \le x + 2$

68. Use the word *half-plane* to describe the feasible region of a linear programming problem. What points of the feasible region are most important in determining the optimal solution?

In Exercises 69 and 70, determine the maximum and minimum values of the objective function subject to the given constraints.

69. $z = 3x + 7y$
Constraints:
$3x + 8y \le 32$
$x - 2y \le 6$
$2x + 3y \ge 12$

70. $z = 2x + 4y$
Constraints:
$2x + 3y \le 24$
$2x - y \le 16$
$5x + 6y \ge 30$
$x \ge 0, y \ge 0$

71. Airplane Seating An airplane can accommodate at most 250 passenger seats. The cost of food service is $3 per person in coach and $8 per person in first class. The airline wants the cost of food service not to exceed $1040. The per-seat profit is $50 for first class and $20 for coach. What configuration of seats will maximize the profit?

72. Refer to Exercise 71. Suppose that the per-seat profit in first class falls to $20. Must the airline eliminate all first-class seats to maximize the profit?

Chapter *6*

Sequences, Series, Probability

Genealogy, the study of family history, is a popular pastime in the United States. In a normal "family tree," you would have 2 parents, 4 grandparents, 8 great-grandparents, and so on.

Starting with yourself, the number of people in each generation can be listed as a sequence 1, 2, 4, 8, ... The sum of the numbers in this sequence is called a series: 1 + 2 + 4 + 8 + · · · . You might be surprised to learn that you have 62 direct ancestors in just the five generations that precede you!

The first half of this chapter is devoted to the study of sequences and series, along with the special notation and formulas that are used to generate terms and to find sums. We then turn to some basic concepts of permutations, combinations, and probability—topics that will be very useful to you in a course in statistics. We conclude with the Binomial Theorem, which makes use of combinations in raising a binomial to a power.

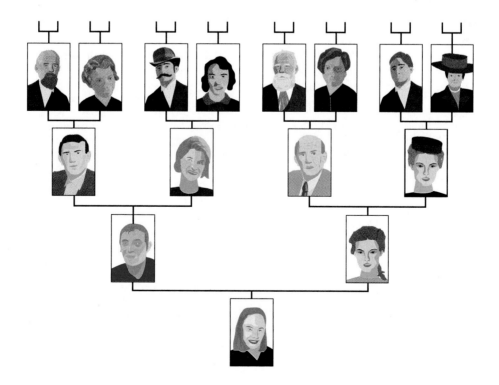

6.1 Sequences and Series

Sequences ■ *Writing the General Term* ■
Recursive Sequences ■ *Series* ■ *Writing a Series*

Sequences

When we refer to a sequence of events, we mean events that occur one after another. In mathematics, a **sequence** is a list or arrangement of numbers called **terms** in which there is a first term, a second term, and so on.

Suppose that the enrollment at a college is currently 1000 and that the enrollment is expected to increase by 5% each year. This means that the projected enrollment for any year is 105% of (or 1.05 times) the enrollment for the preceding year. The following table shows the enrollments for years 1, 2, 3, and so on:

Year	Enrollment	
1	1000	$= 1000(1.05)^0$
2	$1000 \cdot 1.05$	$= 1000(1.05)^1$
3	$1000(1.05) \cdot 1.05$	$= 1000(1.05)^2$
4	$1000(1.05)^2 \cdot 1.05$	$= 1000(1.05)^3$

The list of numbers $1000(1.05)^0$, $1000(1.05)^1$, $1000(1.05)^2$, $1000(1.05)^3$, . . . is an example of a sequence. The enrollments are the terms of the sequence, and the years are the term numbers. The dots . . . indicate that the list continues forever, and so the sequence is called an **infinite sequence.**

In the preceding table, we can see that the exponent on the factor 1.05 is one less than the term number. Thus the **general term** or **nth term** is $1000(1.05)^{n-1}$. We can write the general term as a function $c(n) = 1000(1.05)^{n-1}$ whose domain is all natural numbers. Thus we can think of an infinite sequence as a special type of function whose domain is the set of natural numbers.

If we calculate the enrollment for only a certain number of years—for instance, the first three years—we have an example of a **finite sequence.**

Term number	1	2	3
Term	$1000(1.05)^0$	$1000(1.05)^1$	$1000(1.05)^2$

For this finite sequence, the domain is the first three natural numbers: {1, 2, 3}.

■ *NOTE   We will usually know from the context whether a sequence is finite or infinite, so we will refer to either type simply as a sequence.*

Definition of Infinite and Finite Sequences

An **infinite sequence** is a function whose domain is the set of natural numbers. A **finite sequence** is a function whose domain is the first n natural numbers.

Conventionally, the terms of a sequence are written with subscripts rather than with function notation.

$$c_1 = 1000(1.05)^0$$
$$c_2 = 1000(1.05)^1$$
$$c_3 = 1000(1.05)^2$$
$$c_4 = 1000(1.05)^3$$
$$\cdot$$
$$\cdot$$
$$\cdot$$
$$c_n = 1000(1.05)^{n-1}$$

For a finite sequence, such as 2, 4, 6, 8, we simply restrict n: $a_n = 2n$, where $1 \le n \le 4$.

Mode   To use a calculator to work with sequences, we choose the *Sequence* mode. When you are in this mode, you can display the variable n by pressing the **X** key.

FIGURE 6.1

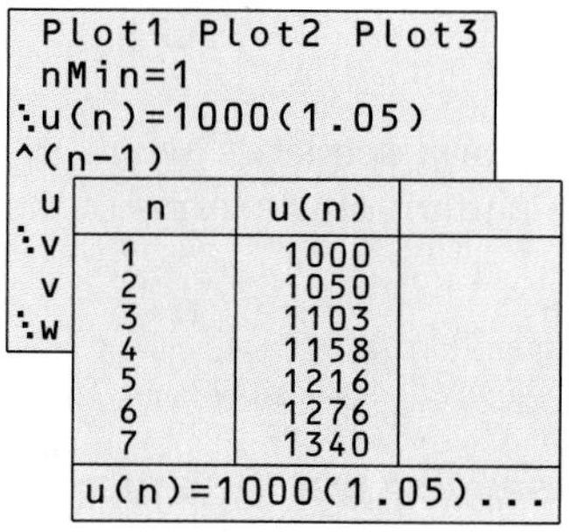

The **Y** screen of a calculator in the *Sequence* mode is quite different from the **Y** screen in the function mode. For simple calculations, you will usually need to enter only the first term number (nMin) and the sequence (u(n)). Doing this will allow you to produce a table of values or a graph.

Figure 6.1 shows the sequence $c_n = 1000(1.05)^{n-1}$ entered on the **Y** screen with the first term number 1. Also shown is a table that lists the term numbers and the terms (rounded to the nearest integer).

If we know the general term of a sequence, we can write any specific term by substituting the term number for n.

EXAMPLE 1   **Writing the Terms of a Sequence**

(a) Write the first four terms of the sequence $a_n = (-1)^n(n + 3)$.

(b) Write the seventh term of the sequence $b_n = \dfrac{n - 1}{n + 1}$.

SOLUTION

(a) Replace n with 1, 2, 3, and 4.

$$a_1 = (-1)^1(1 + 3) = -4 \qquad a_2 = (-1)^2(2 + 3) = 5$$
$$a_3 = (-1)^3(3 + 3) = -6 \qquad a_4 = (-1)^4(4 + 3) = 7$$

(b) $b_7 = \dfrac{7 - 1}{7 + 1} = \dfrac{3}{4}$ **Replace n with 7.**

In Example 1(a), we determined the first four terms of the given sequence by substitution. We can also use a calculator to produce the same results.

FIGURE 6.2

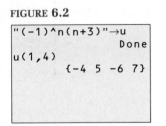

Sequence On the home screen, we can store a sequence in a variable u and then evaluate u (that is, the sequence) by entering the beginning and ending term numbers. The variable u can usually be found on the keypad, possibly with the "2nd" key.

Figure 6.2 shows that the sequence given in Example 1(a) has been stored in u. Then, by entering u(1, 4), we produce the first four terms of the sequence.

We can visualize a sequence by plotting points whose coordinates are of the form (term number, term).

FIGURE 6.3

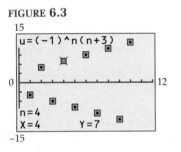

Sequence Plot We can use a calculator to plot the terms of a sequence.

Figure 6.3 shows the results of plotting the first 10 terms of the sequence in Example 1(a). Because the terms alternate in sign, succeeding points are on opposite sides of the x-axis. Note that the tracing cursor is on (4, 7), which indicates that $a_4 = 7$.

Writing the General Term

To determine an expression for the general term of a sequence when the first few terms are given, we must recognize the pattern of the terms of the sequence and express that pattern algebraically.

EXAMPLE 2   Writing the General Term of a Sequence

Predict the next two terms of the sequence. Assuming that the pattern of the sequence continues, write an expression for the general term of the sequence.

(a) 8, 10, 12, 14, . . .
(b) 1, 3, 9, 27, . . .
(c) $-1, \dfrac{1}{4}, \dfrac{-1}{9}, \dfrac{1}{16}, \ldots$

SOLUTION

(a) The next two terms appear to be 16 and 18. Because the terms are even numbers, we might suspect that the general term includes the expression $2n$. Compare $2n$ with the given sequence.

n	1	2	3	4	. . .
$2n$	2	4	6	8	. . .
a_n	8	10	12	14	. . .

Observe that each term of a_n is 6 more than $2n$. Thus the pattern suggests that $a_n = 2n + 6$.

(b) The terms $1, 3, 9, 27, \ldots$ appear to be increasing powers of 3.

n	1	2	3	4	$\ldots$
Term	1	3	9	27	$\ldots$
Powers of 3	3^0	3^1	3^2	3^3	$\ldots$

The next two terms appear to be $3^4 = 81$ and $3^5 = 243$. Observe that the exponent on 3 is 1 less than the term number n. This suggests that the general term is $b_n = 3^{n-1}$.

n	1	2	3	4	$\ldots$	n	$\ldots$
b_n	3^0	3^1	3^2	3^3	$\ldots$	3^{n-1}	$\ldots$

(c) In the following pattern, we see that the signs of the terms alternate and that the denominators are the squares of the term numbers.

n	1	2	3	4	$\ldots$
c_n	$\dfrac{-1}{1^2}$	$\dfrac{1}{2^2}$	$\dfrac{-1}{3^2}$	$\dfrac{1}{4^2}$	$\ldots$

The pattern suggests that the next two terms are $\dfrac{-1}{5^2} = \dfrac{-1}{25}$ and $\dfrac{1}{6^2} = \dfrac{1}{36}$.

Observe that $(-1)^n = -1$ when n is an odd number and that $(-1)^n = 1$ when n is an even number. We can use these facts to write a sequence whose terms alternate in sign.

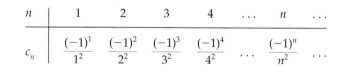

n	1	2	3	4	$\ldots$	n	$\ldots$
c_n	$\dfrac{(-1)^1}{1^2}$	$\dfrac{(-1)^2}{2^2}$	$\dfrac{(-1)^3}{3^2}$	$\dfrac{(-1)^4}{4^2}$	$\ldots$	$\dfrac{(-1)^n}{n^2}$	$\ldots$

Recursive Sequences

Rather than giving the general term of a sequence, we can describe the sequence by giving the first term and stating the relationship between each term and its successor. Such a sequence is said to be defined **recursively**.

For the sequence $8, 10, 12, 14, \ldots$ in Example 2(a), each term after the first term is 2 more than the preceding term. Thus the sequence can be described by $a_1 = 8$, $a_{n+1} = a_n + 2$.

Similarly, for the sequence 1, 3, 9, 27, . . . in Example 2(b), each term after the first term is 3 times the previous term, so we can describe the sequence with $b_1 = 1$, $b_{n+1} = 3b_n$.

EXAMPLE 3   **Writing the Terms of a Recursive Sequence**

Write the first five terms of the sequence with $c_1 = 2$ and $c_{n+1} = c_n + 4$.

SOLUTION

$$c_1 = 2$$
$$c_2 = c_1 + 4 = 2 + 4 = 6$$
$$c_3 = c_2 + 4 = 6 + 4 = 10$$
$$c_4 = c_3 + 4 = 10 + 4 = 14$$
$$c_5 = c_4 + 4 = 14 + 4 = 18$$

In Example 3, the recursive sequence is defined by $c_1 = 2$, $c_{n+1} = c_n + 4$. In words, each term (after the first) is 4 more than the previous term. Another way to define the sequence is $c_1 = 2$, $c_n = c_{n-1} + 4$.

Sequence   You can use a calculator to generate the terms of a recursively defined sequence. The method is similar to that used previously, but for a recursive sequence, you must enter both the general term and the first term.

Sequence Variable   Certain sequence variables, such as u(nMin), are usually located in a variables menu.

FIGURE 6.4

```
"u(n-1)+4"→u
                    Done
2→u(nMin)
                      2
u(1,5)
      {2  6  10  14  18}
```

Figure 6.4 shows the use of a calculator to generate the terms of a recursively defined sequence. The general term $u_{n-1} + 4$ has been stored in u. (Note that we use *u* rather than *c*.) In the next entry, we store 2, the first term of the sequence, in u(nMin). Finally, we enter u(1,5) to generate terms 1 through 5.

Series

Suppose that a couple opens a savings account for a child and makes an initial deposit of $50. Each month thereafter they deposit $10 more than they deposited the previous month.

Month	Deposit
1	$50 = 50 + 10(0)$
2	$60 = 50 + 10(1)$
3	$70 = 50 + 10(2)$
4	$80 = 50 + 10(3)$
.	.
.	.
.	.

The amount d that is deposited in month n is described by the sequence $d_n = 50 + 10(n - 1)$. To determine the total amount of money that the parents had deposited after 4 months, we need to add the terms of the sequence.

$$d_1 + \quad d_2 \quad + \quad d_3 \quad + \quad d_4$$
$$50 + (50 + 10) + (50 + 20) + (50 + 30)$$

The sum of the terms of a sequence is called a **series.**

Definitions of Finite and Infinite Series

A **finite series** is the sum of the terms of a finite sequence. An **infinite series** is the sum of the terms of an infinite sequence.

The following are examples of sequences and their associated series.

Sequence	*Series*	
$3, 6, 12, 24$	$3 + 6 + 12 + 24$	Finite series
$1, \dfrac{1}{3}, \dfrac{1}{9}, \dfrac{1}{27}, \ldots$	$1 + \dfrac{1}{3} + \dfrac{1}{9} + \dfrac{1}{27} + \cdots$	Infinite series

To **evaluate** a series, we perform the indicated addition.

For an infinite series, the sum of the first n terms is a finite series and is called the *nth partial sum*, denoted by S_n.

EXAMPLE 4   **Evaluating *n*th Partial Sums**

Consider the sequence $-3, -1, 1, \ldots, 2n - 5, \ldots$. Evaluate the first four partial sums.

SOLUTION

The first three terms are -3, -1, and 1. The fourth term is the value of $2n - 5$ when $n = 4$. Thus the fourth term is 3.

$$S_1 = -3 \qquad \text{First term}$$
$$S_2 = -3 + (-1) = -4 \qquad \text{Sum of the first two terms}$$
$$S_3 = -3 + (-1) + 1 = -3 \qquad \text{Sum of the first three terms}$$
$$S_4 = -3 + (-1) + 1 + 3 = 0 \qquad \text{Sum of the first four terms}$$

To write a series, we use a notation called *summation notation*. In this notation, the Greek letter Σ (sigma) represents the word *sum.*

Consider the following sequence:

$$2, 5, 8, 11, \ldots, 3i - 1, \ldots$$

The sum of the first four terms is written

$$\sum_{i=1}^{4} (3i - 1)$$

which is read "the sum of $3i - 1$ from $i = 1$ to 4."

The letter i is the **index** of the summation. The lower number 1 is the **lower limit,** and the higher number 4 is the **upper limit.** Letters other than i can also be used, and the lower limit may be a number other than 1.

To **expand** a series, we replace the index variable with its values and write the indicated sum.

EXAMPLE 5   **Expanding a Series**

Expand and evaluate the given series.

(a) $\displaystyle\sum_{i=1}^{5} (2i - 3)$ (b) $\displaystyle\sum_{j=1}^{4} (-1)^j(j^2 + 2)$

SOLUTION

(a) $\displaystyle\sum_{i=1}^{5} (2i - 3) = (2 \cdot 1 - 3) + (2 \cdot 2 - 3) + (2 \cdot 3 - 3) + (2 \cdot 4 - 3) +$
$$(2 \cdot 5 - 3)$$
$$= -1 + 1 + 3 + 5 + 7 = 15$$

(b) $\displaystyle\sum_{j=1}^{4} (-1)^j(j^2 + 2) = (-1)^1(1^2 + 2) + (-1)^2(2^2 + 2) + (-1)^3(3^2 + 2) +$
$$(-1)^4(4^2 + 2)$$
$$= -3 + 6 - 11 + 18 = 10$$

■ **Series**   By entering the general term and the limits of summation, you can use a calculator to evaluate a series.

Figure 6.5 shows the results of evaluating the two series in Example 5. Note that n is used rather than i and j.

FIGURE **6.5**

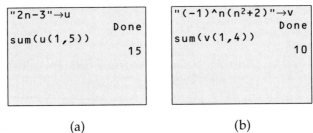

(a)   (b)

Writing a Series

To write a series in summation notation, we determine the general term and the domain of the sequence.

EXAMPLE 6 **Writing a Series in Summation Notation**

Write the given series in summation notation.

(a) $\dfrac{1}{2} + \dfrac{1}{4} + \dfrac{1}{6} + \dfrac{1}{8} + \cdots + \dfrac{1}{20}$

(b) $e^3 + e^4 + e^5 + e^6$

SOLUTION

(a) In each term, the denominator is twice the term number. Thus the general term is $\dfrac{1}{2n}$.

For the last term, $\dfrac{1}{2n} = \dfrac{1}{20}$, so $n = 10$. Thus there are 10 terms in the series, and $1 \le n \le 10$.

$$\frac{1}{2} + \frac{1}{4} + \frac{1}{6} + \frac{1}{8} + \cdots + \frac{1}{20} = \sum_{n=1}^{10} \frac{1}{2n}$$

■ *NOTE   The same series can be represented in different ways. For instance, the series in Example 6(b) can also be described by either of the following:*

$$\sum_{n=2}^{5} e^{n+1} \qquad \sum_{n=3}^{6} e^{n}$$

(b) Note that the exponent on e is 2 more than the term number, so the general term is e^{n+2}. Because there are four terms in the series, $1 \le n \le 4$.

$$e^3 + e^4 + e^5 + e^6 = \sum_{n=1}^{4} e^{n+2}$$

6.1 Quick Reference

Sequences

■ A **sequence** is a list of numbers called **terms** in which there is a first term, a second term, and so on.

■ The **general term** or *n*th **term** is a formula for any term of the sequence.

■ An **infinite sequence** is a function whose domain is the set of natural numbers. A **finite sequence** is a function whose domain is the first n natural numbers.

■ Although a sequence can be written as a function, the conventional notation uses subscripts to indicate the term numbers.

Writing the General Term

■ To determine an expression for the general term of a sequence when the first few terms are given, we must recognize the pattern of the terms of the sequence and express that pattern algebraically.

Recursive Sequences

■ A **recursive** sequence is described by giving the first term and stating the relationship between each term and its successor.

Series ■	A **finite series** is the sum of the terms of a finite sequence. An **infinite series** is the sum of the terms of an infinite sequence.
■	For an infinite series, the sum of the first n terms is a finite series and is called the *nth partial sum*, denoted by S_n.
■	To **evaluate** a series, we perform the indicated addition.
■	To write a series, we use a notation called *summation notation*. In this notation, the Greek letter Σ (sigma) represents the word *sum*.
■	In the summation notation $\sum_{i=1}^{4} c_i$, the letter i is the **index** of the summation. The lower number 1 is the **lower limit,** and the higher number 4 is the **upper limit.**
■	To **expand** a series, we replace the index variable with its values and write the indicated sum.
Writing a Series ■	To write a series in summation notation, we determine the general term and the domain of the sequence.

6.1 Speaking the Language

1. The list 1, 2, 3 is called a(n) ▨▨▨▨▨, whereas the sum of those terms is called a(n) ▨▨▨▨▨ .

2. For 3, 6, 9, 12, . . . , the ▨▨▨▨▨ term appears to be $3n$.

3. If we describe a sequence by giving the first term and stating the relationship between each term and its successor, we say that the sequence is defined ▨▨▨▨▨ .

4. When we write a series with the symbol Σ, we are using ▨▨▨▨▨ notation.

6.1 Exercises

Concepts and Skills

1. Consider the following sequences:
   (i) $b_n = a_n$ (ii) $c_n = (-1)^n a_n$
   How are the two sequences similar? How are they different?

2. Explain how to determine the domain of a finite sequence of n terms.

In Exercises 3–10, predict the next two terms of the given sequence. Then write the general term of the sequence.

3. $12, 14, 16, 18, \ldots$ 4. $3, 2, 1, 0, \ldots$

5. $3, -4, 5, -6, \ldots$ 6. $1, -2, 4, -8, \ldots$

7. $2, \dfrac{3}{2}, \dfrac{4}{3}, \dfrac{5}{4}, \ldots$

8. $\dfrac{3}{2}, -\dfrac{5}{4}, \dfrac{7}{8}, -\dfrac{9}{16}, \ldots$

9. $1 + \dfrac{2}{3}, 1 + \dfrac{3}{4}, 1 + \dfrac{4}{5}, 1 + \dfrac{5}{6}, \ldots$

10. $\dfrac{1}{3} - 1, \dfrac{2}{3} - 1, 0, \dfrac{4}{3} - 1, \ldots$

In Exercises 11–18, for the given sequence, write (a) the first four terms, (b) the seventh term, and (c) the twelfth term.

11. $b_n = 2^n$ 12. $a_n = \left(\dfrac{3}{2}\right)^n$

13. $c_n = 1 - 3n$ 14. $b_n = 3(n + 2)$

15. $a_n = \dfrac{(-1)^n - 1}{n}$

16. $c_n = \dfrac{(-1)^{n+1}n}{n+1}$

17. $b_n = \dfrac{n^2}{n+4}$

18. $c_n = \dfrac{n-3}{n+1}$

In Exercises 19–22, write all terms of the sequence.

19. $a_n = \dfrac{n}{n+1}, 1 \le n \le 5$

20. $b_n = \dfrac{(-1)^n}{n(n+1)}, 1 \le n \le 4$

21. $b_n = \dfrac{1 + (-1)^{n+1}}{n}, 1 \le n \le 8$

22. $c_n = (-1)^n n^2, 1 \le n \le 7$

In Exercises 23–28, use a calculator to write the first four terms of the sequence. (Round answers to four decimal places.)

23. $c_n = \left(1 + \dfrac{1}{n}\right)^n$

24. $a_n = \dfrac{\sqrt{n}}{3 + \sqrt{n}}$

25. $a_n = n^{2/3}$

26. $b_n = \dfrac{\ln n}{n}$

27. $b_n = \sqrt{n+1} - \sqrt{n}$

28. $a_n = \dfrac{1}{n+1} - \dfrac{1}{n}$

29. What information must be given in a recursive definition of a sequence?

30. A sequence is an **increasing sequence** if $a_n \le a_{n+1}$ for each n. From the following list, identify any sequence that is not an increasing sequence and explain why.

(i) $2, -1, 3, -2, 5, -3, \ldots$

(ii) $4, 4, 4, 4, 4, 4, \ldots$

(iii) $\dfrac{1}{2}, \dfrac{2}{3}, \dfrac{3}{4}, \dfrac{4}{5}, \dfrac{5}{6}, \ldots$

In Exercises 31–34, (a) write the first four terms of the given sequence and (b) use a calculator display of the terms of the sequence to find the tenth term.

31. $c_1 = 3, c_n = c_{n-1} + 5$

32. $a_1 = \dfrac{1}{2}, a_n = -2a_{n-1}$

33. $b_1 = 1, b_2 = 3, b_n = b_{n-1} - b_{n-2}$

34. $a_1 = -1, a_2 = 2, a_n = a_{n-1}a_{n-2}$

In Exercises 35–38, use a calculator to write the first four terms of the sequence. (Round answers to four decimal places.)

35. $a_1 = 1, a_n = \sqrt{1 + a_{n-1}}$

36. $a_1 = 2, a_n = \dfrac{2}{1 + a_{n-1}}$

37. $a_1 = 2, a_2 = 3, a_n = \dfrac{1}{a_{n-1}} - \dfrac{1}{a_{n-2}}$

38. $a_1 = 1, a_2 = 1, a_n = (a_{n-1} + a_{n-2})^{1/2}$

39. What do we mean by the fourth partial sum of a sequence?

40. In summation notation, what do the lower and upper limits signify?

In Exercises 41–44, for the given sequence, evaluate the indicated partial sums.

41. $2, 4, 6, 8, \ldots;$ S_4, S_9

42. $1, -3, 5, -7, \ldots;$ S_3, S_7

43. $-1, 4, -7, 10, \ldots;$ S_5, S_8

44. $\dfrac{2}{1}, \dfrac{2}{2}, \dfrac{2}{3}, \dfrac{2}{4}, \ldots;$ S_4, S_6

In Exercises 45–48, for the given sequence, use a calculator to evaluate the indicated partial sums.

45. $a_n = 1 + n^2;$ S_6, S_{10} **46.** $b_n = 2^{-n} - 3;$ S_7, S_{11}

47. $c_n = \dfrac{-n}{n+1};$ S_5, S_9 **48.** $b_n = \dfrac{n^2}{n+2};$ S_8, S_{12}

In Exercises 49–54, for the given sequence, use a calculator to evaluate the indicated partial sums.

49. $a_1 = -5, a_n = 2a_{n-1} + 1;$ S_5, S_8

50. $b_1 = 3, b_n = 3 - b_{n-1};$ S_7, S_{12}

51. $a_1 = 2, a_n = \dfrac{n-1}{a_{n-1}};$ S_6, S_9

52. $c_1 = 8, c_n = \sqrt{1 + c_{n-1}};$ S_{11}, S_{16}

53. $c_1 = 1, c_2 = 2, c_n = c_{n-1}c_{n-2};$ S_7, S_{10}

54. $b_1 = 1, b_2 = 2, b_n = b_{n-1} - b_{n-2};$ S_{12}, S_{15}

In Exercises 55–62, evaluate the given series.

55. $\displaystyle\sum_{i=1}^{5} (2i - 5)$ **56.** $\displaystyle\sum_{i=1}^{6} (i+1)(i-1)$

57. $\displaystyle\sum_{i=1}^{4} (-2)^i$ **58.** $\displaystyle\sum_{i=1}^{5} \dfrac{(-1)^i}{i}$

59. $\displaystyle\sum_{i=1}^{7} 3i$ **60.** $\displaystyle\sum_{i=1}^{10} 5$

61. $\displaystyle\sum_{i=1}^{4} \left(1 - \dfrac{2}{i}\right)$ **62.** $\displaystyle\sum_{i=1}^{4} \left(\dfrac{i-2}{i^2}\right)$

In Exercises 63–68, write the series in summation notation.

63. $3 - 4 + 5 - 6 + 7 - 8$

64. $\dfrac{5}{3} - \dfrac{5}{9} + \dfrac{5}{27} - \dfrac{5}{81} + \dfrac{5}{243}$

65. $\dfrac{2}{1+2} + \dfrac{2}{2+2} + \dfrac{2}{3+2} + \dfrac{2}{4+2} + \cdots + \dfrac{2}{15+2}$

66. $\dfrac{3}{2} + \dfrac{5}{4} + \dfrac{7}{6} + \dfrac{9}{8} + \cdots + \dfrac{21}{20}$

67. $2 + 5 + 8 + 11 + \cdots + 29$

68. $\dfrac{3}{1+1} + \dfrac{5}{4+1} + \dfrac{7}{9+1} + \dfrac{9}{16+1} + \cdots + \dfrac{21}{100+1}$

Concept Extension

In Exercises 69–72, for the given sequence write (a) a general term of the sequence and (b) a recursive relation for the sequence.

69. $5, 3, 1, -1, \ldots$

70. $4, 2, 1, \dfrac{1}{2}, \ldots$

71. $7, -7, 7, -7, \ldots$

72. $-1, 3, -9, 27, \ldots$

73. Consider the sequence of odd whole numbers. Evaluate the partial sums S_n for $n = 1, 2, 3, 4, 5$. What pattern do you observe?

74. Consider the sequence $a_n = \dfrac{1}{n} - \dfrac{1}{n+1}$. Write a formula for the sequence of partial sums S_n.

Applications

A well-known sequence is the **Fibonacci sequence,** named for the Italian mathematician Leonardo Fibonacci.

$$1, 1, 2, 3, 5, 8, \ldots$$

The terms of the sequence are called *Fibonacci numbers*. These numbers often occur in nature. For instance, pine cones, daisies, and ears of corn exhibit growth patterns that can be modeled by the Fibonacci sequence. Exercises 75 and 76 involve the Fibonacci sequence.

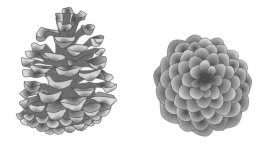

75. In the Fibonacci sequence, $a_1 = 1$ and $a_2 = 1$. Write a recursive formula for a_{n+2}.

76. Use the recursive formula found in Exercise 75 to determine the first twelve terms of the Fibonacci sequence.

Rabbit Population Exercises 77 and 78 refer to the following information.

Suppose that a pair of rabbits will produce a new pair of rabbits in their second month and will continue to produce a new pair each month thereafter. Assuming that each pair of rabbits will do the same and that none of the rabbits die, we can determine the number of rabbits at the end of a given number of months.

In the accompanying table, gray rabbits represent pairs of newborn rabbits, and blue rabbits represent pairs of rabbits that are able to reproduce.

Number of Months	Status of Rabbit Pairs	Number of Pairs
Start		1
1		1
2		2
3		3
4		5

Each reproducing pair in one month produces a pair of newborns in the next month, and each newborn pair in one month becomes a reproducing pair in the next month.

77. Extend the table to 11 months and complete the entries for the added months.

78. What evidence do you see that these birth patterns of rabbits can be modeled by the Fibonacci sequence?

Honeybees Exercises 79–84 refer to the following information.

A male honeybee has no father because it comes from an unfertilized egg. A female honeybee comes from a fertilized egg, and so it has a mother and a father. The accompanying figure shows a *tree diagram* of six generations for a male honeybee and its ancestors.

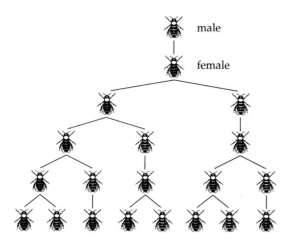

Observe that the male bee (first generation) has only a mother (second generation), which has both a father and a mother (third generation).

79. Extend the tree diagram to a seventh generation. Then, using M (male) and F (female), list the entries in the seventh row.

80. Write a sequence in which each term a_n is the number of bees in generation n, where $1 \le n \le 7$.

81. Compare your sequence in Exercise 80 with the Fibonacci sequence. What can you conclude?

82. Use a calculator to display the terms of the sequence in Exercise 80. In what generation would a single male honeybee have 144 ancestors?

83. Suppose that the figure had begun with a female bee in the first generation. Then $a_1 = 1$ and $a_2 = 2$. Write a formula for a_{n+2}. Is this sequence the same as the Fibonacci sequence?

84. Refer to the diagram described in Exercise 83. Use a calculator to evaluate the series that represents the total number of bees in 12 generations.

6.2 *Arithmetic Sequences and Series*

Arithmetic Sequences ■ *Arithmetic Series* ■
Real-Life Applications

Arithmetic Sequences

Suppose that a physical therapist recommends that a patient walk 10 minutes the first day and that the walking time be increased by 3 minutes each day thereafter. We can describe the time the patient walks on days 1, 2, 3, 4, . . . with the sequence 10, 13, 16, 18, ... Because the difference between any two consecutive terms is always 3, this sequence is an example of an **arithmetic sequence**.

> *Definition of an Arithmetic Sequence*
>
> An **arithmetic sequence** is a sequence in which the difference between any term and the preceding term is a constant called the **common difference**.

In the following sequences, we see that any two consecutive terms can be used to determine the common difference.

	Common	
Sequence	*difference*	
$-4, -1, 2, 5, \ldots$	3	$-1 - (-4) = 3, 2 - (-1) = 3, 5 - 2 = 3, \ldots$
$8, 2, -4, -10, \ldots$	-6	$2 - 8 = -6, -4 - 2 = -6, -10 - (-4) = -6, \ldots$

In general, $a_n - a_{n-1} = d$, or $a_n = a_{n-1} + d$.

EXAMPLE 1   **Writing the Terms of an Arithmetic Sequence**

Write the first five terms of the arithmetic sequence whose first term is 8 and whose common difference is -5.

SOLUTION

Each term is obtained by adding -5 to the previous term. The first five terms are $8, 3, -2, -7$, and -12.

In general, the terms of an arithmetic sequence with a common difference d can be written

n	1	2	3	4	$\ldots$
Term	$a_1,$	$a_1 + d,$	$a_1 + 2d,$	$a_1 + 3d,$	$\ldots$

Because the coefficient of d is 1 less than the term number n, we can write $a_n = a_1 + (n - 1)d$.

> **The General Term of an Arithmetic Sequence**
>
> The general term of an arithmetic sequence whose first term is a_1 and whose common difference is d is $a_n = a_1 + (n - 1)d$.

EXAMPLE 2   **Writing the General Term of an Arithmetic Sequence**

Determine whether the sequence is an arithmetic sequence. For each arithmetic sequence, write the general term.

(a) $-2, 2, 6, 10, \ldots$
(b) $2, 4, 7, 11, 16, \ldots$
(c) $\dfrac{8}{3}, 1, -\dfrac{2}{3}, -\dfrac{7}{3}, \ldots$

SOLUTION

(a) Each term is 4 more than the previous term. Thus the sequence is an arithmetic sequence. Because we know that $a_1 = -2$ and $d = 4$, we can write the general term.

$$a_n = a_1 + (n - 1)d$$
$$a_n = -2 + (n - 1)4 = -2 + 4n - 4 = 4n - 6$$

(b) The differences $4 - 2 = 2$ and $7 - 4 = 3$ are not the same. Because the differences between consecutive terms are not constant, the sequence is not an arithmetic sequence.

(c) Calculate the differences of consecutive terms.

$$1 - \frac{8}{3} = -\frac{5}{3} \qquad -\frac{2}{3} - 1 = -\frac{5}{3} \qquad -\frac{7}{3} - \left(-\frac{2}{3}\right) = -\frac{5}{3} \qquad \cdots$$

Because the differences are the same, the sequence is arithmetic. We use $a_1 = \frac{8}{3}$ and $d = -\frac{5}{3}$ to write the general term.

$$a_n = a_1 + (n - 1)d$$
$$a_n = \frac{8}{3} + (n - 1)\left(-\frac{5}{3}\right) = \frac{13}{3} - \frac{5}{3}n = \frac{13 - 5n}{3}$$

■ *NOTE   It can be shown that the general term of the sequence in Example 2(b) is $a_n = \frac{1}{2}n^2 + \frac{1}{2}n + 1$. The function defined by the general term is not a linear function, and the sequence is not arithmetic.*

For the arithmetic sequences in Example 2, we can graph the sequences by plotting points whose coordinates are of the form (term number, term) or (n, a_n). Figure 6.6 shows the graphs for the first 10 terms in parts (a) and (c) of Example 2. Note that the terms correspond to points of a line. In general, an arithmetic sequence is a linear function whose domain is restricted to the natural (term) numbers.

FIGURE **6.6**

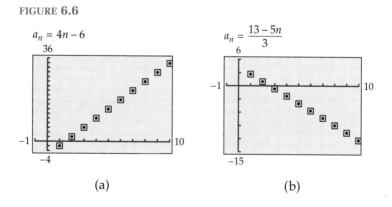

$a_n = 4n - 6$ (a)

$a_n = \dfrac{13 - 5n}{3}$ (b)

EXAMPLE **3**   **Finding a Specific Term of an Arithmetic Sequence**

Two terms of an arithmetic sequence are $a_1 = 3$ and $a_7 = -9$. Determine a_{12}.

SOLUTION

We can use a_1 and a_7 to determine d.

$$a_n = a_1 + (n-1)d$$
$$-9 = 3 + (7-1)d \qquad a_1 = 3, n = 7, a_7 = -9$$
$$-9 = 3 + 6d$$
$$-12 = 6d$$
$$d = -2$$

Now we know a_1 and d, so we can write the general term.

$$a_n = a_1 + (n-1)d$$
$$a_n = 3 + (n-1)(-2) = 5 - 2n$$

Finally, $a_{12} = 5 - 2(12) = -19$.

■ *NOTE   We can use a calculator to determine the general term of an arithmetic sequence. For instance, to determine the general term of the sequence described in Example 3, enter (1, 3) (for $a_1 = 3$) and (7, −9) (for $a_7 = −9$) in the statistics list. Then produce the linear regression equation. The expression ax + b is the general term of the sequence.*

EXAMPLE 4   **Writing the General Term of an Arithmetic Sequence**

Write the general term of an arithmetic sequence with $a_6 = 10$ and $a_{16} = 25$.

SOLUTION

We use the information about each term in the formula $a_n = a_1 + (n-1)d$.

$n = 6, a_6 = 10$	$n = 16, a_{16} = 25$
$10 = a_1 + (6-1)d$	$25 = a_1 + (16-1)d$
$10 = a_1 + 5d$ \qquad (1)	$25 = a_1 + 15d$ \qquad (2)

We use equations (1) and (2) to form a system of two equations in two variables. After writing both equations in standard form, we solve the system with the addition method.

$$-a_1 - \ 5d = -10 \qquad \text{Multiply equation (1) by } -1.$$
$$\underline{a_1 + 15d = \ \ 25} \qquad \text{Equation (2)}$$
$$10d = \ \ 15 \qquad \text{Add the equations.}$$
$$d = \ \ \frac{3}{2} \qquad \text{Solve for } d.$$

To determine a_1, substitute $d = \frac{3}{2}$ into either equation.

$$a_1 + 5d = 10 \qquad \text{Equation (1)}$$
$$a_1 + 5\left(\frac{3}{2}\right) = 10 \qquad d = \frac{3}{2}$$
$$a_1 = \frac{5}{2}$$

Using a_1 and d, we can write the general term.

$$a_n = a_1 + (n-1)d$$
$$a_n = \frac{5}{2} + (n-1)\left(\frac{3}{2}\right) = \frac{5}{2} + \frac{3}{2}n - \frac{3}{2} = \frac{3}{2}n + 1$$

EXAMPLE 5   **Finding the Number of Terms in an Arithmetic Sequence**

Determine the number of terms in the finite sequence

$$5, 2, -1, -4, \ldots, -88$$

SOLUTION

Let n represent the number of terms. The common difference is -3, $a_1 = 5$, and $a_n = -88$.

$$a_n = a_1 + (n - 1)d$$
$$-88 = 5 + (n - 1)(-3) \qquad a_n = -88, a_1 = 5, d = -3$$
$$-88 = -3n + 8 \qquad \text{Simplify.}$$
$$-3n = -96$$
$$n = 32 \qquad \text{Solve for } n.$$

The given sequence has 32 terms.

Arithmetic Series

An arithmetic series is a specific example of a series, as defined in the previous section.

> **Definition of an Arithmetic Series**
>
> An **arithmetic series** is the sum of the terms of an arithmetic sequence.

Recall that the nth partial sum of a series is the sum of the first n terms of the series.

EXPLORATION   *The nth Partial Sum of an Arithmetic Series*

Consider the series $1 + 3 + 5 + 7 + 9 + 11 + \cdots$. The fourth partial sum is $S_4 = 1 + 3 + 5 + 7$. Because the terms can be added in any order, we can also write the sum as $S_4 = 7 + 5 + 3 + 1$.

Suppose we add these two representations.

$$S_4 = 1 + 3 + 5 + 7$$
$$S_4 = 7 + 5 + 3 + 1$$
$$\overline{2S_4 = 8 + 8 + 8 + 8 = 4(8) = 4(1 + 7)}$$

(a) Solve the last equation for S_4.
(b) Let $n =$ the number of terms in the series, $a_1 =$ the first term, and $a_n =$ the last term. What does your result in part (a) suggest as a general formula for the nth partial sum of an arithmetic series?

DEVELOPING THE CONCEPT

The nth Partial Sum of an Arithmetic Series

Let S_n represent the sum of the first n terms of an arithmetic sequence. We can write the sum in both increasing and decreasing order and add the terms.

$$S_n = a_1 + (a_1 + d) + (a_1 + 2d) + \cdots + [a_1 + (n-1)d]$$
$$S_n = a_n + (a_n - d) + (a_n - 2d) + \cdots + [a_n - (n-1)d]$$

$$2S_n = (a_1 + a_n) + (a_1 + a_n) + (a_1 + a_n) + \cdots + (a_1 + a_n)$$
$$2S_n = n(a_1 + a_n) \qquad \text{The sum contains } n \text{ terms } (a_1 + a_n).$$
$$S_n = \frac{n(a_1 + a_n)}{2} \qquad \text{Divide both sides by 2.}$$

The result is a formula for the nth partial sum of an arithmetic series.

The nth Partial Sum of an Arithmetic Series

For an arithmetic sequence whose first term is a_1 and whose nth term is a_n, the nth partial sum S_n is given by

$$S_n = \frac{n(a_1 + a_n)}{2}$$

In words, the nth partial sum is found by averaging the first and last terms of the sequence and multiplying the result by n.

EXAMPLE 6   Evaluating a Finite Arithmetic Series

Determine the sum $2 + 7 + 12 + 17 + \cdots + 117$.

SOLUTION

We already know that the first term is $a_1 = 2$ and that the last term is $a_n = 117$. However, to determine the sum, we must also know n, the number of terms in the sum.

$$a_n = a_1 + (n-1)d$$
$$117 = 2 + (n-1)(5) \qquad a_n = 117, a_1 = 2, \text{ and } d = 5$$
$$117 = 5n - 3 \qquad \text{Simplify.}$$
$$5n = 120$$
$$n = 24 \qquad \text{Solve for } n.$$

Now determine the sum.

$$S_n = \frac{n(a_1 + a_n)}{2}$$

$$S_n = \frac{24(2 + 117)}{2} = 1428 \qquad n = 24,\, a_1 = 2,\, \text{and } a_{24} = 117$$

EXAMPLE 7   **Evaluating a Finite Arithmetic Series**

Evaluate the series $\sum\limits_{i=1}^{100} (7 - 2i)$.

SOLUTION

The series has 100 terms ($n = 100$). The first term is $a_1 = 7 - 2(1) = 5$ and the last term is $a_{100} = 7 - 2(100) = -193$.

$$S_n = \frac{n(a_1 + a_n)}{2} = \frac{100(5 - 193)}{2} = -9400$$

EXAMPLE 8   **Evaluating an Arithmetic Series**

Determine the sum of the first 100 terms of an arithmetic sequence with $a_1 = -20$ and $d = 5$.

SOLUTION

We know that the first term is $a_1 = -20$ and that $n = 100$. To determine the sum, we must know the last term a_{100}.

$$a_n = a_1 + (n - 1)d$$
$$a_{100} = -20 + (100 - 1)(5) = 475 \qquad a_1 = -20,\, d = 5,\, \text{and } n = 100$$

Now calculate the sum.

$$S_n = \frac{n(a_1 + a_n)}{2} = \frac{100(-20 + 475)}{2} = 22{,}750$$

Real-Life Applications

We often find that quantities increase or decrease by a constant amount over a limited domain. In such cases, an arithmetic sequence is an appropriate model.

EXAMPLE 9   **Career Earnings**

Suppose that the average starting salary for a liberal arts graduate is $30,000 per year and that a graduate can expect to receive a $1600 raise each year.

(a) Write an arithmetic sequence to describe the annual salary.
(b) Write and evaluate an arithmetic series to determine the person's total earnings for a 30-year career.

SOLUTION

(a) The salaries 30,000, 31,600, 33,200, 34,800, . . . form an arithmetic sequence with $a_1 = 30{,}000$ and $d = 1600$.

$$a_n = 30{,}000 + (n - 1)(1600)$$
$$= 30{,}000 + 1600n - 1600$$
$$= 28{,}400 + 1600n$$

(b) The total earnings is the sum of the first 30 terms of the sequence: $\sum_{n=1}^{30} (28{,}400 + 1600n)$. The first term is 30,000, and the 30th term is $28{,}400 + 1600(30) = 76{,}400$.

$$S_n = \frac{n(a_1 + a_n)}{2} = \frac{30(30{,}000 + 76{,}400)}{2} = 1{,}596{,}000$$

For the given assumptions, the person's total career earnings will be $1,596,000.

6.2 Quick Reference

Arithmetic Sequences
■ An **arithmetic sequence** is a sequence in which the difference between any term and the preceding term is a constant called the **common difference.**

■ The general term of an arithmetic sequence whose first term is a_1 and whose common difference is d is $a_n = a_1 + (n - 1)d$.

■ An arithmetic sequence can be regarded as a linear function whose domain is restricted to the natural numbers.

Arithmetic Series
■ An **arithmetic series** is the sum of the terms of an arithmetic sequence.

■ For an arithmetic sequence whose first term is a_1 and whose nth term is a_n, the nth partial sum S_n is given by

$$S_n = \frac{n(a_1 + a_n)}{2}$$

In words, the nth partial sum is found by averaging the first and last terms of the sequence and multiplying the result by n.

6.2 Speaking the Language

1. For the arithmetic sequence $2, 7, 12, \ldots$, we say that 5 is the ▨▨▨▨▨ .

2. An arithmetic sequence is a(n) ▨▨▨▨▨ function whose domain is the set of natural numbers.

3. If the terms of a series are such that $a_n - a_{n-1}$ is constant, then we call the series a(n) ▨▨▨▨▨ series.

4. The sum of the first six terms of $\sum_{i=1}^{20} (2i + 3)$ is called the sixth ▨▨▨▨▨ .

6.2 Exercises

Concepts and Skills

1. Explain how to determine the common difference of an arithmetic sequence.

2. What information is needed to write the general term of an arithmetic sequence?

In Exercises 3–10, determine whether the given sequence is an arithmetic sequence. For each arithmetic sequence, (a) identify the first term and common difference and (b) write the general term.

3. $-5, -2, 1, 4, \ldots$

4. $8, 2, -4, -10, \ldots$

5. $3, 1, -1, -3, \ldots$

6. $-\dfrac{1}{2}, 1, \dfrac{5}{2}, 4, \ldots$

7. $0, 2, 6, 12, \ldots$

8. $2, -1, 2, -1, \ldots$

9. $0.2, 0.6, 1.0, 1.4, \ldots$

10. $7, 7, 7, 7, \ldots$

In Exercises 11–18, use the given information about arithmetic sequences to write

(a) the first four terms.
(b) the general term.
(c) the tenth term.

11. $c_1 = -3, d = 7$

12. $a_1 = 15, d = 2$

13. $a_1 = 18, d = -6$

14. $b_1 = \dfrac{5}{3}, d = \dfrac{5}{6}$

15. $k_5 = 24, d = -4$

16. $a_4 = -8, d = 5$

17. $c_4 = -3, d = 0.3$

18. $k_7 = -2, d = -1.2$

19. Describe the difference between the graph of the equation $y = 2x + 1$ and the graph of the arithmetic sequence $a_n = 2n + 1$.

20. Suppose that the first term of an arithmetic sequence is 1 and the last term is 50. Explain why determining the number of terms in the sequence is not possible.

In Exercises 21–26, two terms of an arithmetic sequence are given. Write the general term.

21. $c_1 = 4, c_6 = -11$

22. $k_1 = -6, k_9 = 10$

23. $a_3 = 1, a_{11} = 3$

24. $b_8 = 12, b_{12} = 18$

25. $t_4 = 6, t_8 = 10.8$

26. $a_7 = -17, a_{10} = -23$

In Exercises 27–30, determine the number of terms in the given finite arithmetic sequence.

27. $-10, -3, 4, 11, \ldots, 158$

28. $103, 98, 93, 88, \ldots, -112$

29. $\dfrac{5}{3}, \dfrac{7}{3}, 3, \dfrac{11}{3}, \ldots, 99$

30. $1.1, 0.8, 0.5, 0.2, \ldots, -3.7$

In Exercises 31–36, for the given arithmetic sequence, find the sum of the indicated number of terms.

31. $9, 13, 17, 21, \ldots; \quad n = 24$

32. $6, 6, 6, 6, \ldots; \quad n = 20$

33. $9, 2, -5, -12, \ldots; \quad n = 30$

34. $-10, -5, 0, 5, \ldots; \quad n = 54$

35. $\dfrac{3}{2}, 2, \dfrac{5}{2}, 3, \ldots; \quad n = 41$

36. $1.7, 1.2, 0.7, 0.2, \ldots; \quad n = 28$

37. Parts (i), (ii), and (iii) represent given information about an arithmetic series. In which part would determining the ninth partial sum of the series be impossible? Why?

(i) $a_1 = 2, a_9 = 30$
(ii) $a_1 = 2, d = 5$
(iii) $a_1 = 2, n = 9$

38. Explain why $\displaystyle\sum_{i=1}^{5} (3i - 1) = \sum_{j=0}^{4} (3j + 2)$.

In Exercises 39–42, evaluate the given arithmetic series.

39. $-1 + 1 + 3 + 5 + \cdots + 45$

40. $2 - 3 - 8 - 13 - \cdots - 173$

41. $\dfrac{1}{6} - \dfrac{1}{6} - \dfrac{1}{2} - \dfrac{5}{6} - \cdots - \dfrac{13}{2}$

42. $2 + \dfrac{8}{3} + \dfrac{10}{3} + 4 + \cdots + \dfrac{34}{3}$

In Exercises 43–52, evaluate the given series.

43. $\displaystyle\sum_{i=1}^{24} (8 - 5i)$

44. $\displaystyle\sum_{n=1}^{15} (1 - n)$

45. $\displaystyle\sum_{k=1}^{30} \left(\dfrac{1}{2}k + 4\right)$

46. $\displaystyle\sum_{i=1}^{18} (1 - 3i)$

47. $\displaystyle\sum_{i=1}^{20} \dfrac{3(1 + 2i)}{5}$

48. $\displaystyle\sum_{n=1}^{36} \dfrac{1}{4}n$

49. $\displaystyle\sum_{n=5}^{30} (n - 4)$

50. $\displaystyle\sum_{k=3}^{14} \dfrac{k}{3}$

51. $\displaystyle\sum_{i=12}^{26} 4i$

52. $\displaystyle\sum_{i=15}^{50} \left(\dfrac{1}{2}i + 1\right)$

Concept Extension

In Exercises 53–56, an arithmetic sequence is defined recursively. Write the general term and the 50th term.

53. $k_1 = 14, k_{n+1} = k_n + 6$

54. $a_1 = 8, a_{n+1} = a_n - 4$

55. $c_1 = -7, c_{n+1} = c_n - 3.2$

56. $b_1 = 5, b_{n+1} = b_n$

57. Write a formula for determining the sum of the first n positive even integers.

58. Determine whether $\ln a, \ln a^2, \ln a^3, \ln a^4, \ldots$ is an arithmetic sequence. If it is, determine the common difference and the general term.

59. Determine whether 2400 is a term of the arithmetic sequence $-2, 1, 4, 7, \ldots$. If not, write the first three terms of the sequence that exceed 2400.

60. Write the first three terms of the arithmetic sequence for which $a_{20} = -7$ and $S_{20} = -45$.

Applications

61. Flower Garden   A gardener plants 18 semicircular rows of marigolds. The first row has 12 plants, and every other row has 6 more plants than the preceding row.

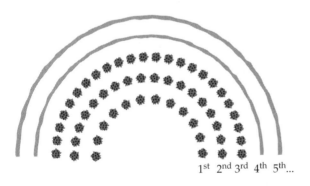

1st 2nd 3rd 4th 5th...

(a) Write an arithmetic sequence to describe the number of plants in each row.

(b) If marigolds are sold in trays of 6 plants, how many trays should the gardener purchase?

62. Amphitheater Seating   The floor plan for an amphitheater is trapezoidal with 36 rows of seats. The first row has 16 seats, and each of the other rows has 2 more seats than the preceding row.

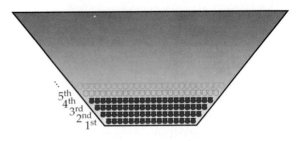

(a) Write an arithmetic sequence to represent the number of seats in row n.

(b) For a sold-out event, the receipts were $22,032. What was the average price of a ticket?

63. Clock Strikes   A clock strikes once on the half hour and once at 1:00, twice at 2:00, and so on. How many times does the clock strike between 12:15 P.M. on Thursday and 12:15 P.M. on Friday?

64. Handshakes   Suppose that 250 politicians attended a reception and that each one shook hands with everyone else exactly once. How many handshakes were there?

65. Volunteer Tutors   A college volunteer coordinator recruited 78 students to participate in a tutoring program for children in the community. Each week 4 fewer students participated than the week before.

(a) Write a sequence to describe the number of student tutors for week n.

(b) Each student volunteer provided 3 hours of tutoring per week. During a 15-week semester, how many student-hours were spent tutoring?

66. Contaminated Water   Because of a malfunction of a water purification facility at a poultry processing plant, contaminated water began leaking into a stream. The state environmental agency levied an initial fine of $1000 per day, with the fine increasing by $100 per day each day until the facility was operational.

(a) Write a sequence to describe the daily fine.

(b) Suppose that the repairs were completed in 18 days. What was the total fine?

67. Skiing   A Super G skier found that for each practice run, she could reduce her time by 0.08 second. For the first run, the skier's time was 1 minute, 20.05 seconds.

(a) Write a sequence for the run times.

(b) What was her total time on the ski slope after 8 practice runs?

68. Running   A runner wants to run a total of at least 1000 miles. He begins by running 1.5 miles the first day and then running $\frac{1}{4}$ mile further each day than he ran the preceding day.

(a) Write a sequence for the distance he runs each day.

(b) After how many days will the total distance first exceed 1000 miles?

69. Cellular Phone Cost   The bar graph shows the average monthly cellular phone bill (in dollars) for the years 1991–1995. (*Source*: Cellular Telecommunications Industry Association.)

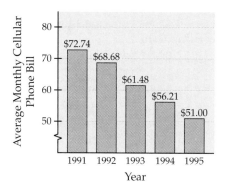

(a) Use a calculator's regression feature to write an arithmetic sequence b_n to model the data. Let n represent the number of years since 1990. (Round coefficients to the nearest tenth.) What does the term b_{11} represent?

(b) Use the bar graph and suppose that a person pays the average bill for the five years 1991–1995. Write and evaluate a series to

give the total expenditure for cell phone service.

(c) Use the regression model and suppose that a person pays the average bill for the seven years 1994–2000. Write a series for the projected expenditures.

70. Women in the Work Force   The bar graph shows the percentage of women in the work force who have children under age 18. (*Source*: Bureau of Labor Statistics.)

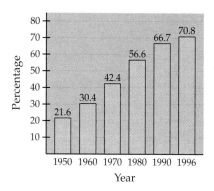

(a) Use your calculator's regression feature to write an arithmetic sequence p_n to model the data. Let n represent the number of years since 1949. (Round coefficients to the nearest hundredth.)

(b) What is the common difference and what does it represent?

(c) Evaluate and interpret $\dfrac{1}{46}\displaystyle\sum_{i=1}^{46} p_n$.

6.3 *Geometric Sequences and Series*

Geometric Sequences   ■   *Geometric Series*   ■
Infinite Geometric Series   ■   *Real-Life Applications*

Geometric Sequences

Your family tree begins with yourself (1 person) and includes your parents (2 people), your grandparents (4 people), your great-grandparents (8 people), and so on.

The number of people in each generation can be described by the sequence 1, 2, 4, 8, 16, ... Note that each term is two times the preceding term. The pattern of multiplying each term by a constant to obtain the next term defines a **geometric sequence.** We call the constant the **common ratio** because it is the ratio of any term to the preceding term.

Definition of a Geometric Sequence

A **geometric sequence** is a sequence in which there is a constant r, called the **common ratio,** such that $\dfrac{a_n}{a_{n-1}} = r$ (or $a_n = ra_{n-1}$) for any integer n.

In the following sequences, we see that the common ratio can be obtained by dividing any term by the preceding term.

Sequence	*Common ratio*	
$3, 6, 12, 24, \ldots$	2	$\dfrac{6}{3} = 2, \dfrac{12}{6} = 2, \dfrac{24}{12} = 2, \ldots$
$-2, 6, -18, 54, \ldots$	-3	$\dfrac{6}{-2} = -3, \dfrac{-18}{6} = -3, \dfrac{54}{-18} = -3, \ldots$
$2, 3, \dfrac{9}{2}, \dfrac{27}{4}, \ldots$	$\dfrac{3}{2}$	$\dfrac{3}{2}, \dfrac{9/2}{3} = \dfrac{3}{2}, \dfrac{27/4}{9/2} = \dfrac{3}{2}, \ldots$

EXAMPLE 1   **Writing the Terms of a Geometric Sequence**

Write the first four terms of the geometric sequence whose first term is 6 and whose common ratio is $-\frac{1}{2}$.

SOLUTION

Because $r = -\dfrac{1}{2}, a_n = -\dfrac{1}{2}a_{n-1}$.

$$a_1 = 6, a_2 = -\frac{1}{2}(6) = -3, a_3 = -\frac{1}{2}(-3) = \frac{3}{2}, a_4 = -\frac{1}{2}\left(\frac{3}{2}\right) = -\frac{3}{4}$$

Note that when $r < 0$, the terms alternate in sign.

The terms of a geometric sequence have the following form:

$$a_1, a_2 = a_1r, a_3 = a_2r = a_1r^2, a_4 = a_3r = a_1r^3, \ldots$$

Note that the exponent on r is 1 less than the term number. This observation leads to the following formula for the general term.

General Term of a Geometric Sequence

The general term of a geometric sequence whose first term is a_1 and whose common ratio is r is $a_n = a_1r^{n-1}$.

EXAMPLE 2   **Finding the General Term of a Geometric Sequence**

Write the general term for the geometric sequence whose first three terms are 8, $-\frac{40}{3}$, and $\frac{200}{9}$.

SOLUTION

The common ratio is $r = \dfrac{\frac{-40}{3}}{8} = -\dfrac{5}{3}$.

$$a_n = a_1 r^{n-1} = 8\left(-\frac{5}{3}\right)^{n-1} \qquad a_1 = 8 \text{ and } r = -\frac{5}{3}.$$

EXAMPLE 3   **Finding a Term of a Geometric Sequence**

Determine the 12th term of the geometric sequence whose first term is -5 and whose common ratio is -2.

SOLUTION

$$a_n = a_1 r^{n-1} = (-5)(-2)^{n-1} \qquad a_1 = -5 \text{ and } r = -2.$$

The 12th term is $a_{12} = (-5)(-2)^{11} = 10{,}240$.

EXAMPLE 4   **Finding the General Term of a Geometric Sequence**

Write the general term for a geometric sequence whose fourth term is 15 and whose seventh term is 405.

SOLUTION

$$\begin{aligned} a_7 = a_1 r^6 = 405 & \qquad \text{The seventh term is 405.} \\ a_4 = a_1 r^3 = 15 & \qquad \text{The fourth term is 15.} \\ \frac{a_7}{a_4} = \frac{a_1 r^6}{a_1 r^3} = \frac{405}{15} & \qquad \text{Divide } a_7 \text{ by } a_4 \text{ and simplify.} \\ r^3 = 27 & \qquad \text{Divide out } a_1 \text{ and subtract exponents.} \\ r = 3 & \qquad \text{Odd Root Property} \end{aligned}$$

Now we use $r = 3$ and $a_4 = 15$ to solve for a_1.

$$\begin{aligned} a_4 = a_1(3)^3 &= 15 \\ 27a_1 &= 15 \\ a_1 &= \frac{5}{9} \end{aligned}$$

The general term is $a_n = \frac{5}{9}(3)^{n-1}$.

Just as an arithmetic sequence is a linear function with a restricted domain, a geometric sequence is an exponential function whose domain is the set of natural (term) numbers. For example, the sequence $a_n = \frac{5}{9}(3)^{n-1}$ in Example 4 can be compared with the exponential function $f(x) = \frac{5}{9}(3)^{x-1}$.

Figure 6.7 shows the graph of f along with three points that represent the sequence values a_3, a_4, and a_5. Every ordered pair of the form (n, a_n) is represented by a point of the graph of f.

FIGURE **6.7**

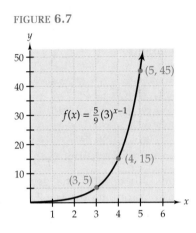

Geometric Series

A **geometric series** is the sum of the terms of a geometric sequence. We can obtain a formula for the nth partial sum of a geometric series.

DEVELOPING THE CONCEPT

The Sum of the First n Terms of a Geometric Series

For the geometric sequence $a_1, a_1r, a_1r^2, a_1r^3, \ldots$, we can write the sum of the first n terms as

$$S_n = a_1 + a_1r + a_1r^2 + a_1r^3 + \cdots + a_1r^{n-1} \quad (1)$$

We multiply both sides of equation (1) by $-r$ and add the result to equation (1).

$$S_n = a_1 + a_1r + a_1r^2 + a_1r^3 + \cdots + a_1r^{n-1}$$
$$-rS_n = \quad\ - a_1r - a_1r^2 - a_1r^3 - \cdots - a_1r^{n-1} - a_1r^n$$

$$S_n - rS_n = a_1 \qquad\qquad\qquad\qquad\qquad - a_1r^n$$

$$(1 - r)S_n = a_1(1 - r^n) \qquad \text{Factor both sides.}$$

$$S_n = \frac{a_1(1 - r^n)}{1 - r} \qquad \text{Divide both sides by } 1 - r, r \neq 1.$$

The result is a formula for the nth partial sum of a geometric series.

■ *NOTE   In the formula for S_n, r cannot be 1. If r = 1, then*

$$S_n = a_1 + a_1 + \cdots + a_1 = na_1$$

$\underbrace{\qquad\qquad\qquad}_{n \text{ terms}}$

> **The nth Partial Sum of a Geometric Series**
>
> The sum of the first n terms of a geometric sequence whose first term is a_1 and whose common ratio is r is given by
>
> $$S_n = \frac{a_1(1 - r^n)}{1 - r}, r \neq 1$$

EXAMPLE 5   **Determining the nth Partial Sum of a Geometric Series**

Determine the sum of the first eight terms of the following sequence:

$$2, -10, 50, -250, \ldots$$

SOLUTION

We use $a_1 = 2$, $n = 8$, and $r = \dfrac{a_2}{a_1} = -5$ to determine the sum.

$$S_n = \frac{a_1(1 - r^n)}{1 - r}$$

$$S_8 = \frac{2[1 - (-5)^8]}{1 - (-5)} = -130{,}208$$

EXAMPLE 6   **Evaluating a Geometric Series**

Evaluate $\displaystyle\sum_{i=1}^{7} 10\left(\frac{2}{5}\right)^{i-1}$.

SOLUTION

Recall that the general term of a geometric sequence is given by $a_1 r^{n-1}$. In this case, the general term is $10\left(\frac{2}{5}\right)^{i-1}$, so $a_1 = 10$ and $r = \frac{2}{5}$. We use these values, along with $n = 7$, to evaluate the series.

$$S_n = \frac{a_1(1 - r^n)}{1 - r}$$

$$S_7 = \frac{10\left[1 - \left(\frac{2}{5}\right)^7\right]}{1 - \frac{2}{5}} = 16.63936$$

Infinite Geometric Series

So far, we have evaluated only finite arithmetic and geometric series—that is, nth partial sums. Now we consider the conditions under which we can evaluate an infinite geometric series, which is represented by the summation notation $\displaystyle\sum_{i=1}^{\infty} a_i$.

Evaluating an Infinite Geometric Series

The number $S = 0.99999\ldots$ can be written as an infinite geometric series:

$$S = \sum_{i=1}^{\infty} 0.9 \left(\frac{1}{10}\right)^{i-1} = 0.9 + 0.09 + 0.009 + 0.0009 + \cdots \quad (1)$$

Suppose we multiply both sides of equation (1) by 10 and then subtract equation (1) from the result.

$$10S = 9 + 0.9 + 0.09 + 0.009 + 0.0009 + \cdots$$
$$S = \quad\quad 0.9 + 0.09 + 0.009 + 0.0009 + \cdots$$
$$\overline{9S = 9}$$

(a) Solve the resulting equation for S. What have we shown about the value of $0.99999\ldots$?

(b) Recall that we wrote S as an infinite geometric series. Does your result in part (a) mean that an infinite geometric series can have a sum?

(c) Consider the series $1 + 2 + 4 + 8 + 16 + 32 + \cdots$. Do you think that this infinite geometric series has a sum? What is your conjecture about how the value of r dictates whether an infinite geometric series has a sum?

DEVELOPING THE CONCEPT

Evaluating an Infinite Geometric Series

Consider the infinite geometric series

$$\sum_{i=0}^{\infty} 2^i = 1 + 2 + 4 + 8 + \cdots$$

and the associated partial sums

$$S_n = \frac{a_1(1 - r^n)}{1 - r} = \frac{1(1 - 2^n)}{1 - 2}$$

To examine the value of S_n as n becomes larger, we can produce a graph of S_n [see Figure 6.8(a)] or a table of values of S_n [see Figure 6.8(b)].

FIGURE **6.8**

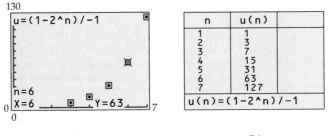

(a)   (b)

Observe that the partial sums S_n appear to increase without bound as n increases.

By simplifying the formula for S_n, we can arrive at the same conclusion:

$$S_n = \frac{1(1 - 2^n)}{1 - 2} = \frac{1 - 2^n}{-1} = 2^n - 1$$

We can see that as n increases, the value of S_n also continues to increase without bound.

Now consider the infinite geometric series

$$\sum_{i=1}^{\infty} \frac{6}{10^i} = 0.6 + 0.06 + 0.006 + \cdots$$

and the associated partial sums

$$S_n = \frac{a_1(1 - r^n)}{1 - r} = \frac{0.6[1 - (0.1)^n]}{1 - 0.1}$$

Again, we produce a graph and a table of values to examine the value of S_n as n becomes larger. (See Figure 6.9.)

FIGURE **6.9**

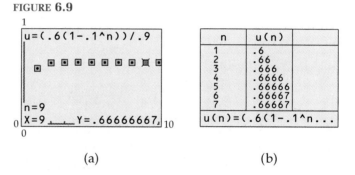

| | (a) | | (b) |

Observe that as n increases, the partial sums S_n appear to approach $\frac{2}{3}$.

We can arrive at the same conclusion by observing that a repeating decimal such as $\frac{2}{3} = 0.666\ldots$ can be expressed as an infinite series.

$$\frac{2}{3} = 0.666\ldots = 0.6 + 0.06 + 0.006 + \cdots$$

$$= \frac{6}{10} + \frac{6}{100} + \frac{6}{1000} + \cdots$$

$$= \sum_{i=1}^{\infty} \frac{6}{10^i}$$

Thus $\displaystyle\sum_{i=1}^{\infty} \frac{6}{10^i} = \frac{2}{3}$.

All of these observations suggest that by making n large enough, we can make S_n as close to $\frac{2}{3}$ as we want. For this reason, we *define* $\displaystyle\sum_{i=1}^{\infty} \frac{6}{10^i}$ to be $\frac{2}{3}$.

For the preceding infinite geometric series, the sum is defined. In general, the value of r dictates whether an infinite sum exists. In the formula for S_n, $r \neq 1$. Moreover, if $|r| > 1$, then the partial sums increase without bound. Thus we can define the sum of an infinite series only if $|r| < 1$, or $-1 < r < 1$.

Sum of an Infinite Geometric Series

The sum S of an infinite geometric series is

$$S = \frac{a_1}{1 - r}, \quad -1 < r < 1$$

For $|r| \geq 1$, the sum does not exist.

EXAMPLE 7   **Evaluating an Infinite Geometric Series**

Determine the sum if it exists.

(a) $6 - 4 + \dfrac{8}{3} - \dfrac{16}{9} + \cdots$ (b) $\displaystyle\sum_{i=1}^{\infty} 0.7(-1.3)^i$ (c) $\displaystyle\sum_{i=1}^{\infty} 3(0.4)^i$

SOLUTION

(a) We see that $a_1 = 6$ and $r = -\frac{2}{3}$. Because $-1 < -\frac{2}{3} < 1$, the sum exists.

$$S = \frac{a_1}{1 - r} = \frac{6}{1 - \left(-\frac{2}{3}\right)} = \frac{6}{\frac{5}{3}} = \frac{18}{5} = 3.6$$

(b) Because the general term is $a_1 r^{n-1} = 0.7(-1.3)^i$, we see that $r = -1.3$. Because $|-1.3| > 1$, the sum does not exist.

(c) Use $a_1 = 3(0.4) = 1.2$ and $r = 0.4$ $(-1 < r < 1)$ to determine the sum.

$$S = \frac{a_1}{1 - r} = \frac{1.2}{1 - 0.4} = 2$$

Real-Life Applications

The following examples illustrate some real-life uses of $S_n = \dfrac{a_1(1 - r^n)}{1 - r}$ for finding the sum of a finite geometric series (see Example 8) and $S = \dfrac{a_1}{1 - r}$ for finding the sum of an infinite geometric series (see Example 9).

EXAMPLE 8   **Retirement Investment**

Each year on January 2, a person plans to invest \$2000 in an IRA account. The person expects a return of 7% compounded annually. What is the expected value of the account after 30 years?

SOLUTION

The table shows the value of each of the yearly deposits after 30 years. The values form a geometric sequence.

Year	1	2	...	30
Value	$2000(1.07)^{30}$	$2000(1.07)^{29}$	...	$2000(1.07)$

The total value of the account is given by a geometric series:
$2000(1.07)^1 + 2000(1.07)^2 + \cdots + 2000(1.07)^{29} + 2000(1.07)^{30}$.

$$\sum_{i=1}^{30} 2000(1.07)^i = \frac{a_1(1 - r^n)}{1 - r}$$

$$= 2000(1.07)\frac{[1 - (1.07)^{30}]}{1 - 1.07} \qquad \begin{array}{l} a_1 = 2000(1.07) \\ r = 1.07 \end{array}$$

$$\approx 202,146.08$$

The expected value of the account after 30 years is $202,146.08.

EXAMPLE 9   Bungee Jump

A bungee jumper falls 100 feet before being pulled back up by the bungee cord. Each time, she rebounds 40% of the last fall and then falls 70% of the previous rebound. Approximately how far does the jumper travel before coming to rest?

SOLUTION

Figure 6.10 shows the first two falls and rebounds along with the numerical expressions that represent the distances.

FIGURE **6.10**

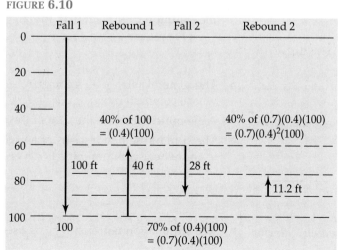

The distances of the falls and the distances of the rebounds form two different geometric sequences.

Term number	Falls	Rebounds
1	100	0.4(100)
2	$(0.7)(0.4)(100)$	$(0.7)(0.4)^2(100)$
3	$(0.7)^2(0.4)^2(100)$	$(0.7)^2(0.4)^3(100)$
4	$(0.7)^3(0.4)^3(100)$	$(0.7)^3(0.4)^4(100)$
.	.	.
.	.	.
.	.	.
i	$(0.7)^{i-1}(0.4)^{i-1}(100)$	$(0.7)^{i-1}(0.4)^i(100)$
.	.	.
.	.	.
.	.	.

The total of the fall distances is $\displaystyle\sum_{i=1}^{\infty}(0.7)^{i-1}(0.4)^{i-1}(100)$, and the total of the rebound distances is $\displaystyle\sum_{i=1}^{\infty}(0.7)^{i-1}(0.4)^i(100)$. Thus the total of the falls and the rebounds is the sum of the two infinite geometric series.

$$\sum_{i=1}^{\infty}(0.7)^{i-1}(0.4)^{i-1}(100) + \sum_{i=1}^{\infty}(0.7)^{i-1}(0.4)^i(100) =$$

$$\frac{100}{1-(0.7)(0.4)} + \frac{40}{1-(0.7)(0.4)} \approx 194.44$$

■ *NOTE   In Example 9, the infinite geometric series is only an approximate model of the situation because the jumper will eventually come to rest. Such modeling is appropriate when the series contains a large number of terms.*

The total distance is approximately 194.44 feet.

6.3 Quick Reference

Geometric Sequences   ■   A **geometric sequence** is a sequence in which there is a constant r called the **common ratio** such that, for any integer n,

$$\frac{a_n}{a_{n-1}} = r \ (\text{or } a_n = ra_{n-1})$$

■ The general term of a geometric sequence whose first term is a_1 and whose common ratio is r is $a_n = a_1 r^{n-1}$.

Geometric Series   ■   A **geometric series** is the sum of the terms of a geometric sequence.

■ The sum of the first n terms of a geometric sequence whose first term is a_1 and whose common ratio is r is given by

$$S_n = \frac{a_1(1-r^n)}{1-r}, r \neq 1$$

Infinite Geometric Series   ■   The summation notation $\displaystyle\sum_{i=1}^{\infty}a_i$ is used to represent an infinite geometric series.

■ The sum S of an infinite geometric series is

$$S = \frac{a_1}{1-r}, \quad -1 < r < 1$$

For $|r| \geq 1$, the sum does not exist.

6.3 Speaking the Language

1. In the geometric sequence 2, 8, 32, 128, ..., the ▬▬▬▬ is 4.

2. For the geometric sequence in Question 1, we call $2 + 8 + 32 + 128$ the fourth ▬▬▬▬.

3. The symbol $\sum\limits_{i=1}^{\infty} a_i$ represents a(n) ▬▬▬▬ geometric series.

4. A geometric sequence is a(n) ▬▬▬▬ function for which the domain is the set of natural numbers.

6.3 Exercises

Concepts and Skills

1. By what method can the common ratio of a geometric sequence be determined?

2. Suppose that the common ratio of a geometric sequence is 2. What other information is needed to write the general term of the sequence?

In Exercises 3–6, determine whether the given sequence is geometric, arithmetic, or neither.

3. (a) $3, -6, 12, -24, \ldots$
   (b) $1, 2, 6, 24, \ldots$
   (c) $-9, -5, -1, 3, \ldots$

4. (a) $2, 4, 6, 8, \ldots$
   (b) $5, -5, 5, -5, \ldots$
   (c) $1, -2, 3, -4, \ldots$

5. (a) $a_n = 1 - 2^n$ (b) $c_n = 2(-5)^{n-2}$
   (c) $k_n = \dfrac{3-n}{5}$

6. (a) $r_n = 0.7n$ (b) $k_n = -\left(\dfrac{7}{2}\right)^{n+3}$
   (c) $c_n = n^2$

In Exercises 7–12, information is given about a geometric sequence. (a) Write the first four terms. (b) Determine the ninth term (rounded to the nearest hundredth).

7. $a_1 = -2, r = -\dfrac{3}{2}$

8. $c_1 = 3, r = \sqrt{2}$

9. $a_n = (-2)^{n+3}$

10. $k_n = \dfrac{3}{4^{n-3}}$

11. $c_n = 0.004(0.2)^{n-6}$

12. $b_n = -\dfrac{4}{9}\left(-\dfrac{3}{2}\right)^{n-1}$

In Exercises 13–16, write the general term of the geometric sequence.

13. $b_4 = \dfrac{9}{7}, r = \dfrac{3}{7}$

14. $a_6 = 250, r = \sqrt{5}$

15. $a_3 = -\dfrac{125}{2}, r = -5$

16. $c_5 = -44.8, r = -2$

In Exercises 17–22, write the general term of the geometric sequence.

17. $c_1 = 3, c_4 = -\dfrac{8}{9}$

18. $b_1 = \dfrac{1}{2}, b_6 = \dfrac{16}{243}$

19. $a_2 = \dfrac{45}{2}, a_5 = -\dfrac{1215}{16}$

20. $c_4 = \dfrac{3}{4}, c_7 = \dfrac{3}{32}$

21. $c_3 = \dfrac{3}{2}, c_6 = 12$

22. $b_2 = -\dfrac{1}{8}, b_9 = -\dfrac{1}{1024}$

23. Suppose that we know only the first two terms of a geometric series. Explain why this is enough information to determine the eighth partial sum of the series.

24. Suppose that for a geometric sequence, $r < 0$. Explain how you know that the terms of the sequence alternate in sign.

In Exercises 25–28, for the given geometric sequence, find the sum of the indicated number of terms.

25. $0.75, 1.5, 3, 6, \ldots; n = 24$

26. $-\dfrac{2}{9}, -\dfrac{2}{3}, -2, -6, \ldots; n = 15$

27. $-\dfrac{\sqrt{3}}{3}, 1, -\sqrt{3}, 3, \ldots; n = 14$

28. $10, 6, 3.6, 2.16, \ldots; n = 8$

In Exercises 29–32, determine the sum.

29. $\dfrac{3}{4} - \dfrac{3}{2} + 3 - 6 + \cdots + 3072$

30. $1 + 5 + 25 + 125 + \cdots + 390{,}625$

31. $8 + 6 + \dfrac{9}{2} + \dfrac{27}{8} + \cdots + \dfrac{6561}{8192}$

32. $\sqrt{3} - 3 + 3\sqrt{3} - 9 + \cdots - 19{,}683$

In Exercises 33–40, evaluate the given geometric series.

33. $\displaystyle\sum_{i=1}^{15} \dfrac{1}{8}(2^{i-1})$

34. $\displaystyle\sum_{i=0}^{12} 3(4^i)$

35. $\displaystyle\sum_{k=0}^{8} 7(0.1)^k$

36. $\displaystyle\sum_{n=1}^{10} 10\left(-\dfrac{4}{5}\right)^{n-1}$

37. $\displaystyle\sum_{i=1}^{9} 25\left(\dfrac{1}{5}\right)^{i-3}$

38. $\displaystyle\sum_{k=1}^{7} 3\left(-\dfrac{2}{3}\right)^{k-2}$

39. $\displaystyle\sum_{i=1}^{10} 5\left(-\dfrac{5}{3}\right)^{i-4}$

40. $\displaystyle\sum_{n=0}^{11} 2\left(\dfrac{4}{3}\right)^{n+2}$

41. Suppose that the first two terms of an infinite geometric series are 0.5 and -1. How can you tell immediately that the sum of the series does not exist?

42. If the sum of an infinite geometric series is 1, what do you know about the nth partial sums of the series as n increases?

In Exercises 43–50, evaluate the given geometric series.

43. $\displaystyle\sum_{i=1}^{\infty} \dfrac{9}{4}\left(\dfrac{2}{3}\right)^i$

44. $\displaystyle\sum_{i=1}^{\infty} (0.8)^i$

45. $\displaystyle\sum_{i=0}^{\infty} 7^{1-i}$

46. $\displaystyle\sum_{i=0}^{\infty} 3(0.5)^{i-1}$

47. $\displaystyle\sum_{i=1}^{\infty} \dfrac{4}{(-3)^{i-1}}$

48. $\displaystyle\sum_{i=1}^{\infty} \dfrac{1}{6^{i+1}}$

49. $\displaystyle\sum_{i=1}^{\infty} 2\left(\sqrt{7}\right)^{1-i}$

50. $\displaystyle\sum_{n=0}^{\infty} 3^{-n/2}$

In Exercises 51–54, evaluate the sum.

51. $4 + 2 + 1 + \dfrac{1}{2} + \cdots$

52. $27 - 9 + 3 - 1 + \cdots$

53. $5 - 1 + \dfrac{1}{5} - \dfrac{1}{25} + \cdots$

54. $49 + 1 + 7^{-2} + 7^{-4} + \cdots$

Concept Extension

55. Consider the sequence $1, r, r^2, r^3, \ldots$, where $r = 0.7, 0.5,$ and 0.1. For each value of r, use a calculator to display the first eight terms of the sequence. What value does r^n approach as n becomes larger and larger?

56. Consider the sequence $1, r, r^2, r^3, \ldots$, where $r = 0.7, 0.5,$ and 0.1. For each value of r, use a calculator to display the first eight partial sums of the sequence. What value does S_n approach as n becomes larger and larger?

57. Evaluate $\displaystyle\sum_{i=1}^{N} 4(-1)^{i-1}$ for (a) any even integer N and (b) any odd integer N.

58. Give an example of a sequence that is both an arithmetic sequence and a geometric sequence.

In Exercises 59–62, write the repeating decimal as a geometric series. Then evaluate the series to write the repeating decimal as a fraction.

59. $0.\overline{8}$

60. $1.\overline{27}$

61. $5.8\overline{3}$

62. $0.\overline{123}$

63. Suppose that $b_1, b_2, \ldots$ is a geometric sequence. Determine whether the following sequences are geometric or arithmetic.

(a) $b_1^2, b_2^2, \ldots$
(b) $\ln b_1, \ln b_2, \ldots$

64. Write a rational function that is equal to

$$1 - x^2 + x^4 - x^6 + \cdots$$

65. Determine whether the given sequence is arithmetic, geometric, or neither.

(i) $\ln 2, \ln 4, \ln 6, \ln 8, \ldots$
(ii) $\ln 2, \ln 4, \ln 8, \ln 16, \ldots$
(iii) $\ln 2, \ln 4, \ln 16, \ln 256, \ldots$

66. Evaluate the following sum:

$$\left(1 + \sqrt{2}\right) + 1 + \left(-1 + \sqrt{2}\right) + \cdots$$

Applications

67. Payroll Impact According to some economists, 70% of all money that is placed in circulation in a certain community is recirculated in that community within a month. Suppose that an industry in the community has a monthly payroll of $200,000.

(a) Write the general term of a sequence that describes the amount of the payroll that is recirculated at the end of month n.
(b) What is the sixth term of the sequence?
(c) The total economic impact of money placed in circulation is the sum of the initial amount and the amounts recirculated. Write a series for the total yearly economic impact of one month's payroll.
(d) Evaluate the series in part (c).

68. Political Campaigning In a county with 30,000 registered voters, a candidate for a commission seat contacted $\frac{1}{5}$ of the remaining uncontacted voters each week.

(a) Write a sequence c_n to model the number of voters that the candidate contacted each week of the campaign.
(b) Use the sequence in part (a) to determine the number of voters contacted in the fifth week of the campaign.
(c) Write and evaluate a series to determine the total number of voters contacted (i) after 3 weeks; (ii) after 6 weeks.

69. Toy Sales A toy company sells beanbag animals. The initial advertising creates a demand that results in sales of $2 million in the first month. Each month thereafter, the sales are 80% of the sales of the preceding month.

(a) Write a geometric sequence to describe the monthly sales.
(b) After how many months will sales drop below $0.5 million?
(c) What are the total sales for the first year?
(d) Explain why the total sales will never exceed $10 million.

70. Family Tree Suppose that you make a family tree listing yourself, your parents, your grandparents, your great-grandparents, and so on.

(a) Write a geometric sequence to describe the number of ancestors in a particular generation.
(b) How many people would be listed in the generation in which the word *great* must be used 25 times to describe the ancestor?
(c) Write a series to determine the total number of people included in the first 15 generations.
(d) How many generations should be included to have at least 1 million people in the family tree?

71. Bouncing Ball A ball is dropped from 20 feet above ground level. Each time the ball hits the ground, it rebounds $\frac{4}{5}$ of its previous height.

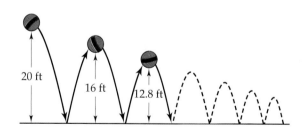

(a) Write a geometric sequence to describe the distances that the ball falls.
(b) Write a geometric sequence to describe the distances that the ball rebounds.
(c) How many times does the ball hit the ground before the rebound is less than 1 inch?
(d) Write and evaluate a series to determine the total distance that the ball travels.

72. Social Security Expenditures The graph shows the total amount (in billions of dollars) paid in social security benefits for selected years. (*Source:* Social Security Administration.)

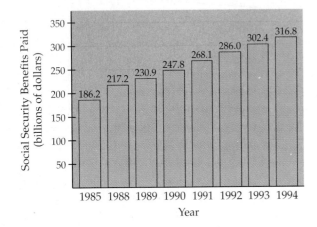

(a) Use a calculator's exponential regression feature to write a geometric sequence a_n to model the data. (Let n represent the number of years since 1980. Round to the nearest hundredth.)

(b) In which year will payments first exceed $500 billion?

(c) What does the series $\sum_{n=1}^{20} a_n$ represent?

(d) Evaluate the series in part (c).

73. Lifetime Earnings In 1995 the average starting salary for a graduate with a bachelor's degree in mathematics was $30,271. (*Source:* National Association of Colleges and Employers.) Suppose that such a graduate anticipates annual raises of 5%.

(a) Write a geometric sequence to model the predicted yearly salaries.

(b) What salary does the person expect for year 20?

(c) Write and evaluate a geometric series to determine the total earnings during a 20-year career.

(d) Determine the number of years the person would need to work to earn over $2 million.

74. National Debt The table shows that although the budget deficit (in billions of dollars) decreased during the period 1991–1998, the national debt (in trillions of dollars) continued to rise. (*Source: USA Today.*)

Year	Deficit	National debt
1991	269.4	3.6
1992	290.4	4.0
1993	255.0	4.4
1994	203.1	4.6
1995	163.9	4.9
1996	107.3	5.2
1997	22.6	5.4
1998	22.0	5.5

(a) Use a calculator to model the budget deficit with (i) a geometric sequence and (ii) an arithmetic sequence. (Let n represent the number of years since 1990. Round results to the nearest hundredth.)

(b) Use a calculator to model the national debt with (i) a geometric sequence and (ii) an arithmetic sequence. (Let n represent the number of years since 1990. Round results to the nearest hundredth.)

6.4 *Permutations and Combinations*

Fundamental Counting Principle ▪ *Permutations* ▪
Combinations

Fundamental Counting Principle

Suppose that the local Chamber of Commerce gives a community service award to 1 man and 1 woman and that there are 3 female nominees and 2 male nominees. The following *tree diagram* shows the number of ways in which the two recipients can be selected.

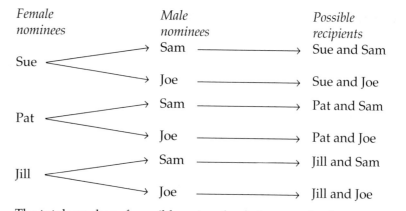

The total number of possible pairs of recipients can be determined by multiplying the number of female nominees by the number of male nominees: $3 \cdot 2 = 6$. This illustrates the Fundamental Counting Principle, in which the selection of a recipient is called an *event*.

> **Fundamental Counting Principle**
>
> Suppose that one event E_1 can occur in m ways and a second event E_2 can occur in n ways. Then the number of ways in which E_1 followed by E_2 can occur is mn.

EXAMPLE 1   **Using the Fundamental Counting Principle**

A man has 3 sport coats and 11 ties. In how many ways can the man dress in a coat and tie?

SOLUTION

There are 3 ways to select a coat (E_1) and 11 ways to select a tie (E_2). Thus there are $3 \cdot 11 = 33$ coat and tie arrangements.

The Fundamental Counting Principle can be extended to any number of events.

EXAMPLE 2   **Extending the Fundamental Counting Principle**

Suppose that a state's license plates consist of a three-digit number followed by three letters. Assuming that repetitions of digits or letters is permitted and that any letter or digit is permitted, how many license plates are possible?

SOLUTION

Because any digit from 0 to 9 can be selected, there are 10 choices for each of the three digits, and there are 26 choices for each of the three letters. Thus the possible number of license plates is the product of the ways in which each number and letter can be chosen.

$$10 \cdot 10 \cdot 10 \cdot 26 \cdot 26 \cdot 26 = 17{,}576{,}000$$

Permutations

In the preceding examples, each event was independent of the previous event. The number of ways in which an event can occur may depend on events that have already occurred.

The following table shows the number of code words that can be created from the given letters with no letter being repeated.

Letters	A	A, B	A, B, C
Words	A	AB	ABC, ACB, BAC
		BA	BCA, CAB, CBA
Number	1	$2 = 2 \cdot 1$	$6 = 3 \cdot 2 \cdot 1$

We can use the Fundamental Counting Principle to determine the number of code words that can be formed with 4 letters. There are 4 ways to select the first letter, then 3 ways to select the second letter, then 2 ways to select the third letter, and then only 1 way to select the fourth letter. Thus the number of words is

$$4 \cdot 3 \cdot 2 \cdot 1 = 24$$

Each of the 24 different code words is called a **permutation** of the given four letters.

Definition of Permutation

A **permutation** is an arrangement of distinct objects in a particular or sequential order.

EXAMPLE 3   **Permutations of Sales Calls**

A sales representative must meet with 8 clients. In how many ways can the representative arrange the call schedule?

SOLUTION

The first appointment can be with any one of the 8 clients. Then the second appointment can be with any of the 7 remaining clients, and so on. The number of permutations of the 8 clients is

$$8 \cdot 7 \cdot 6 \cdot 5 \cdot 4 \cdot 3 \cdot 2 \cdot 1 = 40{,}320$$

The call schedule can be arranged in 40,320 ways.

In Example 3, the product $8 \cdot 7 \cdot 6 \cdot 5 \cdot 4 \cdot 3 \cdot 2 \cdot 1$ can be written in the more compact form 8! (read 8 factorial).

> ### *Definition of Factorial*
>
> For a positive integer n, the symbol $n!$ (which is read as "n factorial") is the product of the integers 1 through n.
>
> $$n! = n \cdot (n-1) \cdot (n-2) \cdot \cdots \cdot 2 \cdot 1$$
>
> The symbol $0!$ is defined to be 1: $0! = 1$.

■ **Factorial**   We can use a calculator to evaluate factorials.

FIGURE **6.11**

```
7!
              5040
10!
           3628800
```

Figure 6.11 shows the results of using a calculator to evaluate $7!$ and $10!$.

In Example 3, the sales representative arranged to meet with all 8 of her clients. We say that the possible number of call schedules is the number of permutations of 8 clients taken 8 at a time.

In general, an arrangement of n distinct objects in sequential order is called a permutation of n objects taken n at a time. The symbol for all such permutations is $_nP_n$, where the first subscript indicates the number of objects, and the second subscript indicates the number of objects used. When we say "n objects taken n at a time," we mean that all objects are used. The first object can be chosen in n ways, the second object can be chosen in $n-1$ ways, and so on.

> ### *Permutations of n Objects Taken n at a Time*
>
> The number of permutations of n objects taken n at a time is given by $_nP_n = n!$.

■ **Permutations**   You can use your calculator to compute the number of permutations of n objects taken n at a time. Although the menu selection is probably $_nP_r$, you can let $r = n$.

EXAMPLE **4**   **Seating Arrangements**

FIGURE **6.12**

```
6 nPr 6
               720
```

Suppose that 6 students attend a seminar and that they arrange themselves in the 6 chairs across the front of the room. How many seating arrangements are possible?

SOLUTION

This is an example of 6 objects (students) taken 6 at a time, so we calculate $_6P_6$. (See Figure 6.12.)
There are 720 possible seating arrangements.

Sometimes we are interested in permutations in which we do not use all the objects. For example, if the sales representative in Example 3 needs to meet with only 3 of the 8 clients, the first appointment could be with any of the 8 clients, then the second appointment could be with any of the remaining 7 clients, and then the third appointment could be with any of the remaining 6 clients. According to the Fundamental Counting Principle, the number of permutations is $8 \cdot 7 \cdot 6 = 336$.

The notation $_8P_3$ is used to represent the number of permutations of 8 objects taken 3 at a time.

$$_8P_3 = 8 \cdot 7 \cdot 6$$

Multiplying the numerator and denominator of $\dfrac{8 \cdot 7 \cdot 6}{1}$ by 5! results in the equivalent expression

$$_8P_3 = \frac{8 \cdot 7 \cdot 6 \cdot 5 \cdot 4 \cdot 3 \cdot 2 \cdot 1}{5 \cdot 4 \cdot 3 \cdot 2 \cdot 1} = \frac{8!}{5!} = \frac{8!}{(8 - 3)!}$$

This form is preferred for purposes of computation.

> **Permutations of n Objects Taken r at a Time**
>
> The number of permutations of n objects taken r at a time, $_nP_r$, is given by
>
> $$_nP_r = \frac{n!}{(n - r)!}, \quad r \le n$$

■ NOTE   *If n objects are taken 0 at a time, then*

$$_nP_0 = \frac{n!}{(n - 0)!} = \frac{n!}{n!} = 1$$

Also, because we defined 0! = 1,

$$_0P_0 = \frac{0!}{(0 - 0)!} = \frac{0!}{0!} = \frac{1}{1} = 1$$

 Permutations   A calculator can be used to calculate permutations of n objects taken r at a time.

EXAMPLE **5**   **Batting Order**

Suppose that a baseball team has 15 players and the manager must assign the batting order for the 9 players who start the game. How many possible batting orders are there?

FIGURE **6.13**

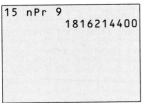

SOLUTION

$$_{15}P_9 = \frac{15!}{(15 - 9)!} = \frac{15!}{6!}$$
$$= 15 \cdot 14 \cdot 13 \cdot 12 \cdot 11 \cdot 10 \cdot 9 \cdot 8 \cdot 7$$
$$= 1,816,214,400$$

Figure 6.13 shows how you can use a calculator to compute $_{15}P_9$.

Sometimes not all of the objects are distinguishable. Consider the following arrangement of 2 identical orange marbles and 3 identical blue marbles:

●   ●   ●   ●   ●

If the first and last orange marbles were exchanged, the arrangement would appear to be exactly the same and would therefore not be considered a distinguishable permutation.

> **Distinguishable Permutations**
>
> Suppose a collection of n objects has n_1 of one type, n_2 of a second type, and so on, where $n = n_1 + n_2 + \cdots + n_k$. The number of distinguishable permutations of the n objects is given by
>
> $$\frac{n!}{n_1! \cdot n_2! \cdots n_k!}$$

Returning to the five marbles, we can apply this formula to determine the number of distinguishable ways in which they can be arranged. There are $n = 5$ marbles, of which $n_1 = 2$ are orange and $n_2 = 3$ are blue. Thus, as shown in Figure 6.14, there are $\dfrac{n!}{n_1! \cdot n_2!} = \dfrac{5!}{2! \cdot 3!} = 10$ distinguishable permutations.

FIGURE 6.14

●●●●●   ●●●●●   ●●●●●   ●●●●●   ●●●●●
●●●●●   ●●●●●   ●●●●●   ●●●●●   ●●●●●

EXAMPLE **6**   **Distinguishable Permutations**

Kokomo is a city in Indiana. Determine the number of distinguishable permutations of the letters in the word KOKOMO.

SOLUTION

Of the 6 letters, there are 3 O's, 2 K's and 1 M. Thus there are $\dfrac{6!}{3!2!1!} = 60$ distinguishable permutations of the letters.

EXAMPLE **7**   **Fund-Raising Volunteers**

From a group of 9 volunteers for fundraising, 4 will be assigned to telephone solicitation, 3 will be assigned to mailings, and 2 will be assigned to door-to-door visitations. In how many distinguishable ways can the assignments be made?

SOLUTION

In effect, the question asks for the number of permutations of 4 T's (telephone), 3 M's (mailings), and 2 D's (door-to-door visitations).

$$\frac{9!}{4!3!2!} = 1260$$

Combinations

Suppose that a bakery makes apple (A), blueberry (B), chocolate (C), and date (D) muffins. The bakery sells trays containing two varieties of muffins. If the order in which the two varieties of muffins are displayed in the tray is important, then the possible displays would be as follows:

AB	AC	AD	BC	BD	CD
BA	CA	DA	CB	DB	DC

The possible number of displays is the number of permutations of 4 varieties of muffins taken 2 at a time:

$$_4P_2 = \frac{4!}{(4-2)!} = \frac{4!}{2!} = 12$$

If the order in which the muffins are displayed is *not* important, then the possible displays would be as follows:

AB	AC	AD	BC	BD	CD

Because order is unimportant, we have simply eliminated all the permutations of each display. We refer to this as the number of **combinations** of 4 objects taken 2 at a time, which we represent with the symbol $_4C_2$.

Because each display of two muffin varieties has two permutations, we can determine $_4C_2$ by dividing $_4P_2$ by 2!.

$$_4C_2 = \frac{_4P_2}{2!} = \frac{12}{2} = 6$$

In general, $_nC_r = \dfrac{_nP_r}{r!} = \dfrac{n!}{(n-r)!r!}$.

■ *NOTE   Be sure to distinguish between permutations, for which the order of an arrangement is meaningful, and combinations, for which the order is not important.*

> ### *Combinations of n Objects Taken r at a Time*
> The number of combinations of n items taken r at a time is given by
>
> $$_nC_r = \frac{n!}{(n-r)!r!}, \quad r \leq n$$

> **Combinations**   You can use a calculator to evaluate $_nC_r$.

EXAMPLE **8**   **Combinations of Cards**

A standard deck of cards contains 52 cards, 13 of each suit. What is the possible number of 5-card hands that contain only diamonds?

SOLUTION

Because the order of the cards in each 5-card hand is not important, we need to determine the number of combinations of 5 cards from the 13 cards that are diamonds.

$$_{13}C_5 = \frac{13!}{5!8!} = 1287$$

EXAMPLE **9**   **Window Displays**

FIGURE **6.15**

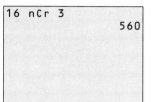

A bridal shop has 16 different bridal gowns, but the front window area is large enough to display only 3 gowns. What is the possible number of bridal gown displays available to the shop owner?

SOLUTION

Figure 6.15 shows the number of combinations of 16 gowns displayed 3 at a time.

We can also use the Fundamental Counting Principle and combinations together.

EXAMPLE **10**   **Committee Formation**

From 8 men and 10 women, in how many ways can a committee consisting of 2 men and 3 women be formed?

FIGURE **6.16**

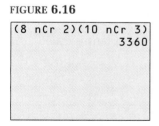

SOLUTION

The number of ways in which the 2 men can be chosen is $_8C_2$, and the number of ways in which the 3 women can be chosen is $_{10}C_3$. By the Fundamental Counting Principle, the number of ways in which the committee can be formed is the product of those two numbers. (See Figure 6.16.) The committee can be formed in 3360 ways.

6.4 Quick Reference

Fundamental Counting Principle   ■   If one event E_1 can occur in m ways and a second event E_2 can occur in n ways, then the number of ways in which E_1 followed by E_2 can occur is mn.

Permutations   ■   A **permutation** is an arrangement of distinct objects in a particular or sequential order.

■ For a positive integer n, the symbol $n!$ (which is read "n factorial") is the product of the integers 1 through n.

$$n! = n \cdot (n - 1) \cdot (n - 2) \cdot \cdots \cdot 1$$

The symbol 0! is defined to be 1: $0! = 1$.

■ The number of permutations of n objects taken n at a time is given by $_nP_n = n!$.

■ The number of permutations of n objects taken r at a time is given by
$$_nP_r = \frac{n!}{(n - r)!}.$$

■ If a collection of n objects has n_1 of one type, n_2 of a second type, and so on, where $n = n_1 + n_2 + \cdots + n_k$, then the number of distinguishable permutations of the n objects is given by $\dfrac{n!}{n_1! \cdot n_2! \cdots \cdots n_k!}$.

Combinations ■ The number of **combinations** of n items taken r at a time is the number of ways in which n items can be selected r at a time, where the order of the arrangements is unimportant.

■ The number of combinations of n items taken r at a time is given by
$$_nC_r = \frac{n!}{(n - r)!r!}.$$

6.4 Speaking the Language

1. If event A can occur in a ways and event B can occur in b ways, then, according to the ▒▒▒▒▒▒ Principle, event A followed by event B can occur in ab ways.

2. An arrangement of objects in a particular order is known as a(n) ▒▒▒▒▒▒.

3. The number 8! is read as 8 ▒▒▒▒▒▒.

4. An arrangement of objects in which order is not important is called a(n) ▒▒▒▒▒▒.

6.4 Exercises

Concepts and Skills

1. Explain the difference between 4^4 and $4!$.

2. Explain why there are more ways to arrange the letters A, B, and C if repetitions of letters are allowed than if repetitions are not allowed.

In Exercises 3–8, evaluate.

3. $7!$

4. $5!$

5. $\dfrac{12!}{8!}$

6. $\dfrac{15!}{12!}$

7. $\dfrac{16!}{10!6!}$

8. $\dfrac{20!}{15!5!}$

In Exercises 9–12, evaluate.

9. $_6P_3$

10. $_8P_5$

11. $_{10}P_4$

12. $_9P_6$

In Exercises 13–16, evaluate.

13. $_7C_5$

14. $_9C_4$

15. $_{15}C_7$

16. $_{11}C_6$

Concept Extension

17. Show that $\sum_{i=0}^{3} {}_3C_i = 8$.

18. Show that $\sum_{i=0}^{3} {}_3P_i = 16$.

19. Show that $\dfrac{{}_nP_r}{{}_nC_r} = r!$.

20. Show that $\sum_{i=0}^{2} {}_5C_i = \sum_{j=3}^{5} {}_5C_j$.

Applications: Counting Principle

21. Applicant Selection A firm plans to hire an accountant and a computer operator. If the company has 24 applicants for the accounting position and 15 applicants for the computer position, in how many ways can the two positions be filled?

22. Interior Design An interior designer showed a client samples of 8 types of wallpaper and 5 types of borders. How many possible selections of one wallpaper and one border can the client make?

Wallpaper

Borders

23. Core Course Requirements To complete a college's core requirements, a student must take 1 of 5 possible history courses, 1 of 4 science courses, 1 of 6 humanities courses, and 1 of 2 math courses. In how many ways can the student complete the core requirements?

24. Party Plans A company plans and caters children's birthday parties. The basic party plan includes 1 of 5 types of cake, 1 of 3 types of entertainment, 1 of 8 theme decorations, and 1 of 4 types of ice cream. How many different party plans are available?

Cake

Entertainment

Decorations

Ice Cream

25. Telephone Numbers Within a particular area code, how many 7-digit phone numbers are possible if (a) any digit can be used? (b) the first digit cannot be 0 or 1?

26. Computer Passwords A person must choose a password for a computer system. The password consists of two letters followed by three digits, and repetitions are allowed.

(a) In how many ways can the password be formed?

(b) Suppose that the letter O and the digit 0 cannot be used because they might be confused. In how many ways can the password be formed?

27. Multiple-Choice Tests A test consists of 10 multiple-choice questions with 4 choices for each question. In how many ways can the test be answered?

28. True/False Tests In how many ways can a 25-question true/false test be answered?

29. Field Trip A kindergarten teacher takes groups of 10 students (4 girls and 6 boys) on field trips. For each trip, the teacher lines up the students in a different arrangement. How many ways can this be done if

(a) only girls are at the head of the line?

(b) only boys are at the head of the line?

(c) the students can be arranged in any order?

30. **Baseball Lineups** A baseball coach has 15 players on the team. In how many ways can he field a team if the pitching position can be filled by only 2 players (and they cannot play any other position) and the remaining players can play any position?

31. **Sandwiches** A deli wants to advertise the number of sandwiches it offers. Sandwiches are made from 3 varieties of bread, 4 kinds of meat, and a choice of 7 extras, such as lettuce or tomato. How many different sandwiches can be made with one item from each category?

32. **Pizzas** A pizza comes in three sizes, with thin or thick crust, a choice of 3 meats, and a choice of 5 vegetables. In how many different ways can a pizza be made with one topping from each category?

Applications: Permutations

33. What is the difference between a permutation and a combination?

34. Suppose that 5 colored blocks are to be arranged in a row. Which of the following has the fewer number of arrangements? Why?
   (i) There are 3 red blocks and 2 blue blocks.
   (ii) There are 2 red blocks, 2 blue blocks, and 1 green block.

35. **Figure Skating** The 6 finalists in a figure-skating competition draw for the order of skating. In how many ways can the final order be drawn?

36. **Track Meet** In a track meet with 8 runners, in how many possible orders can the runners finish?

37. **Parade Lineup** A small town's centennial parade had 18 entries. In how many ways can the organizer line up the parade?

38. **Airport Shuttle Routes** An airport shuttle bus transports passengers from BWI Airport to 6 hotels in the inner harbor area of Baltimore. In how many ways can the shuttle arrange the drop-offs?

39. **Film Festival** The organizer of a film festival plans to show 8 films. In how many ways can the program be arranged?

40. **Soccer Tickets** Two families of four purchased 8 tickets for a soccer game.

(a) In how many ways can they be seated in the 8 seats?

(b) In how many ways can they be seated if each family sits together?

41. **Flags** The government of a newly formed country decides to create a flag with three horizontal stripes. Using the colors red, green, black, yellow, blue, white, and orange, in how many ways can a flag be created?

42. **Fraternity Names** The Greek alphabet consists of 24 letters. How many fraternity names can be formed using three different letters?

43. **Teaching Assignments** A college schedules a college algebra course to begin each hour from 8:00 A.M. through 2:00 P.M. There are 10 mathematics professors available to teach the classes, but no professor will teach more than one college algebra class. In how many ways can professors be assigned to classes?

44. **Bobsled Teams** In how many ways can a coach fill the seats in the four-man bobsled if

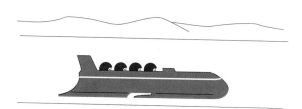

(a) any of the 6 team members can fill any seat?

(b) only 2 of the team members can be the driver, but any other team member can fill the other 3 positions?

45. **PINs** A person's personal identification number (PIN) for a certain ATM machine consists of 5 different letters.

(a) How many different PINs are possible?

(b) How many different PINs would have been possible if repetitions of letters had been allowed?

46. **Flag Signals** Signals are sent from a ship by raising three flags up the mast.

(a) If 8 different flags are available, how many different signals can be sent?

(b) If a signal consists of 4 flags, how many different signals can be sent?

47. Determine the number of distinguishable permutations of the letters in MISSISSIPPI.

48. Determine the number of distinguishable permutations of the letters in CINCINNATI.

49. Job Assignments A grocery store has 9 new employees and wants to assign 4 as cashiers, 3 to stock shelves, and 2 for customer service. In how many distinguishable ways can this be done?

50. Recall Petition To recall a public official, a group of 30 volunteers must collect signatures from each district of a county. The coordinator plans to assign 5 people to District A, 8 people to District B, 10 people to District C, 5 people to District D, and 2 people to District E. In how many distinguishable ways can this be done?

51. Shrub Arrangements A homeowner buys 6 laurels, 5 azaleas, and 4 nandinas to plant around a patio. How many distinguishable arrangements of the shrubs are possible?

52. Parade Lineups A small Independence Day parade has 3 bands, 12 floats, and 7 cars with politicians. In how many ways can the organizer line up the parade if the bands, floats, and politicians are regarded as (a) distinguishable and (b) indistinguishable?

Applications: Combinations

53. How does the number of combinations of 5 items taken 3 at a time compare with the number of combinations of 5 items taken 2 at a time? Why?

54. Calculate and compare $_nC_n$ and $_nC_1$. Interpret the results in your own words.

55. Basketball Tournament Semifinals A basketball tournament begins with 32 teams. In how many ways can 4 teams reach the semifinal round?

56. Student Government From a class of 24 students, 3 are elected to serve on the student government. In how many ways can the representatives be chosen?

57. Jury Selection For a civil case, 24 potential jurors are questioned. How many possible 12-person juries can be chosen?

58. Softball Schedules A recreation director must schedule games for a softball league with 12 teams. If each team must play every other team, how many games must be scheduled?

59. Ice Cream Cones An ice cream shop advertises 36 different flavors of ice cream. How many three-dip cones are possible if each dip is a different flavor?

60. Scholarship Recipients A foundation has received 35 applications for 5 scholarships. If all the applicants are equally qualified, how many different groups of students can be selected as recipients?

61. Poker Hands A standard deck consists of 52 cards. How many 5-card hands of the same suit are possible?

62. Bridge Hands A bridge hand has 13 cards. How many possible bridge hands can be dealt from a standard deck of 52 cards?

63. Computerized Tests A computerized testing system has 100 test items in its data bank, from which test questions are randomly selected. How many 20-question tests can be generated?

64. Lottery Numbers To win a certain state lottery, a person must select the correct 6 numbers (in any order) from the integers 1 through 46. In how many ways can a person select the 6 numbers?

65. Senate Committees Suppose that the U.S. Senate has 54 Democrats and 46 Republicans. In how many ways can a judiciary committee of 12 be chosen if the committee must have 7 Democrats and 5 Republicans?

66. Finance Committees A city council consists of 28 members, of whom 12 are Democrats, 10 are Republicans, and 6 are Independents. How many finance committees of 10 can be formed if the membership must have 2 Independents, 4 Republicans, and 4 Democrats?

67. Advisory Groups An apartment residents' advisory group consists of 5 men and 5 women chosen from the 42 men and 52 women who live in the complex. How many possible advisory groups can be formed?

68. Variety Programs The director of a variety show wants to put together a program with 5 music groups, 3 comedy routines, and 4 dance performances. If 12 music groups, 8 comedians, and 10 dance groups audition, in how many ways can performers be chosen for the program?

6.5 *Probability*

Experimental Probability ■ *Theoretical Probability* ■
Mutually Exclusive Events ■ *Odds*

Experimental Probability

An **experiment** is an activity in which the outcome is uncertain. Any particular outcome of an experiment is called an **event.** For example, when you toss a coin, the events "heads" and "tails" are the only possible outcomes.

Suppose that you toss a coin 1000 times and record the numbers of heads and tails. Your final tally might be 495 heads and 505 tails. Based on the results of your experiment, we would say the experimental probability of obtaining heads is $\frac{495}{1000}$, or 0.495.

Many probabilities that you are familiar with are based on **experimental probability.** For instance, the probability of rain or the probability that you will be involved in an automobile accident are both based on collected data.

> *Definition of Experimental Probability*
>
> If an experiment was performed n times and an event E occurred m times, then the **experimental probability** of the event is
>
> $$P(E) = \frac{m}{n}$$

EXAMPLE 1 **Defective Computer Chips**

A manufacturer of computer chips tested 1200 chips and found that 50 were defective.

(a) What is the experimental probability that a chip is defective?
(b) How many defective chips would you expect to find in a shipment of 360 chips?

SOLUTION

(a) There were 1200 chips tested, where the event E is a defective chip. Because event E occurred 50 times, the experimental probability of a defective chip is

$$P(E) = \frac{50}{1200} = \frac{1}{24}$$

(b) The expected number of defective chips is the number of times that event E is expected to occur in a shipment of 360 chips. Solving the experimental probability formula for m, we obtain

$$m = n \cdot P(E)$$

$$m = 360 \cdot \frac{1}{24}$$

$$= 15$$

Of the 360 chips, we would expect 15 to be defective.

EXAMPLE **2** **Television Viewing**

The following results were obtained from 600 households in a survey about Sunday afternoon television viewing.

Event	E_1	E_2	E_3	E_4	E_5
Viewing	Sports	TV movie	Rented movie	Other	None
Households	170	65	85	90	190

(a) What is the probability that a household is watching sports?
(b) What is the probability that a household is not watching sports or movies?
(c) In a city of 24,000 households, how many households would you expect to be watching sports?

SOLUTION

(a) $P(E_1) = \frac{170}{600} = \frac{17}{60}$
(b) If a household is not watching sports or movies, then either event E_4 or event E_5 occurred. These two events occurred a total of 280 times.

$$P(E_4 \text{ or } E_5) = \frac{280}{600} = \frac{7}{15}$$

(c) The expected number of households watching sports is the product of the total number of households in the city and the probability that a household is watching sports.

$$24{,}000 \cdot \frac{17}{60} = 6800$$

Theoretical Probability

When you toss a fair coin, there are two equally likely outcomes: heads or tails. Thus the **theoretical probability** of obtaining heads is $\frac{1}{2}$.

When the experimental probability of an event is based on large sets of data, the result should be approximately the same as the theoretical probability. If a coin is tossed 1000 times and 495 heads are obtained, the difference between the experimental result and the theoretical probability that 500 heads would be obtained is attributed to "chance." Statistical methods can be used to determine whether the difference is too great to be explained by chance alone.

We will focus on using mathematical reasoning to determine theoretical probabilities.

We call the set of all possible outcomes of an experiment the **sample space.** If each outcome occurs with the same frequency, we say that the outcomes are **equally likely.** An event is a subset of a sample space.

Consider again the experiment of tossing a fair coin. The experiment has two equally likely outcomes: heads (H) and tails (T). Thus the sample space is {H, T}. The two possible events are {H} and {T}.

■ *NOTE   If a particular event can never occur, then P(E) = 0. If a particular event is certain to occur, then P(E) = 1. In all cases, $0 \le P(E) \le 1$.*

> ### Theoretical Probability of an Event
>
> If S is a sample space with n equally likely outcomes and if an event E can occur in m ways, then the **theoretical probability** of event E is defined by
>
> $$P(E) = \frac{m}{n}$$

In words, the probability of a particular event is the number of equally likely ways in which the event can occur divided by the number of equally likely outcomes of the experiment.

EXAMPLE 3   ### Sample Spaces, Events, and Probability

Write the sample space for each experiment. Then describe each event as a subset of the sample space and compute the probability of the event's occurring.

(a) Experiment: Two coins are tossed.
   Event: Tails appear on both coins.
(b) Experiment: Three coins are tossed.
   Event: Tails appear on exactly one coin.
(c) Experiment: One die is rolled.
   Event: The result is an odd number.

SOLUTION

(a) Sample space: {HH, HT, TH, TT}
Event: {TT}
$$P(E) = \frac{1}{4}$$

(b) Sample space: {HHH, HHT, HTH, THH, HTT, THT, TTH, TTT}
Event: {HHT, HTH, THH}
$$P(E) = \frac{3}{8}$$

(c) Sample space: {1, 2, 3, 4, 5, 6}
Event: {1, 3, 5}
$$P(E) = \frac{3}{6} = \frac{1}{2}$$

Sample spaces can be relatively large, and events can contain many elements of the sample space.

EXAMPLE 4 **Two Dice Events**

If you roll two dice, what is the probability that

(a) the sum is 7?
(b) the sum is 8?
(c) the sum is no more than 3?

SOLUTION

The following is the sample space, which consists of the 36 equally likely outcomes from rolling two dice. The first number of each pair is the number showing on the first die, and the second number is the number showing on the second die.

$$\{(1, 1), (1, 2), (1, 3), (1, 4), (1, 5), (1, 6)$$
$$(2, 1), (2, 2), (2, 3), (2, 4), (2, 5), (2, 6)$$
$$(3, 1), (3, 2), (3, 3), (3, 4), (3, 5), (3, 6)$$
$$(4, 1), (4, 2), (4, 3), (4, 4), (4, 5), (4, 6)$$
$$(5, 1), (5, 2), (5, 3), (5, 4), (5, 5), (5, 6)$$
$$(6, 1), (6, 2), (6, 3), (6, 4), (6, 5), (6, 6)\}$$

(a) Let E_1 represent the event of rolling a sum of 7. As shown by the highlighted pairs in the sample space, E_1 can occur in 6 ways.

$$E_1 = \{(1, 6), (2, 5), (3, 4), (4, 3), (5, 2), (6, 1)\}$$

Thus the probability of rolling a sum of 7 is

$$P(E_1) = \frac{6}{36} = \frac{1}{6}$$

■ *NOTE   In Example 4, the number of elements in the sample space could have been determined without actually listing them. Because each die has 6 possible outcomes, the Fundamental Counting Principle states that there are $6 \cdot 6 = 36$ equally likely outcomes from rolling the two dice.*

(b) Let E_2 represent the event of rolling a sum of 8.

$$E_2 = \{(2, 6), (3, 5), (4, 4), (5, 3), (6, 2)\}$$

Thus the probability of rolling a sum of 8 is

$$P(E_2) = \frac{5}{36}$$

(c) Let E_3 represent the event of rolling a sum of 2 or 3.

$$E_3 = \{(1, 1), (1, 2), (2, 1)\}$$

Thus the probability that the sum is no more than 3 is

$$P(E_3) = \frac{3}{36} = \frac{1}{12}$$

Sometimes listing all outcomes is not practical. In such cases, we can use counting principles to determine the number of possible outcomes of an experiment and the number of ways in which an event can occur.

EXAMPLE 5   **Using Counting Principles**

Suppose that 5 cards are selected from a standard deck. What is the probability that all 5 cards will be of the same suit?

SOLUTION

The sample space consists of all possible 5-card combinations. The number of such combinations is

$$_{52}C_5 = 2,598,960$$

■ *NOTE   In poker, 5 cards of the same suit is called a flush. Example 5 shows that the probability of being dealt a flush is very low.*

Let E represent the event "5 cards of the same suit." We can choose any of the 4 suits and then choose any 5 of the 13 cards in that suit. Thus the number of ways in which event E can occur is

$$4 \cdot {}_{13}C_5 = 4 \cdot 1287 = 5148$$

Thus $P(E) = \frac{5148}{2,598,960} \approx 0.002$.

Mutually Exclusive Events

If we select a card from a deck, the card cannot be both a club and a diamond. The two events "the card is a club" and "the card is a diamond" are said to be **mutually exclusive.**

Two events E_1 and E_2 are mutually exclusive if there is no possibility that both can occur at the same time. Symbolically, $E_1 \cap E_2 = \emptyset$. (Recall that the symbol $\cap$ means *intersection*.) The probability of one or the other of two mutually exclusive events occurring is the sum of the probabilities of the two events.

> ### *Probability of Mutually Exclusive Events*
> If E_1 and E_2 are mutually exclusive events, then
> $$P(E_1 \text{ or } E_2) = P(E_1) + P(E_2)$$

EXAMPLE **6**   **Mutually Exclusive Events**

A container holds 2 white marbles, 3 red marbles, and 5 blue marbles. If a single marble is drawn from the container, what is the probability that the marble is white or blue?

SOLUTION

Because the marble cannot be both white and blue, the events are mutually exclusive.

$$P(\text{white or blue}) = P(\text{white}) + P(\text{blue}) = \frac{2}{10} + \frac{5}{10} = \frac{7}{10}.$$

EXAMPLE **7**   **Mutually Exclusive Events**

Suppose that a card is drawn from a standard deck. What is the probability that the card will be either a club or a diamond face card (jack, queen, or king)?

SOLUTION

Let C represent the event of a club, and let D represent the event of a diamond face card. Because C and D cannot both occur, the events are mutually exclusive.

The sample space consists of the 52 possible cards that can be drawn.

The probability of drawing a club is $P(C) = \frac{13}{52}$.

The probability of drawing a diamond face card is $P(D) = \frac{3}{52}$.

Thus $P(C \text{ or } D) = \frac{13}{52} + \frac{3}{52} = \frac{16}{52} = \frac{4}{13}$.

> ### *Definition of the Complement of an Event*
> The **complement** of an event E, denoted by $\overline{E}$, is the set of all outcomes in the sample space that are not in E. Because E and $\overline{E}$ are mutually exclusive events,
> $$P(E \text{ or } \overline{E}) = P(E) + P(\overline{E}) = 1$$
> Thus $P(\overline{E}) = 1 - P(E)$.

EXAMPLE **8**   **Complement of an Event**

If two dice are rolled, what is the probability that the sum is *not* 7?

SOLUTION

In Example 4, we found that the probability of rolling a sum of 7 is $\frac{1}{6}$. Because a sum of 7 and a sum that is not 7 are mutually exclusive,

$$P(\text{sum is 7}) + P(\text{sum is not 7}) = 1$$

$$P(\text{sum is not 7}) = 1 - P(\text{sum is 7}) = 1 - \frac{1}{6} = \frac{5}{6}$$

Suppose that we select a card from a deck and consider the events "the card is a number card" and "the card is a heart." Because the card could be both a heart *and* a number card, the events are not mutually exclusive.

If two events are not mutually exclusive, then we must account for the possibility that both events can occur.

> *Probability of Non-Mutually Exclusive Events*
>
> If E_1 and E_2 are not mutually exclusive events, then
>
> $$P(E_1 \text{ or } E_2) = P(E_1) + P(E_2) - P(E_1 \text{ and } E_2)$$

EXAMPLE **9**   **Non-Mutually Exclusive Events**

Suppose that you draw one card from a standard deck. What is the probability of obtaining a number card or a heart?

SOLUTION

The sample space consists of 52 possible outcomes. Let H represent the event "heart drawn" and N represent the event "number card" drawn.

A standard deck contains 13 hearts and 36 number cards, but 9 of the number cards are also hearts.

$$P(H \text{ or } N) = P(H) + P(N) - P(H \text{ and } N)$$

$$= \frac{13}{52} + \frac{36}{52} - \frac{9}{52} = \frac{40}{52} = \frac{10}{13}$$

Odds

If a team's probability of winning a game is $\frac{2}{3}$, then the probability that the team loses the game is $\frac{1}{3}$. Therefore, the team is twice as likely to win as to lose. We say that the odds in favor of winning are 2 to 1.

■ *NOTE   Keep in mind that the odds favoring an event are not the same thing as the probability of the event.*

> **Odds of an Event's Occurring or Not Occurring**
>
> If E is an event, then the **odds in favor of** E are defined as $\dfrac{P(E)}{P(\overline{E})}$. The **odds against** E are defined as $\dfrac{P(\overline{E})}{P(E)}$.

EXAMPLE **10**   **Odds in Dice Rolling**

If two dice are rolled, what are the odds in favor of rolling a sum of 7? What are the odds against rolling a sum of 7?

SOLUTION

We have seen that $P(\text{sum is } 7) = \frac{1}{6}$, so $P(\text{sum is not } 7) = \frac{5}{6}$. Thus the odds in favor of rolling a sum of 7 are

$$\frac{\frac{1}{6}}{\frac{5}{6}} = \frac{1}{5}$$

The odds in favor of rolling a sum of 7 are 1 to 5.

The odds against rolling a sum of 7 are

$$\frac{\frac{5}{6}}{\frac{1}{6}} = \frac{5}{1}$$

The odds against rolling a sum of 7 are 5 to 1.

EXAMPLE **11**   **Odds of Contracting the Flu**

If you have not had a flu shot, the odds in favor of your contracting the flu are 3 to 8. What is the probability that you will contract the flu?

SOLUTION

The odds in favor of contracting the flu are 3 to 8, which means that

$$\frac{P(E)}{P(\overline{E})} = \frac{3}{8}$$

We solve for $P(E)$ by cross-multiplying and replacing $P(\overline{E})$ with $1 - P(E)$.

$$8 \cdot P(E) = 3 \cdot P(\overline{E})$$
$$8 \cdot P(E) = 3[1 - P(E)]$$
$$8 \cdot P(E) = 3 - 3 \cdot P(E)$$
$$11 \cdot P(E) = 3$$
$$P(E) = \frac{3}{11}$$

The probability of your contracting the flu is $\frac{3}{11}$.

6.5 Quick Reference

Experimental Probability

■ An **experiment** is an activity in which the outcome is uncertain. Any particular outcome is called an **event.**

■ If an experiment was performed n times and an event E occurred m times, then the **experimental probability** of the event is $P(E) = \dfrac{m}{n}$.

Theoretical Probability

■ The **sample space** is the set of all possible outcomes of an experiment. If each outcome occurs with the same frequency, we say the outcomes are **equally likely.**

■ An event is a subset of a sample space.

■ If S is a sample space with n equally likely outcomes and if an event E can occur in m ways, then the **theoretical probability** of event E is defined by $P(E) = \dfrac{m}{n}$.

Mutually Exclusive Events

■ Two events are **mutually exclusive** if there is no possibility that both events can occur at the same time.

■ If E_1 and E_2 are mutually exclusive events, then
$$P(E_1 \text{ or } E_2) = P(E_1) + P(E_2)$$

■ The **complement** of an event E, denoted by $\bar{E}$, is the set of all outcomes in the sample space that are not in E. Because E and $\bar{E}$ are mutually exclusive events,
$$P(E \text{ or } \bar{E}) = P(E) + P(\bar{E}) = 1$$

■ If events E_1 and E_2 are not mutually exclusive events, then
$$P(E_1 \text{ or } E_2) = P(E_1) + P(E_2) - P(E_1 \text{ and } E_2)$$

Odds

■ If E is an event, then the **odds in favor of** E are defined by $\dfrac{P(E)}{P(\bar{E})}$. The **odds against** E are defined by $\dfrac{P(\bar{E})}{P(E)}$.

6.5 Speaking the Language

1. A(n) ▨▨▨▨ is an outcome of an experiment or a subset of a sample space.

2. If there is no possibility that two events can occur at the same time, we say that the events are ▨▨▨▨ .

3. If the probability of an event is $\frac{1}{3}$, then the probability of the ▨▨▨▨ of the event is $\frac{2}{3}$.

4. If E is an event, then we call the ratio $\dfrac{P(E)}{P(\bar{E})}$ the ▨▨▨▨ in favor of E.

6.5 Exercises

Concepts and Skills

1. Suppose that you toss a fair coin 24 times and the result is heads every time. What is the probability of obtaining a heads on the 25th toss? Explain.

2. In a certain lottery, players are to match three numbers from 0 through 9. Compare the likelihood of winning with a selection of 000 with that of winning with a selection of 345.

In Exercises 3–16, an experiment is described. Determine the probability that the indicated event occurs.

3. A single die is rolled.
   (a) The result is 5.
   (b) The result is a number no larger than 3.
   (c) The result is 2 or 6.
   (d) The result is an even number.

4. One coin is tossed and then a second coin is tossed.
   (a) Both are heads.
   (b) Both are tails.
   (c) The first is heads and the second is tails.
   (d) One is heads and the other is tails.

5. Four coins are tossed at once.
   (a) Exactly two are heads.
   (b) At least two are heads.
   (c) All are heads or all are tails.
   (d) One is heads and three are tails.

6. Five coins are tossed at once.
   (a) At least four are heads.
   (b) At most two are tails.
   (c) All are heads or all are tails.
   (d) Exactly three are heads or exactly three are tails.

7. A pair of dice is rolled.
   (a) The sum is 9.
   (b) The sum is at least 5.
   (c) The sum is 3 or 11.
   (d) Both show the same number.

8. A pair of dice is rolled.
   (a) The sum is 6.
   (b) The sum is less than 8.
   (c) The sum is 4 or 8.
   (d) Both are even numbers.

9. One card is selected from a standard deck of 52 cards.
   (a) The card is a club.
   (b) The card is an ace.
   (c) The card is a face card (jack, queen, or king).
   (d) The card is a heart or a diamond, but it is not a face card.
   (e) The card is a club or a face card.

10. One card is selected from a standard deck of 52 cards.
    (a) The card is a number card (2–10).
    (b) The card is a heart or a diamond.
    (c) The card is a diamond face card (jack, queen, or king).
    (d) The card is a 7 or an 8.
    (e) The card is a number card or a heart.

11. Two cards are selected (without replacement) from a standard deck of 52 cards.
    (a) Both cards are aces.
    (b) Neither card is a face card.
    (c) At least one card is a face card.
    (d) The cards are different colors.
    (e) Both cards are face cards or both cards are red cards.

12. Two cards are selected (without replacement) from a standard deck of 52 cards.
    (a) Both cards are face cards.
    (b) Both cards are clubs.
    (c) One card is a number card and the other is a face card.
    (d) The cards are the same color.
    (e) Both cards are number cards or both are diamonds.

13. A jar contains 6 red marbles, 3 blue marbles, 4 white marbles, and 1 green marble. One marble is selected at random.
    (a) The marble is white.
    (b) The marble is not red.
    (c) The marble is blue or red.

14. There are 6 keys that open door A, 5 keys that open door B, and 4 keys that open door C. A key is chosen at random.
    (a) The key opens door C.
    (b) The key opens door B.
    (c) The key opens either door A or door B.

15. There are 7 keys that open door A, 3 keys that open door B, and 6 keys that open door C. Two keys are chosen at random.

(a) Both keys open door C.

(b) Both keys open the same door.

(c) Both keys open either door A or door B.

16. A jar contains 7 red marbles, 4 blue marbles, 6 white marbles, and 1 green marble. Suppose two marbles are selected at random.

(a) Both marbles are blue.

(b) One marble is blue and the other is white.

(c) Both marbles are the same color.

17. Explain whether events E_1 and E_2 are mutually exclusive.

(i) E_1: A vehicle is blue.
    E_2: A vehicle is a van.

(ii) E_1: A computer is on.
    E_2: A computer is off.

18. Suppose that on an interstate, the probability that a car is exceeding the speed limit is 0.6 and the probability that the car has two or more passengers is 0.5. Is the probability of a speeding car with two or more passengers $0.6 + 0.5 = 1.1$? Why?

Applications: Probability

19. Dormitory Room Assignments A college assigns dormitory rooms based on a random selection. If the college has space for 300 students and 420 apply for a room, what is the probability that a student receives a room?

20. Telephone Number What is the probability that the last four digits of your telephone number are identical? (Assume that any digit 0 through 9 is permissible.)

21. Seating Arrangements The names of four guests at a dinner party are written on four place cards. If the cards are randomly placed in a row at the head table, what is the probability that the guests are seated in alphabetical order? (Assume that each person has a different name.)

22. A container has five balls with numbers 1 through 5. The balls are drawn one at a time. What is the probability that they are drawn in numerical order?

23. Random Number Suppose the first digit of a two-digit number is chosen at random from the numbers 1 through 5, and the second digit is chosen at random from the numbers 1 through 9. What is the probability that the two-digit number is less than 30?

24. Suppose that 2 people are selected at random from the first 8 people to enter a room. Of the first 8 people to enter, 3 are women. What is the probability that

(a) both people selected are men?

(b) one man and one woman are selected?

25. Jury Composition A jury of 12 is to be selected from a group of 36 people, of which 20 are women. What is the probability that the jury will consist of an equal number of men and women?

26. Boy/Girl Assuming that the probability of the birth of a child of a particular sex is 0.5, what is the probability that a family with 4 children has

(a) 4 boys?

(b) 2 boys and 2 girls?

(c) at least 2 girls?

27. Lottery Numbers A lottery draws 6 numbers from the numbers 1 through 25. What is the probability of winning the lottery if

(a) a person must match all 6 numbers to win?

(b) a person must match at least 4 numbers to win?

28. Scratch-and-Win A scratch-and-win card at a fast food restaurant has 20 spots, and the customer scratches off 4 spots. Under the spots are

Scratch-and-Win!

Scratch off any 4 spots. If at least 3 spots match, you win that prize!

5 burger symbols, 6 fries symbols, 5 soft drink symbols, and 4 money symbols. If at least 3 match, the customer wins the item. What is the probability of winning each of the following?

(a) Money (b) Fries (c) A food item

29. A container has 8 balls with numbers 1 through 8. If 3 balls are selected without replacement, what is the probability that each is less than 4?

30. **Scholarship Recipients** From a group of 15 equally qualified students, 3 are to be selected at random to receive $1000 scholarships. If the group includes 6 male students and 9 female students, what is the probability that

(a) a particular student wins?
(b) all winners are female?
(c) only one male wins?

31. A container has 8 balls with numbers 1 through 8. A ball is drawn, the number is recorded, and the ball is returned to the container. If 3 balls are drawn in this manner, what is the probability that each is less than 4?

32. **Essay Test** A history instructor gives 6 possible topics for essay questions, of which 3 will be included on the test. A student knows the answer to 4 questions. What is the probability that the student can answer all 3 questions on the test?

33. **Thank-you Notes** A mother addressed envelopes for the 5 thank-you notes her child wrote for birthday gifts. Unfortunately, the child randomly inserted the thank-you notes into the envelopes. What is the probability that

(a) all notes were in the correct envelopes?
(b) the first envelope contains the correct note?

34. **Bacteria Carrier** Of 80 soldiers returning from an assignment in the Middle East, 5 are carriers of a certain bacteria. The army selects 10 to be tested and will test all 80 if at least 1 of the 10 is infected. What is the probability that all will be tested?

35. **Spoiled Milk** Of the 20 cartons of milk on the grocery shelf, 3 are spoiled. Suppose that a shopper selects 4 cartons at random. What is the probability that

(a) none are spoiled?
(b) at least 1 is spoiled?

36. **Defective Monitors** A computer store received a shipment of 50 monitors, of which 8 were defective. A company purchases 6 monitors from the store. What is the probability that

(a) all are defective?
(b) at least 1 is defective?

37. **U.S. Work Force** The table shows the 1998 data for the number (in millions) of full-time wage and salary workers in the United States. (*Source*: U.S. Bureau of Labor Statistics.)

Male		Female	
16–24	25 and over	16–24	25 and over
6	45	4	34

Suppose a worker is chosen at random. What is probability that the worker is

(a) age 16–24?
(b) male?
(c) age 25 and over?
(d) female and under age 25?

38. **Marital Status** The table shows the 1998 data for the number (in millions) of adults (at least 18 years of age) in the United States according to marital status. (*Source*: U.S. Bureau of the Census.)

Never Married	Married	Widowed	Divorced
44	117	13	18

Suppose a person at least 18 years of age is chosen at random. What is probability that the person is

(a) married?
(b) not currently married?
(c) widowed or divorced?

39. **Education Level** The table shows the 1998 data for the number (in millions) of people age 25 and over in the United States according to highest level of education. (*Source*: U.S. Bureau of the Census.)

Not a High School Graduate	High School Graduate	Some College	At Least a Bachelor's Degree
31	58	42	40

If a person age 25 or over is chosen at random, what is the probability that the person

(a) has no more than a high school education?
(b) has at least some college education?
(c) is a college graduate?

40. **Popular Forms of Exercise** The table shows the number (in millions) of people in the United States who participate in selected forms of exercise. (*Source*: National Sporting Goods Association.)

Aerobics	Biking	Walking	Jogging
23	50	71	21

Assume that the population of the United States is 260 million. Suppose that a person is chosen at random. What is the probability that the person

(a) jogs?
(b) walks?
(c) does not bike?

41. **Blood Pressure and Weight** A hospital conducted a study relating to blood pressure and weight. The table shows the probabilities for adult patients.

Condition	Probability
High blood pressure	0.25
Overweight	0.35
Both	0.10

What is the probability that a patient has high blood pressure or is overweight?

42. **Health Habits** A clinic has compiled information based on patients' background information sheets. Of 80 patients selected at random, 40 are regular smokers or drink two or more drinks per day, 20 are regular smokers, and 30 drink two or more alcoholic drinks per day. What is the probability that a patient drinks and smokes?

43. **Student Demographics** At a certain college, a survey finds that the probability that a student is a female is 60%, the probability that a student lives more than 10 miles from campus is 30%, and the probability that a student is female or lives more than 10 miles from campus is 70%. What is the probability that a

student is female and lives more than 10 miles from campus?

44. **Weather Forecast** The weather report indicates that the probability of sunshine is 70%, the probability that the temperature will be above 50°F is 50%, and the probability of both is 30%. What is the probability of at least one of these conditions?

45. Compare the odds in favor of event E to the odds in favor of event $\overline{E}$.

46. Compare the probability of event E to the probability of event $\overline{E}$.

Applications: Computing Odds

47. **Rolling Dice** A pair of dice is rolled. What are the odds in favor of rolling a 5? What are the odds against rolling a 5?

48. **Tossing Coins** Four coins are tossed at once. What are the odds in favor of at most one tail? What are the odds against at most one tail?

49. **Busy Signal** The probability of a busy signal when you dial a repair service is 15%. What are the odds in favor of completing the call?

50. **Snow Forecast** The probability of snow is 70%. What are the odds in favor of snow?

51. **Football Prediction** The odds that the Packers will win at least 10 games are 5 to 2. What is the probability that the Packers will win at least 10 games?

52. **Election Winner** The odds that a candidate will win an election are 3 to 7. What is the probability that the candidate will win?

53. **Matching Socks** Suppose that a person randomly selects 2 socks from a drawer that contains 10 white socks, 8 black socks, and 4 orange socks. What are the odds in favor of the person's selecting

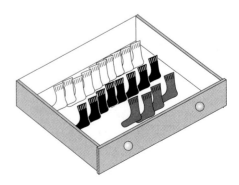

(a) a matching pair?
(b) a pair of black socks?
(c) one orange and one white sock?

54. A box contains 3 white balls, 7 blue balls, and 4 orange balls. If a ball is chosen at random, what are the odds in favor of selecting

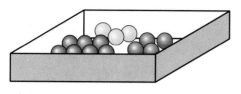

(a) a white ball?
(b) a blue ball?
(c) either a white or an orange ball?

55. Jury Selection   A jury of 12 is to be selected from a jury pool of 24. Of the prospective jurors, 8 have served on a previous jury. What are the odds in favor of having a jury with no one having previous jury experience?

56. Lottery Winner   To win a lottery, a person must match three of five numbers between 1 and 30. (Repetitions are not allowed.) What are the odds in favor of winning the lottery?

6.6   *The Binomial Theorem*

Pascal's Triangle   ■   *The Binomial Theorem*   ■
Specified Terms   ■   *Applications*

Pascal's Triangle

Writing $(a + b)^2 = a^2 + 2ab + b^2$ is called *expanding* the binomial. By observing patterns, we can determine a general method for expanding $(a + b)^n$ for any nonnegative integer n.

EXPLORATION

Expansions of $(x + y)^n$

(a) Use multiplication to expand $(x + y)^4$.
(b) Look at the terms in your result and describe the pattern of exponents on x and y.
(c) Calculate $_4C_r$ for $r = 0, 1, 2, 3,$ and 4. How do your results compare to the numerical coefficients in your expansion in part (a)?

DEVELOPING THE CONCEPT

Expansions of $(a + b)^n$

Consider the following results of expanding $(a + b)^n$.

$$(a + b)^0 = 1$$
$$(a + b)^1 = a + b$$
$$(a + b)^2 = a^2 + 2ab + b^2$$
$$(a + b)^3 = a^3 + 3a^2b + 3ab^2 + b^3$$
$$(a + b)^4 = a^4 + 4a^3b + 6a^2b^2 + 4ab^3 + b^4$$

These expansions reveal certain patterns.

1. The sum has one more term than the value of the exponent n.
2. The exponents on a begin with n and decrease to 0.
3. The exponents on b begin with 0 and increase to n.
4. The sum of the exponents in each term is n.

Furthermore, the coefficients follow a definite pattern.

			1			$(a + b)^0 = 1$
		1		1		$(a + b)^1 = a + b$
	1		2		1	$(a + b)^2 = a^2 + 2ab + b^2$

$$1 \quad\quad\quad\quad\quad\quad 1 \quad\quad\quad\quad 1 \quad\quad\quad\quad\quad\quad (a+b)^0 = 1$$

$$\begin{array}{ccccccc}
& & & 1 & & & (a+b)^0 = 1 \\
& & 1 & & 1 & & (a+b)^1 = a + b \\
& 1 & & 2 & & 1 & (a+b)^2 = a^2 + 2ab + b^2 \\
1 & & 3 & & 3 & & 1 \quad (a+b)^3 = a^3 + 3a^2b + 3ab^2 + b^3 \\
\end{array}$$

1   1 $(a + b)^0 = 1$

1   1 $(a + b)^1 = a + b$

1   2   1 $(a + b)^2 = a^2 + 2ab + b^2$

1   3   3   1 $(a + b)^3 = a^3 + 3a^2b + 3ab^2 + b^3$

1   4   6   4   1 $(a + b)^4 = a^4 + 4a^3b + 6a^2b^2 + 4ab^3 + b^4$

1   5   10   10   5   1   Coefficients of $(a + b)^5$

This triangular array of numbers is called **Pascal's triangle.** Note that each entry in the triangle is the sum of the two numbers above it. For example, in the bottom row, 5 is the sum of the 1 and 4 above it, and 10 is the sum of the 4 and 6 above it. By continuing the pattern, we can write additional lines that contain the coefficients of expansions of $(a + b)^n$ for higher values of n.

We can number the rows of Pascal's triangle starting with row 0, then row 1, row 2, and so on. In this way, the row number corresponds to the power of the binomial. For example, to find the coefficients of the expansion of $(a + b)^3$, we look in row 3.

Suppose that we wish to expand $(a + b)^6$. To determine the coefficients, we add a row to Pascal's triangle.

1   Row 0

1   1   Row 1

1   2   1   Row 2

1   3   3   1   Row 3

1   4   6   4   1   Row 4

1   5   10   10   5   1   Row 5

1   6   15   20   15   6   1   Row 6

Now we can write the expansion of $(a + b)^6$. The exponents on a begin with 6 and decrease to 0, and the exponents on b begin with 0 and increase to 6.

$$(a + b)^6 = 1a^6b^0 + 6a^5b^1 + 15a^4b^2 + 20a^3b^3 + 15a^2b^4 + 6a^1b^5 + 1a^0b^6$$
$$= a^6 + 6a^5b + 15a^4b^2 + 20a^3b^3 + 15a^2b^4 + 6ab^5 + b^6$$

EXAMPLE 1   **Using Pascal's Triangle to Expand a Binomial**

Expand $(x - 3)^5$.

SOLUTION

From Pascal's Triangle, we know that the coefficients of the expansion of $(a + b)^5$ are

$$1 \quad 5 \quad 10 \quad 10 \quad 5 \quad 1$$

Thus the model for the expansion is

$$(a + b)^5 = a^5 + 5a^4b + 10a^3b^2 + 10a^2b^3 + 5ab^4 + b^5$$

■ *NOTE   In Example 1, the binomial x − 3 is a difference, and the terms of the expansion alternate in sign. This is true for all expansions of the form $(a − b)^n$.*

Replace a with x and b with -3.

$$(a + b)^5 = a^5 + 5a^4b + 10a^3b^2 + 10a^2b^3 + 5ab^4 + b^5$$
$$(x - 3)^5 = x^5 + 5x^4(-3) + 10x^3(-3)^2 + 10x^2(-3)^3 + 5x(-3)^4 + (-3)^5$$
$$= x^5 - 15x^4 + 90x^3 - 270x^2 + 405x - 243$$

The Binomial Theorem

When we expand $(a + b)^n$, the variable factors of each term are easy to determine from the exponent patterns on a and b. The coefficients (called **binomial coefficients**) of the terms can be found with Pascal's triangle. However, this is not a practical method when n is large. The Binomial Theorem offers an alternative method for determining the coefficients. In this theorem, the coefficients are calculated with combinations, which we introduced in the section of this chapter on permutations and combinations.

The Binomial Theorem

For any positive integer n,

$$(a + b)^n = {}_nC_0a^n + {}_nC_1a^{n-1}b + {}_nC_2a^{n-2}b^2 + \cdots + {}_nC_{n-1}ab^{n-1} + {}_nC_nb^n$$

$$= \sum_{r=0}^{n} {}_nC_ra^{n-r}b^r$$

where ${}_nC_r = \dfrac{n!}{(n-r)!\,r!}$.

Now that we have a formula for the binomial coefficients, we can use the Binomial Theorem to expand any binomial. We usually begin by determining the binomial coefficients with the formula

$$_nC_r = \frac{n!}{(n-r)!r!}$$

EXAMPLE **2**   **Using the Binomial Theorem to Expand a Binomial.**

Expand $(3x - 2y)^5$.

SOLUTION

We can calculate the coefficients one at a time with the formula $_nC_r = \dfrac{n!}{(n-r)!r!}$ for $r = 0, 1, 2, 3, 4, 5$. Alternatively, we can use a calculator to compute all the binomial coefficients at once (see Figure 6.17). The binomial coefficients are

$$_5C_0 = 1 \qquad _5C_1 = 5 \qquad _5C_2 = 10 \qquad _5C_3 = 10 \qquad _5C_4 = 5 \qquad _5C_5 = 1$$

Now we know that the expansion has the following form:

$$(a + b)^5 = a^5 + 5a^4b + 10a^3b^2 + 10a^2b^3 + 5ab^4 + b^5$$

Replace a with $3x$ and b with $-2y$.

$$(3x - 2y)^5 = (3x)^5 + 5(3x)^4(-2y) + 10(3x)^3(-2y)^2 + 10(3x)^2(-2y)^3 +$$
$$5(3x)(-2y)^4 + (-2y)^5$$
$$= 243x^5 - 810x^4y + 1080x^3y^2 - 720x^2y^3 + 240xy^4 - 32y^5$$

Note again that the binomial is a difference and that the signs of the terms of the expansion alternate.

FIGURE 6.17

```
5 nCr {0,1,2,3,4,5}
    {1 5 10 10 5 1}
```

EXAMPLE **3**   **Using the Binomial Theorem to Expand a Binomial**

Expand $\left(x^2 + \dfrac{2}{y}\right)^6$.

SOLUTION

We begin by determining the first four coefficients.

$$_6C_0 = 1 \qquad _6C_1 = 6 \qquad _6C_2 = 15 \qquad _6C_3 = 20$$

By symmetry, the remaining coefficients are 15, 6, and 1. Now we know that the expansion has the following form:

$$(a + b)^6 = a^6 + 6a^5b + 15a^4b^2 + 20a^3b^3 + 15a^2b^4 + 6ab^5 + b^6$$

Replace a with x^2 and b with $\dfrac{2}{y}$.

$$\left(x^2 + \frac{2}{y}\right)^6 = (x^2)^6 + 6(x^2)^5\left(\frac{2}{y}\right) + 15(x^2)^4\left(\frac{2}{y}\right)^2 + 20(x^2)^3\left(\frac{2}{y}\right)^3 +$$
$$15(x^2)^2\left(\frac{2}{y}\right)^4 + 6(x^2)\left(\frac{2}{y}\right)^5 + \left(\frac{2}{y}\right)^6$$
$$= x^{12} + \frac{12x^{10}}{y} + \frac{60x^8}{y^2} + \frac{160x^6}{y^3} + \frac{240x^4}{y^4} + \frac{192x^2}{y^5} + \frac{64}{y^6}$$

■ *NOTE   Pascal's triangle and the Binomial Theorem reveal a symmetry in the pattern of coefficients. For example, if there are six coefficients, then the last three coefficients are the same as the first three, but in reverse order. If there are seven coefficients, we need to calculate only the first four coefficients; we can then use symmetry for the remaining three.*

Specified Terms

The Binomial Theorem allows us to find a specific term of an expansion. Observe the pattern of the terms.

Term Number	1	2	3	$\ldots$	$k + 1$
Term	$_nC_0a^nb^0$	$_nC_1a^{n-1}b^1$	$_nC_2a^{n-2}b^2$	$\ldots$	$_nC_ka^{n-k}b^k$

> ### The (k + 1)st Term of a Binomial Expansion
> When $(a + b)^n$ is expanded, the $(k + 1)$st term is given by $_nC_ka^{n-k}b^k$.

Note that the formula includes the variable k, not $k + 1$. You will need to let $k + 1$ equal the given term number and solve for k.

EXAMPLE 4   **Finding a Specified Term of a Binomial Expansion**

Write the seventh term of the expansion of $(5x - y)^9$.

SOLUTION

To write the seventh term, note that $k + 1 = 7$, so $k = 6$. Substitute $n = 9$, $k = 6$, $a = 5x$, and $b = -y$ into the formula for the $(k + 1)$st term.

$$_nC_ka^{n-k}b^k = {_9C_6}(5x)^3(-y)^6 = 10,500x^3y^6$$

Applications

Recall that a set B is a *subset* of a set A if every element of B is also an element of A. Set A and the empty set $\varnothing$ are both considered subsets of set A.

Consider the set $A = \{1, 2, 3\}$. In the following table, we list the subsets of A, and we use combination notation to indicate the number of subsets with a given number of elements.

Subset of A

$\varnothing$ Number of subsets with 0 elements: $_3C_0 = 1$

$\{1\}$
$\{2\}$ Number of subsets with 1 element: $_3C_1 = 3$
$\{3\}$

$\{1, 2\}$
$\{1, 3\}$ Number of subsets with 2 elements: $_3C_2 = 3$
$\{2, 3\}$

$\{1, 2, 3\}$ Number of subsets with 3 elements: $_3C_3 = 1$

Total number of subsets: 8

The total number of subsets of set A is ${}_3C_0 + {}_3C_1 + {}_3C_2 + {}_3C_3$, which is equal to 8.

Note that $8 = 2^3$, where the exponent on 2 is the number of elements in set A. We can show that this is not a coincidence by writing 2^3 in a different way: $(1 + 1)^3$. Now we can use the Binomial Theorem to expand $(1 + 1)^3$.

$$(1 + 1)^3 = {}_3C_0 \cdot 1^3 + {}_3C_1 \cdot 1^2 \cdot 1 + {}_3C_2 \cdot 1 \cdot 1^2 + {}_3C_3 \cdot 1^3$$
$$= {}_3C_0 + {}_3C_1 + {}_3C_2 + {}_3C_3$$

Expanding $(1 + 1)^3$ results in ${}_3C_0 + {}_3C_1 + {}_3C_2 + {}_3C_3$, which is the total number of subsets of set A. Thus a set with 3 elements has a total of $(1 + 1)^3$ or 2^3 subsets.

This result can be generalized: A set with n elements has a total of 2^n subsets.

EXAMPLE 5 **Choices of Sundae Toppings**

A customer can select any, all, or none of the following toppings for an ice cream sundae.

{chocolate, butterscotch, strawberry, nuts, marshmallows, whipped cream}

How many different kinds of sundaes can be served?

SOLUTION

Each subset of toppings corresponds to a different sundae.

Because the set of toppings has 6 elements, the total number of subsets is $2^6 = 64$. Thus the store can serve 64 different sundaes.

This chapter contains many formulas. Some of the formulas may appear complicated and, because they are not used frequently, may be difficult to remember for very long. Memorizing the formulas is not as important as knowing that the formulas exist and being able to retrieve and use them when they are needed.

6.6 Quick Reference

Pascal's Triangle
- Writing $(a + b)^n$ as a sum is called *expanding* the binomial.
- **Pascal's triangle** is a triangular array of numbers in which each entry is the sum of the two entries above it. We number the first row as row 0.
- The coefficients of the expansion of $(a + b)^n$ can be found in the nth row of **Pascal's triangle.**
- The following patterns can be found in the expansion of $(a + b)^n$:
  1. The sum has one more term than the value of the exponent n.
  2. The exponents on a begin with n and decrease to 0.
  3. The exponents on b begin with 0 and increase to n.
  4. The sum of the exponents in each term is n.

The Binomial Theorem　■　In the expansion of $(a - b)^n$, the terms of the expansion alternate in sign.

■　The Binomial Theorem is a formula for the expansion of $(a + b)^n$.

$$(a + b)^n = \sum_{r=0}^{n} {}_nC_r a^{n-r} b^r, \text{ where } {}_nC_r = \frac{n!}{(n-r)!r!}$$

■　In the Binomial Theorem, the numbers ${}_nC_r$ are called the **binomial coefficients.**

Specified Terms　■　When $(a + b)^n$ is expanded, the $(k + 1)$st term is given by ${}_nC_k a^{n-k} b^k$.

6.6 Speaking the Language

1. The triangular array of numbers in which each number is the sum of the two numbers above it is known as ▨▨▨▨▨ triangle.

2. Writing a binomial to the nth power as a sum is known as ▨▨▨▨▨ the binomial.

3. The ▨▨▨▨▨ Theorem states that $(a + b)^n = \sum_{r=0}^{n} {}_nC_r a^{n-r} b^r$, where the numbers ${}_nC_r$ are called ▨▨▨▨▨.

4. If a set has 7 elements, then the set has 2^7 ▨▨▨▨▨.

6.6 Exercises

Concepts and Skills

1. Suppose that you have a chart showing the first 12 lines of Pascal's triangle. Would this chart give you the coefficients that you need to expand $(x + y)^{12}$? Explain.

2. Compare the sign patterns in the expansions of $(a + b)^n$ and $(a - b)^n$.

In Exercises 3–6, use Pascal's triangle to expand the binomial.

3. $(2x - y)^4$
4. $(t + 3)^5$
5. $(3y + 2)^5$
6. $(5 - 2z)^4$

In Exercises 7–16, use the Binomial Theorem to expand the binomial.

7. $(y - 2)^6$
8. $(x + 1)^5$
9. $(1 + 2x)^5$
10. $(5 - 3y)^4$
11. $\left(2x - 3\sqrt{y}\right)^4$
12. $\left(\sqrt{x} + 2y\right)^6$
13. $(t^2 + 4w)^5$
14. $(x^2 - y^3)^4$
15. $\left(\frac{x}{y} - \frac{z}{2}\right)^6$
16. $\left(\frac{2}{t} + \sqrt{t}\right)^5$

In Exercises 17–24, write the first four terms of the expanded binomial.

17. $(x + 4y)^{10}$
18. $(2x - 5y)^8$
19. $(a^2 - 2b^4)^9$
20. $(x^3 + 3y^2)^{10}$
21. $(4y + 5z)^8$
22. $(ac + 3b)^{11}$
23. $\left(\frac{1}{y} - \frac{z}{2}\right)^{12}$
24. $\left(\frac{a}{3} - \frac{3b}{4}\right)^7$

25. The expansions of both $(a + b)^5$ and $(a - b)^5$ have 6 terms. Does this imply that the expansion of the expression $(a + b)^5 + (a - b)^5$ has 12 terms? Why?

26. If you expand $(a + b)^6$, the coefficient of the fourth term is ${}_6C_3$. If you expand and simplify $(2x + 3y)^6$, will ${}_6C_3$ also be the coefficient of the fourth term? Explain.

In Exercises 27–30, write the first two terms in the expansion of the given expression.

27. $(x + 2y)^{50}$
28. $\left(w - \frac{2}{y}\right)^{90}$
29. $(z - 3)^{200}$
30. $(x + 5)^{100}$

In Exercises 31–38, find the indicated term of the expansion of the given expression.

31. $(z + 4w)^{12}$; second term

32. $(5y - z)^{10}$; eighth term

33. $(x - 3y)^9$; last term

34. $(2a - b^2)^7$; fourth term

35. $(2t^2 + s)^{10}$; seventh term

36. $(3xy + z)^{11}$; last term

37. $(3 - x^2)^8$; third term

38. $(2a^2 + 3b^4)^9$; fifth term

Concept Extension

39. Suppose that you draw a vertical line through the center of Pascal's triangle. Note that the entries on the left and right of the line are symmetric. In general, show that $_nC_j = {}_nC_{n-j}$.

40. Determine the sum of the numbers in each row of Pascal's triangle, where the first row is row 0. Write a formula for the sum of the numbers in row n.

41. Show that the first binomial coefficient of the expansion of any binomial is always 1.

42. Show that the last binomial coefficient of the expansion of any binomial is always 1.

In Exercises 43–50, the expansion of the given expression will include a term with the indicated variable part. Write the coefficient of that term.

43. $(y + 4)^{10}$; y^3

44. $(x - 2)^{12}$; x^4

45. $(2x + 5y)^{11}$; x^7y^4

46. $(3t + z)^7$; t^3z^4

47. $(x - 2y)^8$; x^2y^6

48. $(a^2 - 3)^9$; a^8

49. $(b - 3a^2)^5$; a^4b^3

50. $\left(y^2 + \dfrac{4}{z}\right)^{15}$; $\dfrac{y^{24}}{z^3}$

In Exercises 51 and 52, use the Binomial Theorem to raise the given complex number to the indicated power.

51. $(1 - i)^{10}$

52. $(2 + 3i)^8$

53. The accompanying figure shows the first six rows of Pascal's triangle. Add the numbers along the indicated diagonal lines and write the results as a sequence. What have you discovered?

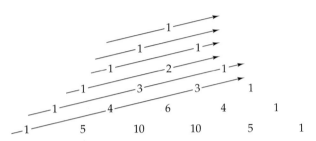

54. Use your calculator to compute 11^n for $n = 0, 1, 2, 3,$ and 4. What do you observe about the digits in the results? Generalize the observations of all n.

55. Use the Binomial Theorem to expand $(2 - y^{-3})^6$. Express your result with positive exponents.

56. (a) Use the Binomial Theorem to show that

$$(x + 2)^n = \sum_{j=0}^{n} {}_nC_j x^{n-j} \cdot 2^j$$

(b) Now let $x = 1$ and show that

$$\sum_{j=0}^{n} {}_nC_j \cdot 2^j = 3^n$$

Applications

In Exercises 57 and 58, assume that none, any, or all of the given options may be chosen.

57. **Truck Options** A customized truck dealer offers the following options for pickup trucks:

{bed liners, power steering, air conditioning, radio, mud flaps, oversize tires, and automatic transmission}

How many different option packages are available?

58. **Benefits Packages** A company offers the following benefits from which employees can choose:

{health insurance, dental insurance, child care, stock options, retirement savings}

How many possible benefit packages does the company offer?

59. Baldness Medication Effectiveness Suppose that a certain medication is successful in treating baldness in 80% of the men who take the medication. If 10 men take the medication, the probability that the medication is successful for exactly 7 of the men is the eighth term of the expansion of $\left(\frac{1}{5} + \frac{4}{5}\right)^{10}$. The probability that the medication is successful for at least 7 of the men is the sum of the last four terms of the expansion.

  (a) What is the probability of success for exactly 7 men?

  (b) What is the probability of success for at least 7 men?

60. Test Score On a 20-question multiple-choice test with 4 choices for each question, a student decides to guess randomly. To pass the test, the student must score at least 70%. The probability that the student scores exactly 70% is the 15th term of the expansion of $\left(\frac{3}{4} + \frac{1}{4}\right)^{20}$. The probability that the student scores at least 70% is the sum of the last 7 terms of the expansion.

  (a) What is the probability that the student scores exactly 70%?

  (b) What is the probability that the student scores at least 70%?

Chapter Review Exercises

Section 6.1

1. Predict the next two terms of the given sequence. Then write the general term of the sequence.

  (a) $-5, 6, -7, 8, -9, \ldots$ (b) $\dfrac{3}{2}, \dfrac{5}{4}, \dfrac{7}{6}, \dfrac{9}{8}, \dfrac{11}{10}, \ldots$

2. Write the first four terms of the following sequence.

$$a_1 = 5, a_{n+1} = 1 - a_n$$

In Exercises 3 and 4, for the given sequence, write (a) the first four terms, (b) the sixth term, and (c) the tenth term.

3. $b_n = \dfrac{2n - 1}{n + 3}$

4. $k_n = \dfrac{(-1)^n + 1}{n}$

5. If $b_1 = -1$, $b_2 = 2$, $b_n = b_{n-1} + 2b_{n-2}$, what is b_5?

6. Explain the difference between a sequence and a series.

In Exercises 7 and 8, for the given sequence, evaluate the indicated partial sums.

7. $1, 2, 0, 3, -1, 4, \ldots$; S_4, S_7

8. $a_n = \dfrac{n + 1}{n}$; S_2, S_5

9. Evaluate $\displaystyle\sum_{i=1}^{5} (1 - 2i)$.

10. Write the following series in summation notation.

$$\frac{3}{2} - \frac{3}{3} + \frac{3}{4} - \frac{3}{5} + \cdots + \frac{3}{10}$$

11. A person borrowed $1500 on a credit card account and planned to pay it back in monthly payments of $250 plus the interest charge of 1.5% of the unpaid balance.

  (a) Write a sequence p_n for the payments on the balance each month.

  (b) Write a sequence I_n for the interest payments each month.

  (c) Determine p_6 and I_6.

  (d) What does the sequence $p_n + I_n$ represent?

12. Refer to Exercise 11.

  (a) What does the series $\displaystyle\sum_{n=1}^{6} p_n$ represent?

  (b) Write and evaluate a series for the total amount of interest paid for 6 months.

Section 6.2

In Exercises 13 and 14, determine whether the given sequence is an arithmetic sequence.

13. $-1, 4, -7, 10, -13, \ldots$ **14.** $b_n = \dfrac{1}{2}n$

In Exercises 15–18, use the given information to write the general term of the arithmetic sequence. Determine the seventh term.

15. $\dfrac{5}{2}, 3, \dfrac{7}{2}, 4, \ldots$ **16.** $a_1 = 2, d = 5$

17. $k_3 = 8, d = -2$ **18.** $b_3 = -\dfrac{1}{2}, b_{10} = 10$

19. Suppose that you know the common difference and the number of terms of a finite arithmetic sequence. Can you write the general term? Why?

20. Determine the sum of the first 30 positive multiples of 5.

21. Evaluate the following arithmetic series.

$$42 + 33 + 24 + \cdots - 192$$

22. Evaluate $\displaystyle\sum_{i=1}^{20} (2 - 3i)$.

23. Because of rapid growth in the area, a school system expects to increase the number of teachers from 130 to 142 next year. For the next 9 years after that, the increase is projected to be 4 more than the preceding year.

(a) Write the general term for the sequence that models the number of new teachers each year.

(b) In which year will the system add 40 teachers?

24. Refer to Exercise 23.

(a) Write and evaluate a series to determine the total number of new teachers hired in the next four years.

(b) Determine the total number of teachers hired during the 10-year period.

Section 6.3

25. Classify each sequence as arithmetic, geometric, or neither. Write a formula for the general term of each sequence.

(a) $\dfrac{3}{2}, 1, \dfrac{1}{2}, 0, \ldots$

(b) $3, -4, 5, -6, \ldots$

(c) $\dfrac{1}{8}, -\dfrac{1}{4}, \dfrac{1}{2}, -1, \ldots$

26. Determine the sum of the first 25 terms of the following geometric sequence.

$$6, 3\sqrt{2}, 3, \dfrac{3}{2}\sqrt{2}, \ldots$$

In Exercises 27–29, write the general term of the geometric sequence.

27. $c_1 = 25, r = \dfrac{1}{5}$ **28.** $a_1 = -\dfrac{3}{2}, a_6 = \dfrac{1}{162}$

29. $b_3 = -54, b_8 = -\dfrac{6561}{16}$

30. Evaluate the following geometric series.

$$\dfrac{8}{5} + \dfrac{12}{5} + \dfrac{18}{5} + \dfrac{27}{5} + \cdots + \dfrac{531{,}441}{2560}$$

31. Evaluate the following infinite geometric series.

$$-32 + 16 - 8 + 4 + \cdots$$

32. Evaluate the geometric series.

$$\sum_{n=1}^{12} 0.25\left(-\dfrac{3}{2}\right)^{n-3}$$

33. Evaluate the infinite geometric series.

$$\sum_{i=1}^{\infty} \dfrac{2}{3^{i-2}}$$

34. Write $2.\overline{15}$ as a quotient of integers.

35. Explain why you can evaluate one of the following series but not the other.

(i) $\displaystyle\sum_{i=1}^{10} 5(-2)^i$ (ii) $\displaystyle\sum_{i=1}^{\infty} 5(-2)^i$

36. Suppose that you are offered a job with a starting annual salary of $30,000 per year and a 4% raise each year.

(a) Write a sequence for your salaries for the first five years.

(b) Write the general term for the sequence of salaries for any year n.

Section 6.4

In Exercises 37–39, evaluate.

37. $\dfrac{18!}{14!}$ **38.** $_8P_5$ **39.** $_{12}C_4$

40. Explain why a *combination lock* actually should be named a *permutation lock*.

41. **Computer Packages** A store sells 8 different types of computers, 5 different types of monitors, and 3 different types of printers. The company plans to offer a special for customers who buy a computer, monitor, and printer. In how many different ways can the company sell a package of a computer, monitor, and printer?

42. **Banquet Seating** Suppose that 5 men and 4 women are to be seated in a row at the head table at a banquet. In how many ways can the host arrange the head table if men and women must alternate seats?

43. UPS Deliveries A UPS delivery person has 12 packages that must be delivered (each to a different address) by 10 A.M. In how many ways can the deliveries be arranged?

44. Baseball Lineups A little league coach has 12 players on the team. How many lineups (batting orders of 9 players) are possible?

45. Emergency Management After a tornado, the Hall County emergency management director had 20 workers, of whom 4 were needed for traffic control, 10 for rescue work, and the remaining workers for first aid. In how many ways can the jobs be assigned?

46. Poker Hands A poker hand consists of 5 cards. In how many ways can a hand of all clubs be dealt?

47. Interview Selections A personnel director has 15 applicants for a payroll clerk position and must select 6 of the applicants for interviews. In how many ways can the 6 finalists be chosen?

48. Civic Club Directors A civic organization has 36 members (20 men and 16 women). The bylaws require that the organization have 5 directors. If the organization plans to have 3 men and 2 women as directors, in how many ways can the directors be chosen?

Section 6.5

College Students Exercises 49 and 50 refer to the following table, which shows college and university full-time and part-time enrollments (in millions). (*Source:* U.S. Center for Education Statistics.)

Male		Female	
Full-time	Part-time	Full-time	Part-time
4.5	2.5	4.7	3.8

49. Suppose a college student is chosen at random. What is the probability that the student is

(a) a female full-time student?
(b) a male part-time student?

50. Suppose a college student is chosen at random. What is the probability that the student is

(a) a female?
(b) a part-time student?

51. Rolling Dice Suppose two dice are rolled. What is the probability that

(a) the sum is 9?
(b) the sum is at least 9?
(c) both numbers are even?

52. Tossing Coins Suppose 3 coins are tossed. What is the probability that

(a) exactly 2 are heads?
(b) at least 2 are heads?

53. Selecting Cards Suppose 2 cards are selected from a standard deck of 52 cards. What is the probability that

(a) both are clubs?
(b) both are either face cards or aces?
(c) both are red cards?

54. Candy Dish A candy dish contains 6 chocolate pieces, 8 peppermints, and 2 gumdrops. Suppose a piece is selected at random. What is the probability that

(a) the piece is a gumdrop?
(b) the piece is not a chocolate?
(c) the piece is a gumdrop or a chocolate?

55. Matching Socks A person has 8 white socks, 6 blue socks, and 4 brown socks in a drawer. If the person randomly selects 2 socks, what is the probability that

(a) both socks are white?
(b) the socks match?

56. Store Grand Opening To celebrate the grand opening of a discount store, the first 20 customers to enter the store are given an envelope containing a gift certificate. Of these envelopes, 2 contain $100 gift certificates, 4 contain $50 certificates, 6 contain $25 certificates, and 8 contain $10 certificates. What is the probability that a person receives

(a) $100? (b) less than $50? (c) $50 or more?

57. Auto Parts An auto parts store has 15 alternators, but 3 are defective. Suppose a mechanic purchases 2 alternators. Assuming that the alternators are selected at random, what is the probability that

(a) both are defective?
(b) both are good?
(c) exactly one is good?

58. Fleas and Tapeworms A veterinarian's records of cat checkups indicate that 70% had fleas,

40% had tapeworms, and 90% had fleas or tapeworms. What is the probability that a cat had both fleas and tapeworms?

59. Suppose that a coin is tossed. Explain the difference between the probability that the coin is heads and the odds in favor of the coin being heads.

60. Lottery In a certain lottery, three numbers from 0 to 9 are chosen at random. What are the odds in favor of the three numbers being the same?

Section 6.6

61. Explain how to determine a row of Pascal's triangle if you know the preceding row.

62. Write the first five terms of the expansion of $\left(x - \dfrac{2}{y}\right)^{11}$.

In Exercises 63–65, use the Binomial Theorem to expand the binomial.

63. $(x + 5)^4$

64. $(2x - y)^6$

65. $\left(\sqrt{3} - y\right)^5$

66. Determine the seventh term of the expansion of $(3x^2 + y)^{12}$.

67. The expansion of $(a^2 - 2b)^9$ will include a term with $a^{12}b^3$ as the variable part. Write the coefficient of the term.

68. The third term of the expansion of a binomial $x + c$ is $40x^3$. What is the binomial?

69. Use the Binomial Theorem to write $(1 - 3i)^5$ in the form $a + bi$.

70. Mother's Day Flowers For a Mother's Day basket, a florist advertises that the customer can order the basic basket with no flowers or choose any or all of the following flowers to be added to the basket.

{carnations, daisies, roses, lilies, chrysanthemums}

In how many ways can the baskets be filled?

Conic Sections

A.1 Introduction

Circle ■ *Parabola* ■ *Ellipse* ■ *Hyperbola*

A **conic section** is obtained from the intersection of a plane with an **infinite right circular cone.** Figure A.1 shows that altering the angle of the plane produces four types of curves: the parabola, the circle, the ellipse, and the hyperbola.

FIGURE **A.1**

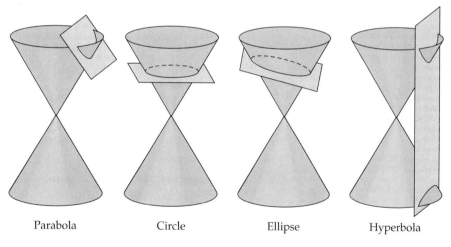

| Parabola | Circle | Ellipse | Hyperbola |

Conic sections can also be defined as the graph of second-degree equations of the form $Ax^2 + By^2 + Cx + Dy + E = 0$. We will use a third approach by defining conic sections geometrically in terms of distances.

Circle

> ### *Definition of a Circle*
> A **circle** is the set of points in a plane that are equidistant from a fixed point. The distance is called the **radius,** and the fixed point is called the **center.**

521

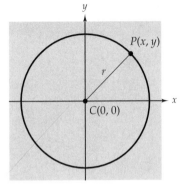

Figure A.2 shows a circle with center $C(0, 0)$ and radius r and a point $P(x, y)$ of the circle.

To write the equation of the circle, we apply the Distance Formula.

$$\sqrt{(x - 0)^2 + (y - 0)^2} = r \qquad \text{The distance between } C(0, 0) \text{ and } P(x, y) \text{ is } r.$$

$$\sqrt{x^2 + y^2} = r \qquad \text{Simplify.}$$

$$x^2 + y^2 = r^2 \qquad \text{Square both sides of the equation.}$$

> **Equation of a Circle Whose Center Is the Origin**
>
> The standard form of the equation of a circle with center $C(0, 0)$ and radius $r > 0$ is
>
> $$x^2 + y^2 = r^2$$

EXAMPLE **1** **Writing an Equation of a Circle**

Write an equation of a circle whose center is $C(0, 0)$ and that contains the point $P(3, -4)$.

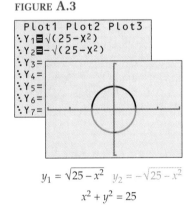

$y_1 = \sqrt{25 - x^2} \quad y_2 = -\sqrt{25 - x^2}$

$x^2 + y^2 = 25$

SOLUTION

We use the distance formula to determine the radius.

$$r = CP = \sqrt{(3 - 0)^2 + (-4 - 0)^2} = \sqrt{9 + 16} = \sqrt{25} = 5$$

The equation is $x^2 + y^2 = 25$.

A circle does not represent a function because it does not pass the Vertical Line Test. To produce the circle in Example 1, we solve $x^2 + y^2 = 25$ for y to obtain $y_1 = \sqrt{25 - x^2}$ and $y_2 = -\sqrt{25 - x^2}$. Then we enter both functions on the Y screen and produce the graphs. In Figure A.3, we use the *Square* setting in order to prevent a distortion in the shape of the circle.

Parabola

> **Definition of a Parabola**
>
> A **parabola** is the set of all points that are equidistant from a fixed point called the **focus** and a fixed line called the **directrix.**

As shown in Figure A.4, the **vertex** is the midpoint between the focus and the directrix. The **axis of symmetry** is the line containing the vertex and the focus.

FIGURE **A.4**

FIGURE **A.5**

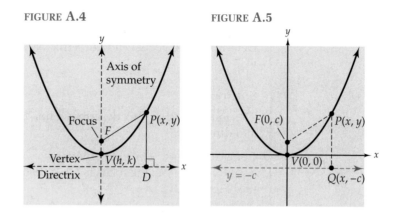

Figure A.5 shows a parabola with focus $F(0, c)$, vertex $V(0, 0)$, and directrix $y = -c$. To write the equation of this parabola, we use the fact that for any point $P(x, y)$ of the parabola, $FP = PQ$.

FIGURE **A.6**

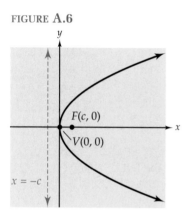

$$\sqrt{(x - 0)^2 + (y - c)^2} = \sqrt{(x - x)^2 + [y - (-c)]^2} \qquad FP = FQ$$

$$\sqrt{x^2 + (y - c)^2} = \sqrt{(y + c)^2} \qquad \text{Simplify.}$$

$$x^2 + (y - c)^2 = (y + c)^2 \qquad \text{Square both sides.}$$

$$x^2 + y^2 - 2cy + c^2 = y^2 + 2cy + c^2 \qquad \text{Simplify.}$$

$$x^2 = 4cy \qquad \text{Solve for } x^2.$$

A similar derivation gives the equation $y^2 = 4cx$ for a parabola with focus $F(c, 0)$, vertex $V(0, 0)$, and directrix $x = -c$. (See Figure A.6.)

When the axis of symmetry is vertical, as in Figure A.5, we say that the parabola is *vertical*; when the axis of symmetry is horizontal, as in Figure A.6, we say that the parabola is *horizontal*.

> ### Equations and Properties of a Parabola Whose Vertex Is the Origin
>
> | | Vertical Parabola | Horizontal Parabola |
> |---|---|---|
> | Equation | $x^2 = 4cy$ | $y^2 = 4cx$ |
> | Focus | $F(0, c)$ | $F(c, 0)$ |
> | Directrix | $y = -c$ | $x = -c$ |
> | Axis of symmetry | y-axis | x-axis |
> | Opens | Upward if $c > 0$ | Right if $c > 0$ |
> | | Downward if $c < 0$ | Left if $c < 0$ |

FIGURE **A.7**

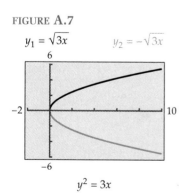

Because a horizontal parabola, such as $y^2 = 3x$, does not pass the Vertical Line Test, it does not represent a function. However, by solving the equation for y and entering the two functions $y_1 = \sqrt{3x}$ and $y_2 = -\sqrt{3x}$ on the Y screen, we can produce a horizontal parabola. (See Figure A.7.)

EXAMPLE **2**   **Determining the Focus and Directrix of a Parabola**

Determine the focus and directrix of the parabola whose equation is $y^2 = -6x$.

SOLUTION

The equation $y^2 = -6x$ describes a horizontal parabola with focus $F(c, 0)$ and directrix $x = -c$.

Comparing $y^2 = -6x$ to the model $y^2 = 4cx$, we see that $4c = -6$, and so $c = -\frac{3}{2}$. Therefore, the focus is $F\left(-\frac{3}{2}, 0\right)$ and the directrix is $x = -\left(-\frac{3}{2}\right) = \frac{3}{2}$. (See Figure A.8.)

FIGURE **A.8**

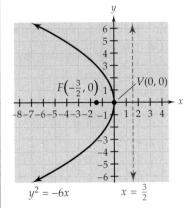

$$y^2 = -6x \qquad x = \frac{3}{2}$$

Note that because $c < 0$, the parabola opens to the left.

When we are given an equation of a parabola, as in Example 2, we can use it to determine the characteristics of the curve. Conversely, if we have enough information about a parabola, then we can use one of the models to write its equation.

EXAMPLE **3**   **Writing an Equation of a Parabola**

Write an equation of the parabola whose focus is $F(0, -3)$ and whose vertex is $V(0, 0)$.

SOLUTION

The focus is $F(0, -3)$, so we know that $c = -3$. Because the focus is a point of the y-axis and c is negative, the parabola is vertical and opens downward.

$$x^2 = 4cy \qquad \text{Model for vertical parabola}$$
$$x^2 = -12y \qquad c = -3$$

Figure A.9 shows the parabola. Note that the equation of the directrix is $y = -c = -(-3) = 3$.

FIGURE **A.9**

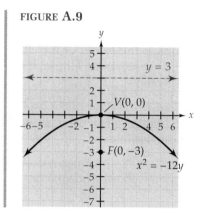

Ellipse

A circle has just one center, and a parabola has just one focus. The shape of the next conic section that we will discuss, the ellipse, is governed by two points, each called a focus. (The plural of focus is *foci*, pronounced fō-sī.)

> ### *Definition of an Ellipse*
>
> An **ellipse** is the set of all points in a plane such that the sum of the distances from each point to two fixed points is a constant. Each of the two fixed points is called a **focus** of the ellipse, and the two together are called the **foci** of the ellipse.

Figure A.10(a) shows an ellipse with foci $F_1(-c, 0)$ and $F_2(c, 0)$. Figure A.10(b) shows an ellipse with foci $F_1(0, -c)$ and $F_2(0, c)$. For any point $P(x, y)$ of the ellipse, by definition, $PF_1 + PF_2 = k$, where k is a constant.

FIGURE **A.10**

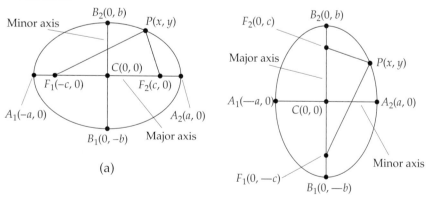

(a)

(b)

The line containing the foci intersects the ellipse at points called the **vertices** of the ellipse. The line segment connecting the vertices is the **major axis**, and its midpoint is the **center** of the ellipse. If the major axis is horizontal, we call the ellipse a *horizontal* ellipse; if the major axis is vertical, we call the ellipse a *vertical* ellipse.

The line perpendicular to the major axis at the center of the ellipse intersects the ellipse at points called **co-vertices.** The line segment connecting the co-vertices is the **minor axis.**

Using the definition of an ellipse and the Distance Formula, we can derive the equation for an ellipse whose center is the origin.

Equations and Properties of an Ellipse Whose Center Is the Origin

The following properties hold for $a > 0$ and $b > 0$.

	Horizontal Ellipse	Vertical Ellipse
Equation	$\dfrac{x^2}{a^2} + \dfrac{y^2}{b^2} = 1, a > b$	$\dfrac{x^2}{a^2} + \dfrac{y^2}{b^2} = 1, a < b$
Vertices	$A_1(-a, 0)$ and $A_2(a, 0)$	$B_1(0, -b)$ and $B_2(0, b)$
Co-vertices	$B_1(0, -b)$ and $B_2(0, b)$	$A_1(-a, 0)$ and $A_2(a, 0)$
Foci	$F_1(-c, 0)$ and $F_2(c, 0)$, where $c^2 = a^2 - b^2$	$F_1(0, -c)$ and $F_2(0, c)$, where $c^2 = b^2 - a^2$

To sketch an ellipse, the intercepts $A(\pm a, 0)$ and $B(0, \pm b)$ are the key points to plot.

EXAMPLE **4**   **Sketching an Ellipse**

Sketch the ellipse whose equation is $9x^2 + 16y^2 = 144$.

SOLUTION

$$\frac{9x^2}{144} + \frac{16y^2}{144} = \frac{144}{144} \qquad \text{Divide both sides by 144 to write the equation in standard form.}$$

$$\frac{x^2}{16} + \frac{y^2}{9} = 1 \qquad \text{From this form, we can see that } a = 4 \text{ and } b = 3.$$

Because $a > b$, the ellipse is a horizontal ellipse. The center is $C(0, 0)$, and the intercepts are $(\pm 4, 0)$ and $(0, \pm 3)$. (See Figure A.11.)

FIGURE A.11

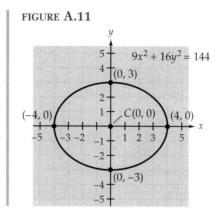

Like a circle, an ellipse does not represent a function. To produce the ellipse in Example 4 with a calculator, we must solve for y and then graph the two associated functions: $y_1 = \frac{3}{4}\sqrt{16 - x^2}$ and $y_2 = -\frac{3}{4}\sqrt{16 - x^2}$. (See Figure A.12.)

FIGURE A.12

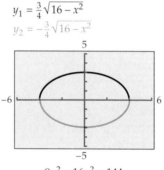

$y_1 = \frac{3}{4}\sqrt{16 - x^2}$

$y_2 = -\frac{3}{4}\sqrt{16 - x^2}$

$9x^2 + 16y^2 = 144$

EXAMPLE **5** **Writing an Equation of an Ellipse**

Write an equation of the ellipse whose foci are $(0, \pm 4)$ and whose x-intercepts are $(\pm 3, 0)$.

SOLUTION

Because the foci are points of the y-axis, the ellipse is a vertical ellipse. From the foci, we see that $c = 4$, and from the x-intercepts, we see that $a = 3$. We use those values to determine b.

$$c^2 = b^2 - a^2$$
$$16 = b^2 - 9$$
$$b^2 = 25$$
$$b = 5$$

We use $a = 3$ and $b = 5$ to write an equation of the ellipse.

$$\frac{x^2}{9} + \frac{y^2}{25} = 1$$

Hyperbola

FIGURE A.13

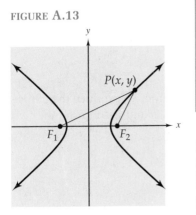

> ### Definition of a Hyperbola
>
> A **hyperbola** is the set of all points in a plane such that the absolute value of the difference of the distances from each point to two fixed points is a constant. The fixed points are called **foci,** and the line containing the foci is the **transverse axis.**

Figure A.13 shows a hyperbola whose center is the origin and whose foci F_1 and F_2 are points of the x-axis. From the definition of a hyperbola, $|PF_1 - PF_2| = k$, where k is a constant.

If the transverse axis is horizontal, we say that the hyperbola is *horizontal* [see Figure A.14(a)]. If the transverse axis is vertical, we say that the hyperbola is *vertical* [see Figure A.14(b)].

FIGURE A.14

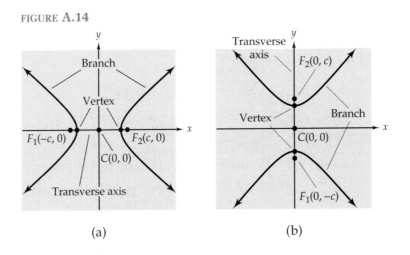

(a)   (b)

The two separate parts of the graph are called **branches.** Each branch intersects the transverse axis at a point called the **vertex.** The midpoint of the line segment connecting the vertices is the **center** of the hyperbola.

As shown in Figure A.15, as the two branches are extended, they become closer to (but never touch) two lines called **asymptotes.** The equations of the asymptotes are $y = \pm\dfrac{b}{a}x$. The asymptotes of the hyperbola are the extended diagonals of the **central rectangle.** Observe that the midpoints of the sides of the central rectangle are $(\pm a, 0)$ and $(0, \pm b)$.

FIGURE **A.15**

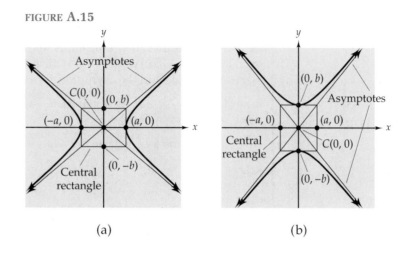

(a)   (b)

To write an equation of a hyperbola centered at the origin with foci on the x- or y-axis, we use the definition and the Distance Formula.

> ### *Equations and Properties of a Hyperbola Whose Center Is the Origin*
>
> The following properties hold for $a > 0$ and $b > 0$.
>
> | | *Horizontal Hyperbola* | *Vertical Hyperbola* |
> |---|---|---|
> | Equation | $\dfrac{x^2}{a^2} - \dfrac{y^2}{b^2} = 1$ | $\dfrac{y^2}{b^2} - \dfrac{x^2}{a^2} = 1$ |
> | Vertices | $V_1(-a, 0)$ and $V_2(a, 0)$ | $V_1(0, -b)$ and $V_2(0, b)$ |
> | Foci | $F_1(-c, 0)$ and $F_2(c, 0)$, where $c^2 = a^2 + b^2$ | $F_1(0, -c)$ and $F_2(0, c)$, where $c^2 = a^2 + b^2$ |

To sketch a hyperbola, we begin by using the values of a and b to draw the central rectangle and asymptotes.

EXAMPLE **6**   **Sketching a Hyperbola**

Describe the hyperbola whose equation is $9x^2 - 25y^2 = 225$.

SOLUTION

$$\frac{9x^2}{225} - \frac{25y^2}{225} = \frac{225}{225}$$ Divide both sides by 225 to write the equation in standard form.

$$\frac{x^2}{25} - \frac{y^2}{9} = 1$$ Standard form of a horizontal hyperbola

With the equation in standard form, we can see that a is 5, which means that the vertices are $V_1(-5, 0)$ and $V_2(5, 0)$. We can also see that b is 3, which means that the midpoints of the other sides of the central rectangle are $(-3, 0)$ and $(3, 0)$. With this information, we can draw the central rectangle and the asymptotes. [See Figure A.16(a).]

Using Figure A.16(a) as a template, we can sketch the hyperbola, as shown in Figure A.16(b).

To determine the foci, we use the relationship

$$c^2 = a^2 + b^2$$
$$c^2 = 5^2 + 3^2 = 25 + 9 = 34$$
$$c = \pm\sqrt{34} \approx \pm 5.83$$

Thus the foci are $F_1\left(-\sqrt{34}, 0\right)$ and $F_2\left(\sqrt{34}, 0\right)$.

The equations of the asymptotes are $y = \pm\dfrac{b}{a}x = \pm\dfrac{3}{5}x$.

FIGURE **A.16**

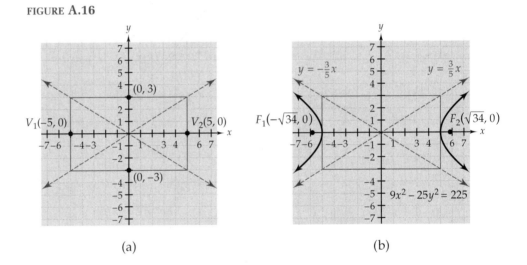

(a)   (b)

FIGURE **A.17**

$y_1 = \frac{3}{5}\sqrt{x^2 - 25}$
$y_2 = -\frac{3}{5}\sqrt{x^2 - 25}$

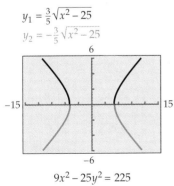

$9x^2 - 25y^2 = 225$

Observe that neither horizontal nor vertical hyperbolas represent functions. To produce a hyperbola with a calculator, we solve the given equation for y and graph the associated functions. For instance, to graph the equation in Example 6, we solve $\dfrac{x^2}{25} - \dfrac{y^2}{9} = 1$ for y to obtain $y = \pm\frac{3}{5}\sqrt{x^2 - 25}$. Then we enter $y_1 = \frac{3}{5}\sqrt{x^2 - 25}$ and $y_2 = -\frac{3}{5}\sqrt{x^2 - 25}$ on the Y screen and produce the graphs of the two functions. (See Figure A.17.)

EXAMPLE 7 **Writing an Equation of a Hyperbola**

Write an equation of a hyperbola with vertices $V(0, \pm2)$ and foci $F(0, \pm3)$.

SOLUTION

Because the vertices are points of the y-axis, the hyperbola is vertical and the model equation is

$$\frac{y^2}{b^2} - \frac{x^2}{a^2} = 1$$

From the given vertices, we know that $b = 2$, and from the given foci, we know that $c = 3$. To determine a^2, we use the relationship $c^2 = a^2 + b^2$ or $a^2 = c^2 - b^2$.

$$a^2 = c^2 - b^2$$
$$a^2 = 3^2 - 2^2 = 9 - 4 = 5$$

Now we use the values of b^2 and a^2 to write the equation.

$$\frac{y^2}{4} - \frac{x^2}{5} = 1 \qquad b^2 = 4 \text{ and } a^2 = 5$$

A.1 Exercises

In Exercises 1–4, determine the radius of the circle. Then sketch the graph.

1. $x^2 + y^2 = 16$ **2.** $x^2 + y^2 = 25$

3. $4x^2 + 4y^2 = 40$ **4.** $2x^2 + 2y^2 = 16$

In Exercises 5–8, write an equation of the circle whose center is the origin and that contains the given point.

5. $P(3, 4)$ **6.** $P(3, -5)$

7. $P(0, 7)$ **8.** $P(-2, -6)$

In Exercises 9–16, determine the focus and directrix of the parabola. Then sketch the graph.

9. $x^2 = y$ **10.** $x^2 = -4y$

11. $y = -\frac{1}{2}x^2$ **12.** $y = 3x^2$

13. $y^2 = -20x$ **14.** $y^2 = 3x$

15. $x = 2y^2$ **16.** $x = -y^2$

In Exercises 17–24, write an equation of the parabola whose vertex is the origin and that satisfies the given conditions.

17. Focus $(0, 4)$ **18.** Focus $(0, -2)$

19. Focus $(-6, 0)$ **20.** Focus $(0.5, 0)$

21. Directrix $y = -2$ **22.** Directrix $x = 3$

23. Vertical parabola containing $P(6, -3)$

24. Horizontal parabola containing $P(10, 5)$

In Exercises 25–32, determine the vertices and co-vertices of the ellipse. Then sketch the graph.

25. $\frac{x^2}{4} + \frac{y^2}{9} = 1$ **26.** $\frac{x^2}{16} + \frac{y^2}{4} = 1$

27. $x^2 + \frac{y^2}{36} = 1$ **28.** $\frac{x^2}{10} + \frac{y^2}{25} = 1$

29. $25x^2 + y^2 = 25$ **30.** $3x^2 + 4y^2 = 12$

31. $4x^2 + y^2 = 16$ **32.** $x^2 + 9y^2 = 9$

In Exercises 33–40, write an equation of an ellipse whose center is the origin and that satisfies the given conditions.

33. Vertices $(0, \pm 3)$, length of minor axis is 2

34. Vertices $(\pm 4, 0)$, co-vertices $(0, \pm 3)$

35. Vertices $(\pm 4, 0)$, foci $(\pm 3, 0)$

36. Vertices $(0, \pm 5)$, foci $(0, \pm 3)$

37. Horizontal ellipse, length of major axis is 10, length of minor axis is 6

38. Vertical ellipse, length of major axis is 12, foci $(0, \pm 5)$

39. Vertices $(\pm 7, 0)$, contains $P(0, -6)$

40. Horizontal ellipse, length of major axis is 10, contains $P(-4, 2.4)$

In Exercises 41–48, determine the vertices of the hyperbola and the equations of the asymptotes. Then sketch the graph.

41. $\dfrac{x^2}{9} - y^2 = 1$

42. $\dfrac{x^2}{16} - \dfrac{y^2}{25} = 1$

43. $y^2 - x^2 = 36$

44. $4x^2 - y^2 = 16$

45. $9x^2 - 4y^2 = 4$

46. $y^2 - 8x^2 = 8$

47. $4y^2 - 9x^2 = 36$

48. $4x^2 - y^2 = 36$

In Exercises 49–54, write an equation of the hyperbola whose center is the origin and that satisfies the given conditions.

49. Vertices $(\pm 4, 0)$, foci $(\pm 5, 0)$

50. Vertices $(0, \pm 2)$, foci $(0, \pm 3)$

51. Vertices $(0, \pm 3)$, asymptotes $y = \pm \dfrac{3}{4}x$

52. Vertices $(\pm 5, 0)$, asymptotes $y = \pm 2x$

53. Horizontal hyperbola, midpoints of the central rectangle are $(\pm 6, 0)$ and $(0, \pm 4)$

54. Vertical hyperbola, midpoints of the central rectangle are $(\pm 5, 0)$ and $(0, \pm 2)$

A.2 *Translations*

Standard Forms of Conic Sections ▪ *Writing Equations in Standard Form* ▪ *Writing an Equation of a Conic Section*

Standard Forms of Conic Sections

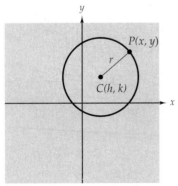

By using our knowledge of translation, we can write the equation of a conic section whose vertex or center has been shifted to a point (h, k).

> ***Equation of a Circle Whose Center Is C(h, k)***
>
> The standard form of the equation of a circle with center $C(h, k)$ and radius $r > 0$ is
>
> $$(x - h)^2 + (y - k)^2 = r^2$$

Figure A.18 shows a circle whose center is $C(h, k)$.

Equations and Properties of a Parabola Whose Vertex Is V(h, k)

	Vertical Parabola	Horizontal Parabola
Equation	$(x - h)^2 = 4c(y - k)$	$(y - k)^2 = 4c(x - h)$
Focus	$F(h, c + k)$	$F(c + h, k)$
Directrix	$y = -c + k$	$x = -c + h$
Axis of symmetry	$x = h$	$y = k$
Opens	Upward if $c > 0$	Right if $c > 0$
	Downward if $c < 0$	Left if $c < 0$

The parabolas described in the preceding summary are shown in Figure A.19.

FIGURE **A.19**

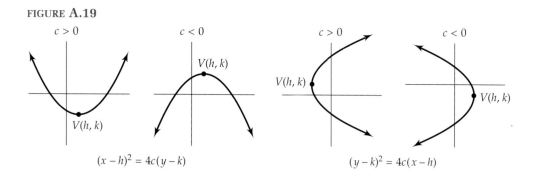

$$(x - h)^2 = 4c(y - k)$$

$$(y - k)^2 = 4c(x - h)$$

Equations and Properties of an Ellipse Whose Center Is C(h, k)

The following properties hold for $a > 0$ and $b > 0$.

HORIZONTAL ELLIPSE

Equation: $\dfrac{(x - h)^2}{a^2} + \dfrac{(y - k)^2}{b^2} = 1, a > b$

Vertices: $A_1(h - a, k)$ and $A_2(h + a, k)$

Co-vertices: $B_1(h, k - b)$ and $B_2(h, k + b)$

Foci: $F_1(h - c, k)$ and $F_2(h + c, k)$, where $c^2 = a^2 - b^2$

VERTICAL ELLIPSE

Equation: $\dfrac{(x - h)^2}{a^2} + \dfrac{(y - k)^2}{b^2} = 1, a < b$

Vertices: $B_1(h, k - b)$ and $B_2(h, k + b)$

Co-vertices: $A_1(h - a, k)$ and $A_2(h + a, k)$

Foci: $F_1(h, k - c)$ and $F_2(h, k + c)$, where $c^2 = b^2 - a^2$

Figure A.20 shows a horizontal ellipse whose center is $C(h, k)$.

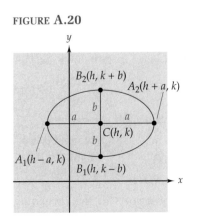

<hr/>

Equations and Properties of a Hyperbola Whose Center Is C(h, k)

HORIZONTAL HYPERBOLA

Equation: $\dfrac{(x - h)^2}{a^2} - \dfrac{(y - k)^2}{b^2} = 1$, where $a > 0$ and $b > 0$

Vertices: $V_1(h - a, k)$ and $V_2(h + a, k)$

Foci: $F_1(h - c, k)$ and $F_2(h + c, k)$, where $c^2 = a^2 + b^2$

VERTICAL HYPERBOLA

Equation: $\dfrac{(y - k)^2}{b^2} - \dfrac{(x - h)^2}{a^2} = 1$, where $a > 0$ and $b > 0$

Vertices: $V_1(h, k - b)$ and $V_2(h, k + b)$

Foci: $F_1(h, k - c)$ and $F_2(h, k + c)$, where $c^2 = a^2 + b^2$

In both cases, the asymptotes are $y - k = \pm\dfrac{b}{a}(x - h)$.

<hr/>

Figure A.21 shows a horizontal hyperbola whose center is $C(h, k)$.

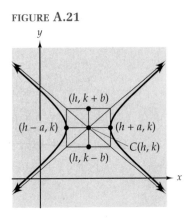

EXAMPLE **1**   **Properties of the Graphs of Conic Sections**

Describe the graph of the given equation.

(a) $x^2 + (y + 3)^2 = 36$ (b) $4(x + 1) = (y + 3)^2$

(c) $\dfrac{(x - 2)^2}{4} + (y + 3)^2 = 1$ (d) $\dfrac{(y - 7)^2}{16} - \dfrac{(x + 1)^2}{25} = 1$

SOLUTION

(a) *What is the model?*

We can write the equation as $(x - 0)^2 + [y - (-3)]^2 = 6^2$, which corresponds to $(x - h)^2 + (y - k)^2 = r^2$, the model equation of a circle.

Center and radius

Because $h = 0$, $k = -3$, and $r = 6$, the center is $C(0, -3)$ and the radius is 6. (See Figure A.22.)

(b) *What is the model?*

Writing the equation as $[y - (-3)]^2 = 4[x - (-1)]$, we see that the equation corresponds to $(y - k)^2 = 4c(x - h)$, which is the model for a horizontal parabola.

Vertex

We see that $h = -1$ and $k = -3$, so the vertex is $V(-1, -3)$.

Focus and directrix

Because $4c = 4$, $c = 1$. Thus the focus is $F(0, -3)$, and the directrix is $x = -2$.

Orientation

Because $c > 0$, the parabola opens to the right. (See Figure A.23.)

(c) *What is the model?*

The equation can be written $\dfrac{(x - 2)^2}{2^2} + \dfrac{[y - (-3)]^2}{1^2} = 1$, which corresponds to $\dfrac{(x - h)^2}{a^2} + \dfrac{(y - k)^2}{b^2} = 1$, the model for an ellipse.

FIGURE **A.22**   FIGURE **A.23**

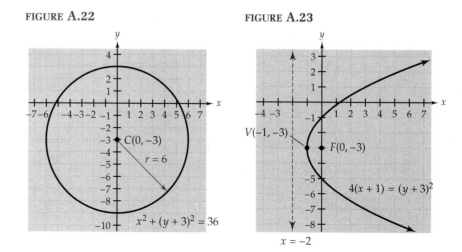

Center
Because $h = 2$ and $k = -3$, the center of the ellipse is $C(2, -3)$.

Vertices and co-vertices
Because $a = 2$ and $b = 1$, the vertices are 2 units to the left and right of the center, and the co-vertices are 1 unit above and below the center.

Foci
Note that $a > b$, which means that the ellipse is horizontal. Thus $c^2 = a^2 - b^2 = 4 - 1 = 3$, so $c = \sqrt{3}$. The foci are $F_1\left(2 - \sqrt{3}, -3\right)$ and $F_2\left(2 + \sqrt{3}, -3\right)$. The graph is shown in Figure A.24.

(d) *What is the model?*

By writing the equation as $\dfrac{(y - 7)^2}{4^2} - \dfrac{[x - (-1)]^2}{5^2} = 1$, we can compare the equation to the model for a vertical hyperbola: $\dfrac{(y - k)^2}{b^2} - \dfrac{(x - h)^2}{a^2} = 1$.

Center
Because $h = -1$ and $k = 7$, the center is $C(-1, 7)$.

Vertices
Because $b = 4$, the vertices are 4 units above and below the center; that is, $V_1(-1, 3)$ and $V_2(-1, 11)$.

Foci
Because $c^2 = a^2 + b^2 = 25 + 16 = 41$, $c = \sqrt{41}$. Thus the foci are $F_1\left(-1, 7 - \sqrt{41}\right)$ and $F_2\left(-1, 7 + \sqrt{41}\right)$.

Central rectangle and asymptotes
Because $a = 5$, the vertical sides of the central rectangle are 5 units to the left and right of the center. The equations of the asymptotes are $y - 7 = \pm\frac{4}{5}(x + 1)$. After we draw the central rectangle and the asymptotes, we can sketch the hyperbola. (See Figure A.25.)

▪ *NOTE   As discussed in the previous section, if a conic section does not represent a function, then producing the graph with a calculator requires solving the given equation for y and graphing each of the two resulting functions separately.*

FIGURE **A.24**

FIGURE **A.25**

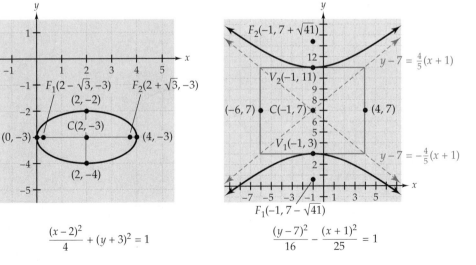

$$\frac{(x - 2)^2}{4} + (y + 3)^2 = 1$$

$$\frac{(y - 7)^2}{16} - \frac{(x + 1)^2}{25} = 1$$

Writing Equations in Standard Form

An equation of the form $Ax^2 + By^2 + Cx + Dy + E = 0$ is a second-degree equation in two variables. Its graph is a conic section. To determine the characteristics of the graph, we use the method of completing the square to write the equation in standard form.

EXAMPLE 2   **Writing Equations in Standard Form**

(a) Determine the center and radius of the circle whose equation is $x^2 + y^2 - 2x + 8y + 8 = 0$.
(b) Find the center and vertices of the hyperbola whose equation is $7x^2 - 9y^2 + 42x + 36y - 36 = 0$.

SOLUTION

(a)

$x^2 + y^2 - 2x + 8y + 8 = 0$	The given equation
$(x^2 - 2x + \ \rule{1cm}{0.4pt}\ ) +$	Group the x- and
$\qquad (y^2 + 8y + \ \rule{1cm}{0.4pt}\ ) = -8$	y-terms.
$(x^2 - 2x + 1) + (y^2 + 8y + 16) = -8 + 1 + 16$	Complete the square in each group.
$(x - 1)^2 + (y + 4)^2 = 9$	Factor the perfect square trinomials.

The center is $C(1, -4)$, and the radius is 3. (See Figure A.26.)

FIGURE **A.26**

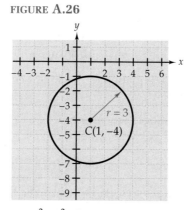

$x^2 + y^2 - 2x + 8y + 8 = 0$

(b)

$7x^2 - 9y^2 + 42x + 36y - 36 = 0$	The given equation
$7x^2 + 42x - 9y^2 + 36y = 36$	Rearrange terms.
$7(x^2 + 6x + \ \rule{1cm}{0.4pt}\ ) -$	Factor out the leading
$\qquad 9(y^2 - 4y + \ \rule{1cm}{0.4pt}\ ) = 36$	coefficient from each group.
$7(x^2 + 6x + 9) -$	Complete the square
$\qquad 9(y^2 - 4y + 4) = 36 + 63 - 36$	in each group.
$7(x + 3)^2 - 9(y - 2)^2 = 63$	Factor the perfect square trinomials.

$$\frac{(x+3)^2}{9} - \frac{(y-2)^2}{7} = 1 \qquad \text{**Divide both sides by 63.**}$$

The graph is a horizontal hyperbola with center $C(-3, 2)$ and vertices $V_1(-6, 2)$ and $V_2(0, 2)$. (See Figure A.27.)

FIGURE A.27

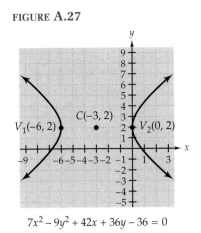

$$7x^2 - 9y^2 + 42x + 36y - 36 = 0$$

Writing an Equation of a Conic Section

Given enough information about a conic section, we can write its equation.

EXAMPLE **3**   **Writing an Equation of a Conic Section**

Write an equation of the conic section that satisfies each of the given conditions.

(a) Parabola whose focus is $F(-2, 0)$ and whose vertex is $V(-2, 1)$
(b) Ellipse with center $C(2, 1)$, focus $F(2, 4)$, and minor axis of length 2

SOLUTION

(a) As we see in Figure A.28, the focus $F(-2, 0)$ is below the vertex $V(-2, 1)$, so the parabola is vertical and opens downward.

FIGURE A.28

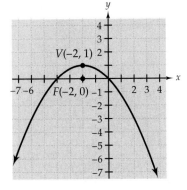

What is the model?
The standard form of the equation is $(x - h)^2 = 4c(y - k)$.

What are h and k?
The vertex is $V(-2, 1)$, so $h = -2$ and $k = 1$.

What is c?
The distance from the vertex to the focus is 1 unit. However, the parabola opens downward, so $c = -1$.

What is the equation?
The equation is $(x + 2)^2 = -4(y - 1)$.

FIGURE **A.29**

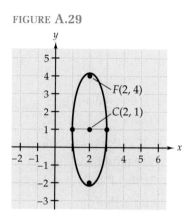

(b) Figure A.29 shows that the described ellipse is vertical.

What is the model?

The standard form of the equation is $\dfrac{(x-h)^2}{a^2} + \dfrac{(y-k)^2}{b^2} = 1$.

What are h and k?
Because the center is $C(2, 1)$, we know that $h = 2$ and $k = 1$.

What are a^2 and b^2?
The length of the minor axis is 2, so $a = 1$ and $a^2 = 1$. The distance from the center to the focus is 3, so $c = 3$. We now use the values of a and c to compute b^2.

$$b^2 = c^2 + a^2 = 9 + 1 = 10$$

What is the equation?
The equation is $(x-2)^2 + \dfrac{(y-1)^2}{10} = 1$.

A.2 Exercises

In Exercises 1–6, determine the center and radius of the circle whose equation is given. Then sketch the graph.

1. $(x + 5)^2 + (y - 2)^2 = 16$
2. $(x - 3)^2 + (y - 6)^2 = 9$
3. $x^2 + y^2 + 2x - 4y + 1 = 0$
4. $x^2 + y^2 - 2x - 6y - 15 = 0$
5. $x^2 - 8x + y^2 = 0$
6. $x^2 + 4x + y^2 + 8y + 15 = 0$

In Exercises 7–12, write an equation of a circle that satisfies the given conditions.

7. Center $C(5, 8)$, $r = 7$
8. Center $C(-1, -6)$, $r = \sqrt{10}$
9. Center $C(4, -7)$, contains $P(2, -3)$
10. Center $C(-2, 4)$, contains $P(1, 2)$
11. Endpoints of the diameter are $P(-1, -3)$ and $Q(5, 7)$.
12. Endpoints of the diameter are $P(4, -2)$ and $Q(-2, 6)$.

In Exercises 13–20, determine the vertex, focus, and directrix of the parabola whose equation is given. Then sketch the graph.

13. $(x - 3)^2 = 8(y + 1)$ 14. $(y + 5)^2 = 12(x + 2)$
15. $(y - 2)^2 = -6(x - 1)$

16. $(x + 4)^2 = -4(y - 6)$
17. $x^2 - 4x - y + 3 = 0$ 18. $x + y^2 - 4y - 5 = 0$
19. $x - y^2 + 6y + 16 = 0$ 20. $x^2 + 5x + y - 6 = 0$

In Exercises 21–26, write an equation of a parabola that satisfies the given conditions.

21. Vertex $V(2, 4)$, focus $F(2, 6)$
22. Vertex $V(-1, 3)$, focus $F(-1, 0)$
23. Vertex $V(3, 0)$, directrix $x = 4$
24. Vertex $V(-2, -3)$, directrix $x = -3$
25. Focus $F(3, -1)$, directrix $x = -1$
26. Focus $F(2, 5)$, directrix $y = 3$

In Exercises 27–34, determine the center, the vertices, the co-vertices, and the foci of the ellipse whose equation is given. Then sketch the graph.

27. $\dfrac{(x+2)^2}{9} + \dfrac{(y-1)^2}{4} = 1$

28. $\dfrac{x^2}{16} + \dfrac{(y+3)^2}{36} = 1$

29. $4(x - 3)^2 + 25(y + 4)^2 = 100$
30. $16(x + 4)^2 + 9(y + 6)^2 = 144$
31. $x^2 + 49y^2 - 294y + 392 = 0$
32. $4x^2 + 9y^2 + 16x + 18y - 11 = 0$
33. $25x^2 - 200x + 9y^2 - 36y + 420 = 0$

34. $4x^2 - 40x + 9y^2 + 99 = 0$

In Exercises 35–40, write an equation of an ellipse that satisfies the given conditions.

35. Center $C(4, -1)$, vertex $A(4, 4)$, co-vertex $B(2, -1)$

36. Center $C(3, -3)$, vertex $A(9, -3)$, co-vertex $B(3, -2)$

37. Vertices $A_1(-1, 3)$, $A_2(5, 3)$; minor axis of length 2

38. Vertices $A_1(-1, 8)$, $A_2(-1, 0)$; co-vertex $B(1, 4)$

39. Foci $F_1(0, 2)$, $F_2(0, 8)$; major axis of length 12

40. Foci $F_1(-1, 1)$, $F_2(3, 1)$; major axis of length 8

In Exercises 41–48, determine the center, the vertices, the foci, and the equations of the asymptotes of the hyperbola whose equation is given. Then sketch the graph.

41. $\dfrac{(x + 2)^2}{9} - \dfrac{(y + 1)^2}{25} = 1$

42. $\dfrac{(y - 2)^2}{36} - \dfrac{(x - 5)^2}{4} = 1$

43. $4y^2 - 49(x - 7)^2 = 196$

44. $16(x - 6)^2 - 25(y + 2)^2 = 400$

45. $y^2 - x^2 - 2x - 4y + 2 = 0$

46. $x^2 - y^2 + 2y - 2 = 0$

47. $x^2 - 16y^2 + 6x - 160y - 407 = 0$

48. $4y^2 - 36x^2 + 360x - 909 = 0$

In Exercises 49–54, write an equation of a hyperbola that satisfies the given conditions.

49. Vertices $A_1(0, 2)$, $A_2(2, 2)$; foci $F_1(-1, 2)$, $F_2(3, 2)$

50. Vertices $A(3, \pm 2)$; foci $F(3, \pm 4)$

51. Center $C(-3, 1)$; vertex $A(-3, 3)$; central rectangle of width 2

52. Center $C(4, -2)$; vertex $A(7, -2)$; central rectangle is square

53. Center $C(5, 2)$; vertex $A(7, 2)$; slopes of asymptotes are $\pm\frac{1}{2}$

54. Vertex $A(0, 2)$; equations of asymptotes are $y = \pm\frac{2}{3}x + 4$

Calculator Guide for the TI-83 Plus

Absolute Value   (*page 4*)

PURPOSE: To report the absolute value of a number or expression.

EXAMPLE: Calculate $|-5|$.

[MATH] [NUM] [abs(] [(−)] [5] [)] [ENTER]

```
abs(-5)
                5
```

Activate   (*page 286*)

PURPOSE: When a function is "active," its graph will be displayed; when a function is "inactive," its graph will not be displayed.

EXAMPLE: Activate $Y_1=2X$ and deactivate $Y_2=X+1$. The black box around the = symbol indicates that Y_1 is active. To deactivate Y_2, use the arrow key to place the cursor over the = symbol and press [ENTER]. Pressing [ENTER] again will activate Y_2.

```
Plot1  Plot2  Plot3
\Y1■2X
\Y2=X+1
```

Alpha   (*page 19*)

PURPOSE: To enter or display variables other than X.

EXAMPLE: Enter the expression $a + b$.

[ALPHA] [A] [+] [ALPHA] [B]

```
A+B
```

Base *e*   (*page 311*)

PURPOSE: To enter or display e^x.

EXAMPLE: Enter the expression $e^{x + 1}$.

[2nd] [eˣ] [x] [+] [1] [)]

```
e^(X+1)
```

Clear   (*page 2*)

PURPOSE: To clear the home screen or a function on the Y screen.

EXAMPLE: To clear the home screen, press [Clear].

To clear a function on the Y screen, place the cursor anywhere on the function and press [Clear].

Combinations *(page 490)*

PURPOSE: To calculate the number of combinations of n items taken r at a time.

EXAMPLE: Calculate $_7C_4$.

7 MATH PRB nCr 4 ENTER

```
7 nCr 4
                35
```

Compose *(page 286)*

PURPOSE: To enter the composition of two functions.

EXAMPLE: Suppose function f is entered as Y_1 and function g is entered as Y_2. Enter $f \circ g$.

VARS Y-VARS Function Y_1 (

VARS Y-VARS Function Y_2)

```
Y₁(Y₂)
```

To evaluate $(f \circ g)(3)$, use $Y_1(Y_2(3))$.

Correlation *(page 149)*

PURPOSE: To view the correlation coefficient r for a regression equation.

EXAMPLE: After data have been entered in the statistics list, press

STAT CALC

and select the regression equation option. The value of r or r^2 is displayed with the regression equation.

The correlation coefficient is displayed only if the "Diagnostic" function is on. To do this, press

2nd CATALOG

and scroll to  DiagnosticOn . Press ENTER .

Cursor *(page 63)*

PURPOSE: To highlight points on the graph screen and display their x- and y-coordinates.

EXAMPLE: To highlight a point in the viewing window, use the arrow keys to move the general cursor to the desired point.

Decimal   *(page 63)*

PURPOSE: To create a viewing window near the origin with coordinates reported in tenths.

EXAMPLE: Press [ZOOM] [ZDecimal] to create a viewing window with

$$\text{Xmin} = -4.7 \qquad \text{Xmax} = 4.7 \qquad \text{Xscl} = 1$$
$$\text{Ymin} = -3.1 \qquad \text{Ymax} = 3.1 \qquad \text{Yscl} = 1$$

Draw Inverse   *(page 296)*

PURPOSE: To produce the graph of an inverse function.

EXAMPLE: If $f(x) = 3x + 1$ is entered on the Y screen as Y_1, produce the graph of f^{-1} by

[2nd] [DRAW] [DrawInv]

[VARS] [Y-VARS] [Function] [Y_1] [ENTER]

```
DrawInv Y₁
```

Exponent   *(page 4)*

PURPOSE: To raise a number or expression to a power.

EXAMPLE: Calculate 2^9.

[2] [^] [9] [ENTER]

```
2^9
            512
```

Factorial   *(page 487)*

PURPOSE: To calculate $n! = n \cdot (n - 1) \cdot (n - 2) \ldots \ldots 2 \cdot 1$.

EXAMPLE: Calculate 6!.

[6] [MATH] [PRB] [!] [ENTER]

```
6!
            720
```

Fraction   *(page 2)*

PURPOSE: To report numerical results in simplified fraction form.

EXAMPLE: Simplify $\frac{21}{14}$.

[2] [1] [÷] [1] [4] [MATH] [Frac] [ENTER]

```
21/14▶Frac
            3/2
```

Graph   *(page 63)*

PURPOSE: To display a graph in the viewing window.

EXAMPLE: To display the graphs of all active functions on the Y screen, press [GRAPH]. If no functions are entered or active, only a coordinate system with the current window settings will be displayed.

Home *(page 2)*

PURPOSE: To display the home screen.

EXAMPLE: Display the home screen from any other screen.

$$\boxed{\text{2nd}}\ \boxed{\text{QUIT}}$$

Integer *(page 63)*

PURPOSE: To create a viewing window in which at least one coordinate is reported as an integer.

EXAMPLE: Press $\boxed{\text{ZOOM}}$ $\boxed{\text{ZInteger}}$ $\boxed{\text{ENTER}}$ to create a viewing window with

| Xmin = -47 | Xmax = 47 | Xscl = 10 |
| Ymin = -31 | Ymax = 31 | Yscl = 10 |

Intersect *(page 74)*

PURPOSE: To determine the coordinates of the point of intersection of two graphs.

EXAMPLE: Display the graphs of the two functions. The point of intersection must be visible in the viewing window.

1. Press $\boxed{\text{2nd}}$ $\boxed{\text{CALC}}$ $\boxed{\text{intersect}}$.
2. Place the tracing cursor on the graph of Y_1 and press $\boxed{\text{ENTER}}$.
3. Place the tracing cursor on the graph of Y_2 and press $\boxed{\text{ENTER}}$.
4. Move the tracing cursor as close as possible to the point of intersection and press $\boxed{\text{ENTER}}$.

The coordinates of the point of intersection are displayed.

Inverse *(page 424)*

PURPOSE: To determine the inverse of a matrix.

EXAMPLE: Determine the inverse of matrix A.

$$\boxed{\text{2nd}}\ \boxed{\text{MATRX}}\ \boxed{\text{NAMES}}\ \boxed{\text{[A]}}\ \boxed{x^{-1}}\ \boxed{\text{ENTER}}$$

```
[A]⁻¹
```

Log, Ln *(page 322)*

PURPOSE: To determine the common (log) or natural (ln) logarithm of a number or expression.

EXAMPLE: Calculate log 5 and ln 3.

$$\boxed{\text{LOG}}\ \boxed{5}\ \boxed{)}\ \boxed{\text{ENTER}}\qquad\boxed{\text{LN}}\ \boxed{3}\ \boxed{)}\ \boxed{\text{ENTER}}$$

```
log(5)
             .6989700043
ln(3)
            1.098612289
```

Matrix Product   *(page 412)*

PURPOSE: To determine the product of matrices.

EXAMPLE: If A and B are matrices, determine AB.

[2nd] [MATRX] [NAMES] [[A]]

[2nd] [MATRX] [NAMES] [[B]] [ENTER]

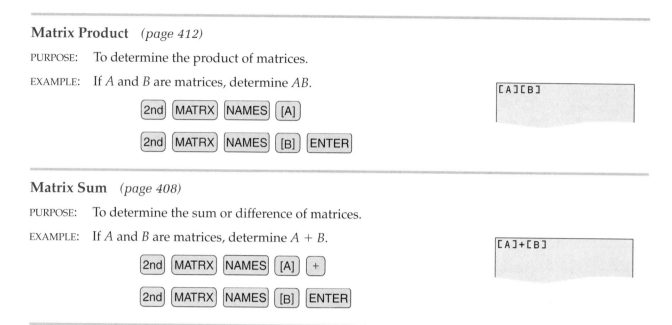

Matrix Sum   *(page 408)*

PURPOSE: To determine the sum or difference of matrices.

EXAMPLE: If A and B are matrices, determine $A + B$.

[2nd] [MATRX] [NAMES] [[A]] [+]

[2nd] [MATRX] [NAMES] [[B]] [ENTER]

Maximum/Minimum   *(page 173)*

PURPOSE: To determine the coordinates of a maximum or minimum point of a graph.

EXAMPLE: Determine the relative minimum of $y = (x - 5)^2 + 3$.

1. Display the graph and press [2nd] [CALC] [minimum].
2. Move the tracing cursor just to the left of the minimum point and press [ENTER].
3. Move the tracing cursor just to the right of the minimum point and press [ENTER].
4. Move the tracing cursor as close as possible to the minimum point and press [ENTER].

The coordinates of the minimum point are displayed.

Mode   *(pages 2, 23 199, 268, 451)*

PURPOSE: (Line 1) To display numbers in decimal notation (Normal) or in scientific notation (Sci).
(Line 2) To set the number of decimal places to which a number is displayed.
(Line 4) To set the function (Func) mode or sequence (Seq) mode.
(Line 5) To set the display of graphs in connected mode or in dot mode.
(Line 7) To perform operations with real numbers (Real) or complex numbers (a + bi).

EXAMPLE: Press [MODE]. Use the up and down arrow keys to move between lines. Use the left and right arrow keys to select a mode within a line. Press [ENTER].

Negative *(page 4)*

PURPOSE: To enter a negative number or the opposite of a variable or expression.

EXAMPLE: Enter $-5 - (-x)$.

(−) 5 − ((−) x) ENTER

```
-5-(-X)
```

Permutations *(pages 487, 488)*

PURPOSE: To calculate the number of permutations of n items taken r at a time.

EXAMPLE: Calculate $_9P_3$.

9 MATH PRB nPr 3 ENTER

```
9 nPr 3
              504
```

Piecewise *(page 120)*

PURPOSE: To produce the graph of a piecewise function.

EXAMPLE: Produce the graph of $f(x) = \begin{cases} x^2, x < 0 \\ x + 1, x \geq 0 \end{cases}$

Y= (x ^ 2) (x 2nd TEST < 0)

ENTER (x + 1) (x 2nd TEST ≥ 0)

```
 Plot1 Plot2 Plot3
\Y1◻(X²)(X<0)
\Y2◻(X+1)(X≥0)
```

Quit *(page 2)*

PURPOSE: To display the home screen.

EXAMPLE: Press 2nd QUIT.

Regression *(pages 147, 184, 229, 355)*

PURPOSE: To produce a regression equation for data entered in the statistics list.

EXAMPLE: To access the regression equation menu, press STAT CALC.

1. For a linear regression equation, select LinReg.
2. For a quadratic regression equation, select QuadReg.
3. For a third-degree regression equation, select CubicReg.
4. For a fourth-degree regression equation, select QuartReg.
5. For an exponential regression equation, select ExpReg.
6. For a logarithmic regression equation, select LnReg.
7. For a logistic regression equation, select Logistic.

RREF   *(page 399)*

PURPOSE: To produce the reduced row-echelon form of a matrix.

EXAMPLE: Display the reduced row-echelon form of a matrix *A*.

[2nd] [MATRX] [MATH] [rref(] [2nd] [MATRX] [NAMES] [[A]] [ )] [ENTER]

> rref([A])

Scalar   *(page 409)*

PURPOSE: To multiply a matrix by a scalar (constant).

EXAMPLE: Multiply a matrix *A* by the scalar 3.

[3] [2nd] [MATRX] [NAMES] [[A]]

> 3[A]

Scatterplot   *(page 146)*

PURPOSE: To produce a scatterplot of data.

EXAMPLE: Enter the data as ordered pairs in the statistics list. Clear the Y screen of unwanted functions, and decide on an appropriate window setting. To display the scatterplot, press

[2nd] [STAT PLOT] [ENTER] [On]

Use the arrow key to select the first option (scatterplot) beside Type. Then press [GRAPH].

Scientific   *(page 23)*

PURPOSE: To enter numbers in scientific notation.

EXAMPLE: Enter $5.3 \cdot 10^4$ in scientific notation.

[5] [.] [3] [2nd] [EE] [4] [ENTER]

> 5.3E4
>
>   53000

Sequence   *(pages 452, 454)*

PURPOSE: To store a sequence in a variable so that the variable can be used to generate terms of the sequence. You must be in sequence mode (see Mode). In the sequence mode, the X key is used to display the variable *n*.

EXAMPLE: Generate the first three terms of $a_n = n + 1$. First, store the sequence in the variable u.

[ALPHA] ["] [n] [+] [1] [ALPHA] ["]

[STO→] [2nd] [u] [ENTER]

> "n+1"→u
>
>   Done
> u(1,3)
>
>   {2 3 4}

(Continued on next page)

Then generate the first three terms.

[2nd] [u] [(] [1] [,] [3] [)] [ENTER]

For a recursive sequence, you must also store the value of the first term in u(nMin). (See Sequence Variable.)

Sequence Plot   *(page 452)*

PURPOSE: To plot points representing the terms of a sequence. You must be in sequence mode (see Mode).

EXAMPLE: Graph the first seven terms of the sequence $a_n = n + 1$.

1. On the Y screen, enter $n + 1$ as u(n).
2. On the window screen, enter the beginning term number, 1, as nMin and the ending term number, 7, as nMax. Also, enter appropriate values for the window settings.
3. Press [GRAPH] to display the graph.

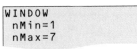

```
WINDOW
 nMin=1
 nMax=7
```

Sequence Variable   *(page 454)*

PURPOSE: To access sequence variables. You must be in sequence mode (see Mode).

EXAMPLE: For the recursive sequence $b_1 = 3$, $b_n = b_{n-1} + 2$, store 3, the value of the first term, in the sequence variable u(nMin).

[3] [STO→] [VARS] [WINDOW] [U/V/W]

[u(nMin)] [ENTER]

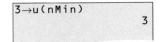

Series   *(page 456)*

PURPOSE: To evaluate a series. You must be in sequence mode (see Mode).

EXAMPLE: Evaluate $\sum\limits_{n=1}^{6} (n + 3)$.

After storing $n + 3$ in u (see Sequence), press

[2nd] [LIST] [MATH] [sum(]

[2nd] [u] [(] [1] [,] [6] [)] [)] [ENTER]

Shade   *(page 435)*

PURPOSE: To produce the graph of an inequality in two variables.

EXAMPLE: Graph the inequality $y \le x + 2$.

1. Enter the boundary line equation $y = x + 2$ on the Y screen as Y_1.
2. Use the arrow key to move the cursor over the icon to the left of the symbol Y_1.

(Continued on next page)

3. Press |ENTER| until the ◣ icon (shade below) is displayed.

4. Press |GRAPH| to display the graph of the inequality.

To shade above, select the ◤ icon, as shown in the second line of the figure.

Solve   *(page 77)*

PURPOSE: To use the calculator to solve an equation.

EXAMPLE: Solve $5x + 1 = 3x - 7$.

1. Enter $5x + 1$ as Y_1 and $3x - 7$ as Y_2.
2. Press |MATH| |Solver|.
3. Enter $Y_1 - Y_2$ beside eqn: 0= and press |ENTER|.
4. Enter a guess of the solution in the line X=. (Using the graphing method first will help you make a guess.)
5. Press |ALPHA| |SOLVE|. (SOLVE is above the ENTER key.)
6. The solution, -4, is displayed next to X=. For equations with more than one solution, repeat steps 4 through 6.

Square   *(page 155)*

PURPOSE: To create a viewing window in which the actual distances between tick marks on the x- and y-axes are the same. This setting prevents distortions in the shapes of graphs.

EXAMPLE: Press |ZOOM| |ZSquare| to create a viewing window with

| Xmin ≈ −15.16 | Xmax ≈ 15.16 | Xscl = 1 |
| Ymin = −10 | Ymax = 10 | Yscl = 1 |

Square Root   *(page 4)*

PURPOSE: To calculate the square root of a number or expression.

EXAMPLE: Calculate $\sqrt{576}$.

 |2nd| |√| |5| |7| |6| |)| |ENTER|

```
√(576)
            24
```

Standard   *(page 63)*

PURPOSE: To create a standard viewing window.

EXAMPLE: Press |ZOOM| |ZStandard| to create a viewing window with

| Xmin = −10 | Xmax = 10 | Xscl = 1 |
| Ymin = −10 | Ymax = 10 | Yscl = 1 |

Store *(page 5)*

PURPOSE: To assign a value to a variable.

EXAMPLE: Let $x = 5$.

$\boxed{5}$ $\boxed{\text{STO}\rightarrow}$ $\boxed{x}$ $\boxed{\text{ENTER}}$

```
5→X
               5
```

Table *(page 6)*

PURPOSE: To display the values of an expression for selected values of the independent variable.

EXAMPLE: To display the values of the expression $x^2 - 4x$, which has been entered as Y_1 on the Y screen, press

$\boxed{\text{2nd}}$ $\boxed{\text{TABLE}}$

Values of x are displayed in the X column, and the corresponding values of $x^2 - 4x$ are displayed in the Y_1 column.

Table Set *(page 6)*

PURPOSE: To set up a table by selecting the initial x-value, the increments of the x-values, and the method by which table values are displayed.

EXAMPLE: To access the table setup menu, press $\boxed{\text{2nd}}$ $\boxed{\text{TBL SET}}$. Use the up and down arrow keys to select options.

```
TABLE SETUP
 TblStart=-3
 △Tbl=2
Indpnt: Auto Ask
Depend: Auto Ask
```

1. TblStart=−3 means that −3 is the first displayed x-value.
2. Δ Tbl=2 means that the x-values will be listed in increments of 2.
3. The Auto and Ask options can be set in four possible ways. The most common is to set both to Auto, which provides a complete table for both variables.

Setting Indpnt: Ask and Depend: Auto allows you to enter an x-value of your choice; the corresponding value of the expression will be reported.

Test *(page 3)*

PURPOSE: To access the symbols $=$, $\neq$, $>$, $\geq$, $<$, and $\leq$, usually for the purpose of testing the truth of a relationship.

EXAMPLE: Determine whether $\frac{1}{3} > 0.33$ is true.

$\boxed{1}$ $\boxed{\div}$ $\boxed{3}$ $\boxed{\text{2nd}}$ $\boxed{\text{TEST}}$ $\boxed{>}$

$\boxed{.}$ $\boxed{3}$ $\boxed{3}$ $\boxed{\text{ENTER}}$

```
1/3>.33
              1
```

The returned value of 1 means that the given inequality is true. If an entered relationship is false, then the returned value is 0.

Trace   *(page 66)*

PURPOSE: To move a tracing cursor to the displayed points of a graph and to report their coordinates.

EXAMPLE: To display the tracing cursor, press TRACE . Use the left and right arrow keys to move the tracing cursor along the graph.

If more than one graph is displayed, use the up and down arrow keys to move the tracing cursor from one graph to the other.

Window   *(page 63)*

PURPOSE: To set the minimum and maximum values on the x- and y-axes that are to be displayed in the viewing window, and to set the distance between tick marks on each axis.

EXAMPLE: To access the window settings menu, press Window . Use the up and down arrow keys to enter values.

$$\text{Xmin} = \text{minimum } x\text{-value that will be displayed}$$
$$\text{Xmax} = \text{maximum } x\text{-value that will be displayed}$$
$$\text{Xscl} = \text{distance between tick marks on the } x\text{-axis}$$
$$\text{Ymin} = \text{minimum } y\text{-value that will be displayed}$$
$$\text{Ymax} = \text{maximum } y\text{-value that will be displayed}$$
$$\text{Yscl} = \text{distance between tick marks on the } y\text{-axis}$$

```
WINDOW
 Xmin=0
 Xmax=10
 Xscl=2
 Ymin=-5
 Ymax=15
 Yscl=3
 Xres=1
```

In the sequence mode, the options in this menu will be different. See the "Sequence" keywords.

Y Screen   *(page 5)*

PURPOSE: To enter functions, which can then be, for example, evaluated or graphed.

EXAMPLE: Enter the function $y = 3x + 1$. To access the Y screen, press Y= .

In the sequence mode, the options in this menu will be different. See the "Sequence" keywords.

```
 Plot1 Plot2 Plot3
\Y₁■3X+1
\Y₂=
\Y₃=
```

Y Var   *(page 5)*

PURPOSE: To use a Y variable in place of the expression that it represents.

EXAMPLE: Suppose that $3x - 2$ has been entered as Y_1 on the Y screen and that 5 has been stored in X. On the home screen, the current value of the expression $3x - 2$ can be determined simply by entering Y_1.

```
Y₁
          13
```

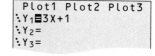

Zero   *(page 227)*

PURPOSE: To determine the zero of a function (or the *x*-intercept of a graph).

EXAMPLE: Display the graph of the function. The *x*-intercept must be visible in the viewing window.

1. Press [2nd] [CALC] [zero].
2. For the left bound, move the tracing cursor just to the left of the *x*-intercept. Then press [ENTER].
3. For the right bound, move the tracing cursor just to the right of the *x*-intercept. Then press [ENTER].
4. For "guess," press [ENTER].

The coordinates of the *x*-intercept are displayed. The *x*-coordinate is a zero of the function.

Zoom In   *(page 67)*

PURPOSE: To magnify a portion of a graph near the current location of the tracing cursor.

EXAMPLE: Display the graph and move the tracing cursor to the point where you wish to magnify the graph. Then press

[ZOOM] [Zoom In] [ENTER]

APPENDIX **C**

Calculator Programs

Friendly Window   *(page 64)*

PURPOSE: The *Decimal* (or *ZDecimal*) window settings are as follows:

Xmin: -4.7	Ymin: -3.1
Xmax: 4.7	Ymax: 3.1
Xscl: 1	Yscl: 1

This program multiplies the decimal window settings by a factor that the user selects. For example, if the user selects a factor of 3, then each of the decimal settings is multiplied by 3. The window is thereby expanded, but coordinates are still reported in tenths.

Program

Disp "SET FACTOR"	Ask the user to select a factor.
Input F	Accept the selected factor and store it in **F**.
$-4.7*F{\rightarrow}$Xmin	Multiply -4.7 by the factor and store the result in **Xmin**.
$4.7*F{\rightarrow}$Xmax	Multiply 4.7 by the factor and store the result in **Xmax**.
$-3.1*F{\rightarrow}$Ymin	Multiply -3.1 by the factor and store the result in **Ymin**.
$3.1*F{\rightarrow}$Ymax	Multiply 3.1 by the factor and store the result in **Ymax**.
$F{\rightarrow}$Xscl	Store the factor in **Xscl**.
$F{\rightarrow}$Yscl	Store the factor in **Yscl**.
Stop	End the program.

Quadratic Formula   *(page 95)*

PURPOSE: This program uses the Quadratic Formula to calculate the solution(s), either real numbers or complex numbers, of a quadratic equation that is written in standard form $ax^2 + bx + c = 0$.

Program

Disp "ENTER A,B,C"	Ask the user to enter the coefficients a, b, and c.
Input A	Accept the value of a and store it in **A**.
Input B	Accept the value of b and store it in **B**.
Input C	Accept the value of c and store it in **C**.
B²−4AC→D	Calculate the discriminant and store it in **D**.
If D<0	Determine whether the discriminant is negative.
Goto P	If it is, skip to the part of the program labeled **P**.
If D=0	Determine whether the discriminant is 0.
Goto Q	If it is, skip to the part of the program labeled **Q**.
(−B+√(D))/(2A)→M	Use the Quadratic Formula with the + symbol to calculate the first solution. Store the result in **M**.
(−B−√(D))/(2A)→N	Use the Quadratic Formula with the − symbol to calculate the second solution. Store the result in **N**.
Disp "TWO REAL ROOTS"	Display this message.
Disp M	Report the first solution.
Disp N	Report the second solution.
Stop	End the program.
Lbl Q	The program skips to here if the discriminant is 0.
(−B)/(2A)→M	Calculate the one solution (double root).
Disp "DOUBLE ROOT"	Display this message.
Disp M	Report the one solution.
Stop	End the program.
Lbl P	The program skips to here if the discriminant is negative.
(−B)/(2A)→R	Calculate the real part of the complex number solution and store the result in **R**.
√(−D)/(2A)→I	Calculate the imaginary part of the complex number solution and store the result in **I**.
Disp "COMPLEX ROOTS"	Display this message.
Disp "REAL PART"	Display this message.
Disp R	Report the real part of the complex number solution.
Disp "IMAGINARY PART"	Display this message.
Disp I	Report the imaginary part of the complex number solution.
Stop	End the program.

Vertex   *(page 176)*

PURPOSE: The graph of a quadratic function is a parabola. If the function has the form $f(x) = ax^2 + bx + c$, then this program will calculate the coordinates of the vertex of the parabola.

Program

Disp "ENTER A,B,C"	Ask the user to enter the coefficients a, b, and c.
Input A	Accept the value of a and store it in **A**.
Input B	Accept the value of b and store it in **B**.
Input C	Accept the value of c and store it in **C**.
$(-B)/(2A)\rightarrow X$	Calculate the x-coordinate of the vertex and store it in **X**.
$(4AC-B^2)/(4A)\rightarrow Y$	Calculate the y-coordinate of the vertex and store it in **Y**.
Disp "X-COORDINATE"	Display this message.
Disp X	Report the x-coordinate of the vertex.
Disp "Y-COORDINATE"	Display this message.
Disp Y	Report the y-coordinate of the vertex.
Stop	End the program.

ANSWERS to Odd-Numbered Exercises

Preamble

1. An algebraic expression contains a variable and a numerical expression does not.

3. b 5. f 7. a 9. -16 11. 3 13. 1

15. (a) (i) 0.43 (ii) 1.05 (iii) 7.91 (b) (i) $\frac{3}{7}$

17. $\frac{89}{420}$ 19. 0.79 21. True

23. False, $|-10| = 10$ 25. -5 27. 0

29. The expression would need to be entered only once.

31. $-5, 13.75, 32.95, 16.49$

33. $201.06, 105.10, 159.13, 465.46$

35. Use Table Start 0.2 and Δ Table 0.3.

37. (a) -1.25 (b) 0 (c) 0.75 (d) 1

39. (a) 0 (b) 0 (c) 10 (d) 18.8

Chapter R

■ Section R.1

1. Because the decimal number does not have a repeating block of numbers, it is not a rational number.

3. (a) $\sqrt{64}$ (b) $\sqrt{64}$ (c) $-15, \sqrt{64}$ (d) $\frac{\pi}{2}, \sqrt{10}, \sqrt[3]{36}$
   (e) $-15, 4.\overline{52}, -\frac{9}{5}, -8.2, \sqrt{64}, 6\frac{3}{7}$

5. $1.\overline{24}$ 7. 6.875

9. Irrational numbers are members of **R** but not of Q.

11. 1 13. $2.\overline{23}, \sqrt{5}, \frac{3\pi}{4}, \frac{26}{11}, \sqrt{15} - \sqrt{2}, \frac{1319}{500}$ 15. $-8, 2$

17. $1, 5$ 19. 11 21. 0.9 23. 0.86

25. The coordinates of points P and Q would be increased by 5, but the distance between P and Q would be unchanged.

27. -6 29. 9 31. $>$ 33. $>$

37. $2 - 5 \neq 5 - 2; 2 - (3 - 7) \neq (2 - 3) - 7$

39. Distributive Property

41. Associative Property of Multiplication

43. Distributive Property

45. Commutative Property of Addition

47. $\frac{1}{3x}$; Multiplicative Inverse Property

49. 1; Multiplicative Identity Property

51. $-x + 4$ 53. $x + (5 + 6)$ 55. $15a + 5b$

57. $-8x + 20y$ 59. $5(3x + 4y)$ 61. $a(3b + 1)$

■ Section R.2

1. Algebraic expressions are equivalent if they have the same value for all permissible replacements for the variable.

3. -5 5. 15 7. 1 9. 18 11. $\frac{13}{4}$ 13. 5

15. -3 17. -48

19. (a) The condition is true if n is odd.
    (b) Both 0^0 and division by 0 are undefined.

21. -128 23. -1 25. $\frac{1}{36}$ 27. $\frac{81}{16}$ 29. $-\frac{64}{3}$ 31. $-18x^6$

33. $64y^3$ 35. $-32a^{30}b^{15}$ 37. $\frac{1}{z^2}$ 39. $\frac{-3a^5}{4b^5}$ 41. $9x^{12}y^{14}$

43. $2x^{-1} > \frac{1}{2x}$ 45. $\frac{1}{x^5}$ 47. $\frac{3}{x^9}$ 49. $\frac{y^6}{5}$ 51. $\frac{b^6}{8c^9}$

53. $\frac{y^2}{3x^8}$ 55. $\frac{1}{9xy^5}$ 57. (a) \$184.80 (b) \$355.81

59. \$19,759.29 61. 30,000,000 63. 4360

65. 0.00000334 67. 0.00042 69. $7.03 \cdot 10^6$

71. $5.32 \cdot 10^5$ 73. $4.37 \cdot 10^{-4}$ 75. $2.0 \cdot 10^{-3}$

77. $1.426 \cdot 10^4$ 79. $2.06 \cdot 10^{38}$ 81. $6 \cdot 10^{-5}$

83. $\$5.00 \cdot 10^{11}$ 85. \$23,055.56

87. (a) Kilo means thousand and mega means million.
    (b) $6.18 \cdot 10^9$ (c) 927 tons

■ Section R.3

1. Statement (ii) is false, because, for example, $(4 - x^2) + (x^2 - x) = 4 - x$. Statement (iii) is false because a trinomial must have three nonzero terms.

3. Yes 5. No 7. Yes 9. 5 11. 0 13. 5

15. $x^2 - x - 3$ 17. $2x - y - 3$ 19. $6x^3 - 2$

21. $5a + b - 6c + 1$

23. Statement (i) is false: $(x + y)^2 = x^2 + 2xy + y^2$, not $x^2 + y^2$. Statement (iii) is false because, for example, $(x + 2)(x - 2) = x^2 - 4$.

25. $-6xy^5 + 10xy^4 - 12xy^3$ 27. $a^3 - 14a + 15$

29. $2x^3 - x^2 - 9x - 4$

31. $4x^2 + 13xy - 9x + 3y^2 - 5y + 2$

33. Because $(a + b)^2 = a^2 + 2ab + b^2, a^2 + b^2 = (a + b)^2 - 2ab = 7 - 2(-5) = 17$.

35. $y^2 + 2y - 35$ 37. $6x^2 + 13x + 6$ 39. $2y^2 - xy - 15x^2$

41. $10 + 9z - 9z^2$ 43. (iii) 45. $49x^2 - 4$ 47. $y^4 - 16z^2$

49. $25 - 10y + y^2$ 51. $a^2 + 6ab + 9b^2$

53. $y^3 - 6y^2 + 12y - 8$ 55. $8y^3 + 36y^2z + 54yz^2 + 27z^3$

57. $18x^4 + 3x^3 - 6x^2$ 59. $x^3 - 7x + 6$ 61. $x^2 + x + 1$

63. $y^2 - y + 1$ 65. $6x - 2$ 67. -35

69. $[(t + 3)(t + 3)](3 - t)$ or $[(3 + t)(3 - t)](t + 3) = 27 + 9t - 3t^2 - t^3$

71. $4x^2 + 4xy + y^2 - 25$ 73. $x^4 - 16$

75. $a^2 + 2ab + b^2 + 6a + 6b + 9$

■ *Section R.4*

1. Choose the smallest exponent on a and the smallest exponent on b. The GCF is a^3b^2.

3. $3(2a + 3)$ **5.** $y^2(y - 1)$ **7.** $(x - 2)(3x + y)$

9. $(6x + 1)(6x - 1)$ **11.** $(7 + 5c)(7 - 5c)$

13. $(x - 3)^2$ **15.** $(5 + 3z)^2$

17. Produce a table of values for $36x^2 - 1$ and its factored form. The tables should be the same.

19. $(4 - w)(16 + 4w + w^2)$ **21.** $(3x + 2)(9x^2 - 6x + 4)$

23. Each product is $12 - x - x^2$. Both are correct.

25. $(y - 3)(y + 2)$ **27.** $(2y + 3)(6y - 5)$

29. $(4 - x)(1 + x)$ **31.** $(8x - 3)(x - 1)$

33. $(x^2 + 3)(x^2 + 2)$

35. Because the middle term is $4x$, not $8x$, (i) is false. Because $x^2 - 4$ can be factored, (ii) is false.

37. $(a + 2b)(x - y)$ **39.** $(x^2 + 2)(x - 5)$

41. $(b + 7)(b + 6)$ **43.** $(11m + 2n)(11m - 2n)$

45. $(x + 5)(5x^2 + 1)$ **47.** $10y^{30}(4y^{10} - 3)$

49. $(4a^2 - 3b)(5a^2 + 4b)$ **51.** $(4x - 21y)(x - 2y)$

53. $(9 - 2y)^2$ **55.** $4x(x + 1)(x^2 - x + 1)$

57. $-x(3x - 4)(x + 2)$ **59.** $(9a^2 + 4)(3a + 2)(3a - 2)$

61. $(x + 3y)(x - 10y)$ **63.** $(1 + 2x^2)(1 - 2x^2 + 4x^4)$

65. $(2a + 7)^2$ **67.** Prime   **69.** $(x + 3)(x + 2)(x - 2)$

71. $(x - 3)(y + 4z)$ **73.** $(2t^3 + 3)(t^3 + 4)$

75. $-3x(x - 3)^2$ **77.** $2(7 + 2y)(7 - 2y)$

79. $x^3y^2(2x - y)(x + 3y)$ **81.** $2y^2(2 - 3y)(4 + 6y + 9y^2)$

83. $(16 - 3x^2)^2$ **85.** $(2y + z + 10)(2y + z - 10)$

87. $2(3y - 2)(3y - 1)$ **89.** $(x + 2)^3(x + 3)(x + 1)$

91. $(x + 3 + y)(x + 3 - y)$

95. (a) $\left(x + \sqrt{10}\right)\left(x - \sqrt{10}\right)$ (b) $\left(\sqrt{15} + y\right)\left(\sqrt{15} - y\right)$
    (c) $\left(x\sqrt{6} + 1\right)\left(x\sqrt{6} - 1\right)$

■ *Section R.5*

1. The tables are the same except for $x = 1$. The table for $x + 1$ indicates that the expression has a value of 2 when x is 1. Because $\dfrac{x^2 - 1}{x - 1}$ is not defined for $x = 1$, the table for this fraction displays an error message when x is 1.

3. $-5, 4$ **5.** $-5, 2$ **7.** $-3, -1, 1$ **9.** $\dfrac{4}{c - 2}$

11. $-(x + 3)$ **13.** $\dfrac{x^2 + 2x + 4}{x + 4}$ **15.** $\dfrac{m - 4}{2m - 5}$ **17.** $\dfrac{x - 2}{x + 3}$

19. $\dfrac{6(a + 1)}{a(2a - 1)}$ **21.** $\dfrac{x - 2}{x + 3}$ **23.** $\dfrac{1}{x^2 - 16}$ **25.** $\dfrac{10}{3x}$

27. $\dfrac{x + 4}{x + 1}$ **29.** $\dfrac{4}{a + 1}$ **31.** $\dfrac{2}{5}$ **33.** $(x + 3)(x - 1)$

35. The tables would not be the same because the expressions are not equivalent. The student made a sign error. $\dfrac{2}{x + 1} - \dfrac{x + 3}{x + 1} = \dfrac{2 - x - 3}{x + 1} = \dfrac{-x - 1}{x + 1} = -1$

37. $\dfrac{3x - y}{x - y}$ **39.** $\dfrac{x + 2}{x}$ **41.** 6   **43.** $\dfrac{n - 2}{n + 2}$

45. $(x + 3)(x - 3)(x + 2)$ **47.** $(x + 1)(x - 1)(x^2 + x + 1)$

49. $(x + 1)^3(x - 1)$ **51.** $\dfrac{3x - 2}{x - 5}$ **53.** $\dfrac{2 - b}{a - b}$

55. $\dfrac{x^2 + 7x + 6}{(x + 3)(x + 5)}$ **57.** $\dfrac{4mn}{(m - n)(m + n)}$ **59.** $\dfrac{x + 8}{x + 5}$

61. $\dfrac{x}{(x - 2)(x - 3)}$ **63.** 0   **65.** $\dfrac{-4}{n - 2}$ **67.** $\dfrac{5}{x + 5}$

69. $\dfrac{1 + a^2}{a}$ **71.** $\dfrac{1}{x(x + h)}$ **73.** $\dfrac{1 + x}{1 - x}$

■ *Section R.6*

1. The number 16 has two square roots, -4 and 4, but 64 has only one cube root, 4.

3. 7   **5.** $5|y^3|$ **7.** -2 **9.** $-2x^5$ **11.** $|x - 5|$

13. $5\sqrt{3}$ **15.** $2\sqrt[3]{5}$ **17.** $x^2y^3\sqrt{x}$ **19.** $2x^5\sqrt{7x}$

21. $-2x\sqrt[3]{3x}$ **23.** $3\sqrt{x^2 + 4}$ **25.** $m < n$

27. $-12a\sqrt{3}$ **29.** $7a^2b^4\sqrt{2}$

31. Produce a table of values for the product and for the simplified expression. The tables should be the same for all nonnegative values of the variable.

33. $\dfrac{2x^5\sqrt{3}}{y^3}$ **35.** $\dfrac{2x^2\sqrt{2x}}{y}$ **37.** $\dfrac{3}{5t^5}$ **39.** $\dfrac{t}{2}$ **41.** $4x\sqrt{7}$

43. $8\sqrt{5}$ **45.** $-4t\sqrt{3}$ **47.** $30 - 3\sqrt{6}$

49. $\sqrt{10a} - 2\sqrt{2} + 5a - 2\sqrt{5a}$

51. The expression $\dfrac{1}{\sqrt{x} + 1}$ is defined for all nonnegative values of x. However, the rationalized expression $\dfrac{\sqrt{x} - 1}{x - 1}$ is not defined for $x = 1$.

53. $4\sqrt{3}$ **55.** $\dfrac{5\sqrt{3}}{6}$ **57.** $\dfrac{7\sqrt[3]{x^2}}{x}$ **59.** $6 - 2\sqrt{6}$

61. $\dfrac{5 + 2\sqrt{5x} + x}{5 - x}$ **63.** $\dfrac{7\left(\sqrt{a} - a\right)}{1 - a}$

65. (i) -1;   (iv) -1;   (ii) and (iii) are even roots of negative numbers.

67. $\frac{1}{5}$ **69.** $-\frac{1}{3}$ **71.** $\frac{16}{81}$ **73.** 32

75. (a) $y^{1/3}$ (b) $\sqrt[4]{2a}$ (c) $2\sqrt[4]{a^3}$ (d) $-x^{2/3}$

77. y **79.** $\dfrac{n^8}{m^4}$ **81.** x^2 **83.** $x^{1/3}y$

85. The calculator result is 256: $\sqrt[0.5]{16} = 16^{1/0.5} = 16^2 = 256$.

87. $\sqrt[3]{y}$ **89.** $2\sqrt[3]{50}$ **91.** $\sqrt[12]{6}$ **93.** $\sqrt[6]{a^5}$ **95.** $3\sqrt[6]{243}$

97. 8.82 **99.** -1.51 **101.** 0.98 **103.** $\dfrac{y + 3\sqrt{y}}{y}$

105. $\dfrac{3\sqrt{2}}{2}$ **107.** 12.1 **109.** 3.2 **111.** 7.2 miles **113.** 2.5

■ **Chapter Review Exercises**

1. (a) $\sqrt{9}$ **(b)** $0, \sqrt{9}$ **(c)** $0, -6, -\sqrt{25}, \sqrt{9}$

   (d) $\dfrac{2}{3}, 0, -6, -\sqrt{25}, 0.\overline{6}, -0.29, \sqrt{9}$ **(e)** $\dfrac{\pi}{3}$ **3.** $0.0\overline{3}$

5. $-\sqrt{3}, \sqrt{3}$ **7. (a)** $>$ **(b)** $<$ **(c)** $=$

9. (a) $y^2 + 4$; Multiplicative Inverse Property

   (b) $-$; Property of the Opposite of a Difference

11. (a) 12 **(b)** -1 **13. (a)** 70 **(b)** $\frac{4}{5}$

15. The expression -5^{-2} means $-1 \cdot 5^{-2} = -1 \cdot \dfrac{1}{5^2} =$

   $-\dfrac{1}{25}$, whereas $(-5)^{-2}$ means $\dfrac{1}{(-5)^2} = \dfrac{1}{25}$.

17. (a) $\dfrac{1}{x}$ **(b)** $\dfrac{y^8}{9x^2}$ **19.** $1.2 \cdot 10^{10}$

21. Because a monomial cannot have a negative exponent, (iv) is not a monomial.

23. $3xy^2 - 2x^2y$ **25.** $a^2 - 4ab + 5b^2$ **27.** $x^3 - xy^2 - 6y^3$

29. (a) $x^2y^2 - 9$ **(b)** $4x^2 - 20x + 25$

31. Each of the factors contains a common factor. The complete factorization is $4(x + 2)(x - 2)$.

33. $-2(2x^2 + x - 4)$

35. (a) $(2x - y)^2$ **(b)** $2c(a + 5)^2$

37. (a) $(a + 10)(a^2 - 10a + 100)$

   (b) $(xy + 1)(x^2y^2 - xy + 1)$

39. (a) $(4x - 5)(x + 1)$ **(b)** $(3x - 2)(4x + 5)$

41. (a) 1 **(b)** 0 **(c)** $-1, -2$ **43.** $\dfrac{1}{2(x - 1)}$

45. (a) $\dfrac{1}{x + 3}$ **(b)** $-\dfrac{x + 1}{2}$ **47. (a)** $\dfrac{1}{2x + 1}$ **(b)** 1

49. $\dfrac{x}{x - 1}$ **51. (a)** $-7, 7$ **(b)** 7

53. (a) $5x^5y^2\sqrt{2x}$ **(b)** $\dfrac{2x^4\sqrt{x}}{y^2}$ **55.** $\dfrac{2(3 - \sqrt{x})}{9 - x}$

57. (a) -4 **(b)** $\frac{1}{4}$ **(c)** $\frac{1}{4}$ **59.** 76.18

Chapter 1

■ **Section 1.1**

1. Any point $P(2, y)$ is a point of the vertical line located 2 units to the right of the y-axis.

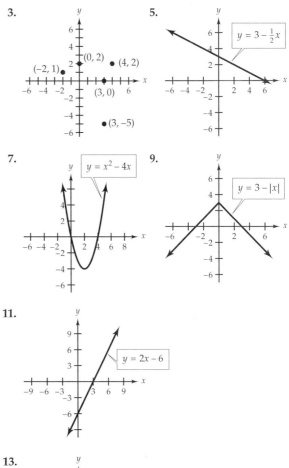

3.

5. $y = 3 - \frac{1}{2}x$

7. $y = x^2 - 4x$

9. $y = 3 - |x|$

11. $y = 2x - 6$

13. $y = x^2 + 3x$

15. Decimal **17.** Integer **19.** Standard

21. When x is 5, the value of the expression is 3.

23. (a) 1 **(b)** 15 **(c)** -6 **(d)** -1

25. (a) 0.75 **(b)** 2 **(c)** -0.24 **(d)** -0.09

27. (B) Xmin -3, Xmax 5, Xscl 1
    Ymin -4, Ymax 6, Yscl 1

29. (A) Xmin -6, Xmax 6, Xscl 3
    Ymin -6, Ymax 6, Yscl 3

31. (F) Xmin -4, Xmax 12, Xscl 4
    Ymin -2, Ymax 6, Yscl 2

33. (a) 3.5 **(b)** $(3.5, 0)$ **35. (a)** $-1, 4$

   (b) $(-1, 0), (4, 0)$ **37. (a)** -2 **(b)** $(-2, 0)$

39. The first coordinate is the value of the variable for which the expressions have the same value.

41. $(2, 0), (0, -1)$ **43.** $(-8, 0), (8, 0), (0, -16)$

45. $(-12, 0), (12, 0), (0, 12)$ **47.** $(-2, 0), (0, 0), (2, 0)$

49. $(15, 0), (0, 5)$ **51.** $(-3.3, 0), (4.3, 0), (0, -14)$

53. $(2, \frac{2}{3})$ **55.** $(-4, 4), (3, 11)$ **57.** $(-4, 7), (24, 21)$

59. $(-0.7, 10.3)$ **61.** $(0.3, 4.7), (11.7, -6.6)$

■ *Section 1.2*

1. The linear equation $Ax + B = 0$ has one solution, $-\dfrac{B}{A}$.

3. 4   **5.** $-9, 12$ **7.** 1   **9.** 2   **11.** 1.3   **13.** $-1.6, 0.6$

15. The solution of (ii) and (iii) is $-\frac{8}{3}$, so they are equivalent. The solution of (i) is $\frac{8}{3}$, so it is not equivalent to the other two.

17. -14 **19.** 1   **21.** 0   **23.** -24 **25.** 2   **27.** $-\frac{1}{2}$

29. $-\frac{109}{83}$ **31.** $\frac{5}{3}$ **33.** $-\frac{87}{14}$ **35.** $-\frac{15}{11}$

37. **(a)** The graphs coincide.

(b) The graphs never intersect.

39. Contradiction   **41.** Conditional   **43.** Identity

45. Ø   **47.** -1 **49.** R

■ *Section 1.3*

1. Statement (i) is false. Multiplication by a negative number reverses the inequality symbol: $ac < bc$.

3. $(-\infty, 5)$ **5.** $x > 1$ **7.** $[-8, \infty)$ **9.** $(-5, -2)$

11. $-5 \le x < 3$ **13.** $[0, 1)$ **15.** $t \ge 4$ **17.** $x < 0$

19. $t \le -3$ **21.** R

23. **(a)** The first solution set is Ø and the second solution set is $(-\infty, -5) \cup (3, \infty)$.

(b) The first solution set is $(-5, 3)$ and the second solution set is **R**.

25. $[-4, 5]$ **27.** $(-\infty, -1) \cup (5, \infty)$ **29.** $(-\infty, -2)$

31. $(-\infty, -3] \cup [\frac{1}{2}, \infty)$ **33.** $[-4, 2)$ **35.** $(-6, -\frac{2}{3})$

37. The solution of (i) is $[3, \infty)$ and the solution of (ii) is $(-\infty, 3]$.

39. $-3, 4$ **41.** $\frac{1}{3}, 7$ **43.** Ø   **45.** $-\frac{7}{3}$

47. $(-\infty, -\frac{5}{2}) \cup (\frac{11}{2}, \infty)$ **49.** $(-\infty, -3] \cup [2, \infty)$

51. $[-\frac{11}{3}, \frac{19}{3}]$ **53.** Ø   **55.** $[-7, 3]$ **57.** Ø

59. $[2, \infty)$ **61.** $(-4.5, 3.5)$

63. The solution of (i) is **R** and the solution of (ii) is Ø.

■ *Section 1.4*

1. All of these methods can be used.   **3.** $-7, 3$ **5.** $0, 7$

7. $-\frac{5}{4}, 4$ **9.** 3   **11.** $\pm\sqrt{3} \approx \pm1.73$ **13.** $\pm\sqrt[6]{42} \approx \pm1.86$

15. No real solution   **17.** $-3 \pm \sqrt{10} \approx -6.16, 0.16$

19. The discriminant $b^2 + 4$ is never negative. The equation has real number solutions for any value of b.

21. $-3 \pm \sqrt{5} \approx -5.24, -0.76$ **23.** No real solution

25. $\dfrac{-1 \pm \sqrt{2}}{3} \approx -0.80, 0.14$ **27.** $\dfrac{1 \pm \sqrt{61}}{6} \approx -1.14, 1.47$

29. $\dfrac{-2 \pm \sqrt{5}}{3} \approx -1.41, 0.08$ **31.** No real solution

33. $\dfrac{-1 \pm \sqrt{7}}{3} \approx -1.22, 0.55$ **35.** $\dfrac{-3 \pm \sqrt{149}}{10} \approx -1.52, 0.92$

37. For a quadratic inequality, either both endpoints are solutions or neither endpoint is a solution.

39. $(-\infty, -5] \cup [3, \infty)$ **41.** $(-\infty, -3) \cup (-\frac{2}{3}, \infty)$ **43.** R

45. $[-3.56, 0.56]$ **47.** One rational   **49.** No real solution

51. Two rational   **53.** Two irrational   **55.** ±2

57. $\pm2, \pm\sqrt{5} \approx \pm2.24$ **59.** $\pm\sqrt{5} \approx \pm2.24$

61. Determine the intervals for which the graph is above the x-axis.

63. $\sqrt[3]{4} \approx 1.59$ **65.** $\sqrt[3]{-50} \approx -3.68$ **67.** $0, \pm\frac{5}{2}$

69. $-3, \pm2$ **71.** ±2 **73.** $(-\frac{5}{2}, 0) \cup (\frac{5}{2}, \infty)$

75. $(-\infty, -3] \cup [-2, 2]$ **77.** $\sqrt{2} \pm 1 \approx 0.41, 2.41$

79. $y = -1 \pm \sqrt{2 + x}$

■ *Section 1.5*

1. In (i), we write each term with the LCD. In (ii), we multiply both sides of the equation by the LCD to clear the fractions.

3. $-5, 1$ **5.** 2   **7.** $-2, \frac{5}{2}$ **9.** $-4, -1$ **11.** $-\frac{15}{4}$

13. -2 **15.** 4

17. In (i), the principal square root is never negative. In (ii) the domains of the two radical expressions have no elements in common.

19. $-\frac{20}{3}$ **21.** 9   **23.** -7 **25.** 4   **27.** $5, 3$ **29.** 8   **31.** $\frac{81}{16}$

33. $\pm4\sqrt{2} \approx \pm5.66$ **35.** $-13, 3$ **37.** $b_2 = \dfrac{2A}{h} - b_1$

39. $P = \dfrac{A}{1 + rt}$ **41.** $H = \dfrac{A - 2LW}{2L + 2W}$ **43.** $-27, 1$

■ *Chapter Review Exercises*

1. Use the Window menu. Xmin would be negative and Ymax positive.

3.

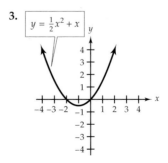

5. Integer   **7.** $(-5, 0), (2, 0)$

9. $(0, 6), (-6, 0), (20, 0)$ **11.** 1

13. **(a)** None   **(b)** One   **(c)** Infinitely many

15. $-\frac{3}{4}$ **17.** 2   **19.** 2.1875   **21.** $\left[-\frac{9}{2}, \infty\right)$

23. The double inequality is equivalent to the conjunction $-3 < 2x + 1$ and $2x + 1 < 5$.

25. $[-3, 4]$ **27.** 1, 6   **29.** $(0, 8)$

31. The terms $2x^2$ and $3x$ cannot be combined.

33. 3   **35.** **(a)** ± 2 **(b)** $-\sqrt[3]{9} \approx -2.08$

37. $\dfrac{-1 \pm \sqrt{19}}{3} \approx -1.79, 1.12$

39. **(a)** $(-\infty, -3) \cup (1, \infty)$ **(b)** $[-3, 1]$

41. An apparent solution may be extraneous.

43. 1   **45.** -3 **47.** 5   **49.** 4.008, 3.992

Chapter 2

■ Section 2.1

1. The domain is the set of all first coordinates of the ordered pairs of the relation, and the range is the set of all second coordinates.

3. Domain = $\{-4, 0, 3, 5\}$
Range = $\{-1, 5, 6\}$
Function

5. Domain = $\{0, 1, 4, 9\}$
Range = $\{-3, -2, -1, 0, 1, 2, 3\}$

7. If a vertical line intersects the graph at more than one point, then two pairs have the same first coordinate and the graph does not represent a function.

9. Yes   **11.** No   **13.** Yes   **15.** Yes

17. Domain: $[-2, \infty)$; range: **R**; no

19. Domain: **R**; range: **R**; yes

21. Domain: **R**; range: $(-3, \infty)$; yes

23. Domain: $(-\infty, 4]$; range: $(-\infty, 3]$; yes

25. No; $(3, 4)$ and $(3, 5)$ satisfy the relation and have the same first coordinate.

27. **(a)** 5   **(b)** -7 **(c)** 3   **(d)** $4a - 3$

29. **(a)** -2 **(b)** 1   **(c)** $-t^2 - 4t - 2$ **(d)** 1.75

31. **(a)** Not a real number   **(b)** 0   **(c)** 4   **(d)** $3\sqrt{t^2 - 1}$

33. The value of the function evaluated at 2 is -6.

35. **(a)** -1.225 **(b)** 1.4   **(c)** -3.7

37. **(a)** -1.5 **(b)** 1.6   **(c)** -0.8 **39.** 1, 5   **41.** 3

43. **(a)** 10   **(b)** -2 **(c)** $t^2 + 3t$

45. **(a)** 2   **(b)** -2 **(c)** $2t^2 - 3$ **47.** -1

49. 4   **51.** $6x + 3h - 1$

53. Domain = **R**
Range = **R**

55. Domain = **R**
Range = $(-\infty, 5]$

57. Domain = $[-4, 4]$
Range = $[-4, 0]$

59. **R**   **61.** $[-4, \infty)$ **63.** $\{x \mid x \neq 0, 1\}$

65. The two pairs $(5, 8)$ and $(5, 25)$ belong to the relation. Thus the relation is not a function.

67. **(a)** 0   **(b)** 0   **(c)** 9   **(d)** -5

69. **(a)** -2 **(b)** 4   **(c)** 6   **(d)** 8

71. **(a)** Because $(1, 75)$ and $(1, 80)$ are both pairs of the relation, the relation is not a function.

   (b) Each score corresponds to only one student number. The set is a function.

73. **(a)** 14 gallons   **(b)** 1990

75. **(a)** $x + 1, x(x + 1)$ **(b)** $C(x) = 45x(x + 1)$
   (c) $\{1, 2, 3, 4, 5\}$ **(d)** $900

77. **(a)** $R(n) = 1.1 + 0.2n$ **(b)** $8.8 billion   **79.** **(b)** 1976

■ Section 2.2

1. **(a)** The slope is undefined.   **(b)** The slope is 0.

3. $-\frac{1}{2}$ **5.** 0   **7.** $\frac{5}{3}$ **9.** $(0, -9), \left(\frac{9}{2}, 0\right)$ **11.** $(0, -12), (3, 0)$

13. $f(x) = x - 5; 1; (0, -5)$ **15.** $f(x) = -\frac{4}{3}x + 1; -\frac{4}{3}; (0, 1)$

17. $f(x) = \frac{1}{2}; 0; \left(0, \frac{1}{2}\right)$

19.

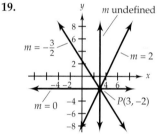

21. The range of a linear function is **R**, whereas the range of a constant function $y = b$ is a single number, $\{b\}$.

23. $y = 3x - 7$ **25.** $x = -6$ **27.** $f(x) = \frac{1}{2}x + 1$

29. $f(x) = -3$ **31.** $y = -\frac{3}{5}x + \frac{3}{5}$ **33.** $y = 7$

35. $f(x) = -2x - 1$ **37.** $f(x) = -3$

39. **(b)** Not collinear   **41.** 3   **43.** -4

45. The slope and the rate of change are the same.

47. a decreased by 10   **49.** b decreased by 4

51. -1 **53.** $-\frac{3}{2}$ **57.** $f(x) = \frac{41}{350}x + \frac{239}{70}$

59. $f(x) = -\frac{19}{15}x + \frac{268}{3}$

61. **(a)** $f(h) = 300$; constant function

   (b) $g(h) = 40h + 100$; the slope is the hourly rate, and the y-intercept corresponds to the annual charge.

63. (a) $N(x) = 25 - x$

(b) The y-intercept, $(0, 25)$, corresponds to no workers belonging to the union, and the x-intercept, $(25, 0)$, corresponds to all workers belonging to the union.

65. (a) $C(x) = 5000 + 2x$

(b) The y-intercept represents the fixed cost.

(c) The slope is the variable cost.

67. (a) 16π (b) $D(t) = \dfrac{4\pi t}{15}$ (c) 29.32 inches

69. (a) $f(t) = 840t + 20{,}000$

(b) The purchasing power would be $2600 less than that in 1985.

71. (a) $P(x) = \frac{5}{2}x$

(b) The slope, $\frac{5}{2}$, is the profit per subscriber.

(c) The y-intercept, $(0, 0)$, indicates that if the newspaper has no subscribers, the profit is 0.

73. (a) $y = 75x$ (b) 32

■ **Section 2.3**

1. Because the three points are not collinear, the linear regression line could contain at most one of the points.

3. Yes   **5.** No

7. (a) $y = -x + 3$ (b) $y = -x + 3$

(c) The equations are the same.

(d) The slope is negative.

9. (a) Yes; $y = \frac{7}{5}x + 4$

(b) $r = 1$; the points are collinear, and the slope is positive.

(c) $y = 1.4x + 4$

11. (a) $y = -1.04x + 18.43$

(b) The negative slope indicates that the number of cases is decreasing.

(c) 12.2

13. Because the points belong to a vertical line, the slope is undefined, and so the calculator gives an error message.

15. Not linear

17. (a) $y = 3.05x + 1.6$

(b) Because $r = 0.97$, a linear model is reasonable.

19. (a) $y = -0.19x + 43.65$ (b) 41.75 seconds

(c) Because $r = -0.93$, a linear model is reasonable.

(d) The winning times are decreasing.

21. (a) (i) $y = 25x + 71$ (ii) $y = 30x + 66$
(iii) $y = 35x + 51$

(b) (ii)

(c) $y = 30x + 64.33$; the regression equation

23. (a) $y = 7.55x + 101.05$
Because $r = 0.98$, the model is a good fit.

(b) According to the model, the number of calls in 1992 was about 101 million.

(c) Because $r = 0.99$, the prediction is reasonable.

■ **Section 2.4**

1. Two lines that are not in the same plane may not be parallel even though they do not intersect.

3. Perpendicular   **5.** Neither   **7.** Parallel   **9.** Parallel

11. Neither   **13.** Parallel

15. (a) $f(x) = -x - 3$ (b) $f(x) = x - 7$

17. (a) $f(x) = \frac{3}{4}x + 4$ (b) $f(x) = -\frac{4}{3}x + 4$

19. (a) $f(x) = 6$ (b) Not a function; the equation is $x = 4$.

21. $y = -2x + 2$ **23.** $y = -3x + 13$ **25.** $y = x$

27. (a) -6 (b) $\frac{3}{2}$

29. Both (i) and (ii) are possible. Because the endpoints have integer coordinates, the coordinates of the midpoint must be rational numbers, so (iii) is not possible.

31. (a) 13   (b) $\left(\frac{1}{2}, -2\right)$

33. (a) $\sqrt{145} \approx 12.04$ (b) $\left(\frac{1}{2}, -4\right)$

35. (a) $2\sqrt{a^2 + b^2}$ (b) (a, b)

37. $(5, -12)$ **39.** $(-3, 4)$ **41.** $(1, 3)$ **45.** Yes   **47.** No

49. $A = 25\pi \approx 78.54, C = 10\pi \approx 31.42$

51. $A = 21, P = 20$ **53.** $A = 20, P \approx 21.78$

55. $A = 20, P \approx 20.47$ **57.** $y = -x - 1$

59. $y = -\frac{5}{12}x + \frac{31}{8}$ **65.** $3\sqrt{5} \approx 6.71$ **67.** $\dfrac{\sqrt{5}}{5} \approx 0.45$

69. The midpoint (1993, 7.2) indicates that in 1993 the median age of a car was 7.2 years.

71. 2383 feet

■ **Section 2.5**

1. The x-intercepts correspond to the zeros of the function.

3. The graph of f is the graph of $y = x^2$ reflected across the x-axis and shifted upward 9 units.

5. The graph of f is the graph of $y = x^2$ shifted to the left 2 units and downward 5 units.

7. The graph of f is the graph of $y = x^2$ reflected across the x-axis, and then shifted to the left 4 units and upward 2 units.

9. B   **11.** D   **13.** C   **15.** $V(0, -8); x = 0$

17. $V(4, 1); x = 4$ **19.** $V(-1, 3); x = -1$ **21.** $\left(\frac{2}{3}, 0\right), (0, 4)$

23. $(-1, 0), (-7, 0), (0, 7)$ **25.** $(1, 5); x = 1$

27. $\left(\frac{1}{4}, \frac{7}{8}\right)$; $x = \frac{1}{4}$

29. The vertex is the highest (or lowest) point of the graph, so the y-coordinate is the largest (or smallest) value of the elements of the range of the function.

31. Maximum: $\frac{33}{4}$; range: $\left(-\infty, \frac{33}{4}\right]$

33. Minimum: $-\frac{49}{4}$; range: $\left[-\frac{49}{4}, \infty\right)$

35. Maximum: $\frac{227}{8}$; range: $\left(-\infty, \frac{227}{8}\right]$ **37.** $y = (x + 3)^2 - 2$

39. $y = 2(x + 2)^2 - 8$ **41.** $y = -2(x + 4)^2 - 5$

43. $[-3, \infty)$; 2   **45.** $(-\infty, -1]$; 0   **47.** $[0, \infty)$; 1   **49.** 0

51. 2   **53.** 2   **55.** 3   **57.** 2   **59.** $k = -\frac{10}{3}, 2$

61. If the discriminant is positive, the function has two zeros and the graph has two x-intercepts. If the discriminant is zero, the function has one zero and the graph has one x-intercept. If the discriminant is negative, the function has no zeros and the graph has no x-intercept.

63. **(a)** The store's highest sales were $9000 on December 13.

   (b) $6000

   (c) November 28, December 27

65. 64 feet, 4 seconds   **67.** 47 million travelers in 1993

69. **(a)** $R(x) = 45x - 0.5x^2$

   (b) $P(x) = -0.5x^2 + 40.5x - 500$

   (c) $320.13   **(d)** 40 or 41   **(e)** 15

71. 1875 square feet

■ **Section 2.6**

1. $y = 2x^2 - 15x + 27$ **3.** $y = 0.4x^2 - 1.2x - 5$

5. Quadratic   **7.** Linear

9. $y = 0.63x^2 - 5.89x + 24.71$; $r^2 = 0.95$

11. Not quadratic

13. $y = -21.88x^2 + 242.43x - 469.65$; $r^2 = 0.97$

15. **(a)** Negative   **(b)** (1991, 43)

   (c) $y = -0.34x^2 + 3.48x + 33.2$; $V(5, 42)$; the coordinates indicate the year in which the rates were highest and the rate for that year.

17. **(a)** Quadratic   **(b)** $y = -193x^2 + 2900x - 6800$

   (c) 1996, 1997

19. **(a)** $y = -0.04x^2 + 0.51x + 11.03$

   (b) Because $r^2 = 0.87$, the model is reasonable.

   (c) (24, 0); in 2003, funding will be 0.

21. **(a)** $y = 26.6x + 201$; $r^2 = 0.75$;
   $y = -9.57x^2 + 84.03x + 134$; $r^2 = 0.89$

   (b) Quadratic

■ **Section 2.7**

1. The first coordinate of a y-intercept is 0, and the second coordinate of an x-intercept is 0.

3. $(0, -7)$, $\left(\frac{7}{2}, 0\right)$ **5.** $(0, -2)$

7. $(0, -6)$, $(-2, 0)$, $(-1, 0)$, $(3, 0)$

9. $(0, -1)$, $(-3, 0)$, $(3, 0)$

11. The function g has a relative minimum at $x = -1$.

13. Increasing: $(-2, 2)$; decreasing: $(2, 4)$; constant: $(-4, -2)$

15. Intercepts: $(0, 3)$, $(1, 0)$, $(5, 0)$; relative minimum: $(3, -2)$; increasing: $(3, \infty)$; decreasing: $(-\infty, 3)$

17. Relative minimum: $(-1, -9)$; increasing: $(-1, \infty)$; decreasing: $(-\infty, -1)$

19. Relative maximum: $(-1, 20)$; relative minimum: $(5, -88)$; increasing: $(-\infty, -1)$, $(5, \infty)$; decreasing: $(-1, 5)$

21. Relative maxima: $(-1.41, 2)$, $(1.41, 2)$; relative minimum: $(0, -2)$; increasing: $(-\infty, -1.41)$, $(0, 1.41)$; decreasing: $(-1.41, 0)$, $(1.41, \infty)$

23. Relative maximum: $(1, 3)$; increasing: $(-\infty, 1)$; decreasing: $(1, \infty)$

25. Relative maxima: $(-0.16, 3.08)$, $(6.16, -0.08)$; increasing: $(-\infty, -0.16)$, $(6.16, \infty)$; decreasing: $(-0.16, 6.16)$

27. Relative maxima: $(-4, 0)$, $(2.83, 8)$; relative minima: $(4, 0)$, $(-2.83, -8)$; increasing: $(-2.83, 2.83)$; decreasing: $(-4, -2.83)$, $(2.83, 4)$

29. An even function is symmetric with respect to the y-axis. An odd function is symmetric with respect to the origin.

31. **(a)** $(2, -3)$, $(-3, 1)$, $(2, 1)$ **(b)** $(-2, 3)$, $(3, -1)$, $(-2, -1)$

33. **(a)** $(1, 4)$, $(3, 2)$, $(4, -6)$, $(-6, 4)$

   (b) $(-1, -4)$, $(-3, -2)$, $(-4, 6)$, $(6, -4)$

35. **(a)**   **(b)**

   (c)

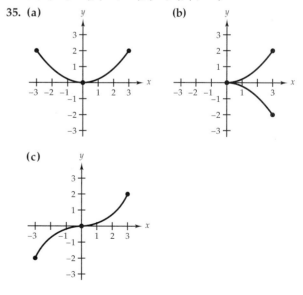

37. (a) **(b)**

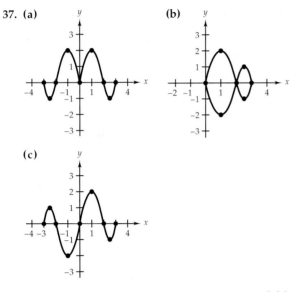

(c)

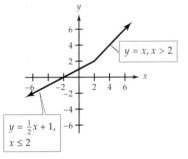

39. Odd **41.** Even **43.** Neither **45.** Even **47.** Odd

49. Yes **51.** x-axis **53.** Origin **55.** x-axis, y-axis, origin

57. Both f and h are always increasing. Function g has a relative minimum at $(0, 0)$.

59. $f(x) = \begin{cases} 3 & x \le 0 \\ x & x > 0 \end{cases}$ **61.** $f(x) = \begin{cases} |x| & x \le 0 \\ \sqrt{x} & x > 0 \end{cases}$

63.

$y = x, x > 2$

$y = \frac{1}{2}x + 1, x \le 2$

65.

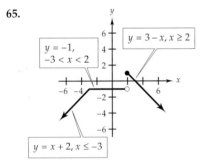

$y = -1, -3 < x < 2$

$y = 3 - x, x \ge 2$

$y = x + 2, x \le -3$

67.

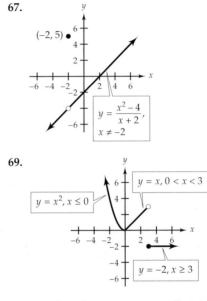

$(-2, 5)$

$y = \dfrac{x^2 - 4}{x + 2},\ x \ne -2$

69.

$y = x, 0 < x < 3$

$y = x^2, x \le 0$

$y = -2, x \ge 3$

71. (a) The relative maximum, $(5.6, 1.7)$, approximates the data for 1991.

(b) Increasing, decreasing

73. $y = x$ **75.** $y = \dfrac{1}{x}$

77. (a) $C(x) = \begin{cases} 0.32 & 0 < x \le 1 \\ 0.55 & 1 < x \le 2 \\ 0.78 & 2 < x \le 3 \\ 1.01 & 3 < x \le 4 \\ 1.24 & 4 < x \le 5 \\ 1.47 & 5 < x \le 6 \end{cases}$

(b) (i) $C(0.6) = 0.32$ (ii) $C(2.9) = 0.78$
(iii) $C(5.1) = 1.47$

■ **Section 2.8**

1. A rigid transformation preserves the shape of the graph, whereas a nonrigid transformation distorts the graph.

3. (a) The graph of g is steeper than the graph of f.

(b) The graph of h is the graph of f shifted vertically.

(c) $y = 3x - 3$

5. (a) The graph of g is the graph of f shifted vertically.

(b) The graph of h is the graph of f shifted horizontally.

(c) $y = (x + 2)^2 + 3$

7. (a) The graph of g is the graph of f shifted horizontally.

(b) The graph of h is the reflection of the graph of f across the x-axis and then shifted vertically.

(c) $y = -|x - 2| - 3$

9. $f(x) = x$. The graph of g is the reflection of the graph of f across the x-axis. The graph of h is the graph of g shifted upward 2 units.

11. $f(x) = |x|$. The graph of g is the graph of f shifted upward 2 units. The graph of h is the graph of g shifted to the left 3 units.

13. $f(x) = \sqrt{x}$. The graph of g is the graph of f shifted to the left 9 units. The graph of h is the graph of g shifted downward 4 units.

15. $f(x) = x^3$. The graph of g is the graph of f shifted downward 2 units. The graph of h is the reflection of the graph of g across the x-axis.

17. $f(x) = \dfrac{1}{x}$. The graph of g is the reflection of the graph of f across the x-axis. The graph of h is the graph of g shifted to the right 6 units.

19. $f(x) = \sqrt{x}$. The graph of g is the reflection of the graph of f across the y-axis. The graph of h is the graph of g shifted to the right 3 units.

21. $-5, 2$

23. **25.**

27. **29.**

31.

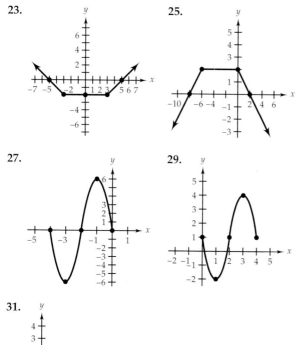

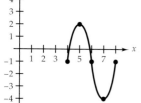

33. The graph becomes steeper. **35.** F **37.** A **39.** E

41. $g(x) = (x + 6)^2 + 5$ **43.** $g(x) = -\dfrac{1}{x}$

45. $g(x) = \sqrt{2 - x}$ **47.** $y = (x - 2)^2$ **49.** $y = \sqrt{x} + 3$

51. $g(x) = f(x) - 3$ and $g(x) = f(x - 3)$ **53.** $y = \sqrt{5 + x}$

55. $y = -\sqrt{16 - x^2}$ **57.** $y = \frac{1}{2}x - 3$

59. **(a)** $f(x) = 16 - x$

  (b) Reflect the graph of $y = x$ across the x-axis and then shift the graph upward 16 units.

61. $g(x) = -f(x) + 60$

■ *Chapter Review Exercises*

1. **(a)** Yes **(b)** No **3.** Yes

5. **(a)** $-\frac{1}{2}$ **(b)** Undefined **(c)** $\dfrac{a + 4}{2a^2 - 8}$

7. Domain: **R**; range: $[-8, 8]$ **9.** $\{x \mid x \neq 0, 5\}$

11. The x-intercepts **13.** $-\frac{3}{2}$ **15.** $y = -4x + 1$

17. $f(x) = -2x$ **19.** $(0, 5), (-15, 0)$

21. **(a)** -3 **(b)** $\left(-\frac{1}{2}, \infty\right)$ **23.** $y = 3.25x + 22.75$; 2002

25. **(a)** $y = 0.95x + 59.73$ **(b)** Fits well **(c)** 2023

27. Parallel **29.** Perpendicular **31.** $f(x) = -\frac{3}{2}x - 4$

33. Determine the lengths of the sides of the triangle and show that the Pythagorean Theorem is satisfied. Determine the slopes of the sides and show that two of the sides are perpendicular.

35. $y = \frac{3}{5}x - \frac{7}{5}$ **37.** $\frac{9}{10}$, \$9.8 billion

39. The graph of f is the graph of y shifted to the right 5 units and upward 1 unit.

41. $(0, -10), (-1.36, 0), (7.36, 0); V(3, -19)$

43. $(-\infty, -1]$; maximum: -1 **45.** 0 **47.** On the x-axis

49. $y = \frac{2}{3}x^2 - 4x + 8$

51. **(a)** Quadratic **(b)** $y = -0.53x^2 + 3.30x + 14.62$

  (c) $V(3.1, 19.8)$; a maximum of 19,800 complaints were filed in 1993.

53. $(0, -2), (8, 0), (-1, 0)$

55. Relative maximum: $(-3.57, 8.21)$; relative minimum: $(2.24, -4.06)$; increasing: $(-\infty, -3.57), (2.24, \infty)$; decreasing: $(-3.57, 2.24)$

57. Origin **59.** None

61. No; the pairs (x, y) and $(x, -y)$ belong to the relation.

63. $y = |x|$; the graph of h is the graph of y shifted to the right 3 units.

65. $y = x^3$; the graph of c is the graph of y shifted to the left 4 units and downward 1 unit.

67. $y = \dfrac{1}{x}$; the graph of r is the graph of y shifted to the right 3 units and upward 5 units.

69. $f(x) = -\dfrac{1}{x - 7}$

71.

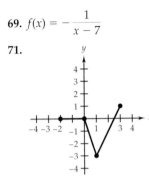

73. $y = |x|; f(x) = |x + 5|$

Chapter 3

■ Section 3.1

1. For even integers n, the graph is symmetric with respect to the y-axis. For odd integers n, the graph is symmetric with respect to the origin.

3. B   5. A

7. The graph of f is the graph of y shifted to the right 4 units.

9. The graph of f is the graph of y shifted downward 7 units.

11. The graph of f is the graph of y shifted to the left 1 unit and downward 7 units.

13. $-2x^3$; rises on the left, falls on the right

15. x^6; rises on the left, rises on the right

17. $-x^4$; falls on the left, falls on the right

19. Because the graph is continuous and rises on one side and falls on the other, it must cross the x-axis. Thus the function must have at least one zero.

21. -3 (multiplicity 1), 0 (multiplicity 1), 2 (multiplicity 1)

23. 0 (multiplicity 1), -2 (multiplicity 4), 1 (multiplicity 3)

25. 2 (multiplicity 2), 4 (multiplicity 2)

27. -4 (multiplicity 1), $-\sqrt{2} \approx -1.41$ (multiplicity 1), $\sqrt{2} \approx 1.41$ (multiplicity 1)

29. -1 (multiplicity 1), 2 (multiplicity 1)

31. 0 (multiplicity 3), $\dfrac{-3 - \sqrt{33}}{2} \approx -4.37$ (multiplicity 1), $\dfrac{-3 + \sqrt{33}}{2} \approx 1.37$ (multiplicity 1)

33. Counting multiplicity, the polynomial has at most four zeros. Because three zeros are distinct, no zero has a multiplicity greater than 2.

35. $(0, 0)$, crosses; $(-2, 0)$, crosses; $(3, 0)$, tangent

37. $\left(\frac{4}{3}, 0\right)$, tangent; $(-5, 0)$, crosses

39. $(0, 0)$, crosses; $(-5, 0)$, crosses; $(-2, 0)$, tangent

41. $(1.33, 0), (1.5, 0), (1.67, 0)$

43. $(-1.41, 0), (-1.33, 0), (1.41, 0), (1.5, 0)$

45. The graph might be close to the point $(4, 0)$ but not contain the point. Zooming in gives a more accurate estimate.

47. $(0, 0), (3, 0)$; relative minimum: $(0, 0)$; relative maximum: $(2, 4)$

49. $(-2, 0), (-1, 0), (1, 0), (2, 0), (0, 4)$; relative minima: $(-1.58, -2.25), (1.58, -2.25)$; relative maximum: $(0, 4)$

51. $(-2, 0), (0, 0), (4, 0)$; relative minimum: $(-1.10, -5.05)$; relative maximum: $(2.43, 16.90)$

53. Relative minimum: $(-2, -20)$; relative maximum: $(6, 236)$

55. Relative minima: $(-1, 3.8), (1, 2.2)$; relative maxima: $(-2, 7.6), (0, 6)$

57. $(2.57, 0)$; range: $\mathbf{R}$

59. $(-2.37, 0), (-0.70, 0), (0.70, 0), (2.37, 0)$; range: $[-29.60, \infty)$

61. $y = -x^3 + 2x^2 + 14x - 5$

63. $y = x^4 - 3x^2 - 12x + 7$ 65. $-4, -2, 3$

67. **(a)** 53 feet   **(b)** 6 minutes, 100 feet   **(c)** 28 minutes

69. **(a)** $w^2 + 4wh = 100$ **(b)** $h = \dfrac{100 - w^2}{4w}$

   (c) $V(w) = 25w - \frac{1}{4}w^3$ **(d)** 5.77 feet

71. **(a)**

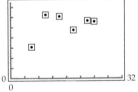

   (b) $y = 0.096x^3 - 4.67x^2 + 70.36x - 211.88$

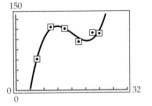

   (c) $y = -0.0083x^4 + 0.595x^3 - 15.165x^2 + 160.746x - 475.438$

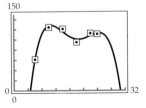

(e) The cubic model indicates a trend of increasing prices, whereas the quartic model indicates a trend of decreasing prices.

73. (a) 259.5 million barrels, 316.57 million barrels

(b) Production increased to 362 million barrels in 1983 and then decreased.

(c) Production decreased to 263 million barrels in 1990 and then increased.

■ *Section 3.2*

1. Evaluate $P(7)$. **3.** $(x^2 + x + 4)(x - 2) + 12$

5. $(x^2 - 6x + 5)(2x + 1) + 0$

7. $(x^4 - x^3 + 6x^2 - 6x + 6)(x + 1) - 16$

9. $(x^2 + x + 5)(x^2 - 5) + 4x + 17$

11. $Q: 4x^2 - 11x + 14; R: -24$ **13.** $Q: 3x^2 - 12x + 9; R: 0$

15. $Q: 2x^3 + x^2 + 3x + 8; R: 9$ **17.** 67 **19.** 3

21. (a) No **(b)** Yes **(c)** Yes

23. (a) No **(b)** Yes **(c)** Yes

25. $-2, -1, 4$ **27.** $-\frac{5}{2}, 1, 2$ **29.** $-\frac{3}{2}, \frac{1}{4}, \frac{5}{2}$

31. If c is a solution of the equation $P(x) = 0$, then c is a zero of P and $(c, 0)$ is an x-intercept of the graph of P.

33. -6 (multiplicity 1), -1 (multiplicity 2)

35. -4 (multiplicity 1), -1 (multiplicity 2), 1 (multiplicity 2)

37. $4, \pm\sqrt{11} \approx \pm 3.32$ **39.** $-0.25, 1.5, \pm\sqrt{7} \approx \pm 2.65$

41. 7 **43.** $G(x) = (x + 1)(x - 1)(x - 2)(x - 4)$

45. $f(x) = (x + 2)(3x - 2)(2x + 3)$

47. $R(x) = (x + 2)(x^2 - 2x + 3)$

49. $P(x) = (5x - 2)(x^2 - x + 4)$

51. The only possible rational zeros are factors of 7 divided by factors of 1: ± 1 and ± 7.

53. $-4, -2, 1, 3$ **55.** $-\frac{1}{5}, \frac{3}{2}, 4$

57. $-5, -1 \pm \sqrt{5} \approx -3.24, 1.24$ **59.** $-\frac{3}{4}$

61. $-\frac{5}{2}, -\frac{4}{3}, \pm\sqrt{3} \approx \pm 1.73$ **63.** $(-4, -1) \cup (5, \infty)$

65. $(-\infty, -3) \cup \left[0, \frac{7}{2}\right]$ **67.** $[-4, 4]$

69. $(-\infty, -4) \cup \left(-\frac{3}{4}, 0\right) \cup (1, \infty)$ **71.** $(-\infty, -1] \cup \{0\}$

73. -2 **75.** $P(x) = -3(x + 2)(x - 1)(x - 4)$

77. $P(x) = 5x(x + 4)^3$

81. (a) $-5, 1, 4$ **(b)** $-1, 2, \frac{7}{2}$

83. (a) $y = 2.\overline{3}x^3 - 17.5x^2 + 41.1\overline{6}x - 26$ **(b)** 1994 data

■ *Section 3.3*

1. A vertical asymptote occurs at a value of x for which the function is not defined.

3. Intercept: $\left(0, -\frac{7}{9}\right)$; asymptotes: $x = \frac{9}{2}, y = 0$

5. Intercepts: $\left(0, -\frac{1}{5}\right), \left(-\frac{1}{2}, 0\right)$; asymptotes: $x = 5, y = 2$

7. Intercepts: $(0, 0), \left(-\frac{5}{3}, 0\right)$; asymptotes: $x = -\frac{1}{2}, x = 1$, $y = \frac{3}{2}$

9. C **11.** D **13.** Vertical asymptote

15.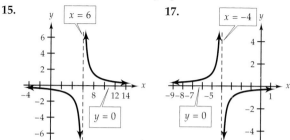

17.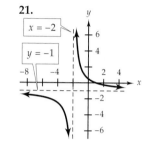

19.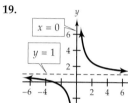

21.

23.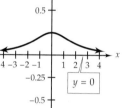

25.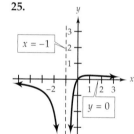

27.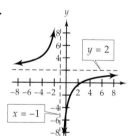

29.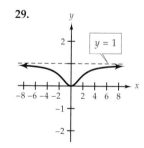

31.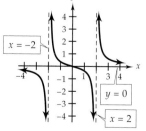

33.

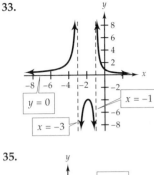

35.

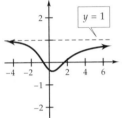

37. Part (iii) is correct. In (i) the vertical asymptotes are $x = -2$ and $x = 3$. In (ii) the horizontal asymptote is $y = 0$.

39. 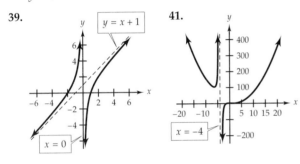 **41.**

43.

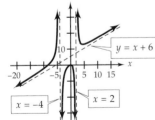

45.

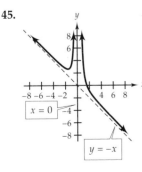

47. The horizontal asymptote is $y = 1$. **49.** $(5, \infty)$

51. $\left(-\infty, \frac{11}{4}\right] \cup (4, \infty)$ **53.** $(-\infty, -5) \cup \left[-\frac{1}{5}, 1\right)$

55. $(-\infty, -1) \cup (0, 2) \cup (4, \infty)$

57. $(-\infty, -1.62) \cup (0, 0.62)$ **59.** $(-2, -1) \cup (2, 5)$

61. $\left(-\infty, -\frac{16}{3}\right) \cup (-4, -1)$ **63.** $(-\infty, -3] \cup [2, 4)$

65. (a) Simplifying $R(x)$ results in $R(x) = x, x \neq 0$, and the graph of $y = x$ is a line.

(b) The function is not defined for $x = 10$.

(c) The domain of R is $\{x \mid x \neq 10\}$ and the domain of f is **R**.

67. **69.**

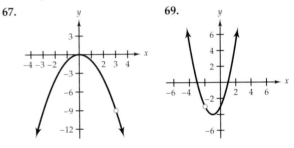

71. (a) 7.6 feet

(b) The model indicates that the height is negative.

(c) 11 feet

73. (a) $x = 0, y = 0$

(b) The function is not defined for $x = 0$.

(c) 16.67%

(d) Because the probability approaches 0, the model appears to be reasonable.

75. (a) $r_2 = 100, r_1 = 100$ **(b)** $(100, \infty)$

(c) As $r_2 \to 100, r_1 \to \infty$. As $r_2 \to \infty, r_1 \to 100$.

77. (a) $h = y + 8, w = x + 6$ **(b)** $xy = 600$

(c) $h = \dfrac{600}{x} + 8$ **(d)** $A(x) = (x + 6)\left(\dfrac{600 + 8x}{x}\right)$

(e) 21 inches

79. 10

■ *Section 3.4*

1. The definition requires that n be positive, so $-n$ is negative.

3. $6i$ **5.** $-3 - 5i\sqrt{3}$ **7.** $3 - i$ **9.** $-11 + 4i$

11. $3 + 3i$ **13.** $-12 - 6i$ **15.** $5 + 5i$ **17.** $-41i$

19. 34 **21.** 7 **23.** $-8 + 6i$ **25.** i **27.** $-i$ **29.** $-i$

31. -3 **33.** $-4 + 2i$ **35.** 7 **37.** $-2 + i\sqrt{3}$

39. The product $(a + b)(a - b)$ is $a^2 - b^2$ and the product $(a + bi)(a - bi)$ is $a^2 + b^2$.

41. $3 - 2i; 13$ **43.** $-4 - i\sqrt{5}; 21$ **45.** $\frac{1}{5} + \frac{3}{5}i$

47. $\frac{1}{2} - \frac{1}{2}i$ **49.** $2 - 3i$ **51.** $\frac{3}{2} + \frac{1}{2}i$ **53.** $-i$

55. Divide n by 4 to determine the remainder r. Then evaluate i^r.

57. $\pm 5i$ **59.** $3 \pm 2i$ **61.** $\frac{1}{2} \pm \frac{1}{4}i$ **63.** $4 \pm i\sqrt{2} \approx 4 \pm 1.41i$

65. $\frac{1}{2} \pm i$ **67.** 4 **69.** $a = 5, b = 1$ **71.** $a = 3, b = 6$

73. 0 **75.** 0 **77.** $x^2 + 4$ **79.** $x^2 - 6x + 10$

81. $\dfrac{a}{a^2 + b^2} - \dfrac{b}{a^2 + b^2}i$ **83.** $5 + 3i$ **85.** $\frac{1}{17} + \frac{4}{17}i$

87. **(a)** i **(b)** -1 **(c)** $\frac{1}{5} - \frac{2}{5}i$ **(d)** $-\frac{5}{169} + \frac{12}{169}i$

■ *Section 3.5*

1. No. Complex roots of a polynomial function occur in conjugate pairs, so the function has an even number of complex roots. Thus a polynomial whose degree is odd has at least one real number zero.

3. $x^3 - 8x^2 + 11x + 6$

5. $x^4 + x^2 - 12$ **7.** $x^3 - 7x^2 + 17x - 15$

9. $x^4 + x^3 - x^2 + x - 2$ **11.** $x^4 - 2x^3 + 2x^2 - 6x - 3$

13. The conjugate of the zero is also a zero of multiplicity 3. So the lowest possible degree is 6.

15. $1 - \sqrt{2}, \pm 3i$ **17.** $3 - 2i, \pm\sqrt{5}$

19. $-1 + i, -2 \pm \sqrt{3}$ **21.** $-1 - \sqrt{3}, 1 \pm i\sqrt{2}$

23. $-3i, -2 \pm 3i$ **25.** $-2 - i, 1 \pm i$

27. No, $-i$ is not a zero. Because the polynomial does not have real coefficients, the result does not violate the Conjugate Zero Theorem.

29. $\frac{3}{2}, -3 \pm i$ **31.** -1 (multiplicity 2), $4 \pm 3i$

33. 1 (multiplicity 2), -1 (multiplicity 2), $\dfrac{1}{2} \pm \dfrac{\sqrt{2}}{2}$

35. The function has two irrational zeros and two imaginary zeros: $\pm\sqrt{2}, \pm i$

37. $4, 3 \pm i\sqrt{2} \approx 3 \pm 1.41i$

39. $-1, \frac{1}{2}, 2 \pm \sqrt{6} \approx -0.45, 4.45$

41. $-1, 1, 2, 1 \pm i\sqrt{3} \approx 1 \pm 1.73i$

43. The zeros of $P(x)$ are the same as the solutions of the equation.

45. $F(x) = \left(x - 3 + \sqrt{5}\right)\left(x - 3 - \sqrt{5}\right)\left(x - 1 + \sqrt{2}\right) \cdot \left(x - 1 - \sqrt{2}\right)$

47. $R(x) = (x + 2 + 4i)(x + 2 - 4i)\left(x + 1 + \sqrt{3}\right) \cdot \left(x + 1 - \sqrt{3}\right)$

49. $G(x) = \left(x - 2 + i\sqrt{5}\right)\left(x - 2 - i\sqrt{5}\right)(x - 1 + 3i) \cdot (x - 1 - 3i)$

51. $f(x) = (x - 5)\left(x + 1 + \sqrt{2}\right)\left(x + 1 - \sqrt{2}\right)$

53. $p(x) = (2x + 1)(3x - 1)(x + 1 + i)(x + 1 - i)$

55. $q(x) = (x + 1)(x + 4)\left(x - \dfrac{1}{2} + \dfrac{\sqrt{2}}{2}i\right)\left(x - \dfrac{1}{2} - \dfrac{\sqrt{2}}{2}i\right)$

57. $2i, 3i$

■ *Chapter Review Exercises*

1. The graph of f is the graph of y shifted upward 4 units.

3. $-x^4$; falls on the left, falls on the right

5. $\frac{1}{2}$ (multiplicity 1), -1 (multiplicity 3), 2 (multiplicity 2)

7. If n is odd, the graph crosses the x-axis at $(c, 0)$. If n is even, the graph is tangent to the x-axis at $(c, 0)$.

9. $y = x^3 - 3x^2 - 4x$

11. Relative maximum: $(-0.79, 7.21)$; relative minimum: $(2.12, -5.06)$

13. $P(x) = (2x - 1)(2x^2 + 4x + 3) + 8$ **15.** $-\frac{1}{2}, \frac{2}{3}, 4$

17. $-6, \dfrac{-3 \pm \sqrt{17}}{2} \approx -3.56, 0.56$

19. $H(x) = x^3(2x + 1)^2(x - 4)$ **21.** $-\frac{1}{2}, 3$

23. $(-\infty, -5] \cup [-3, \infty)$

25. Intercept: $(0, 0)$; asymptote: $y = 0$

27. Intercepts: $(3, 0), (-1, 0)$; asymptotes: $x = 0, x = -2, y = 1$

29.   **31.**

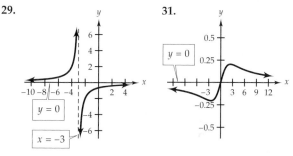

33. $(-\infty, -3) \cup [-2, \infty)$

35. The function simplifies to $y = x$. So as x approaches 2, y approaches 2, not ∞.

37. **(a)** 4 **(b)** -7 **(c)** $4 + 7i$

39. $4i$ **41.** $-14 + i$ **43.** $-7 - 24i$ **45.** $\frac{1}{2} + \frac{1}{2}i$

47. $-1 \pm 2i$ **49.** The conjugate $-4 + 5i$ is also a zero.

51. $f(x) = x^4 - 2x^3 + 24x^2 - 50x - 25$

53. $-i\sqrt{7}, 2 \pm \sqrt{5} \approx -0.24, 4.24$

55. $\frac{5}{3}, -2 \pm i\sqrt{2} \approx -2 \pm 1.41i$

57. $f(x) = (2x - 3)(x + 2)(x - 1 + i)(x - 1 - i)$

59. $\frac{5}{2}, \pm 5i$

Chapter 4

■ *Section 4.1*

1. Either (1) determine the product of $f(x)$ and $g(x)$ and evaluate the resulting function for $x = -2$ or (2) evaluate $f(-2)$ and $g(-2)$ and multiply the results. Method (2) is easier.

3. (a) 3 **(b)** 9 **(c)** 0 **(d)** 0

5. (a) $\frac{3}{2}$ **(b)** 5 **(c)** $-\frac{1}{8}$ **(d)** -3

7. (a) 0 **(b)** 0 **(c)** $\frac{3}{7}$ **(d)** 5

9. $(f+g)(x) = x^2 + 3x + 2$

$(f-g)(x) = -x^2 - x$

$(fg)(x) = x^3 + 3x^2 + 3x + 1$

$\left(\dfrac{f}{g}\right)(x) = \dfrac{1}{x+1}$

Domain of $\dfrac{f}{g} = \{x \mid x \neq -1\}$

11. $(f+g)(x) = \dfrac{4x-1}{(x-1)(x+2)}$

$(f-g)(x) = \dfrac{2x^2+1}{(x-1)(x+2)}$

$(fg)(x) = -\dfrac{x}{x+2}$

$\left(\dfrac{f}{g}\right)(x) = -\dfrac{x^2+2x}{(x-1)^2}$

Domain of $\dfrac{f}{g} = \{x \mid x \neq 1, -2\}$

13. Parts (ii) and (iii) are possible definitions of f and g. In (i), both functions are defined for $x = -1$, so $(fg)(-1)$ is also defined.

15. (a) R **(b)** R **(c)** R **(d)** R

17. (a) $\{x \mid x \neq 0, 3\}$ **(b)** $\{x \mid x \neq 0, 3\}$ **(c)** $\{x \mid x \neq 0, 3\}$
(d) $\{x \mid x \neq -2, 0, 3\}$

19. (a) $\{x \mid x \geq 3\}$ **(b)** $\{x \mid x \geq 3\}$ **(c)** $\{x \mid x \geq 3\}$
(d) $\{x \mid x > 3\}$

21. Because $g(0) = -1$ and -1 is not in the domain of f, 0 is not in the domain of $f \circ g$.

23. (a) -3 **(b)** 0 **(c)** 13 **(d)** 8

25. (a) 1 **(b)** 4 **(c)** $\sqrt[3]{4} \approx 1.59$ **(d)** -3

27. $(f \circ g)(x) = (3-x)^2$

$(g \circ f)(x) = 3 - x^2$

$(f \circ f)(x) = x^4$

29. $(f \circ g)(x) = x - 1$

$(g \circ f)(x) = \sqrt[3]{x^3 - 1}$

$(f \circ f)(x) = (x^3 - 2)^3 - 2$

31. $(f \circ g)(x) = \dfrac{1}{4 - 2x}$

$(g \circ f)(x) = 3 - \dfrac{2}{x+1}$

$(f \circ f)(x) = \dfrac{x+1}{x+2}$

33. $(f \circ g)(x) = x^6$

$(g \circ f)(x) = x^6$

$(f \circ f)(x) = x^{16}$

35. $(f \circ g)(x) = x^3 + 2x^2 - x + 2$

$(g \circ f)(x) = (x+3)^3 + 2(x+3)^2 - (x+3) - 1$

$(f \circ f)(x) = x + 6$

43. (a) R **(b)** R

45. (a) $\{x \mid x \geq 0\}$ **(b)** $\{x \mid x \leq 3\}$

47. $f(x) = \sqrt{x}; g(x) = 2x + 1$

49. $f(x) = \dfrac{2}{x}; g(x) = x^2 - x$

51. $f(x) = \dfrac{x-4}{x+4}; g(x) = x^2$

53. $f(x) = x^3 + 3x^2 + 3x + 5; g(x) = x - 1$

55. $f(x) = |2 - x|; g(x) = |2 - x|$

57. (a) $4t^2 - 2t - 4$ **(b)** $t^2 - t + 2$ **(c)** $\dfrac{a^2 - 4}{a^4 - 2a^2}$
(d) $-b^3 - 3b^2 + b + 3$

59. (a)

(b) $\{x \mid 1 \leq x \leq 6\}$

61. (a) 3 **(b)** $\{x \mid 1 \leq x < 4\}$

63. (a) 3 **(b)** 0 **65.** Odd

67. (a) $P(x) = 100,000\sqrt{x} - 30,000x$ **(b)** 80,000
(c) 80,000

69. (a) $h(d) = \dfrac{d}{50}$ **(b)** $f(h) = 18 - 2.5h$

(c) The composition $(f \circ h)(d) = 18 - \dfrac{d}{20}$ gives the amount of gasoline remaining after the car has traveled d miles.

(d) After the car has traveled 200 miles, 8 gallons of gasoline remain.

71. (a) $r(t) = 1.5t$ **(b)** $A(r) = \pi r^2$ **(c)** $f(t) = 2.25\pi t^2$
(d) 44.2 square feet

73. (a) The function $(C + N)(x) = 1086x + 10,500$ represents the total number of agencies.
(b) $(C + N)(4) = C(4) + N(4)$

Section 4.2

1. The Vertical Line Test guarantees that the relation is a function, and the Horizontal Line Test guarantees that the function is a one-to-one function.

3. $\{(4, -3), (-1, 0), (9, 5)\}$ **5.** $y = \sqrt{9 - x^2}$

7. $xy + x = 3$

9. The inverse is obtained by reversing the coordinates of the ordered pairs, so its graph is in Quadrant IV.

11. No   **13.** Yes   **15.** Yes

17. (a) $\{x \mid x \neq 4\}, \{y \mid y \neq 2\}$
   (b) $\{x \mid x \neq 2\}, \{y \mid y \neq 4\}$

19. (a) $\{x \mid x \leq 3\}, \{y \mid y \geq 0\}$
   (b) $\{x \mid x \geq 0\}, \{y \mid y \leq 3\}$

21. (a) $\{x \mid x \geq 2\}, \{y \mid y \geq 0\}$
   (b) $\{x \mid x \geq 0\}, \{y \mid y \geq 2\}$

23. (a) R, R **(b)** R, R **25.** Yes   **27.** No

29. Yes; $f^{-1}(x) = -\frac{1}{2}x + \frac{5}{2}$

31. No   **33.** Yes; $f^{-1}(x) = \dfrac{\sqrt[3]{x}}{2}$ **35.** No

37. Yes; $f^{-1}(x) = x^2 - 3, x \geq 0$

39. Yes; $f^{-1}(x) = \dfrac{x + 3}{x - 1}$ **41.** No   **43.** No

49. $f^{-1}(x) = 5x - 2$ **51.** $f^{-1}(x) = (x + 7)^3$

53. $f^{-1}(x) = -\dfrac{3}{x} - 5$ **55.** $f^{-1}(x) = (2 - x)^2, x \leq 2$

57. $f^{-1}(x) = \sqrt{x} + 1, x \geq 0$ **59.** $g^{-1}(x) = \sqrt{4 - x}, x \leq 4$

61. $h^{-1}(x) = -\sqrt{x} - 2, x \geq 0$

63. A constant function is not a one-to-one function.

67. Yes   **69.** No

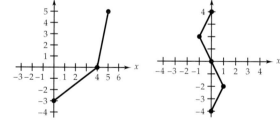

71. Yes   **73.** No

75. Yes; $f^{-1}(x) = \begin{cases} -\sqrt{-x}, & x \leq 0 \\ \sqrt{x}, & x > 0 \end{cases}$

79. (a) No. Two pairs have the same first coordinate: (2, A), (2, B).

   (b) Yes. Each letter is associated with exactly one number.

   (c) Yes

81. (a) $r = 800 - 3m$ **(b)** $0 \leq m \leq 100$

(c) The inverse $m = \dfrac{800 - r}{3}$ gives the number m of members in attendance for total revenue r.

83. (a) $f(x) = \dfrac{3x}{x + 3}$

   (b) The value $f(4.5) = 1.8$ indicates that if the second painter can paint the house in 4.5 days, then the two painters, working together, can paint the house in 1.8 days.

   (c) $f^{-1}(x) = \dfrac{3x}{3 - x}$

   (d) The value $f^{-1}(2) = 6$ indicates that if the painters, working together, can complete the job in 2 days, the second painter could paint the house in 6 days.

85. (a) $f(x) = 0.68x^3 - 5.41x^2 + 15.02x - 6.67$

   (b) The value $f(7) = 66.62$ indicates that in 1995, the number of cards was 66.62 million.

   (c) $f^{-1}(66.62) = 7$ **(d)** $[5, 570]$

Section 4.3

1. The expressions 2^x and $\left(\frac{1}{2}\right)^{-x}$ are equivalent.

3. 172.47   **5.** 0.03   **7.** 0.07   **9.** 20.09   **11.** 0.47

13. 0.74

15. We call f a linear function, g an exponential function, and h a quadratic function.

17. (a) 0   **(b)** 2   **(c)** 64   **(d)** 0.25

19. (a) 1   **(b)** 0   **(c)** 3   **(d)** -1

21. Because e represents a specific number, (i) and (ii) are not variable expressions. Only (iii) is a variable expression.

23. (a) The graph of f is the graph of y shifted to the left 4 units.

   (b) The graph of g is the graph of y shifted upward 4 units.

25. (a) The graph of f is the graph of y shifted upward 5 units.

   (b) The graph of g is the graph of y shifted downward 1 unit.

27. 3   **29. (a)** $(0, 3)$ **(b)** $y = 2$ **(c)** $\{y \mid y > 2\}$

31. (a) $(0, 2)$ **(b)** $y = 0$ **(c)** $\{y \mid y > 0\}$

33. (a) $(0, 6)$ **(b)** $y = 0$ **(c)** $\{y \mid y > 0\}$

35. (a) $(0, -6), (-1.77, 0)$ **(b)** $y = -7$ **(c)** $\{y \mid y > -7\}$

37. (a) $(0, 2)$ **(b)** $y = 1$ **(c)** $\{y \mid y > 1\}$

39. (a) $(0, 20.09)$ **(b)** $y = 0$ **(c)** $\{y \mid y > 0\}$

41. (a) $(0, 2.86), (3.10, 0)$ **(b)** $y = 3$ **(c)** $\{y \mid y < 3\}$

43. (a) $(0, -1), (0.16, 0)$ **(b)** $y = -5$ **(c)** $\{y \mid y > -5\}$

45. B   **47.** F   **49.** C   **51.** 1.46   **53.** -1.25

55. $-9.00, 3.66$

57. The graph of g is the graph of f shifted downward k units. The graph of h is the graph of f shifted to the right k units.

59. (a) Relative maximum: $(1, 1.84)$ **(b)** $y = 0$
(c) $\{y \mid y \le 1.84\}$

61. (a) None **(b)** $y = 3, y = 0$ **(c)** $\{y \mid 0 < y < 3\}$

65. $35,125.39 **67. (a)** $11,264 **(b)** Increase

69. (a) $632 million **(b)** The graph rises very quickly.

71. $3467.36 **73.** $746,202.53 **75.** 9 years

77. If you select the exponential model, your assumption is that the number of strikes will decrease. If you select the quadratic model, your assumption is that the number of strikes will increase.

■ *Section 4.4*

1. Either $\log a = 0$ or $\log b = 0$. Thus either $a = 1$ or $b = 1$.

3. $9^2 = 81$ **5.** $t^3 = x$ **7.** $\log_7 \frac{1}{49} = -2$ **9.** $\ln P = rt$

11. 4 **13.** -2 **15.** 4 **17.** 4 **19.** $\frac{2}{3}$ **21.** $-\frac{2}{3}$ **23.** 1.04

25. 0.22 **27.** 0.71 **29.** 1.20 **31.** 3.96

33. (a) $(1024, 0)$ **(b)** $x = 0$ **(c)** $\{x \mid x > 0\}$

35. (a) $(1, 0)$ **(b)** $x = 0$ **(c)** $\{x \mid x > 0\}$

37. (a) $(0, 3), (-0.95, 0)$ **(b)** $x = -1$ **(c)** $\{x \mid x > -1\}$

39. The vertical asymptote of each is $x = 0$.

41. (a) The graph of f is the graph of y shifted to the left 2 units.
(b) The graph of g is the graph of y reflected across the y-axis and shifted to the right 4 units.

43. (a) The graph of f is the graph of y reflected across the y-axis and then shifted to the right 1 unit and upward 2 units.
(b) The graph of g is the graph of y reflected across the x-axis and then shifted to the left 1 unit and upward 4 units.

45. E **47.** A **49.** F **51.** $f^{-1}(x) = \log_5 x$

53. $g^{-1}(x) = \left(\frac{1}{2}\right)^x - 1$ **55.** $f^{-1}(x) = 2 + \log_4 x$

57. $r^{-1}(x) = e^{1-x}$

59. We can use the Change of Base Formula to write $\log_2 x$ as $\dfrac{\log_4 x}{\log_4 2}$. So the expressions are equivalent.

61. $\{x \mid x > \frac{7}{2}\}$ **63.** $\{x \mid x \ne 2\}$ **65.** $(1.31, 0.27)$; no

67. The domain of $y = \ln e^x$ is $\mathbf{R}$, whereas the domain of $y = e^{\ln x}$ is $\{x \mid x > 0\}$, so the graphs of the two functions are not the same.

69. (a) Relative maximum: $(0, 0)$ **(b)** $(0, 0)$

71. (a) Relative minimum: $(-2.43, -2.29)$
(b) $(0, 0), (-4, 0)$

73. $f^{-1}(x) = 10^{10^x}$

75. (a) (i) 2543 meters **(ii)** 3574 meters
(b) (i) 652.11 mm Hg **(ii)** 610.69 mm Hg

77. (a) (i) 8.6 **(ii)** 5.8 **(iii)** 7.4
(b) (i) $(1.6 \cdot 10^6)I_0$ **(ii)** $(2.5 \cdot 10^8)I_0$ **(iii)** $(4.0 \cdot 10^7)I_0$

79. (a) 58 **(b)** 83.75 **(c)** 6.39

81. (a) $142,000 **(b)** 3.59 **(c)** $397,587

■ *Section 4.5*

1. $\dfrac{\log 100}{\log 10} = \dfrac{2}{1} = 2$ and $\log 100 - \log 10 = 2 - 1 = 1$.
Thus, $\dfrac{\log M}{\log N} \ne \log M - \log N$.

3. $\frac{1}{3}$ **5.** $\frac{1}{125}$ **7.** 3 **9.** 0 **11.** $x - 2$ **13.** $3x$

15. $\ln 3 + \ln (x - 1)$ **17.** $\log x\sqrt{x + 3}$

19. $\log_5 t - \log_5 (t - 2)$ **21.** $\ln \dfrac{t^2 - 1}{\sqrt{t}}$ **23.** $-3 \ln a$

25. $\log_3 \dfrac{1}{w}$ **27.** $\log_9 x + 2 \log_9 y - \log_9 z$

29. $\ln x + 2 \ln (y + 3)$ **31.** $\frac{1}{3} \log a - \frac{1}{3} \log 2 - \frac{1}{3} \log b$

33. $\frac{1}{2} \log_4 (z + 2) - 3 \log_4 x - \log_4 y$

35. $\log_7 (x + 2) + \log_7 (x - 2) - 3 \log_7 x$

37. $2 \ln (x + 3)$ **39.** $1 + \log x + \log (10 - x)$

41. $\log_6 x^3 y^2$ **43.** $\log_3 \dfrac{\sqrt{a}}{b}$ **45.** $\log \dfrac{x^3 z}{\sqrt{y}}$

47. $\log \sqrt[3]{(x + 1)^2(x - 2)}$ **49.** $\ln \dfrac{(a + 1)b^2}{(c - 3)^3}$ **51.** $\log_5 125y^2$

53. $\ln ye^{6x}$

55. The domain of $y = \ln x^2$ is $\{x \mid x \ne 0\}$, whereas the domain of $y = 2 \ln x$ is $\{x \mid x > 0\}$.

57. True **59.** False **61.** False **63.** True **65.** $\frac{3}{2}$

67. 3.5 **69.** $A - B - C$ **71.** $\frac{1}{2} - C - B - 2A$

73. $a + \frac{1}{2}b + \frac{1}{2}c$ **75.** $2 + c$

79. Because each number is 10 times the preceding number, each logarithmic value is 1 more than the preceding value.

83. 10 times

■ *Section 4.6*

1. Equate the logarithms of the expressions.

3. 2 **5.** -5 **7.** 8 **9.** $-3, -1$ **11.** $0, \frac{1}{2}$ **13.** -2

15. 1.26 **17.** -4.82 **19.** 0.03 **21.** 0.51 **23.** 25

25. 0.50 **27.** $\frac{3}{2}$ **29.** 4096 **31.** $-3.30, 0.30$ **33.** 6

35. 5

37. The expression on the left is easy to simplify; however, the Change of Base Formula is needed to evaluate the expression on the right.

39. 1 **41.** 5 **43.** 3 **45.** $\frac{2}{3}$ **47.** 2 **49.** No solution

51. Both -5 and 3 are solutions of (i). In (ii), x must be positive, so 3 is the only solution.

53. ± 16 **55.** $10^4, 10^{-2}$ **57.** $\frac{1}{10}, 1000$ **59.** e

61. $100, 1000$ **63.** e^{e^e} **65.** $1.92, 0.08$ **67.** $0, 1.10$

69. 2 **71.** 0 **73.** $\frac{2}{3}$

75. The solution of (i) and (iii) is 1. The solutions of (ii) are 1 and 10.

77. Because each value of x is $\frac{1}{10}$ the preceding value of x, the value of $\log x$ is 1 less than the value of the preceding logarithm.

79. $-2.29, 1.15$ **81.** $0.11, 2.94$ **83.** $[-1, 1.87]$

85. $[-2.59, 1.53]$

87. (a) Because the base is positive, f is defined on $(0, \infty)$.

(b) The function is defined for some values such as -3 because $(-3)^{-3}$ is a real number. The function is not defined for values such as $-\frac{1}{2}$ because $\left(-\frac{1}{2}\right)^{-1/2}$ is not a real number.

(c) On $(-\infty, 0)$, the graph has "holes" because the function is not defined for all values in the interval.

▪ *Section 4.7*

1. 1.13 **3.** 63.38 minutes **5.** June 14 **7.** 9:24 A.M.

9. 7.70% **11.** 1996 **13.** 1996

15. 10.57 feet/second; 29.173 feet/second; the velocity approaches 40 feet/second.

17. 7.43 minutes **19.** 15 minutes **21.** 57

23. $t = n^k c$; because $n \geq 1$ and $c > 0$, t is never zero.

25. $N = Ak e^{2.834}$ **27.** 14 **29.** 38.45 decibels

31. $1.32 \cdot 10^{10}$ **33.** $62; 6.8$ years **35.** 9.5 years

37. 2123

39. (a) $y = 105.48(0.63)^x$

(b)
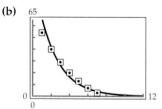

(c) The model function never has a value of 0.

41. (a) Linear: $y = 0.197x - 0.0038$; exponential: $y = 0.274(1.273)^x$

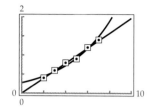

(b) Because the rate of increase is approximately constant, the linear model appears to be better.

(c) Because the expenditures are not likely to increase rapidly, a linear model is more reasonable.

43. (a) $y = 0.034(1.037)^x$ **(b)** 2026

(c) The rate of increase is constant, so a linear model would be appropriate.

45. (a) $y = 11.2 + 15.2 \ln x$ **(b)** 2001

47. (a) The trend is a rapid increase followed by leveling off of the percentage.

(b) $y = \dfrac{69.586}{1 + 154.952e^{-0.215x}}; 67.7\%, 69.4\%$

(c) The limiting value is approximately 69.6%, indicating that the percentage will level off at that level.

49. (a) $y = \dfrac{15.39}{1 + 1597.45e^{-0.82x}}$ **(b)** 2003

(c) $y = -42.97 + 23.1 \ln x$ **(d)** 2002

(e) The logistic model indicates that the number of accounts will level off, and the logarithmic model indicates that the number will continue to increase.

▪ *Chapter Review Exercises*

1. (a) 13 **(b)** $(f - g)(x) = -x^2 - 2x - 1$

3. (a) -3 **(b)** $(g \circ f)(x) = 9x^2 + 3x + 1$

5. $\left(\dfrac{f}{g}\right)(x) = \dfrac{-4}{x + 3}; \quad \{x \mid x \neq -3, 1\}$

7. (a) 1 **(b)** 4 **9.** $(f \circ g)(x) = x; \quad (g \circ f)(x) = x$

11. $f(x) = \dfrac{2}{x}; \quad g(x) = x^2 + x$ **13.** $3xy - x^2 = 7$ **15.** Yes

17. (a) $\{x \mid x \neq -2\}, \{y \mid y \neq 3\}$

(b) $\{x \mid x \neq 3\}, \{y \mid y \neq -2\}$

19. Yes; $f^{-1}(x) = 2 - x^2, x \geq 0$ **21.** Yes; $h^{-1}(x) = \dfrac{1}{x - 1}$

23. $f^{-1}(x) = \sqrt{x + 1}$ **25. (a)** 12.18 **(b)** $\dfrac{1}{e} \approx 0.37$

27. The graph of f is the graph of $y = b^x$ shifted upward k units. The graph of g is the graph of $y = b^x$ shifted downward k units.

29. (a) The graph of f is the graph of y reflected across the x-axis.

(b) The graph of g is the graph of y reflected across the x-axis and then shifted upward 1 unit.

31. (a) $(0, -2)$, $(0.46, 0)$ **(b)** $y = -5$ **(c)** $\{y \mid y > -5\}$

33. 0.79 **35. (a)** $\left(\frac{1}{4}\right)^{-2} = 16$ **(b)** $\ln A = -rt$

37. (a) -3 **(b)** $\frac{2}{3}$

39. Functions f and g are inverses, so their graphs are symmetric with respect to the line $y = x$.

41. (a) The graph of f is the graph of y reflected across the x-axis and shifted downward 1 unit.

(b) The graph of g is the graph of y shifted upward 5 units.

43. (a) 100 **(b)** $10^{6.5} \cdot I_0$ **45.** $g^{-1}(x) = 10^x - 5$

47. (a) $\frac{1}{9}$ **(b)** 7

49. (a) $\log_5 2 + \log_5 a + \log_5 b$ **(b)** $\log t(t + 2)$

51. (a) $-5 \log_2 z$ **(b)** $\ln \sqrt[3]{t}$

53. $\log_7 5 + 2\log_7 a - \log_7 b$

55. $\ln 2 + \ln x + 3 \ln (x + 4)$ **57.** $\log_9 t^2(t^2 + 1)^3$

59. -10 **61.** 0.30 **63.** $-2, 4$ **65.** $\frac{9}{2}$

67. $5, 125$ **69.** 877 years

71. (a) $y = 778.59(0.83)^x$ **(b)** 175

Chapter 5

■ *Section 5.1*

1. Because the y-intercept of both lines is $(0, -3)$, the graphs intersect at $(0, -3)$. Thus $(0, -3)$ is the solution of the system.

3. F; $(4, -2)$

5. E; every point of the line represents a solution of the system.

7. D; $(-6, -2)$ **9.** $(5, 7)$ **11.** $(3, -2)$ **13.** $(14, -12)$

15. $\varnothing$

17. The graphs are straight lines. If the lines do not coincide, they intersect at most once. So the maximum finite number of solutions of the system is 1.

19. $(6, -1)$ **21.** $\left(-\frac{1}{2}, 0\right)$ **23.** $\left\{\left(\frac{3}{2}y - \frac{5}{2}, y\right)\right\}$ **25.** $(-4, 7)$

27. $\varnothing$ **29.** $(3, 1)$

31. $(15, -7)$; consistent; independent

33. $\varnothing$; inconsistent; independent

35. $(-1, -2)$; consistent; independent

37. $(0, -3)$; consistent; independent

39. $\left\{\left(\frac{19}{5} - \frac{3}{5}y, y\right)\right\}$; consistent; dependent

41. The only solution of (i) is $(0, 0)$, whereas (ii) has infinitely many solutions.

43. $a = 5, b = 3$ **45.** $\left(\frac{1}{3}, \frac{1}{5}\right)$ **47.** $(-2, 1)$ **49.** 6 **51.** 18

53. 20% solution: 30 ml; 60% solution: 20 ml

55. Daily rate: \$23; mileage rate: 7 cents

57. Small box: 24 cubic inches; large box: 56 cubic inches

59. Electrician: \$22.50; metal worker: \$20.25

61. Mathematics: 120; English: 90   **63.** 36   **65.** 10

67. 12 feet by 15 feet   **69.** 40 miles

71. 42 faculty, 9 staff   **73.** 6 hours

75. (a) In 1982, the math score was lower than the verbal score, whereas in 1997, it was higher than the verbal score.

(b) Verbal: $y = 0.07x + 503.87$; Mathematics: $y = 1.2x + 490.6$

(c) The solution, approximately $(12, 505)$, indicates that in 1992 both the math and the verbal scores were 505.

■ *Section 5.2*

1. No. In the second equation, substitute 5 for z and solve for y.

3. $(5, 7, 4)$ **5.** $(1, 2, -1)$ **7.** $\left\{\left(\frac{3}{7}z + 1, 2 - \frac{16}{7}z, z\right)\right\}$

9. $(1, 3, -1)$ **11.** $(0, -2, -3)$

13. The planes do not intersect, so the system has no solution.

15. $(3, -2, -1)$ **17.** No solution   **19.** $\left\{\left(z + 2, 1 - \frac{3}{2}z, z\right)\right\}$

21. $(4, -2, 0)$

23. In the solution, $y = 0$ and $x + z = 1$. Because $y = 0$, the solution cannot be written in terms of y. Because $x + z = 1$, we can solve for x and write $(1 - z, 0, z)$, or we can solve for z and write $(x, 0, 1 - x)$.

25. $f(x) = x^2 - 2x - 8$ **27.** $f(x) = -2x^2 + x - 3$

29. $\left\{\left(\frac{1}{5}z + \frac{37}{5}, \frac{3}{5}z + \frac{6}{5}, z\right)\right\}$ **31.** $\left(2, -2, -\frac{1}{2}\right)$ **33.** $(-2, 0, 3, 4)$

35. Small: \$1.20; medium: \$1.80; large: \$3.00   **37.** 2345

39. 1-point shots: 7; 2-point shots: 17; 3-point shots: 9

41. Faculty lot: 380; central lot: 2300; perimeter lot: 3060

43. 2×3: \$1.00; 3×5: \$2.50; 6×8: \$6.00

45. Females: 12 million; males: 3 million; couples: 54 million

47. 20 children, 12 adults, 10 senior citizens

49. Chapman: 575; Bascomb: 891; Booth: 369

51. $T(x) = -0.38x^2 + 7.90x - 17$

■ *Section 5.3*

1. The coefficient matrix contains the coefficients of x and y.

3. 1×2 **5.** 2×3

7. $\begin{bmatrix} 1 & 3 & | & 7 \\ 2 & -5 & | & -8 \end{bmatrix}$ **9.** $\begin{bmatrix} 1 & 0 & 5 & | & 13 \\ 0 & 2 & 1 & | & 4 \\ 1 & 1 & -1 & | & 0 \end{bmatrix}$

11. $\begin{aligned} x - y &= -2 \\ 3x + 2y &= 14 \end{aligned}$ **13.** $\begin{aligned} -2x \quad + z &= 2 \\ 4x + y - 3z &= -2 \\ 2x - 4y + z &= -10 \end{aligned}$

15. For $c = 1$, the operation $c \cdot R_i + R_j \to R_j$ becomes $R_i + R_j \to R_j$.

17. (a) 1 4 -3 **(b)** 0 1 -2 **(c)** 1 0 5

19. (a) 0 1 -3 7 **(b)** 0 -4 13 -31 **(c)** 1 0 -5 16
   (d) 0 0 1 -3 **(e)** 1 0 0 1 **(f)** 0 1 0 -2

21. $\begin{bmatrix} 1 & 0 & | & 9 \\ 0 & 1 & | & -1 \end{bmatrix}$ **23.** Yes

25. The row corresponds to the equation $0 = 0$.
   The system has infinitely many solutions.

27. $(3, -2)$ **29.** No solution **31.** $\left\{ \left(\frac{1}{4}y + \frac{1}{4}, y \right) \right\}$

33. $(2, 0, -6)$ **35.** No solution **37.** $\left(\frac{1}{2}, \frac{5}{2} \right)$

39. $\{(y + 6, y)\}$ **41.** $(6, 4)$ **43.** $(2, -1, 4)$

45. No solution

47. The planes coincide.
   The solution set is $\{(x, y, z) \mid 2x - y + z = 4\}$.

49. $\left(\frac{1}{2}, 1, \frac{3}{2} \right)$ **51.** $\left\{ \left(9 - 3z, \frac{1}{2}z, z \right) \right\}$ **53.** $\left(\frac{1}{2}, 2, 1 \right)$

55. $(1, -1, 1, 2)$ **57.** $(2, 1, -1, 3)$

59. Coffee: \$2.27 billion; sugar: \$0.551 billion;
   wheat: \$0.286 billion

61. (a) Midwest: $y = -0.1x + 24$; West: $y = 0.14x + 21.2$
   (b) 2001

63. (a) Army: $y = -33.5x + 767.2$;
   Navy: $y = -17.15x + 567.3$
   (b) The solution $(12.23, 357.62)$ indicates that in 1997
   the number of enlisted personnel in the Army and
   in the Navy will be the same: 357.62 thousand.

65. (a) $\{(2z - 10, 60 - 3z, z)\}$
   (b) Least: 5 liters; most: 20 liters
   (c) 20% solution: 10 liters; 30% solution: 30 liters;
   50% solution: 10 liters

67. 1 ampere, 2 amperes, 1 ampere

■ *Section 5.4*

1. The corresponding entries of A and B are equal.

3. (a) -3 **(b)** $\frac{5}{3}$ **5. (a)** 5 **(b)** 3 **(c)** 0

7. $\begin{bmatrix} 3 & 2 \\ -2 & 8 \end{bmatrix}$ **9.** $\begin{bmatrix} 8 & 0 \\ -4 & 12 \end{bmatrix}$ **11.** $\begin{bmatrix} 1 & 2 \\ -1 & 5 \end{bmatrix}$

13. $\begin{bmatrix} 0 & 6 \\ -7 & 15 \end{bmatrix}$ **15.** $\begin{bmatrix} -3 & -1 & 4 \\ 2 & -3 & -4 \\ 2 & 0 & 1 \end{bmatrix}$ **17.** $\begin{bmatrix} 6 & 1 & 2 \\ 7 & 0 & 7 \\ 1 & 6 & -1 \end{bmatrix}$

19. $\begin{bmatrix} 3 & 0 & 2 \\ 3 & 1 & 1 \\ 1 & 2 & 2 \end{bmatrix}$ **21.** $\begin{bmatrix} 2 & 5 & -4 \\ 10 & 3 & -12 \\ 6 & 5 & 8 \end{bmatrix}$

23. (a) $\begin{bmatrix} 4 & -2 \\ 3 & 3 \end{bmatrix}$ **(b)** $\begin{bmatrix} 1 & -2 \\ -3 & 6 \end{bmatrix}$ **(c)** $\begin{bmatrix} 15 & -10 \\ 5 & 20 \end{bmatrix}$
   (d) $\begin{bmatrix} -1 & 2 \\ 9 & -4 \end{bmatrix}$ **(e)** $\begin{bmatrix} 3 & -2 \\ 5 & -8 \end{bmatrix}$

25. (a) Not possible **(b)** Not possible
   (c) $\begin{bmatrix} 5 \\ 15 \end{bmatrix}$ **(d)** $\begin{bmatrix} 2 & -6 \\ 6 & -18 \end{bmatrix}$ **(e)** $[-16]$

27. (a) The matrices must have the same dimensions,
   so $m = p$ and $n = q$.
   (b) To determine AB, the number of columns of A
   must be the same as the number of rows of B,
   so $n = p$.

29. $[-5, 5]$ **31.** $\begin{bmatrix} 14 & -3 \\ -11 & 4 \\ 17 & 21 \end{bmatrix}$ **33.** Not possible

35. $\begin{bmatrix} 6 & -3 & -1 \\ 8 & -7 & 0 \\ 4 & 1 & -2 \end{bmatrix}$ **37.** $\begin{aligned} x + 3y &= 10 \\ 2x - y &= -1 \end{aligned}$

39. $\begin{aligned} 2x - 3y + z &= -7 \\ x + y - 4z &= 6 \\ 3x + 2y - z &= 5 \end{aligned}$

41. $\begin{bmatrix} 3 & -1 \\ 4 & 5 \end{bmatrix} \begin{bmatrix} x \\ y \end{bmatrix} = \begin{bmatrix} 10 \\ 26 \end{bmatrix}$ **43.** $\begin{bmatrix} 0 & 1 & 3 \\ 1 & 0 & 1 \\ 1 & -2 & 0 \end{bmatrix} \begin{bmatrix} x \\ y \\ z \end{bmatrix} = \begin{bmatrix} -4 \\ -1 \\ -3 \end{bmatrix}$

45. Matrix I is a square matrix with 1s on the main
   diagonal and 0s elsewhere.

47. $\begin{bmatrix} 0 & -4 & -5 \\ 3 & -9 & 5 \end{bmatrix}$

49. Because $\begin{bmatrix} 1 & 1 \\ 1 & 1 \end{bmatrix} \begin{bmatrix} 1 & 1 \\ -1 & -1 \end{bmatrix} = \begin{bmatrix} 0 & 0 \\ 0 & 0 \end{bmatrix}$, $AB = \mathbf{O}$
   does not imply that $A = \mathbf{O}$ or $B = \mathbf{O}$.

51. (a) $\begin{bmatrix} -27 & -18 & 13 \\ 16 & -9 & -32 \\ 9 & 8 & -44 \end{bmatrix}$ **(b)** $\begin{bmatrix} -24 & -4 & -1 \\ 43 & 10 & -31 \\ 25 & 25 & -66 \end{bmatrix}$

53. (a) $\begin{bmatrix} 26 & 34 & 11 \\ -15 & -26 & 5 \\ 35 & 25 & 4 \end{bmatrix}$ **(b)** $\begin{bmatrix} 23 & 20 & 25 \\ -42 & -45 & 4 \\ 19 & 8 & 26 \end{bmatrix}$

55. (a) $\begin{bmatrix} 38 & 15 \\ 46 & 28 \\ 20 & 42 \end{bmatrix}$ The rows represent the number of
   items made at each location. The columns
   represent the production capabilities at each
   plant location.
   (b) The weekly production capabilities

(c) $S = 0.2P = \begin{bmatrix} 7.6 & 3 \\ 9.2 & 5.6 \\ 4 & 8.4 \end{bmatrix}$ (d) $P + S = \begin{bmatrix} 45.6 & 18 \\ 55.2 & 33.6 \\ 24 & 50.4 \end{bmatrix}$

The matrix represents the daily production capabilities after the equipment upgrade.

57. (a) $C = [67 \quad 22 \quad 58]$

(b) The matrix $CP = [4718 \quad 4057]$ represents the total cost of production at each plant.

(c) $R = [84 \quad 35 \quad 74]$

(d) The matrix $RP = [6282 \quad 5348]$ represents the total revenue from production at each plant.

(e) The matrix $RP - CP = [1564 \quad 1291]$ represents the total profit from each plant.

(f) The matrix $R - C = [17 \quad 13 \quad 16]$ represents the profit per item at each plant.

(g) The matrix $(R - C)P = [1564 \quad 1291]$ represents the total profit at each plant.

59. (a) $C = \begin{bmatrix} 64 & 42 \\ 75 & 52 \\ 93 & 28 \\ 51 & 12 \end{bmatrix}$ **(b)** $\frac{1}{2}C = \begin{bmatrix} 32 & 21 \\ 37.5 & 26 \\ 46.5 & 14 \\ 25.5 & 6 \end{bmatrix}$

(c) The matrix $24C$ represents the daily traffic count.

(d) The weekly traffic count

61. (a) $\begin{bmatrix} 0.75 & 0.34 & 15 \\ 1.25 & 0.68 & 22 \end{bmatrix}$

(b) The matrix $CD = \begin{bmatrix} 100.5 & 50.32 & 1884 \\ 121.25 & 60.86 & 2269 \\ 104.75 & 50.66 & 2011 \\ 53.25 & 25.50 & 1029 \end{bmatrix}$

represents the hourly toll, the hourly number of out-of-town vehicles, and the hourly pollution emitted.

63. (a) $E = \begin{bmatrix} 258 & 180 & 146 \\ 135 & 110 & 94 \end{bmatrix}$ **(b)** $F = [0.45 \quad 0.54]$

(c) Matrix $FE = [189 \quad 140.4 \quad 116.46]$ represents the estimated number of females in each of the courses.

65. (a) $M = [0.55 \quad 0.46]$

(b) Matrix ME represents the estimated number of males in each course.

(c) The matrix $FE + ME = [393 \quad 290 \quad 240]$ gives the total enrollment in each course.

■ *Section 5.5*

1. The matrix is not a square matrix. **3.** Yes **5.** No

7. No **9.** Yes **11.** $\begin{bmatrix} 5 & -7 \\ 2 & -3 \end{bmatrix}$ **13.** $\begin{bmatrix} 2 & -1 \\ -\frac{5}{2} & \frac{3}{2} \end{bmatrix}$

15. Not possible **17.** $\begin{bmatrix} 7 & 3 & -16 \\ 1 & 0 & -2 \\ -3 & -1 & 7 \end{bmatrix}$

19. $\begin{bmatrix} -1 & 1 & 2 \\ \frac{1}{2} & 0 & -\frac{1}{2} \\ 1 & -1 & -1 \end{bmatrix}$ **21.** Not possible

23. $\begin{bmatrix} \frac{2}{5} & -\frac{1}{5} & \frac{4}{5} \\ \frac{1}{5} & -\frac{3}{5} & \frac{2}{5} \\ -\frac{3}{5} & -\frac{1}{5} & -\frac{1}{5} \end{bmatrix}$ **25.** $\begin{bmatrix} -\frac{1}{2} & 1 \\ -\frac{3}{2} & \frac{5}{2} \end{bmatrix}$

27. $\begin{bmatrix} 1 & 0 & 0 \\ 2 & 1 & 0 \\ 0 & 2 & 1 \end{bmatrix}$ **29.** $\begin{bmatrix} 0 & 4 & -1 & -2 \\ 1 & 6 & -3 & -5 \\ 1 & 4 & -3 & -4 \\ -1 & -7 & 4 & 6 \end{bmatrix}$

31. Not possible **33.** $[I \mid A^{-1}]$ **35.** $(-3, 2)$

37. $(1, -1)$ **39.** $(1, 2, -4)$ **41.** $(0, 1, 4)$

43. $(-2, -1, 3)$ **45.** $(-2, 1, 0, 3)$

47. Matrix multiplication is not commutative. The product $A^{-1}AX$ results in X, whereas AXA^{-1} does not.

49. $\begin{bmatrix} 1 & 0 \\ -c & 1 \end{bmatrix}$

53. If A is invertible, then $B = C$.

55. (a) $\begin{bmatrix} 1 & 0 & 0 \\ 2 & 1 & 0 \\ 2 & 0 & 1 \end{bmatrix}$ **(b)** $\begin{bmatrix} 1 & 0 & 0 \\ 3 & 1 & 0 \\ 3 & 0 & 1 \end{bmatrix}$ **(c)** $\begin{bmatrix} 1 & 0 & 0 \\ n & 1 & 0 \\ n & 0 & 1 \end{bmatrix}$

57. $\begin{bmatrix} 44 & 36 & 18 \\ 86 & 40 & 33 \\ 17 & 10 & 6 \\ 80 & 40 & 30 \\ 78 & 40 & 34 \end{bmatrix}$ **59.** SEND MONEY NOW

■ *Section 5.6*

1. The graph is the half-plane above the line.

3. B **5.** D

7.

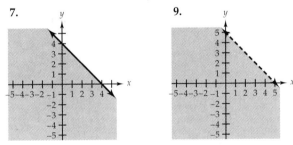

9.

11.

13.

37.

39.

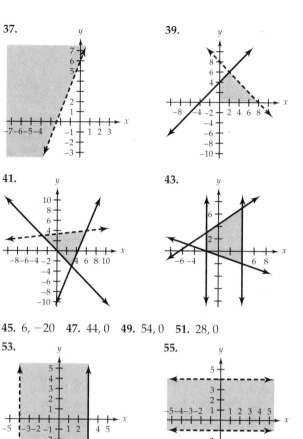

15.

17.

41.

43.

19.

21.

45. 6, −20   **47.** 44, 0   **49.** 54, 0   **51.** 28, 0

53.

55.

23. The graph of the one-variable inequality $x \le -1$ is the interval $(-\infty, -1]$ on the number line. The graph of the two-variable inequality $x \le -1$ is the line $x = -1$ and the half-plane to the left of the line.

25. D   **27.** B

29.

31.

57.

59.

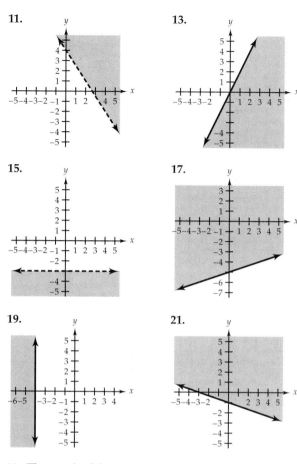

61.

33.

35.

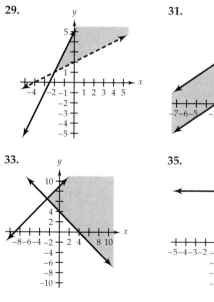

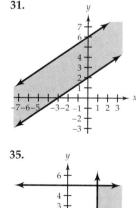

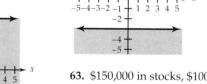

63. $150,000 in stocks, $100,000 in bonds

65. 4 vans, 6 buses   **67.** 6 birdhouses, 12 mailboxes

69. 85 acres of corn, 15 acres of soybeans

71. 7 full-time employees, 35 part-time employees

73. 2400 bleacher seats, 600 reserved seats

75. (a) 50 **(b)** 30 putters, 0 drivers

■ *Chapter Review Exercises*

1. The lines have the same slope but different y-intercepts, so they do not intersect. Thus the system has no solution.

3. $(3, -5)$ **5.** $(4, -6)$ **7.** $\varnothing$; inconsistent

9. $\left\{\left(\frac{1}{3}y + \frac{1}{3}, y\right)\right\}$; dependent equations

11. 15% syrup: 200 ml; 40% syrup: 300 ml

13. The ordered triple corresponds to the point of intersection of three planes.

15. $(2, 0, -4)$ **17.** $\left\{\left(\frac{7}{5}z + \frac{29}{5}, \frac{6}{5}z + \frac{17}{5}, z\right)\right\}$ **19.** $(3, -2, 2)$

21. Solve the second equation for y. Then substitute the value of y into the third equation and solve for z. Finally, substitute the values of y and z into the first equation and solve for x.

23. 0–299: 30%; 300–699: 48%; over 699: 22%

25. (a) 2×2 **(b)** 2×3

27. $\begin{aligned} 2x - 3y &= -7 \\ -5x + y &= -15 \end{aligned}$ **29.** $(4, -1)$ **31.** $(5, -2)$

33. $(1, 3, -1)$ **35.** $E(t) = 0.026t^2 - 1.015t + 30.5$

37. $a = 2, b = -4, c = -\frac{1}{2}, d = 0$

39. (a) $\begin{bmatrix} 1 & -4 & 2 \\ 0 & 0 & -3 \\ 1 & -1 & 5 \end{bmatrix}$ **(b)** $\begin{bmatrix} 5 & 9 & 3 \\ 0 & 8 & 12 \\ 0 & 0 & -1 \end{bmatrix}$

41. (a) Not possible **(b)** Not possible

(c) $\begin{bmatrix} 3 & -6 \\ 9 & -3 \\ 12 & 9 \end{bmatrix}$ **(d)** $\begin{bmatrix} -3 & -2 & -1 \\ 1 & -1 & -3 \\ 10 & 3 & -4 \end{bmatrix}$ **(e)** $\begin{bmatrix} -3 & -5 \\ 5 & -5 \end{bmatrix}$

43. (a) $\begin{bmatrix} 4 & 4 & -1 \\ 8 & -7 & 13 \end{bmatrix}$ **(b)** $\begin{bmatrix} -1 & 2 & -2 \\ 4 & -8 & 5 \end{bmatrix}$

(c) $\begin{bmatrix} 3 & 6 & -3 \\ 12 & -15 & 18 \end{bmatrix}$

(d) Not possible **(e)** Not possible

45. $\begin{aligned} 2x + y &= 1 \\ 3x + 4y &= -1 \end{aligned}$ **47.** $P = \begin{bmatrix} 4 & 8 & 6 \\ 6 & 4 & 4 \\ 6 & 10 & 8 \\ 10 & 12 & 8 \end{bmatrix}$

49. Yes **51.** $\begin{bmatrix} -3 & 2 \\ 5 & -3 \end{bmatrix}$ **53.** $\begin{bmatrix} 1 & -1 & 0 \\ 1 & 0 & 1 \\ 0 & 1 & 2 \end{bmatrix}$

55. $\left(\frac{3}{2}, -\frac{1}{2}\right)$ **57.** $(-1, 0, 4)$ **59.** $(1, -1, 3, -2)$

61.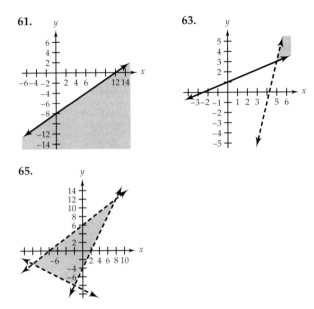

63.

65.

67. In (ii) the points of the line $y = x + 2$ correspond to solutions of the inequalities. In (i), the graphs of the two inequalities have no points in common.

69. 31, 18 **71.** Coach: 192; first class: 58

Chapter 6

■ *Section 6.1*

1. The absolute values of the terms are the same. The signs of some alternating terms differ.

3. 20, 22; $a_n = 2n + 10$ **5.** $7, -8; a_n = (-1)^{n+1}(n + 2)$

7. $\frac{6}{5}, \frac{7}{6}; a_n = \dfrac{n+1}{n}$ **9.** $1 + \frac{6}{7}, 1 + \frac{7}{8}; a_n = 1 + \dfrac{n+1}{n+2}$

11. (a) 2, 4, 8, 16 **(b)** 128 **(c)** 4096

13. (a) $-2, -5, -8, -11$ **(b)** -20 **(c)** -35

15. (a) $-2, 0, -\frac{2}{3}, 0$ **(b)** $-\frac{2}{7}$ **(c)** 0

17. (a) $\frac{1}{5}, \frac{2}{3}, \frac{9}{7}, 2$ **(b)** $\frac{49}{11}$ **(c)** 9 **19.** $\frac{1}{2}, \frac{2}{3}, \frac{3}{4}, \frac{4}{5}, \frac{5}{6}$

21. $2, 0, \frac{2}{3}, 0, \frac{2}{5}, 0, \frac{2}{7}, 0$ **23.** 2, 2.25, 2.3704, 2.4414

25. 1, 1.5874, 2.0801, 2.5198

27. 0.4142, 0.3178, 0.2679, 0.2361

29. The recursive definition must include the first term and a formula describing the nth term in relation to the preceding term.

31. (a) 3, 8, 13, 18 **(b)** 48 **33. (a)** $1, 3, 2, -1$ **(b)** -1

35. 1, 1.4142, 1.5538, 1.5981 **37.** $2, 3, -0.1667, -6.3333$

39. The fourth partial sum is the sum of the first four terms of a sequence.

41. 20, 90 **43.** $-7, 12$ **45.** 97, 395 **47.** $-3.55, -7.07$

49. $-129, -1028$ **51.** 13.52, 28.33 **53.** $305, 1.72 \cdot 10^{10}$

55. 5 **57.** 10 **59.** 84 **61.** $-\frac{1}{6}$ **63.** $\sum_{i=1}^{6} (-1)^{i+1}(i+2)$

65. $\sum_{i=1}^{15} \frac{2}{i+2}$ **67.** $\sum_{i=1}^{10} (3i-1)$

69. (a) $a_n = -2n + 7$ **(b)** $a_1 = 5, a_n = a_{n-1} - 2$

71. (a) $a_n = (-1)^{n+1} \cdot 7$ **(b)** $a_1 = 7, a_n = -a_{n-1}$

73. $S_n = n^2$ **75.** $a_{n+2} = a_{n+1} + a_n$

77. 8, 13, 21, 34, 55, 89, 144 **79.** FMFFMFMFFMFFM

81. The sequence is the Fibonacci sequence.

83. $a_{n+2} = a_{n+1} + a_n$; the sequence is the same except for the first term.

■ Section 6.2

1. Determine the difference between any term and the preceding term.

3. (a) $a_1 = -5, d = 3$ **(b)** $a_n = 3n - 8$

5. (a) $a_1 = 3, d = -2$ **(b)** $a_n = -2n + 5$

7. Not arithmetic

9. (a) $a_1 = 0.2, d = 0.4$ **(b)** $a_n = 0.4n - 0.2$

11. (a) $-3, 4, 11, 18$ **(b)** $c_n = 7n - 10$ **(c)** 60

13. (a) $18, 12, 6, 0$ **(b)** $a_n = -6n + 24$ **(c)** -36

15. (a) $40, 36, 32, 28$ **(b)** $k_n = -4n + 44$ **(c)** 4

17. (a) $-3.9, -3.6, -3.3, -3$ **(b)** $c_n = 0.3n - 4.2$
(c) -1.2

19. The graph of the equation is a solid line. The graph of the sequence contains only those points with a positive integer as a first coordinate.

21. $c_n = -3n + 7$ **23.** $a_n = \frac{1}{4}n + \frac{1}{4}$ **25.** $t_n = 1.2n + 1.2$

27. 25 **29.** 147 **31.** 1320 **33.** -2775 **35.** 471.5

37. Part (iii) does not contain enough information to determine a_9, so we cannot determine the ninth partial sum.

39. 528 **41.** -66.5 **43.** -1308 **45.** 352.5 **47.** 264

49. 351 **51.** 1140 **53.** $k_n = 6n + 8, k_{50} = 308$

55. $c_n = -3.2n - 3.8, c_{50} = -163.8$ **57.** $n(1 + n)$

59. 2401, 2404, 2407

61. (a) $12, 18, 24, \ldots, 114$; $a_n = 6n + 6, 1 \le n \le 18$
(b) 189

63. 180 **65. (a)** $a_n = -4n + 82$ **(b)** 2250

67. (a) $a_n = -0.08n + 80.13$ (in seconds)
(b) 638.16 seconds or 10 minutes, 38.16 seconds

69. (a) $b_n = -5.6n + 78.8$; b_{11} represents the average bill in 2001.
(b) $12(72.74 + 68.68 + 61.48 + 56.21 + 51.00) =$ $3721.32
(c) $12\sum_{i=4}^{10} (-5.6i + 78.8) = 12(277.20) = \$3326.40$

■ Section 6.3

1. The common ratio can be determined by dividing a term by the preceding term.

3. (a) Geometric **(b)** Neither **(c)** Arithmetic

5. (a) Neither **(b)** Geometric **(c)** Arithmetic

7. (a) $-2, 3, -\frac{9}{2}, \frac{27}{4}$ **(b)** -51.26

9. (a) $16, -32, 64, -128$ **(b)** 4096

11. (a) $12.5, 2.5, 0.5, 0.1$ **(b)** $3.2 \cdot 10^{-5}$

13. $b_n = 7\left(\frac{3}{7}\right)^{n-2}$ **15.** $a_n = \frac{(-5)^n}{2}$ **17.** $c_n = 3\left(-\frac{2}{3}\right)^{n-1}$

19. $a_n = -15\left(-\frac{3}{2}\right)^{n-1}$ **21.** $c_n = 3 \cdot 2^{n-4}$

23. Use the two terms to determine r. Then determine the eighth partial sum.

25. 12,582,911.25 **27.** 461.96 **29.** 2048.25 **31.** 29.60

33. 4095.875 **35.** 7.78 **37.** 781.25 **39.** 66.57

41. The common ratio is greater than 1. **43.** 4.5

45. $\frac{49}{6}$ **47.** 3 **49.** 3.22 **51.** 8 **53.** $\frac{25}{6}$ **55.** 0

57. (a) 0 **(b)** 4 **59.** $\frac{8}{9}$ **61.** $\frac{35}{6}$

63. (a) Geometric **(b)** Arithmetic

65. (i) Neither **(ii)** Arithmetic **(iii)** Geometric

67. (a) $a_n = 200,000(0.7)^n$ **(b)** \$23,529.80

(c) $\sum_{i=0}^{12} 200,000(0.7)^i$ **(d)** \$660,207.40

69. (a) $a_n = 2(0.8)^{n-1}$ **(b)** 8 months **(c)** \$9.31 million
(d) The sum of the infinite series is 10.

71. (a) $a_n = 20\left(\frac{4}{5}\right)^{n-1}$ **(b)** $b_n = 20\left(\frac{4}{5}\right)^n$

(c) 25 **(d)** $\sum_{i=1}^{\infty} 20\left(\frac{4}{5}\right)^{i-1} + \sum_{i=1}^{\infty} 20\left(\frac{4}{5}\right)^i = 180$ feet

73. (a) $a_n = 30,271(1.05)^{n-1}$ **(b)** \$76,493.31

(c) $\sum_{i=1}^{20} 30,271(1.05)^{i-1} = 1,000,939.50$ **(d)** 30

■ Section 6.4

1. The notation 4! means $4 \cdot 3 \cdot 2 \cdot 1$, whereas 4^4 means $4 \cdot 4 \cdot 4 \cdot 4$.

3. 5040 **5.** 11,880 **7.** 8008 **9.** 120 **11.** 5040 **13.** 21

15. 6435 **21.** 360 **23.** 240 **25. (a)** 10^7 **(b)** $8 \cdot 10^6$

27. 4^{10} or 1,048,576

29. (a) 17,280 **(b)** 17,280 **(c)** 3,628,800 **31.** 84

33. For a permutation, the order of the items is important, whereas for a combination, the order of the items is not important.

35. 720 **37.** $18! \approx 6.40 \cdot 10^{15}$ **39.** 40,320 **41.** 210

43. 604,800 **45. (a)** 7,893,600 **(b)** 11,881,376

47. 34,650   **49.** 1260   **51.** 630,630

53. Because the values of $_5C_3$ and $_5C_2$ are both $\dfrac{5!}{3! \cdot 2!}$, the number of combinations is the same.

55. 35,960   **57.** 2,704,156   **59.** 7140   **61.** 5148

63. $5.36 \cdot 10^{20}$ **65.** $2.43 \cdot 10^{14}$ **67.** $2.21 \cdot 10^{12}$

■ *Section 6.5*

1. The coin has no memory, so the probability is $\frac{1}{2}$.

3. (a) $\frac{1}{6}$ (b) $\frac{1}{2}$ (c) $\frac{1}{3}$ (d) $\frac{1}{2}$

5. (a) $\frac{3}{8}$ (b) $\frac{11}{16}$ (c) $\frac{1}{8}$ (d) $\frac{1}{4}$

7. (a) $\frac{1}{9}$ (b) $\frac{5}{6}$ (c) $\frac{1}{9}$ (d) $\frac{1}{6}$

9. (a) $\frac{1}{4}$ (b) $\frac{1}{13}$ (c) $\frac{3}{13}$ (d) $\frac{5}{13}$ (e) $\frac{11}{26}$

11. (a) $\frac{1}{221}$ (b) $\frac{10}{17}$ (c) $\frac{7}{17}$ (d) $\frac{26}{51}$ (e) $\frac{188}{663}$

13. (a) $\frac{2}{7}$ (b) $\frac{4}{7}$ (c) $\frac{9}{14}$ **15.** (a) $\frac{1}{8}$ (b) $\frac{13}{40}$ (c) $\frac{3}{8}$

17. Because a van could be blue, the events in (i) are not mutually exclusive. The events in (ii) cannot occur simultaneously, so they are mutually exclusive.

19. $\frac{5}{7}$ **21.** $\frac{1}{24}$ **23.** $\frac{2}{5}$ **25.** 0.25   **27.** (a) $\frac{1}{177,100}$ (b) $\frac{134}{8855}$

29. $\frac{1}{56}$ **31.** $\frac{27}{512}$ **33.** (a) $\frac{1}{120}$ (b) $\frac{1}{5}$ **35.** (a) $\frac{28}{57}$ (b) $\frac{29}{57}$

37. (a) $\frac{10}{89}$ (b) $\frac{51}{89}$ (c) $\frac{79}{89}$ (d) $\frac{4}{89}$

39. (a) $\frac{89}{171}$ (b) $\frac{82}{171}$ (c) $\frac{40}{171}$ **41.** 0.50   **43.** 20%

45. The odds in favor of E is the reciprocal of the odds in favor of $\overline{E}$.

47. 1 to 8; 8 to 1   **49.** 17 to 3   **51.** $\frac{5}{7}$

53. (a) 79 to 152   (b) 4 to 29   (c) 40 to 191

55. 455 to 675,584

■ *Section 6.6*

1. No, the coefficients are in row 12, which is the 13th row of the triangle.

3. $16x^4 - 32x^3y + 24x^2y^2 - 8xy^3 + y^4$

5. $243y^5 + 810y^4 + 1080y^3 + 720y^2 + 240y + 32$

7. $y^6 - 12y^5 + 60y^4 - 160y^3 + 240y^2 - 192y + 64$

9. $1 + 10x + 40x^2 + 80x^3 + 80x^4 + 32x^5$

11. $16x^4 - 96x^3\sqrt{y} + 216x^2y - 216xy\sqrt{y} + 81y^2$

13. $t^{10} + 20t^8w + 160t^6w^2 + 640t^4w^3 + 1280t^2w^4 + 1024w^5$

15. $\dfrac{x^6}{y^6} - \dfrac{3x^5z}{y^5} + \dfrac{15x^4z^2}{4y^4} - \dfrac{5x^3z^3}{2y^3} + \dfrac{15x^2z^4}{16y^2} - \dfrac{3xz^5}{16y} + \dfrac{z^6}{64}$

17. $x^{10} + 40x^9y + 720x^8y^2 + 7680x^7y^3 + \cdots$

19. $a^{18} - 18a^{16}b^4 + 144a^{14}b^8 - 672a^{12}b^{12} + \cdots$

21. $65,536y^8 + 655,360y^7z + 2,867,200y^6z^2 + 7,168,000y^5z^3 + \cdots$

23. $\dfrac{1}{y^{12}} - \dfrac{6z}{y^{11}} + \dfrac{33z^2}{2y^{10}} - \dfrac{55z^3}{2y^9} + \cdots$

25. No, several terms of the expansion are like terms and can be combined.

27. $x^{50} + 100x^{49}y + \cdots$ **29.** $z^{200} - 600z^{199} + \cdots$

31. $48z^{11}w$ **33.** $-19,683y^9$ **35.** $3360t^8s^6$

37. $-20,412x^4$ **43.** 1,966,080   **45.** 26,400,000

47. 1792   **49.** 90   **51.** $-32i$

53. The sequence 1, 1, 2, 3, 5, 8, … is the Fibonacci sequence.

55. $64 - \dfrac{192}{y^3} + \dfrac{240}{y^6} - \dfrac{160}{y^9} + \dfrac{60}{y^{12}} - \dfrac{12}{y^{12}} + \dfrac{1}{y^{18}}$

57. 128   **59.** (a) 0.20   (b) 0.88

■ *Chapter Review Exercises*

1. (a) $10, -11$; $a_n = (-1)^n(n + 4)$ (b) $\frac{13}{12}, \frac{15}{14}$; $a_n = \dfrac{2n + 1}{2n}$

3. (a) $\frac{1}{4}, \frac{3}{5}, \frac{5}{6}, 1$ (b) $\frac{11}{9}$ (c) $\frac{19}{13}$ **5.** 4   **7.** 6, 7   **9.** -25

11. (a) $p_n = 250$ (b) $I_n = 26.25 - 3.75n$
   (c) $p_6 = 250$, $I_6 = 3.75$
   (d) The total monthly payment

13. No   **15.** $a_n = \frac{1}{2}n + 2$; $a_7 = 5.5$

17. $k_n = -2n + 14$; $k_7 = 0$

19. No, you need to know at least one term.

21. -2025 **23.** (a) $t_n = 4n + 8$ (b) 8

25. (a) Arithmetic; $a_n = -\frac{1}{2}n + 2$
   (b) Neither; $b_n = (-1)^{n+1}(n + 2)$
   (c) Geometric; $c_n = \frac{1}{8}(-2)^{n-1}$

27. $c_n = 25\left(\dfrac{1}{5}\right)^{n-1}$ **29.** $b_n = -24\left(\dfrac{3}{2}\right)^{n-1}$ **31.** $-\frac{64}{3}$ **33.** 9

35. Series (i) is finite and can be evaluated. Series (ii) is an infinite series with $r > 1$, so it cannot be evaluated.

37. 73,440   **39.** 495   **41.** 120   **43.** 479,001,600

45. 38,798,760   **47.** 5005   **49.** (a) 0.30   (b) 0.16

51. (a) $\frac{1}{9}$ (b) $\frac{5}{18}$ (c) $\frac{1}{4}$ **53.** (a) $\frac{1}{17}$ (b) $\frac{20}{221}$ (c) $\frac{25}{102}$

55. (a) $\frac{28}{153}$ (b) $\frac{49}{153}$ **57.** (a) $\frac{1}{35}$ (b) $\frac{22}{35}$ (c) $\frac{12}{35}$

59. The probability of heads is $\frac{1}{2}$. The odds in favor of heads is the probability of heads divided by the probability of tails: 1 to 1.

61. Each entry in the triangle is the sum of the two numbers above it.

63. $x^4 + 20x^3 + 150x^2 + 500x + 625$

65. $9\sqrt{3} - 45y + 30\sqrt{3}y^2 - 30y^3 + 5\sqrt{3}y^4 - y^5$

67. -672 **69.** $316 + 12i$

Appendix A

■ Section A.1

1. $r = 4$

3. $r = \sqrt{10} \approx 3.16$

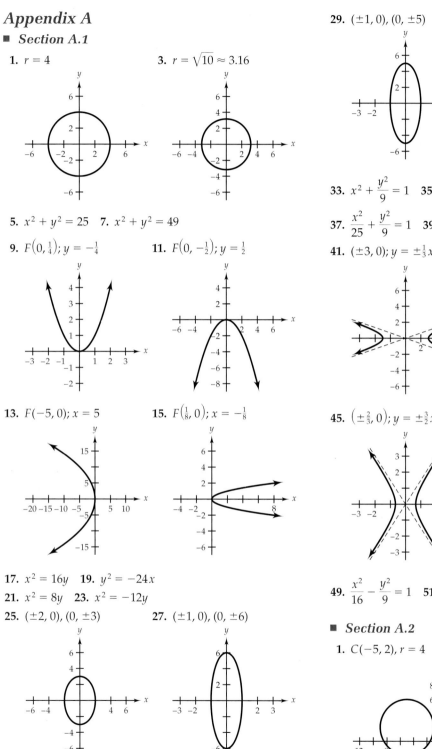

5. $x^2 + y^2 = 25$ **7.** $x^2 + y^2 = 49$

9. $F\left(0, \frac{1}{4}\right); y = -\frac{1}{4}$ **11.** $F\left(0, -\frac{1}{2}\right); y = \frac{1}{2}$

13. $F(-5, 0); x = 5$ **15.** $F\left(\frac{1}{8}, 0\right); x = -\frac{1}{8}$

17. $x^2 = 16y$ **19.** $y^2 = -24x$
21. $x^2 = 8y$ **23.** $x^2 = -12y$

25. $(\pm 2, 0), (0, \pm 3)$ **27.** $(\pm 1, 0), (0, \pm 6)$

29. $(\pm 1, 0), (0, \pm 5)$ **31.** $(\pm 2, 0), (0, \pm 4)$

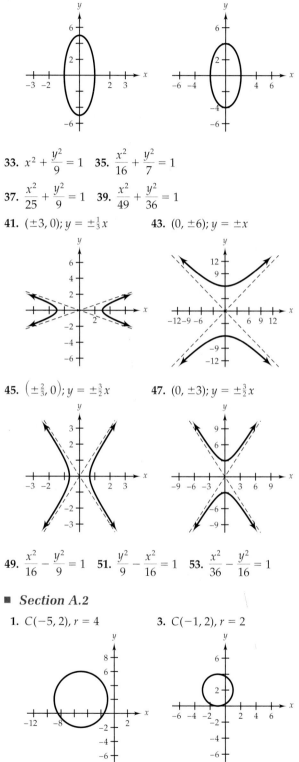

33. $x^2 + \dfrac{y^2}{9} = 1$ **35.** $\dfrac{x^2}{16} + \dfrac{y^2}{7} = 1$

37. $\dfrac{x^2}{25} + \dfrac{y^2}{9} = 1$ **39.** $\dfrac{x^2}{49} + \dfrac{y^2}{36} = 1$

41. $(\pm 3, 0); y = \pm\frac{1}{3}x$ **43.** $(0, \pm 6); y = \pm x$

45. $\left(\pm\frac{2}{3}, 0\right); y = \pm\frac{3}{2}x$ **47.** $(0, \pm 3); y = \pm\frac{3}{2}x$

49. $\dfrac{x^2}{16} - \dfrac{y^2}{9} = 1$ **51.** $\dfrac{y^2}{9} - \dfrac{x^2}{16} = 1$ **53.** $\dfrac{x^2}{36} - \dfrac{y^2}{16} = 1$

■ Section A.2

1. $C(-5, 2), r = 4$ **3.** $C(-1, 2), r = 2$

5. $C(4, 0)$, $r = 4$

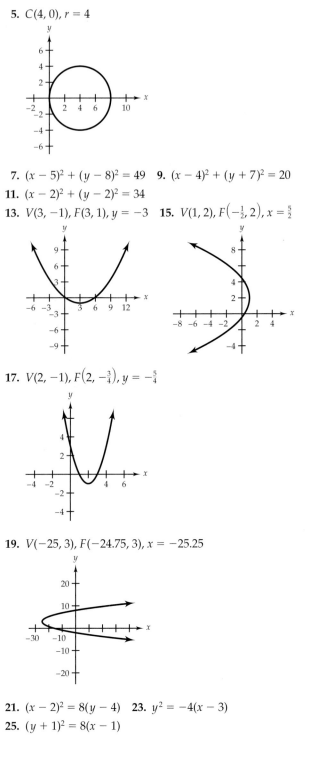

7. $(x - 5)^2 + (y - 8)^2 = 49$ **9.** $(x - 4)^2 + (y + 7)^2 = 20$

11. $(x - 2)^2 + (y - 2)^2 = 34$

13. $V(3, -1)$, $F(3, 1)$, $y = -3$ **15.** $V(1, 2)$, $F\left(-\frac{1}{2}, 2\right)$, $x = \frac{5}{2}$

17. $V(2, -1)$, $F\left(2, -\frac{3}{4}\right)$, $y = -\frac{5}{4}$

19. $V(-25, 3)$, $F(-24.75, 3)$, $x = -25.25$

21. $(x - 2)^2 = 8(y - 4)$ **23.** $y^2 = -4(x - 3)$

25. $(y + 1)^2 = 8(x - 1)$

27. $C(-2, 1)$, $A_1(-5, 1)$, $A_2(1, 1)$, $B_1(-2, -1)$, $B_2(-2, 3)$, $F\left(-2 \pm \sqrt{5}, 1\right)$

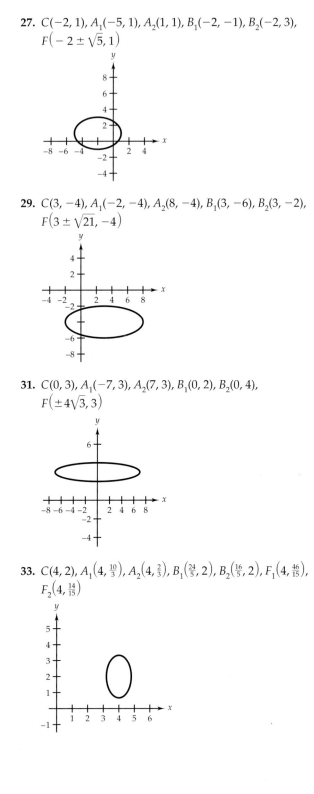

29. $C(3, -4)$, $A_1(-2, -4)$, $A_2(8, -4)$, $B_1(3, -6)$, $B_2(3, -2)$, $F\left(3 \pm \sqrt{21}, -4\right)$

31. $C(0, 3)$, $A_1(-7, 3)$, $A_2(7, 3)$, $B_1(0, 2)$, $B_2(0, 4)$, $F\left(\pm 4\sqrt{3}, 3\right)$

33. $C(4, 2)$, $A_1\left(4, \frac{10}{3}\right)$, $A_2\left(4, \frac{2}{3}\right)$, $B_1\left(\frac{24}{5}, 2\right)$, $B_2\left(\frac{16}{5}, 2\right)$, $F_1\left(4, \frac{46}{15}\right)$, $F_2\left(4, \frac{14}{15}\right)$

35. $\dfrac{(x-4)^2}{4} + \dfrac{(y+1)^2}{25} = 1$ **37.** $\dfrac{(x-2)^2}{9} + (y-3)^2 = 1$

39. $\dfrac{x^2}{27} + \dfrac{(y-5)^2}{36} = 1$

41. $C(-2, -1)$, $V_1(-5, -1)$, $V_2(1, -1)$, $F\left(-2 \pm \sqrt{34}, -1\right)$,
$y_1 = \frac{5}{3}x + \frac{7}{3}$, $y_2 = -\frac{5}{3}x - \frac{13}{3}$

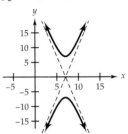

43. $C(7, 0)$, $V_1(7, -7)$, $V_2(7, 7)$, $F\left(7, \pm\sqrt{53}\right)$, $y_1 = \frac{7}{2}x - \frac{49}{2}$,
$y_2 = -\frac{7}{2}x + \frac{49}{2}$

45. $C(-1, 2)$, $V_1(-1, 1)$, $V_2(-1, 3)$, $F\left(-1, 2 \pm \sqrt{2}\right)$,
$y_1 = x + 3$, $y_2 = -x + 1$

47. $C(-3, -5)$, $V_1(-7, -5)$, $V_2(1, -5)$, $F\left(-3 \pm \sqrt{17}, -5\right)$,
$y_1 = \frac{1}{4}x - \frac{17}{4}$, $y_2 = -\frac{1}{4}x - \frac{23}{4}$

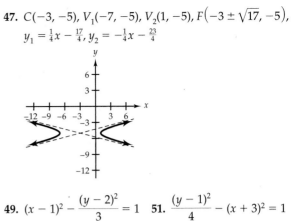

49. $(x-1)^2 - \dfrac{(y-2)^2}{3} = 1$ **51.** $\dfrac{(y-1)^2}{4} - (x+3)^2 = 1$

53. $\dfrac{(x-5)^2}{4} - (y-2)^2 = 1$

INDEX